Advances in Vehicular Technology

Advances in Vehicular Technology

Edited by Mackenzie Ruiz

CLANRYE
INTERNATIONAL
www.clanryeinternational.com

Clanrye International,
750 Third Avenue, 9th Floor,
New York, NY 10017, USA

ISBN: 978-1-64726-126-9

Cataloging-in-Publication Data

Advances in vehicular technology / edited by Mackenzie Ruiz.
p. cm.
Includes bibliographical references and index.
ISBN 978-1-64726-126-9
1. Mobile communication systems. 2. Intelligent transportation systems. 3. Vehicles. I. Ruiz, Mackenzie.
TK5103.2 .A38 2022
621.384--dc23

For information on all Clanrye International publications visit our website at www.clanryeinternational.com

Clanrye International

Contents

Preface

Machines which are used to transport people and cargo are known as vehicles. Some of the most common types of vehicles are motor vehicles, watercraft, amphibious vehicles, aircraft and spacecraft. Vehicles need a source of energy to drive them. Some of the common types of technologies which are used to propel vehicles are internal combustion engines, electric motors and gas turbines. The heat engines where the combustion of a fuel occurs with an oxidizer in a combustion chamber are referred to as internal combustion engines. Gas-turbine engines employ a gas as the working fluid used to turn a turbine. The topics included in this book on vehicular technologies are of utmost significance and bound to provide incredible insights to readers. The book presents researches and studies performed by experts across the globe. It is meant for students who are looking for an elaborate reference text on vehicular technologies.

After months of intensive research and writing, this book is the end result of all who devoted their time and efforts in the initiation and progress of this book. It will surely be a source of reference in enhancing the required knowledge of the new developments in the area. During the course of developing this book, certain measures such as accuracy, authenticity and research focused analytical studies were given preference in order to produce a comprehensive book in the area of study.

This book would not have been possible without the efforts of the authors and the publisher. I extend my sincere thanks to them. Secondly, I express my gratitude to my family and well-wishers. And most importantly, I thank my students for constantly expressing their willingness and curiosity in enhancing their knowledge in the field, which encourages me to take up further research projects for the advancement of the area.

Editor

1

Switched Cooperative Driving Model towards Human Vehicle Copiloting Situation: A Cyberphysical Perspective

Yang Li,[1,2] Dihua Sun,[1,2] Min Zhao,[1,2] Dong Chen,[1,2] Senlin Cheng,[1,2] and Fei Xie[3]

[1]*Key Laboratory of Dependable Service Computing in Cyber Physical Society of Ministry of Education, Chongqing University, Chongqing 400044, China*
[2]*School of Automation, Chongqing University, Chongqing 400044, China*
[3]*China Automotive Engineering Research Institute Co., Ltd., Chongqing 401122, China*

Correspondence should be addressed to Dihua Sun; d3sun@163.com

Academic Editor: Jose E. Naranjo

Development of highly automated and intelligent vehicles can lead to the reduction of driver workload. However, it also causes the out-of-the-loop problem to drivers, which leaves drivers handicapped in their ability to take over manual operations in emergency situations. This contribution puts forth a new switched driving strategy to avoid some of the negative consequences associated with out-of-the-loop performance by having drivers assume manual control at periodic intervals. To minimize the impact of the transitions between automated and manual driving on traffic operations, a switched cooperative driving model towards human vehicle copiloting situation is proposed by considering the vehicle dynamics and the realistic intervehicle communication in a cyberphysical view. The design method of the switching signal for the switched cooperative driving model is given based on the Lyapunov stability theory with the comprehensive consideration of platoon stability and human factors. The good agreement between simulation results and theoretical analysis illustrates the effectiveness of the proposed methods.

1. Introduction

Advanced driver assistance systems (ADASs) assist drivers or take over driving tasks to operate vehicles in a safe, efficient, and comfortable way [1]. Many advanced driver assistance systems are cooperative systems, where vehicles are connected with each other via vehicle-to-vehicle (V2V) communications [2]. The cooperative adaptive cruise control (CACC) system is a promising example of ADAS which enables vehicles to follow one another with short intervehicle distances using intervehicle wireless communication as well as onboard sensors, thus improving road throughput, without compromising safety [3].

The Society of Automotive Engineers (SAE), the National Highway Traffic Safety Administration (NHTSA), and the German Federal Highway Research Institute (BASt) have provided their definitions of "levels of automated driving" [4–6]. Although these definitions are different, the essential criteria that define the different levels of automation are similar in each of the three cases.

The progress of automated driving technology and the development of the ADAS make the problem of human vehicle copiloting a hot topic, which has received a degree of attention in recent years, partly due to the considerable outreach activities undertaken by companies such as Google and Tesla [7].

It is helpful to think of the driving process as a control loop that consists of goal, implementation, and feedback on whether or not the chosen strategy satisfies the desired state. When the driver is in the loop, he/she is actively engaged in all three processes. However, as soon as automation is engaged, the human operator is potentially relegated from one or more of these stages, meaning that they could become out of the loop [8].

Several general conclusions can be derived from previous researches on human-machine cooperation: a greater degree of automation reduces the variability of human performance, but it also leads to "human out-of-the-loop" problems, such as complacency, loss of situation awareness, loss of manual control skills, and behavioral adaptation [9–12]. A literature review by De Winter et al. [13] shows that drivers' workload and situation awareness are vastly different for driving with ACC compared to Highly Automated Driving (HAD).

Such human factors issues resulting from increased automation levels need to be taken into consideration when designing automated driving and ADASs [14, 15]. A driving state model describing the human monitoring level and the allocation of lateral and longitudinal control tasks was introduced. In addition, an authority transition in automated driving is defined as the process of changing from one static state of driving to another static state [16].

Some studies found that the type of functional limitation [17] and drivers' experience of using the automated driving system [18, 19] have an effect on the outcome of the takeover, whereas drivers' trust in the automation has been found to not influence the driver's behavior during the takeover [19, 20]. Further studies have shown that drivers' ability to take over when a transition to manual control is required can be improved by providing information about the limits and reliability of the automation [21, 22].

Previous works that detect driver fatigue or distraction in real time have used both eye tracking data and vehicle-based measures such as speed and lateral position [23–25]. A variety of measures have been used to determine the thresholds of such distraction, ranging from average duration of glances away from the road in a 4.3 s wide sliding window [26] to the percentage of on-road gaze points in a 60 s sliding window [27].

As the above states, the existing studies about human vehicle copiloting and transition between automated and manual driving mainly focus on the specific switching process and conditions. Many studies are subject to expensive devices to detect drivers' states and vehicle running data, and few researches pay close attention to the impact of the transitions on traffic operations.

In level 3 driving automation defined by SAE, the automated driving system performs all aspects of dynamic driving task. The human driver does not need to monitor the driving environment, which will make drivers out of the loop, but should respond appropriately to take over the vehicle in an emergency situation. Many companies and researchers have recognized the potential problem; the likes of Ford and Waymo believe the handoff is fraught with so many problems; they have elected to eschew such development and pursue fully driverless cars.

Meanwhile, many people neglect the issues in level 2. The task of the human driver turns from operating the vehicle to supervising the status of the automated driving system and traffic environment in level 2. However, the studies on human factors show that human beings are not good at supervising tasks. Boredom, distraction, and fatigue will result in the long time supervising task.

Thus, to avoid some of the negative consequences associated with the factors mentioned above, this paper puts forth a periodic switched driving strategy to keep drivers always in the loop and prevent drivers' distraction and fatigue resulting from the long time supervising task. In order to minimize the impact of the periodic transitions between automated and manual driving on traffic operations,

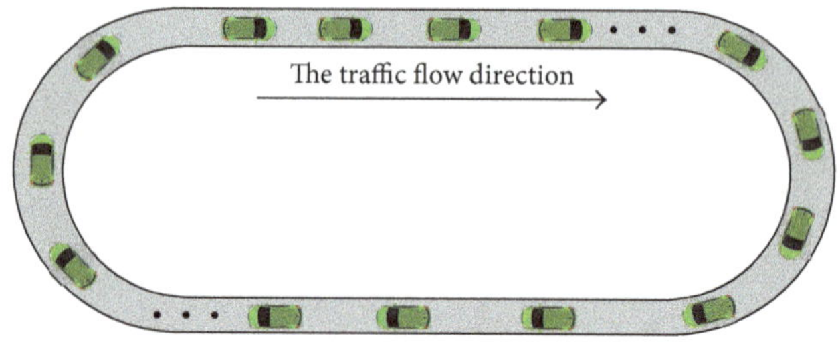

FIGURE 1: The schematic diagram of the vehicle platoon.

a switched cooperative driving model towards human vehicle copiloting situation is proposed by considering the vehicle dynamics and the realistic intervehicle communication in a cyberphysical view. Meanwhile, the design method of the switching signal for the switched cooperative driving model is given based on the Lyapunov stability theory with the comprehensive consideration of platoon stability and human factors.

The remainder of this paper is organized as follows. Section 2 introduces the construction of switched cooperative driving model; stability analysis is given in Section 3; Section 4 presents the design method and suggestion of the switching signal; and numerical examples are provided to illustrate the effectiveness of the proposed methods in Section 5. Section 6 summarizes the main conclusions.

2. The Switched Cooperative Driving Model

Consider a platoon of N vehicles which follow one another along a single-lane ring road; as depicted in Figure 1, the vehicle dynamics can be described by the transfer function $G(s)$ in the Laplace domain, with $s \in \mathbb{C}$, according to

$$G_n(s) = \frac{x_n(s)}{u_n(s)} = \frac{1}{s^2(T_n s + 1)} e^{-\tau_n s}, \tag{1}$$

where T_n and τ_n are the mechanical inertia coefficient and time delay of the nth follower vehicle in the platoon, respectively. u_n is the vehicle input (desired acceleration), whereas the position x_n represents the output. Note that $\cdot(s)$ denotes the Laplace transform of the corresponding time-domain variable $\cdot(t)$. Assume that all vehicles are homogeneous; $G(s)$ is identical for all vehicles; this means $T_1 = T_2 = \cdots = T_N$, $\tau_1 = \tau_2 = \cdots = \tau_N$.

In this paper, we focus on the modeling of the switched cooperative driving system towards human vehicle copiloting situation; the distinguishing feature between automated and manual driving is the CACC system as an extension of the adaptive cruise control with wireless intervehicle communication, which enables vehicles to obtain the control information of the preceding vehicle. Moreover, the vehicle dynamics and the realistic intervehicle communication should be considered in a cyberphysical perspective. Let $a_n(t)$, $v_n(t)$, and $x_n(t)$ denote the acceleration, velocity, and positon of the nth follower vehicle at time t. Thus, the vehicle-to-vehicle cooperative driving model's dynamics equation for automated and manual driving can be formulated as follows:

$$\begin{aligned} a_n(t) = {} & -T_n \dot{a}_n(t) \\ & + k_n^{\mathrm{CACC}} \left[V\left(y_n\left(t-\tau_n\right)\right) - v_n\left(t-\tau_n\right) \right] \\ & + \lambda_n^{\mathrm{CACC}} \Delta v_n\left(t-\tau_n\right) \\ & + \gamma_n^{\mathrm{CACC}} a_{n+1}\left(t-\tau_n^c-\tau_n\right) \\ & \qquad \text{for automated driving} \\ a_n(t) = {} & -T_n \dot{a}_n(t) + k_n^{\mathrm{hum}} \left[V\left(y_n\left(t-\tau_n\right)\right) - v_n\left(t-\tau_n\right) \right] \\ & + \lambda_n^{\mathrm{hum}} \Delta v_n\left(t-\tau_n\right) \quad \text{for manual driving,} \end{aligned} \tag{2}$$

where k_n^{CACC}, $\lambda_n^{\mathrm{CACC}}$, and γ_n^{CACC} are the sensitivities for automated driving and k_n^{hum} and λ_n^{hum} are the sensitivities when vehicles are in manual driving condition. $y_n(t)$ and $\Delta v_n(t)$ denote the headway and the velocity difference between the nth and the $n+1$th vehicle at time t. τ_n^c represents the delay of the realistic communication. In this paper, we adopt the optimal velocity function which is carried out by Helbing and Tilch [28]. The optimal velocity function $V(\cdot)$ is based on the hyperbolic tangent function, so it is a monotonically increasing function and has an upper bound, which is defined as

$$V(x) = V_1 + V_2 \tanh\left[C_1\left(x - l_c\right) - C_2\right], \tag{3}$$

where $V_1 = 6.75\,\mathrm{m/s}$, $V_2 = 7.91\,\mathrm{m/s}$, $C_1 = 0.13\,\mathrm{m}^{-1}$, $C_2 = 1.57$, and l_c denotes the length of the vehicle.

According to the method of stability analysis from [29, 30], some assumptions should be predefined before detailed discussion.

Assumption 1. All vehicles move at an initial constant velocity when $t \le 0$.

Assumption 2. All vehicles follow one another without overtaking.

Assumption 3. The 1th vehicle is the preceding vehicle of the Nth vehicle.

Based on the above assumptions, subject to the optimal velocity function, we can easily obtain the following steady state of model (2):

$$\left[a_n^*(t), v_n^*(t), y_n^*(t)\right]^T = \left[0, V\left(\frac{L}{N}\right), \frac{L}{N}\right]^T, \tag{4}$$

where

L is the length of the ring road;

N is the total number of the vehicles.

This implies that, in the steady state, all vehicles run orderly with velocity $V(L/N)$ and headway distance L/N.

The objective of this paper is to design an adequate switching signal, not only to keep drivers in the loop, but also to guarantee that the cooperative driving system runs orderly with velocity $V(L/N)$ and headway distance L/N by rejecting the effect of the periodic switch, so as to keep drivers in the loop and effectively suppress the traffic congestion.

Aiming at manual driving scenario, we can obtain the error dynamic of the cooperative driving model by linearization of (2) at the steady state (4) as described in

$$\begin{aligned} \delta\dot{a}_n(t) = {} & \frac{k_n^{\mathrm{hum}}}{T_n}\left[\Lambda \delta y_n\left(t-\tau_n\right) - \delta v_n\left(t-\tau_n\right)\right] \\ & + \frac{\lambda_n^{\mathrm{hum}}}{T_n}\left(\delta v_{n+1}\left(t-\tau_n\right) - \delta v_n\left(t-\tau_n\right)\right) \\ & - \frac{1}{T_n}\delta a_n(t) \\ \delta\dot{v}_n(t) = {} & \delta a_n(t) \\ \delta\dot{y}_n(t) = {} & \delta v_{n+1}(t) - \delta v_n(t), \end{aligned} \tag{5}$$

where $\delta a_n(t) = a_n(t) - a_n^*(t)$, $\delta v_n(t) = v_n(t) - v_n^*(t)$, $\delta y_n(t) = y_n(t) - y_n^*(t)$, and Λ is the slope of the optimal velocity function at $y_n(t) = L/N$, $\Lambda = dV(y_n(t))/dy_n(t)|_{y_n(t)=L/N}$.

Let $s_n(t) = [\delta a_n(t), \delta v_n(t), \delta y_n(t)]^T$; the error dynamic (5) is rewritten as

$$\begin{aligned} \dot{s}_n(t) = {} & a_1 s_n(t) + b_1 s_n\left(t-\tau_n\right) + c_1 s_{n+1}(t) \\ & + d_1 s_{n+1}\left(t-\tau_n\right) + e_1 s_{n+1}\left(t-\tau_n^c-\tau_n\right), \end{aligned} \tag{6}$$

where

$$\begin{aligned} a_1 &= \begin{bmatrix} -\dfrac{1}{T_n} & 0 & 0 \\ 1 & 0 & 0 \\ 0 & -1 & 0 \end{bmatrix}, \\ b_1 &= \begin{bmatrix} 0 & -\dfrac{k_n^{\mathrm{hum}} + \lambda_n^{\mathrm{hum}}}{T_n} & \dfrac{k_n^{\mathrm{hum}}\Lambda}{T_n} \\ 0 & 0 & 0 \\ 0 & 0 & 0 \end{bmatrix}, \\ c_1 &= \begin{bmatrix} 0 & 0 & 0 \\ 0 & 0 & 0 \\ 0 & 1 & 0 \end{bmatrix}, \\ d_1 &= \begin{bmatrix} 0 & \dfrac{\lambda_n^{\mathrm{hum}}}{T_n} & 0 \\ 0 & 0 & 0 \\ 0 & 0 & 0 \end{bmatrix}, \\ e_1 &= \begin{bmatrix} 0 & 0 & 0 \\ 0 & 0 & 0 \\ 0 & 0 & 0 \end{bmatrix}. \end{aligned} \tag{7}$$

Same as the manual driving scenario, the error dynamic for automated driving can be obtained as

$$\begin{aligned}\delta\dot{a}_n(t) &= \frac{k_n^{\text{CACC}}}{T_n}\left[\Lambda\delta y_n(t-\tau_n) - \delta v_n(t-\tau_n)\right] \\ &\quad + \frac{\lambda_n^{\text{CACC}}}{T_n}\left(\delta v_{n+1}(t-\tau_n) - \delta v_n(t-\tau_n)\right) \\ &\quad + \frac{\gamma_n^{\text{CACC}}}{T_n}\delta a_{n+1}(t-\tau_n^c-\tau_n) - \frac{1}{T_n}\delta a_n(t) \\ \delta\dot{v}_n(t) &= \delta a_n(t) \\ \delta\dot{y}_n(t) &= \delta v_{n+1}(t) - \delta v_n(t).\end{aligned} \tag{8}$$

Reformulate the error dynamic (8) as

$$\begin{aligned}\dot{s}_n(t) &= a_2 s_n(t) + b_2 s_n(t-\tau_n) + c_2 s_{n+1}(t) \\ &\quad + d_2 s_{n+1}(t-\tau_n) + e_2 s_{n+1}(t-\tau_n^c-\tau_n),\end{aligned} \tag{9}$$

where

$$\begin{aligned}a_2 &= \begin{bmatrix} -\frac{1}{T_n} & 0 & 0 \\ 1 & 0 & 0 \\ 0 & -1 & 0 \end{bmatrix}, \\ b_2 &= \begin{bmatrix} 0 & -\frac{k_n^{\text{CACC}}+\lambda_n^{\text{CACC}}}{T_n} & \frac{k_n^{\text{CACC}}\Lambda}{T_n} \\ 0 & 0 & 0 \\ 0 & 0 & 0 \end{bmatrix}, \\ c_2 &= \begin{bmatrix} 0 & 0 & 0 \\ 0 & 0 & 0 \\ 0 & 1 & 0 \end{bmatrix}, \\ d_2 &= \begin{bmatrix} 0 & \frac{\lambda_n^{\text{CACC}}}{T_n} & 0 \\ 0 & 0 & 0 \\ 0 & 0 & 0 \end{bmatrix}, \\ e_2 &= \begin{bmatrix} \frac{\gamma_n^{\text{CACC}}}{T_n} & 0 & 0 \\ 0 & 0 & 0 \\ 0 & 0 & 0 \end{bmatrix}.\end{aligned} \tag{10}$$

Define the augmented vector $S(t) = [s_N(t)\ \ s_{N-1}(t)\ \ \cdots\ \ s_1(t)]^T$, which consists of the state variables of all vehicles in the platoon; according to (5)–(9), the state space expression of the cooperative platoon can be described as

$$\begin{aligned}\dot{S}(t) &= A_1 S(t) + B_1 S(t-\tau_n) + W_1 S(t-\tau_n^c-\tau_n) \\ &\qquad \text{for manual driving} \\ \dot{S}(t) &= A_2 S(t) + B_2 S(t-\tau_n) + W_2 S(t-\tau_n^c-\tau_n) \\ &\qquad \text{for automated driving,}\end{aligned} \tag{11}$$

where

$$\begin{aligned}A_1 &= \begin{bmatrix} a_1 & 0 & 0 & \cdots & c_1 \\ c_1 & a_1 & 0 & \cdots & 0 \\ 0 & c_1 & a_1 & \cdots & 0 \\ \vdots & \vdots & \vdots & \ddots & \vdots \\ 0 & \cdots & 0 & c_1 & a_1 \end{bmatrix}, \\ B_1 &= \begin{bmatrix} b_1 & 0 & 0 & \cdots & d_1 \\ d_1 & b_1 & 0 & \cdots & 0 \\ 0 & d_1 & b_1 & \cdots & 0 \\ \vdots & \vdots & \vdots & \ddots & \vdots \\ 0 & \cdots & 0 & d_1 & b_1 \end{bmatrix}, \\ W_1 &= \begin{bmatrix} 0 & 0 & 0 & \cdots & 0 \\ 0 & 0 & 0 & \cdots & 0 \\ 0 & 0 & 0 & \cdots & 0 \\ \vdots & \vdots & \vdots & \ddots & \vdots \\ 0 & \cdots & 0 & 0 & 0 \end{bmatrix}, \\ A_2 &= \begin{bmatrix} a_2 & 0 & 0 & \cdots & c_2 \\ c_2 & a_2 & 0 & \cdots & 0 \\ 0 & c_2 & a_2 & \cdots & 0 \\ \vdots & \vdots & \vdots & \ddots & \vdots \\ 0 & \cdots & 0 & c_2 & a_2 \end{bmatrix}, \\ B_2 &= \begin{bmatrix} b_2 & 0 & 0 & \cdots & d_2 \\ d_2 & b_2 & 0 & \cdots & 0 \\ 0 & d_2 & b_2 & \cdots & 0 \\ \vdots & \vdots & \vdots & \ddots & \vdots \\ 0 & \cdots & 0 & d_2 & b_2 \end{bmatrix}, \\ W_2 &= \begin{bmatrix} 0 & 0 & 0 & \cdots & e_2 \\ e_2 & 0 & 0 & \cdots & 0 \\ 0 & e_2 & 0 & \cdots & 0 \\ \vdots & \vdots & \vdots & \ddots & \vdots \\ 0 & \cdots & 0 & e_2 & 0 \end{bmatrix}.\end{aligned} \tag{12}$$

In order to keep drivers in the loop, which means drivers should always monitor the vehicle's running state and supervise the change of driving environment to an adequate level even if the longitudinal and/or lateral driving tasks were conducted by ADAS or automated driving system, the periodic switched driving strategy is put forth creatively to avoid driver fatigue or distraction which may cause serious traffic accidents and fateful consequences. From the perspective of control system, according to (11), the switched cooperative driving model can be regarded as a switched control system with varying time delay, which can be expressed as

$$\begin{aligned}\dot{S}(t) &= A_{\sigma(t_k)}S(t) + B_{\sigma(t_k)}S(t-\tau_n) \\ &\quad + W_{\sigma(t_k)}S(t-\tau_n^c-\tau_n) \quad t \in [t_k, t_{k+1}),\ k \in \mathbb{N},\end{aligned} \tag{13}$$

where $\sigma(t) : [0,+\infty) \to \overline{M} = \{1,2\}$ is a piecewise constant function called the switching signal which specifies the active subsystem at each time; only one subsystem is active at each sampling instant. For a switching sequence $0 \le t_0 < t_1 < \cdots < t_k < t_{k+1} < \cdots, \sigma(t)$ is everywhere continuous from the right; when $t \in [t_k, t_{k+1})$, we say the $\sigma(t_k)$th subsystem is active.

3. Stability Analysis

For all types of vehicle-to-vehicle cooperative driving systems, stability analysis is definitely an important issue. The existing stability analysis for the cooperative driving mainly focuses on the stability condition from the viewpoint of fluid mechanics. In fact, from the perspective of cyberphysical system, the cooperative driving system is a nonlinear and strong coupling complex system. So, it is meaningful to analyze the switched cooperative driving system from the control system point of view.

To obtain the main results, the following definition and lemmas are needed.

Definition 4. System (13) is said to be exponentially stable under switching signal $\sigma(t)$ if the solution of system (13) satisfies

$$\|S(t)\|^2 \le \kappa e^{-\beta(t-t_0)} \|S(t_0)\|_{c1}^2, \quad \forall t \ge 0, \tag{14}$$

for $\kappa \ge 1$ and $\beta > 0$, where $\|S(t)\|_{c1} = \sup_{-\tau_n^c-\tau_n \le \theta \le 0}\{\|S(t+\theta)\|, \|\dot{S}(t+\theta)\|\}$.

Lemma 5. *The matrix inequality*

$$\begin{bmatrix} S_{11} & S_{12} \\ S_{12}^T & -S_{22} \end{bmatrix} < 0, \tag{15}$$

where $S_{11} = S_{11}^T$, $S_{22} = S_{22}^T$ *are equivalent to* $S_{22} > 0$, $S_{11} + S_{12}S_{22}^{-1}S_{12}^T < 0$ *and* $S_{11} < 0$, $S_{22} + S_{12}^T S_{11}^{-1} S_{12} > 0$.

Lemma 6. *For any constant matrix* $Z > 0$, *scalar* $\tau_2 > \tau_1 \ge 0$, $\alpha \ne 0$, *and vector function* $x : [t-\tau_2, t-\tau_1] \to \mathbb{R}^n$, *one has*

$$\begin{aligned}&\frac{e^{\alpha\tau_2} - e^{\alpha\tau_1}}{\alpha} \int_{t-\tau_2}^{t-\tau_1} e^{\alpha(s-t)} x^T(s) Zx(s)\, ds \\ &\quad \ge \int_{t-\tau_2}^{t-\tau_1} x^T(s)\, ds\, Z \int_{t-\tau_2}^{t-\tau_1} x(s)\, ds.\end{aligned} \tag{16}$$

Proof. According to the Schur complement theorem, together with $Z > 0$, it holds that

$$\begin{bmatrix} e^{\alpha(s-t)} x^T(s) Zx(s) & x^T(s) \\ x(s) & e^{\alpha(t-s)} \end{bmatrix} \ge 0, \qquad t-\tau_2 \le s \le t-\tau_1; \tag{17}$$

taking the integral of formula (17) over the closed interval $[t-\tau_2, t-\tau_1]$ yields

$$\begin{bmatrix} \int_{t-\tau_2}^{t-\tau_1} e^{\alpha(s-t)} x^T(s) Zx(s)\, ds & \int_{t-\tau_2}^{t-\tau_1} x^T(s)\, ds \\ \int_{t-\tau_2}^{t-\tau_1} x(s)\, ds & \dfrac{e^{\alpha\tau_2} - e^{\alpha\tau_1}}{\alpha} Z^{-1} \end{bmatrix} \ge 0. \tag{18}$$

Then, by Schur complement theorem, one can get that the matrix inequality (18) is equivalent to (16). The proof of Lemma 6 is complete. □

In this paper, we use $P > 0$ ($\ge, <, \le 0$) to denote a positive definite (semidefinite, negative definite, and seminegative definite) matrix P. $\lambda_{\max}(P)$ and $\lambda_{\min}(P)$ denote the maximum and minimum eigenvalues of P. I and 0 represent identity matrix and zero matrix in the block matrix. The superscript "T" stands for matrix transpose, and the symmetric terms in a matrix are denoted by $*$.

Without loss of generality, we regard the cooperative system with manual driving considering the vehicle dynamics as an unstabilizable system according to the idea of stability analysis from [31, 32], when the drivers' sensitivity coefficient is low. In the research about the platoon of vehicles, the authors of [3] suggest that CACC can guarantee the string stability of the cooperative driving system, which means the effect of disturbances is attenuated along the vehicle string, due to the fact that vehicles can obtain the control information of preceding vehicles beforehand via intervehicle communication before the change of headway and velocity difference. So, the cooperative system with automated driving can be considered as a stabilizable system.

In the following, for the sake of convenience, suppose that M^- corresponds to the stable subsystems of (13) and M^+ corresponds to the unstable subsystems. Obviously, $M^- \cup M^+ = \overline{M}$.

It should be noted that once the unstable subsystem is active in a hybrid system, the stability of the whole system usually cannot be guaranteed. However, if the total activation time of the unstable subsystems is relatively small compared with the stable subsystems, the stability of the system can be achieved.

Theorem 7. *For given positive constants* α_1, α_2, τ_n, *and* τ_n^c, *if there exist symmetric positive definite matrices* P_k, Q_{ik}, R_{ik}, $i \in (1,2)$, $k \in (1,2,\ldots,p)$, *such that*

$$\Phi^k = \begin{bmatrix} \Phi_{11}^k & \Phi_{12}^k & \Phi_{13}^k \\ * & \Phi_{22}^k & \Phi_{23}^k \\ * & * & \Phi_{33}^k \end{bmatrix} < 0 \tag{19}$$

holds, for the Lyapunov-Krasovskii function candidate

$$\begin{aligned} V_k(t) = {} & S^T(t) P_k S(t) \\ & + \int_{t-\tau_n}^{t} e^{\alpha_k(u-t)} S^T(u) Q_{1k} S(u)\, du \\ & + \int_{t-\tau_n^c-\tau_n}^{t} e^{\alpha_k(u-t)} S^T(u) Q_{2k} S(u)\, du \\ & + \int_{-\tau_n}^{0}\int_{t+\theta}^{t} e^{\alpha_k(u-t)} \dot{S}^T(u) R_{1k} \dot{S}(u)\, du\, d\theta \\ & + \int_{-\tau_n^c-\tau_n}^{0}\int_{t+\theta}^{t} e^{\alpha_k(u-t)} \dot{S}^T(u) R_{2k} \dot{S}(u)\, du\, d\theta \end{aligned} \tag{20}$$

along the trajectory of system (13), there holds the following inequality:

$$V_k(t) \le e^{-\alpha_k(t-t_k)} V_k(t_k), \tag{21}$$

where

$$\begin{aligned} \Phi_{11}^k = {} & A_{\sigma(t_k)}^T P_k + P_k A_{\sigma(t_k)} + \alpha_k P_k + Q_{1k} + Q_{2k} \\ & + \tau_n R_{1k} + (\tau_n^c + \tau_n) R_{2k} + \frac{\alpha_k}{1-e^{\alpha_k \tau_n}} R_{1k} \\ & + \frac{\alpha_k}{1-e^{\alpha_k(\tau_n^c+\tau_n)}} R_{2k}, \\ \Phi_{12}^k = {} & P_k B_{\sigma(t_k)} + \frac{\alpha_k}{e^{\alpha_k\tau_n}-1} R_{1k} + \frac{\alpha_k}{e^{\alpha_k(\tau_n^c+\tau_n)}-1} R_{2k}, \\ \Phi_{13}^k = {} & P_k W_{\sigma(t_k)}, \\ \Phi_{22}^k = {} & -e^{\alpha_k\tau_n} Q_{1k} + \frac{\alpha_k}{1-e^{\alpha_k\tau_n}} R_{1k} + \frac{\alpha_k}{1-e^{\alpha_k(\tau_n^c+\tau_n)}} R_{2k}, \\ \Phi_{23}^k = {} & 0, \\ \Phi_{33}^k = {} & -e^{\alpha_k(\tau_n^c+\tau_n)} Q_{2k}, \\ \alpha_k = {} & \begin{cases} \alpha_1 & k \in M^- \\ \alpha_2 & k \in M^+. \end{cases} \end{aligned} \tag{22}$$

Proof. Suppose that system (13) is in the kth mode for $t \in [t_k, t_{k+1})$; one can construct the Lyapunov-Krasovskii function candidate for the switched cooperative driving model (13) as (20).

Taking the derivative of $V_k(t)$ along the trajectory of model (13) yields

$$\begin{aligned} \dot{V}_k(t) = {} & 2S^T(t) P_k \dot{S}(t) + S^T(u) Q_{1k} S(u) \\ & - e^{-\alpha_k\tau_n} S^T(t-\tau_n) Q_{1k} S(t-\tau_n) \\ & - \alpha_k \int_{t-\tau_n}^{t} e^{\alpha_k(u-t)} S^T(u) Q_{1k} S(u)\, du \\ & + S^T(t) Q_{2k} S(t) \\ & - e^{-\alpha_k(\tau_n^c+\tau_n)} S^T(t-\tau_n^c-\tau_n) Q_{2k} S(t-\tau_n^c-\tau_n) \\ & - \alpha_k \int_{t-\tau_n^c-\tau_n}^{t} e^{\alpha_k(u-t)} S^T(u) Q_{2k} S(u)\, du \\ & + \tau_n \dot{S}^T(t) R_{1k} \dot{S}(t) + (\tau_n^c + \tau_n) \dot{S}^T(t) R_{2k} \dot{S}(t) \\ & - \int_{t-\tau_n}^{t} e^{\alpha_k(u-t)} \dot{S}^T(u) R_{1k} \dot{S}(u)\, du \\ & - \alpha_k \int_{-\tau_n}^{0}\int_{t+\theta}^{t} e^{\alpha_k(u-t)} \dot{S}^T(u) R_{1k} \dot{S}(u)\, du\, d\theta \\ & - \int_{t-\tau_n^c-\tau_n}^{t} e^{\alpha_k(u-t)} \dot{S}^T(u) R_{2k} \dot{S}(u)\, du \\ & - \alpha_k \int_{-\tau_n}^{0}\int_{t+\theta}^{t} e^{\alpha_k(u-t)} \dot{S}^T(u) R_{2k} \dot{S}(u)\, du\, d\theta. \end{aligned} \tag{23}$$

Then, direct computation gives

$$\begin{aligned} & \dot{V}_k(t) + \alpha_k V_k(t) \\ & \quad = 2S^T(t) P_k \dot{S}(t) + \alpha_k S^T(t) P_k S(t) \\ & \qquad + S^T(u) Q_{1k} S(u) \\ & \qquad - e^{-\alpha_k\tau_n} S^T(t-\tau_n) Q_{1k} S(t-\tau_n) \\ & \qquad + S^T(t) Q_{2k} S(t) \\ & \qquad - e^{-\alpha_k(\tau_n^c+\tau_n)} S^T(t-\tau_n^c-\tau_n) Q_{2k} S(t-\tau_n^c-\tau_n) \\ & \qquad + \tau_n \dot{S}^T(t) R_{1k} \dot{S}(t) \\ & \qquad - \int_{t-\tau_n}^{t} e^{\alpha_k(u-t)} \dot{S}^T(u) R_{1k} \dot{S}(u)\, du \\ & \qquad + (\tau_n^c + \tau_n) \dot{S}^T(t) R_{2k} \dot{S}(t) \\ & \qquad - \int_{t-\tau_n^c-\tau_n}^{t} e^{\alpha_k(u-t)} \dot{S}^T(u) R_{2k} \dot{S}(u)\, du. \end{aligned} \tag{24}$$

In addition, by Lemma 6, it holds that

$$\begin{aligned} & - \int_{t-\tau_n}^{t} e^{\alpha_k(u-t)} \dot{S}^T(u) R_{1k} \dot{S}(u)\, du \\ & \quad \le \frac{\alpha_k}{1-e^{\alpha_k\tau_n}} \left(S^T(t) - S^T(t-\tau_n)\right) \\ & \quad \cdot R_{1k} \left(S(t) - S(t-\tau_n)\right), \\ & - \int_{t-\tau_n^c-\tau_n}^{t} e^{\alpha_k(u-t)} \dot{S}^T(u) R_{2k} \dot{S}(u)\, du \\ & \quad \le \frac{\alpha_k}{1-e^{\alpha_k(\tau_n^c+\tau_n)}} \left(S^T(t) - S^T(t-\tau_n^c-\tau_n)\right) \\ & \quad \cdot R_{2k} \left(S(t) - S(t-\tau_n^c-\tau_n)\right). \end{aligned} \tag{25}$$

Combining (24)–(25) yields

$$\begin{aligned}\dot{V}_k(t)+\alpha_kV_k(t)&\leq 2S^T(t)P_k\dot{S}(t)+\alpha_kS^T(t)P_kS(t)\\&+S^T(u)Q_{1k}S(u)-e^{-\alpha_k\tau_n}S^T(t-\tau_n)Q_{1k}S(t-\tau_n)\\&+S^T(t)Q_{2k}S(t)-e^{-\alpha_k(\tau_n^c+\tau_n)}S^T(t-\tau_n^c-\tau_n)\\&\cdot Q_{2k}S(t-\tau_n^c-\tau_n)+\tau_n\dot{S}^T(t)R_{1k}\dot{S}(t)\\&+\frac{\alpha_k}{1-e^{\alpha_k\tau_n}}\left(S^T(t)-S^T(t-\tau_n)\right)\\&\cdot R_{1k}(S(t)-S(t-\tau_n))+(\tau_n^c+\tau_n)\dot{S}^T(t)R_{2k}\dot{S}(t)\\&+\frac{\alpha_k}{1-e^{\alpha_k(\tau_n^c+\tau_n)}}\left(S^T(t)-S^T(t-\tau_n^c-\tau_n)\right)\\&\cdot R_{2k}(S(t)-S(t-\tau_n^c-\tau_n)).\end{aligned}\tag{26}$$

Let $\xi(t)=[S(t),S(t-\tau_n),S(t-\tau_n-\tau_n^c)]^T$; inequality (26) can be rewritten as

$$\dot{V}_k(t)+\alpha_kV_k(t)\leq\xi(t)^T\Phi^k\xi(t).\tag{27}$$

If (19) holds, one has

$$\dot{V}_k(t)+\alpha_kV_k(t)\leq 0.\tag{28}$$

Integrating inequality (28) from t_k to t gives $V_k(t)\leq e^{-\alpha_k(t-t_k)}V_k(t_k)$; the proof of Theorem 7 is complete. □

4. Design of the Switching Signal

Now, for any piecewise constant switching signal $\sigma(t)$ and any $0\leq t_0<t$, we let $T^-(t_0,t)$ (resp., $T^+(t_0,t)$) denote the total activation time of stabilizable (resp., unstabilizable) subsystems during (t_0,t). N^- represents the total number of system switching instances from stabilizable subsystems to unstabilizable subsystems during (t_0,t); conversely, N^+ is the number of system switching instances from unstabilizable subsystems to stabilizable subsystems.

Theorem 8. *For given positive constants α_1, α_2, μ_1, μ_2, τ_n, and τ_n^c, if there exist symmetric positive definite matrices P_k, Q_{ik}, R_{ik}, $i\in(1,2)$, $k\in(1,2,\ldots,p)$, such that Theorem 7 and inequalities*

$$\begin{aligned}P_k&\leq\mu_1P_l,\\Q_{ik}&\leq\mu_1Q_{il},\\R_{ik}&\leq\mu_1R_{il},\\&\forall k,l\in M\end{aligned}\tag{29}$$

hold, where $\mu_1\geq 1$, system (13) is exponentially stable for any switching signal satisfying the following condition:

$$\begin{aligned}\frac{T^-}{T^+}&\geq\frac{(\alpha_2+\alpha^*)}{(\alpha_1-\alpha^*)},\quad\alpha^*\in(0,\alpha_1)\\f(t_0,t)&\leq\frac{\alpha}{\ln(\mu_1\mu_2)},\quad\alpha\in(0,\alpha^*),\end{aligned}\tag{30}$$

where $f(T_1,T_2)=N^+/(T_2-T_1)$, $\mu_2=e^{(\alpha_1+\alpha_2)(\tau_n^c+\tau_n)}\mu_1$.

Proof. According to Theorem 7, we have

$$V_k(t)\leq\begin{cases}e^{-\alpha_1(t-t_k)}V_k(t_k), & k\in M^-\\e^{\alpha_2(t-t_k)}V_k(t_k), & k\in M^+.\end{cases}\tag{31}$$

At switching time t_k, based on (20) and (29), one has

$$V_k(t_k^+)\leq\begin{cases}\mu_1V_p(t_k^-), & k\in M^-,\ p\in M^+,\\\mu_2V_p(t_k^-), & k\in M^+,\ p\in M^-,\end{cases}\tag{32}$$

where $\mu_2=e^{(\alpha_1+\alpha_2)(\tau_n^c+\tau_n)}\mu_1$.

Without loss of generality, when $t\in[t_k,t_{k+1})$, $k\in M^+$, combining (31)-(32), along the trajectory of system (13), yields

$$V(t)\leq\mu_1^{N^-}\mu_2^{N^+}e^{-\alpha_1T^-(t_0,t)+\alpha_2T^+(t_0,t)}V(t_0).\tag{33}$$

By (20), it holds that

$$\begin{aligned}V(t)&\geq a\|S(t)\|^2,\\V(t_0)&\leq b\|S(t_0)\|_{c1}^2.\end{aligned}\tag{34}$$

Because of $N^-(t_0,t)\leq N^+(t_0,t)$,

$$\|S(t)\|^2\leq\frac{1}{a}V(t)\leq\frac{b}{a}e^{-(\alpha^*-\alpha)(t-t_0)}\|S(t_0)\|_{c1}^2.\tag{35}$$

According to Definition 4, system (13) is exponentially stable when (30) holds, where

$$\begin{aligned}a&=\min_{\forall k\in M}\lambda(P_k),\\b&=\max_{\forall k\in M}\lambda(P_k)+\tau_n\max_{\forall k\in M}\lambda(Q_{1k})\\&+(\tau_n+\tau_n^c)\max_{\forall k\in M}\lambda(Q_{2k})\\&+\frac{1}{2}(\tau_n)^2\max_{\forall k\in M}\lambda(R_{1k})\\&+\frac{1}{2}(\tau_n+\tau_n^c)^2\max_{\forall k\in M}\lambda(R_{2k}).\end{aligned}\tag{36}$$

The proof of Theorem 8 is complete. □

Theorem 8 provides a sufficient condition to choose a proper switching signal such that the switched cooperative driving system (13) is stable even in the existence of a certain amount of communication delay. The matrix P_k, Q_{ik}, R_{ik}, $i\in(1,2)$, $k\in(1,2,\ldots,p)$, reduces the conservativeness of the proposed method. Additionally, when α_1, α_2, μ_1, μ_2, τ_n, and τ_n^c are given, it is obvious that the condition in Theorems 7 and 8 takes the form of linear matrix inequality, which can be easily determined using the Matlab LMI toolbox.

Reference [7] investigated the driver behavior when resuming control from a highly automated vehicle; the study suggests that drivers require around 30–40 s to resume adequate and stable control of driving from automation. So, single manual driving time should not be less than 30 s when designing the switching signal for the sake of safety.

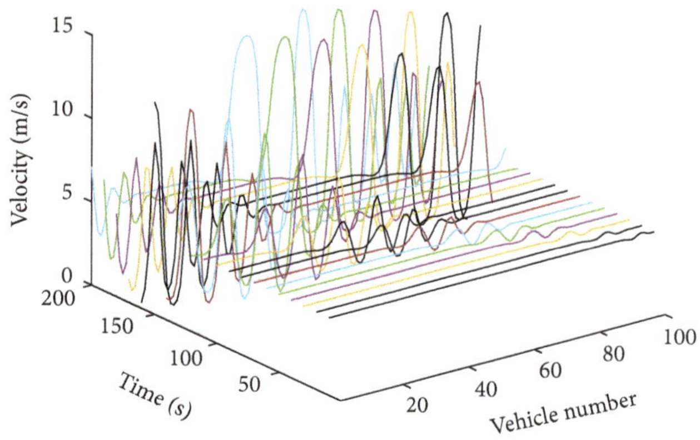

FIGURE 2: Velocity variation of all vehicles for manual driving.

Furthermore, the suggested switching period should not exceed 6 min to keep drivers in the loop according to the previous works about human factors.

According to these conditions and suggestions, a proper switching signal can be easily designed using the Matlab LMI toolbox.

5. Numerical Simulation

In this section, numerical examples are given to illustrate the effectiveness of the proposed methods for the switched cooperative driving system towards human vehicle copiloting situation considering the vehicle dynamic and realistic inter-vehicle communication.

Let us consider a case where 100 vehicles are running on a single lane under a periodic boundary condition, with the length $L = 1500$ m. We set the initial disturbance with reference to [32]. The simulation is conducted for 2000 s with the evolution of time. $x_1(0) = 1$ m, $x_n(0) = (n-1)L/N$ m, for $i = 2, \ldots, 100$, $v_n(0) = V(L/N)$, for $i = 1, \ldots, 100$.

In manual driving situation, the driver's sensitivity is $k_n^{\text{hum}} = 0.41$, and the sensitivity coefficient of the velocity difference is $\lambda_n^{\text{hum}} = 0.5$. Assuming that $T_n = 0.2$ and $\tau_n = 0.2$ s, Figure 2 plots the velocity variations of all vehicles, and Figure 3 shows the snapshots of the velocity of all vehicles at different times.

In Figure 2, the three axes represent the number of the vehicles, the time of the evolution, and the velocity of the vehicles, respectively. This figure depicts the space-time evolution of the platoon when the 1st vehicle suffers an unexpected speed reduction. From this figure, we can see that the speed fluctuation will become severe along the platoon sequence and the initial uniform traffic flow will evolve into a propagating backward traffic congestion under the disturbance.

To further exhibit the effect of the unexpected speed reduction, the velocity profiles were obtained by taking snapshots of the evolution depicted by Figure 2 at time steps $t = 10$ s, 300 s, and 2000 s, respectively, as Figure 3. Figure 3 shows that many vehicles have to decrease their velocities to zero, which means a severe traffic jam occurs at time step $t = 2000$ s. This reveals that the platoon is unstable under

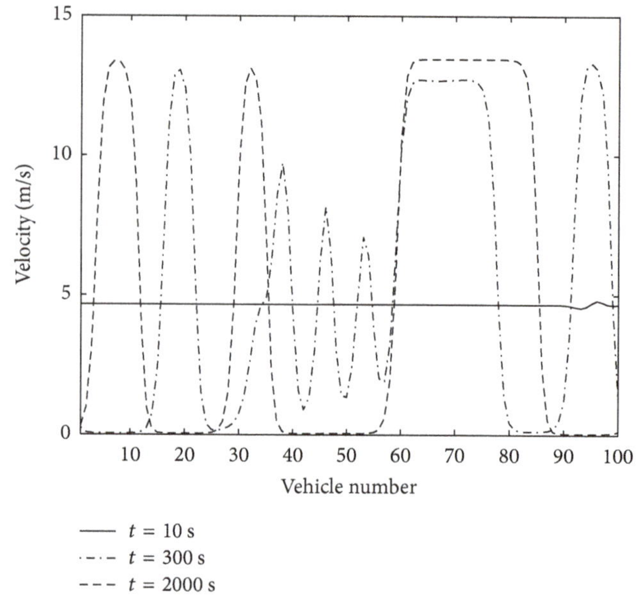

FIGURE 3: The snapshots of the velocity of all vehicles at different times for manual driving.

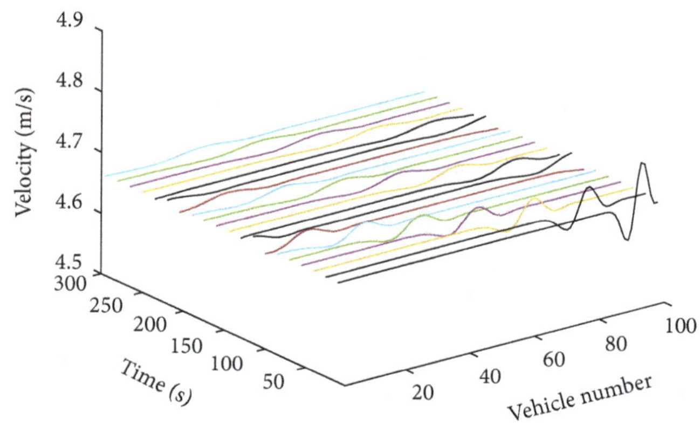

FIGURE 4: Velocity variation of all vehicles for automated driving.

manual driving which is suitable to the practice and the result of the traffic flow research.

Similarly, we set $k_n^{\text{CACC}} = 0.41$, $\lambda_n^{\text{CACC}} = 0.5$, and $\gamma_n^{\text{CACC}} = 0.3$ for the automated driving situation, when $T_n = 0.2$, $\tau_n = 0.2$ s, and $\tau_n^c = 0.2$ s; the simulations are shown in Figures 4 and 5.

In contrast to Figures 2 and 3, we can see clearly from Figures 4 and 5 that the amplitudes of the velocity waves shrink distinctly when intervehicle wireless communication is taken into account and traffic flow becomes much smoother. This reveals that the CACC system can stabilize traffic flow and suppress traffic congestion efficiently, which is consistent with the studies on automated driving system.

From Figures 2–5, it can be seen that the velocity fluctuations will decrease gradually as the simulation time develops under automated driving situation, even if there exists a certain amount of communication delay. However, the platoon is unstable and a traffic jam will occur under the manual driving case. These are in good agreement with the

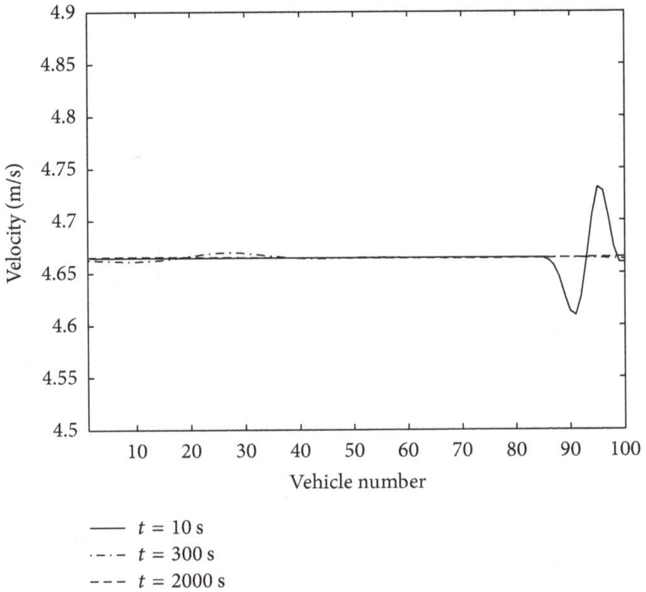

FIGURE 5: The snapshots of the velocity of all vehicles at different times for automated driving.

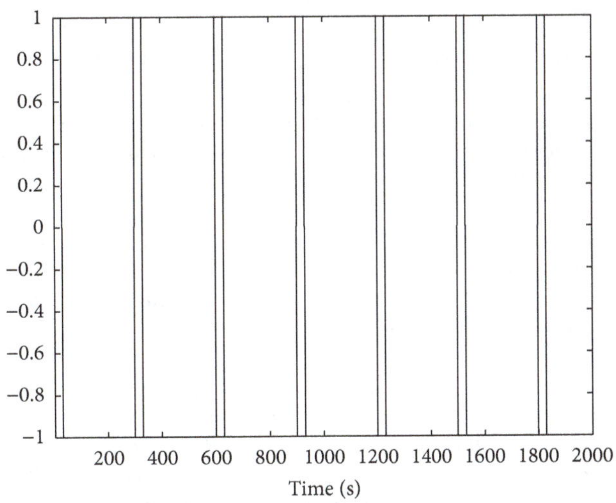

FIGURE 6: Switching signal.

aforementioned theoretical analysis, which can demonstrate indirectly that the constructed manual driving model and automated driving model are reasonable to a certain degree.

To demonstrate the switched cooperative driving model, according to Theorems 7 and 8, we can choose $\alpha_1 = 8$, $\alpha_2 = 2$, and $\mu_1 = 5$ using LMI toolbox, so the switching signal which satisfies condition (30) can be obtained as Figure 6 by selecting the adequate α^* and α; the cycle length is $T = 300$ s, and the duty cycle is Duty $= 0.1$, in which amplitude equal to one means the manual driving mode is active and amplitude equal to -1 means the current mode is automated driving. With the same initial condition and unexpected speed reduction of the 1st vehicle, the platoon revolution under the switching signal depicted by Figure 6 is shown in Figures 7 and 8.

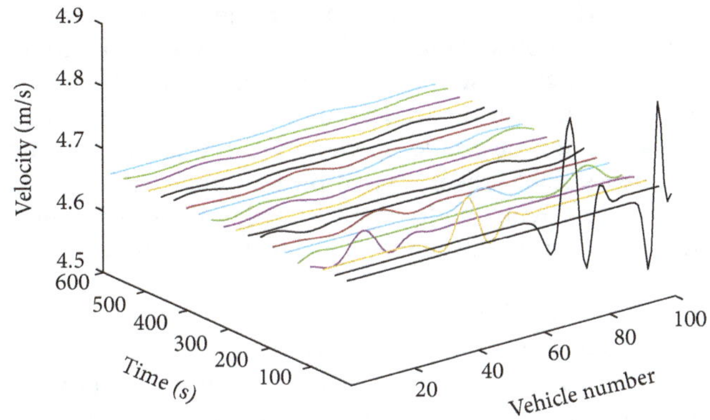

FIGURE 7: Velocity variation of all vehicles for switched cooperative driving model.

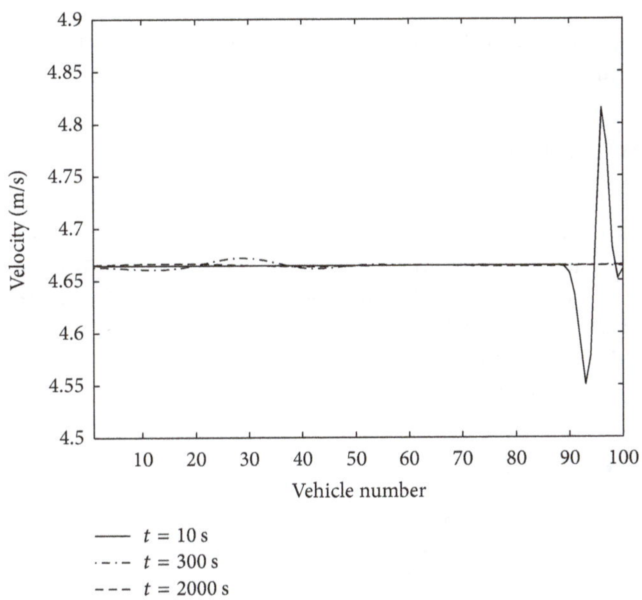

FIGURE 8: The snapshots of the velocity of all vehicles at different times for switched cooperative driving model.

Compared with Figures 4 and 5, we can see clearly from Figures 7 and 8 that the disturbance can also be well suppressed under the switching signal depicted by Figure 6, although the performance of the traffic flow declines slightly due to the periodic transitions between automated driving and manual driving. Fortunately, we can benefit significantly from the periodic switched driving strategy, by which drivers can be kept in the loop to respond to the takeover requests rapidly in L3 situation. Meanwhile, drivers' boredom, distraction, and fatigue caused by long-played supervising task can be prevented in L2 situation. Therefore, we can conclude that the numerical simulation is in good agreement with the analytical results.

6. Conclusion

In order to address the out-of-the-loop problem and distraction problem mentioned in the Introduction, this paper puts forth a periodic switched driving strategy to keep drivers always in the loop and prevent drivers' boredom, distraction, and even fatigue resulting from the supervising task. The potential impact of the transitions between automated and manual driving on traffic operations is revealed using the Lyapunov stability theory based on the proposed switched cooperative driving model towards the human vehicle copiloting situation. The switching signal with the comprehensive consideration of platoon stability and human factors can be easily designed based on the method presented by this paper. Numerical simulations illustrate the effectiveness of the proposed methods. The research results can provide scientific support for the optimization design of human vehicle copiloting.

Conflicts of Interest

The authors declare that there are no conflicts of interest regarding the publication of this paper.

Acknowledgments

This work was supported by the National Key R&D Program (Grant no. 2016YFB0100904), the National Natural Science Foundation of China (Grant no. 61573075), the Natural Science Foundation of Chongqing (Grant no. cstc2017jcyjBX0001), and the Project of Standardization and New Model for Intelligent Manufacture (Grant no. 2016ZXFB06002).

References

[1] P. Varaiya and S. E. Shladover, "Sketch of an IVHS systems architecture," in *Proceedings of the IEEE Vehicle Navigation and Information Systems Conference, VNISC '91*, vol. 2, pp. 909–922, 1991.

[2] M. Wang, W. Daamen, S. P. Hoogendoorn, and B. van Arem, "Rolling horizon control framework for driver assistance systems. Part I: mathematical formulation and non-cooperative systems," *Transportation Research Part C: Emerging Technologies*, vol. 40, pp. 271–289, 2014.

[3] S. Lam, J. Taghia, and J. Katupitiya, "Evaluation of a transportation system employing autonomous vehicles," *Journal of Advanced Transportation*, vol. 50, no. 8, pp. 2266–2287, 2016.

[4] SAE On-Road Automated Vehicle Standards Committee, "Taxonomy and definitions for terms related to on-road motor vehicle automated driving systems," SAE Standard J3016, pp. 1-16, 2014.

[5] N. H. T. S. Administration, Preliminary statement of policy concerning automated vehicles, pp. 1-14, 2013.

[6] T. M. Gasser and D. Westhoff, "BASt-study: definitions of automation and legal issues in Germany," in *Proceedings of the Road Vehicle Automation Workshop*, 2012.

[7] N. Merat, A. H. Jamson, F. C. H. Lai, M. Daly, and O. M. J. Carsten, "Transition to manual: driver behaviour when resuming control from a highly automated vehicle," *Transportation Research Part F: Traffic Psychology and Behaviour*, vol. 27, pp. 274–282, 2014.

[8] N. A. Stanton, M. S. Young, and G. H. Walker, "The psychology of driving automation: a discussion with Professor Don Norman," *International Journal of Vehicle Design*, vol. 45, no. 3, pp. 289–306, 2007.

[9] M. R. Endsley and E. O. Kiris, "The out-of-the-loop performance problem and level of control in automation," *Human Factors: The Journal of the Human Factors and Ergonomics Society*, vol. 37, no. 2, pp. 381–394, 1995.

[10] D. B. Kaber and M. R. Endsley, "Out-of-the-loop performance problems and the use of intermediate levels of automation for improved control system functioning and safety," *Process Safety Progress*, vol. 16, no. 3, pp. 126–131, 1997.

[11] R. Parasuraman and V. Riley, "Humans and automation: use, misuse, disuse, abuse," *Human Factors: The Journal of the Human Factors and Ergonomics Society*, vol. 39, no. 2, pp. 230–253, 1997.

[12] P. A. Hancock, R. J. Jagacinski, R. Parasuraman, C. D. Wickens, G. F. Wilson, and D. B. Kaber, "Human-automation interaction research: past, present, and future," *Ergonomics in Design*, vol. 21, no. 2, pp. 9–14, 2013.

[13] J. C. F. De Winter, R. Happee, M. H. Martens, and N. A. Stanton, "Effects of adaptive cruise control and highly automated driving on workload and situation awareness: a review of the empirical evidence," *Transportation Research Part F: Traffic Psychology and Behaviour*, vol. 27, pp. 196–217, 2014.

[14] N. A. Stanton and M. S. Young, "Vehicle automation and driving performance," *Ergonomics*, vol. 41, no. 7, pp. 1014–1028, 1998.

[15] N. Merat and J. D. Lee, "Preface to the special section on human factors and automation in vehicles: designing highly automated vehicles with the driver in mind," *Human Factors: The Journal of the Human Factors and Ergonomics Society*, vol. 54, no. 5, pp. 681–686, 2012.

[16] Z. J. Lu and J. C. F. de Winter, "A review and framework of control authority transitions in automated driving," in *Proceedings of the 6th International Conference on Applied Human Factors and Ergonomics (Ahfe '15) and the Affiliated Conferences*, pp. 2510–2517, 2015.

[17] B. D. Seppelt, M. N. Lees, and J. D. Lee, "Driver distraction and reliance: adaptive cruise control in the context of sensor reliability and algorithm limits," in *Proceedings of the 3rd International Driving Symposium on Human Factors in Driver Assessment, Training and Vehicle Design*, pp. 255–261, 2005.

[18] A. F. L. Larsson, K. Kircher, and J. A. Hultgren, "Learning from experience: familiarity with ACC and responding to a cut-in situation in automated driving," *Transportation Research Part F: Traffic Psychology and Behaviour*, vol. 27, pp. 229–237, 2014.

[19] G. F. B. Piccinini, C. M. Rodrigues, M. Leitão, and A. Simões, "Reaction to a critical situation during driving with Adaptive Cruise Control for users and non-users of the system," *Safety Science*, vol. 72, pp. 116–126, 2015.

[20] K. Kircher, A. Larsson, and J. A. Hultgren, "Tactical driving behavior with different levels of automation," *IEEE Transactions on Intelligent Transportation Systems*, vol. 15, no. 1, pp. 158–167, 2014.

[21] B. D. Seppelt and J. D. Lee, "Making adaptive cruise control (ACC) limits visible," *International Journal of Human-Computer Studies*, vol. 65, no. 3, pp. 192–205, 2007.

[22] J. Beller, M. Heesen, and M. Vollrath, "Improving the driver-automation interaction an approach using automation uncertainty," *Human Factors: The Journal of the Human Factors and Ergonomics Society*, vol. 55, no. 6, pp. 1130–1141, 2013.

[23] C. Ahlström and K. Kircher, "Review of real-time visual driver distraction detection algorithms," in *Proceedings of the 7th International Conference on Methods and Techniques in Behavioral Research, MB '10*, ACM, Eindhoven, The Netherlands, 2010.

[24] B. Donmez, L. N. Boyle, and J. D. Lee, "Safety implications of providing real-time feedback to distracted drivers," *Accident Analysis & Prevention*, vol. 39, no. 3, pp. 581–590, 2007.

[25] B. Donmez, L. N. Boyle, and J. D. Lee, "Mitigating driver distraction with retrospective and concurrent feedback," *Accident Analysis & Prevention*, vol. 40, no. 2, pp. 776–786, 2008.

[26] H. Zhang, M. Smith, and R. Dufour, "A final report of safety vehicles using adaptive interface technology (phase ii: Task 7c): visual distraction," Tech. Rep., Delphi Electron, Safety, Indiana, Ind, USA, 2008.

[27] T. W. Victor, J. L. Harbluk, and J. A. Engström, "Sensitivity of eye-movement measures to in-vehicle task difficulty," *Transportation Research Part F: Traffic Psychology and Behaviour*, vol. 8, no. 2, pp. 167–190, 2005.

[28] D. Helbing and B. Tilch, "Generalized force model of traffic dynamics," *Physical Review E: Statistical, Nonlinear, and Soft Matter Physics*, vol. 58, no. 1, pp. 133–138, 1998.

[29] Y. Li, H. Zhu, M. Cen, Y. Li, R. Li, and D. Sun, "On the stability analysis of microscopic traffic car-following model: A case study," *Nonlinear Dynamics*, vol. 74, no. 1-2, pp. 335–343, 2013.

[30] K.-T. Chong and D.-H. Roh, "A Lyapunov function approach to longitudinal control of vehicles in a platoon," *IEEE Transactions on Vehicular Technology*, vol. 50, no. 1, pp. 116–124, 2001.

[31] R. Jiang, Q. Wu, and Z. Zhu, "Full velocity difference model for a car-following theory," *Physical Review E: Statistical, Nonlinear, and Soft Matter Physics*, vol. 64, no. 1, Article ID 017101, 2001.

[32] H. Liu, D. Sun, and M. Zhao, "Analysis of traffic flow based on car-following theory: a cyber-physical perspective," *Nonlinear Dynamics*, vol. 84, no. 2, pp. 881–893, 2016.

2

Stability, Bifurcation, and Quenching Chaos of a Vehicle Suspension System

Chun-Cheng Chen[1] and Shun-Chang Chang[2]

[1]Graduate School of Technological and Vocational Education, National Yunlin University of Science and Technology, Yunlin 64002, Taiwan
[2]Department of Mechanical and Automation Engineering, Da-Yeh University, Changhua 51591, Taiwan

Correspondence should be addressed to Shun-Chang Chang; changsc@mail.dyu.edu.tw

Academic Editor: David F. Llorca

This study investigated the dynamics and control of a nonlinear suspension system using a quarter-car model that is forced by the road profile. Bifurcation analysis used to characterize nonlinear dynamic behavior revealed codimension-two bifurcation and homoclinic orbits. The nonlinear dynamics were determined using bifurcation diagrams, phase portraits, Poincaré maps, frequency spectra, and Lyapunov exponents. The Lyapunov exponent was used to identify the onset of chaotic motion. Finally, state feedback control was used to prevent chaotic motion. The effectiveness of the proposed control method was determined via numerical simulations.

1. Introduction

The suspension system of vehicles is meant to isolate the vehicle body and passengers from oscillations imposed by road irregularities and maintain continuous contact between the tires and surface of the road. Suspension systems provide a smoother ride and help the operator to maintain control of the vehicle over rough terrain or during sudden stops [1]. Hysteresis is an important aspect of suspension design, and the development of smart materials for vehicle suspension systems with hysteretic damping has been the subject of considerable studies [2–5]. The fact that the damping force in a vehicle's suspension system is hysteretic means that road excitation can be modeled as a multifrequency harmonic function. Therefore, automotive suspension systems may be considered a nonlinear hysteretic system with multifrequency excitation.

Nonlinear phenomena of codimension bifurcations have been studied extensively, particularly in engineering systems [6–8]. Tseng and Tung [9] examined the bifurcation behavior of a nonlinear structure with a magnetic actuator. Yagasaki [10] investigated nonlinear dynamics and the bifurcation behavior of a pendulum with feedback control. Tian et al. [11] studied the complex nonlinear dynamics of smooth, discontinuous oscillators. Numerous methods have been proposed for nonlinear control. Yin et al. [12] applied an approximated adaptive fuzzy control method to a nonlinear strict-feedback system with unmodeled dynamics. And, Yin et al. [13] developed a fault-tolerant controller based on the adaptive neural network backstepping technique for time-varying delay systems with unmodeled dynamics.

The fact that hysteresis is not a smooth process renders the characteristics of a suspension system with hysteretic damping inherently nonlinear. This can greatly complicate the design of a controller. Accurately predicting and/or controlling the performance of a system require that the effects of nonlinear phenomena be taken into account. The problem of rough road profiles and unwanted vehicle vibrations produced by kinematic excitation make it necessary to form accurate predictions of dynamic behavior over a range of operating conditions in the design of suspension systems. Bifurcation and chaos are crucial concepts in the study of stability in nonlinear systems, and numerous researchers have addressed issues pertaining to nonlinear dynamics in suspension systems [14–17] as well as chaotic responses in nonlinear air suspension systems [18, 19]. However, the issue of codimension bifurcations has largely been overlooked.

In this study, we examined a suspension system presenting codimension-two bifurcation and homoclinic orbits, both of which determine the nature of the dynamic behavior and the existence of strange attractors. We sought to calculate the strange attractor using numerical simulation in which external forces from a road profile are added to the system as a parameter with a specific range of values. We employed bifurcation diagrams, phase portraits, Poincaré maps, frequency spectra, and Lyapunov exponents to explain periodic and chaotic motions in a vehicle suspension system that exerts hysteretic nonlinear damping forces. Algorithms that computes Lyapunov exponents for smooth dynamic systems [20] were used to determine whether the system exhibits chaotic motion.

Chaotic behavior is normally considered undesirable due to the restrictions it imposes on the operating range of electrical and mechanical devices. Many engineering problems can only be resolved by driving a chaotic attractor in a periodic orbit. Numerous methods have been proposed for the control of chaos [21–23], including suspension systems [15, 24, 25]. In this study, we adopted the simple control method proposed by Cai et al. [26], in which state feedback is used to convert chaotic motion into stable motion. Numerical simulations demonstrate the feasibility of the proposed approach.

This paper is organized as follows. Section 2 describes the proposed vehicle suspension system using a quarter-car model under kinematic excitation by a road surface profile. Section 3 outlines the local bifurcation analysis used to identify codimension-two bifurcation and homoclinic orbits. Section 4 elucidates the complex dynamic behavior in the suspension system using numerical analysis, such as a bifurcation diagrams, phase portraits, Poincaré maps, and frequency spectra. Section 5 presents the Lyapunov exponent used to determine whether the system exhibits chaotic motion. A state feedback control technique for controlling chaos in the suspension system is presented in Section 6. Conclusions are drawn in Section 7.

2. Formulation of Problem

Figure 1 presents quarter-car model with a single degree-of-freedom and hysteretic nonlinear damping. The equation of motion for the system was formulated by Li et al. [14]:

$$m\ddot{x} + k_1 (x - x_0) + F_h = 0, \tag{1}$$

where F_h is an additional nonlinear hysteretic suspension damping and stiffness force, which is dependent on the relative displacement and velocity:

$$F_h = k_2 (x - x_0)^3 + C_1 (\dot{x} - \dot{x}_0) + C_2 (\dot{x} - \dot{x}_0)^3, \quad x_0 = \tilde{a}\sin(\Omega t). \tag{2}$$

In defining the relative vertical displacement as $y = x - x_0$, (1) can be written as follows:

$$\ddot{y} + \omega^2 y + B_1 y^3 + B_2 \dot{y} + B_3 \dot{y}^3 = A \sin(\Omega t), \tag{3}$$

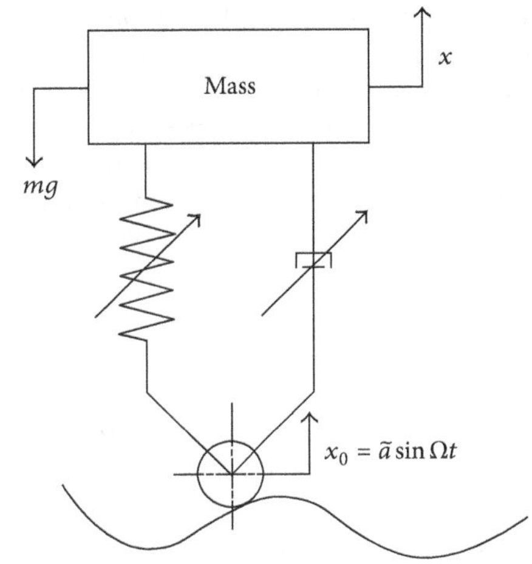

Figure 1: Schematic diagram of the quarter-car model.

Table 1: Nomenclature and parameters.

Notation	Value	Unit	Description
m	240	kg	The body mass
k_1	16×10^4	N/m	The suspension stiffness coefficient
k_2	30×10^4	N/m^3	The nonlinear hysteretic suspension stiffness coefficient
C_1	250	Ns/m	The linear damping coefficient
C_2	−25	Ns^3/m^3	The nonlinear hysteretic suspension viscous damping coefficient
$\tilde{a}$	0.05	m	The amplitude of road excitation
x_0		m	The road excitation
x		m	The body's vertical displacement
Ω		rad/s	The frequency of road excitation

where $\omega^2 = k_1/m$, $B_1 = k_2/m$, $B_2 = C_1/m$, $B_3 = C_2/m$, $A = \tilde{a}\Omega^2$.

The states are chosen such that $y_1 = y$, $y_2 = \dot{y}$. Thus, the state-space model for the suspension system can be written as

$$\dot{y}_1 = y_2, \tag{4a}$$

$$\dot{y}_2 = -\omega^2 y_1 - B_1 y_1{}^3 - B_2 y_2 - B_3 y_2{}^3. \tag{4b}$$

The amplitude of the road profile excitation is $\tilde{a} = 0$. Table 1 lists the values of the parameters in the above equations, which were obtained using the methods outlined by Li et al. [14].

3. Local Codimension-Two Bifurcation Analysis

Our main objective in this analysis is to observe the qualitative behavior of the system and how it changes as system parameters are varied. Changes occurring near the equilibrium point of the system are referred to as local bifurcation. The codimension of bifurcation refers to the minimum number of parameters that must be varied in order to observe a specific type of bifurcation [8].

Nonlinear analysis must be based on linear analysis. Bifurcation analysis is based on the Jacobian matrix in (4a) and (4b). When $(k_1, c_1) = (0, 0)$ and the Jacobian matrix has double-zero eigenvalues of the form of $\left[\begin{smallmatrix} 0 & 1 \\ 0 & 0 \end{smallmatrix}\right]$, we have the most degenerate situation (eigenvalues with zero real part), that is, a double-zero codimension-two bifurcation problem. Bifurcation equations can be written as follows:

$$\dot{y}_1 = y_2, \tag{5a}$$

$$\dot{y}_2 = -B_1 {y_1}^3 + \text{higher order terms}. \tag{5b}$$

Bifurcation equations (5a) and (5b) take the normal form because they are the simplest forms of the equations of motion [8, 27–30].

The blow-up technique is used to check whether the cubic terms in (5a) and (5b) determine local behavior. The blow-up technique involves a singular change to the coordinates that expands degenerate fixed points into circles containing a finite number of fixed points. If these fixed points are hyperbolic following the first change in the coordinates, then the local flow near the circle (and consequently flow near the original fixed point) remains stable with respect to higher-order terms. If these fixed points are all degenerate, then the coordinates must be changed again. If all of the fixed points become hyperbolic within a finite number of coordinate changes, then a system with only lower-order terms will exhibit the same qualitative behavior as the overall system [8]. The steps used in the blow-up method are as follows:

First Blow-Up

$$y_1 = r\cos\theta, \tag{6a}$$

$$y_2 = r\sin\theta. \tag{6b}$$

Equations (5a), (5b) and (6a), (6b) become

$$\dot{r} = r\sin\theta\left(1 - B_1 r^2 \cos^2\theta\right) + \text{h.o.t.}, \tag{7a}$$

$$\dot{\theta} = -\sin^2\theta - B_1 r^2 \cos^4\theta + \text{h.o.t.}. \tag{7b}$$

Stability analysis reveals that the system has two degenerate equilibrium points (0, 0) and $(0, \pi)$, which means that the blow-up procedure must be repeated. In the case of (0, 0), the power series in normal form is

$$\dot{r} = r\theta + \text{h.o.t.}, \tag{8a}$$

$$\dot{\theta} = -\theta^2 - B_1 r^2 + \text{h.o.t.}. \tag{8b}$$

Second Blow-Up. Let

$$r = \rho\sin\phi, \tag{9a}$$

$$\theta = \rho\cos\phi, \tag{9b}$$

so that (8a), (8b) and (9a), (9b) become

$$\dot{\rho} = -\rho^2\cos\phi\left(B_1\sin^2\phi + \cos^2\phi\right) + \text{h.o.t.}, \tag{10a}$$

$$\dot{\phi} = \rho\sin\phi\left(2\cos^2\phi - B_1\sin^2\phi\right) + \text{h.o.t.}. \tag{10b}$$

After dividing by the unaffected common component, ρ, (10a) and (10b) have two hyperbolic fixed points, (0, 0) and $(0, \pi)$. Thus, no further blow-up is required.

For the other equilibrium point, $(0, \pi)$, the power series in the normal form is the same as the power series used for (8a) and (8b), such that the same results are obtained. The degenerate field contains cubic terms of the lowest order and unfolding is performed using the two-parameter family, as follows:

$$\dot{y}_1 = y_2, \tag{11a}$$

$$\dot{y}_2 = K_1 y_1 + K_2 y_2 - B_1 {y_1}^3, \tag{11b}$$

where $K_1 = -k_1/m$, $K_2 = -c_1/m$. The higher-order terms in (11a) and (11b) are disregarded when applied as an alternative for (4a) and (4b).

The term "bifurcation set" refers to a set in parameter space on which the system bifurcates. It is obtained from unfolded equations in the normal form.

There are three equilibrium points in this case:

1. (0, 0) for all K_1, K_2;
2. $(\pm\sqrt{K_1/B_1})$ for $K_1 > 0$.

By checking the stability of these equilibrium points, we obtain the following:

(1) For $K_1 = 0$, a subcritical pitchfork bifurcation occurs if $K_2 > 0$ and a supercritical pitchfork bifurcation occurs if $K_2 < 0$.

(2) For $K_1 > 0$ *and* $K_2 = 0$, a rescaling technique and Melnikov's perturbation method are used to obtain the complete bifurcation set associated with homoclinic orbits in the unfolding bifurcation equations (11a) and (11b) [8].

Let $y_1 = \epsilon u$, $y_2 = \epsilon^2 v$, $K_1 = \epsilon^2 v_1$, $K_2 = \epsilon^3 v_2$, $t \to \epsilon t$.

Thus, after rescaling, (11a) and (11b) are rewritten as

$$\dot{u} = v, \tag{12a}$$

$$\dot{v} = v_1 u + \epsilon^2 v v_2 - B_1 u^3. \tag{12b}$$

Melnikov's method is then used to characterize the relationship between K_1 and K_2.

After a series of operations, the following expression is obtained:

$$K_2 = 0, \tag{13a}$$

$$K_1 = B_1 \epsilon^3. \tag{13b}$$

Equations (13a) and (13b) yield the homoclinic saddle connection, referred to as homoclinic bifurcation.

(3) For $K_1 < 0$ *and* $K_2 = 0$, the Jacobian matrix in (11a) and (11b) is written as

$$\text{DF}(0, 0) = \begin{bmatrix} 0 & 1 \\ K_1 & 0 \end{bmatrix}, \tag{14}$$

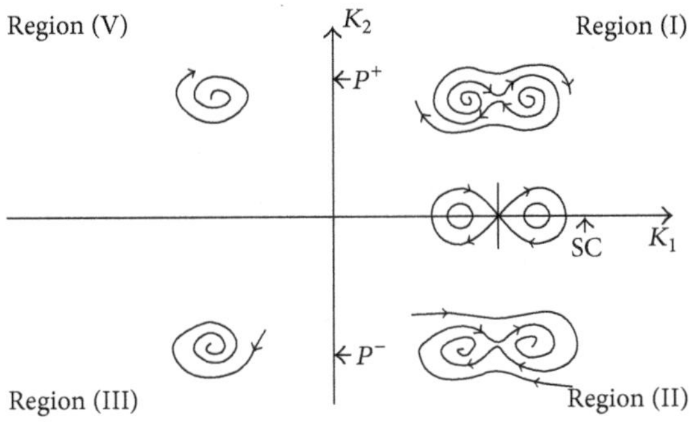

FIGURE 2: The bifurcation sets that are associated with typical phase portraits in each region, where P^+ represents a subcritical pitchfork, P^- represents a supercritical pitchfork, and SC represents a homoclinic bifurcation.

The eigenvalues for DF(0, 0) are

$$\lambda = \pm\sqrt{K_1}i. \tag{15}$$

For the following change in coordinates,

$$\begin{bmatrix} y_1 \\ y_2 \end{bmatrix} = \begin{bmatrix} 0 & 0 \\ -\sqrt{K_1} & 0 \end{bmatrix} \begin{bmatrix} U \\ V \end{bmatrix}. \tag{16}$$

Equations (12a) and (12b) become

$$\begin{bmatrix} \dot{U} \\ \dot{V} \end{bmatrix} = \begin{bmatrix} K_2 & -\sqrt{K_1} \\ \sqrt{K_1} & 0 \end{bmatrix} \begin{bmatrix} U \\ V \end{bmatrix} + \begin{bmatrix} 0 \\ \dfrac{B_1 V^3}{\sqrt{K_1}} \end{bmatrix}. \tag{17}$$

Let $g(U,V) = B_1V^3/\sqrt{K_1}$ and $f(U,V) = 0$. Clearly, it can be seen that

$$\begin{aligned} f_{vvv} &= 0, \\ f_{vuu} &= 0 \\ g_{vvu} &= 0, \\ g_{uuu} &= 0 \\ f_{uu} &= 0, \\ f_{uv} &= 0, \\ f_{vv} &= 0 \\ g_{uu} &= 0, \\ g_{uv} &= 0, \\ g_{vv} &= 0. \end{aligned} \tag{18}$$

According to the Hopf bifurcation theorem [8], coefficient **a** must not equal zero. In this case, the coefficient **a** equals zero; therefore, there is no Hopf bifurcation.

Numerical simulations using (4a) and (4b) were performed to check the validity of this bifurcation analysis. The parameter space was divided into several regions using bifurcation sets, as shown in Figure 2. The commercial package DIVPRK of IMSL in the FORTRAN subroutines for mathematics applications was used to solve these ordinary differential equations [31]. The resulting numerical simulations are presented in Figures 3–7. The results in each parametric region are discussed below.

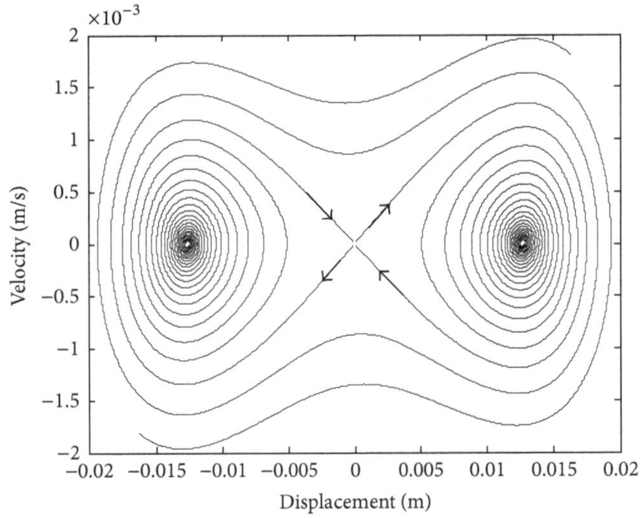

FIGURE 3: Numerical simulation for region (I), for $(K_1, K_2) = (0.02, 0.01)$.

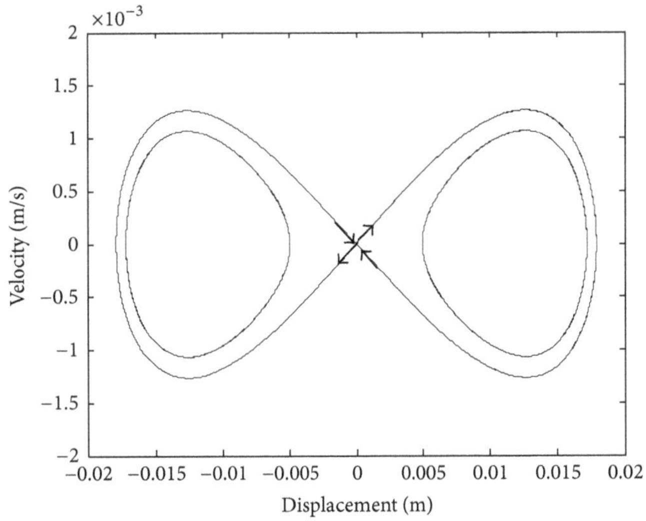

FIGURE 4: Numerical simulation for SC, for $(K_1, K_2) = (0.02, 0.0)$.

Region (I) includes sources and a saddle. If the initial conditions are in the neighborhood of the two sources, then the mass is distant from the equilibrium points (Figure 3). At the boundary between regions (I) and (II) (i.e., in the bifurcation set SC ($K_2 = 0$, $K_1 > 0$)), the system is an integrable system with homoclinic orbits (Figure 4). Therefore, if the damping term is disregarded, then the system is conservative. In region (II), the system is not integrable with two sinks and one saddle. If the initial conditions are in the neighborhood of the saddle, then the mass returns to the two sinks, which are stable points, as time approaches infinity (Figure 5). At the boundary between regions (II) and

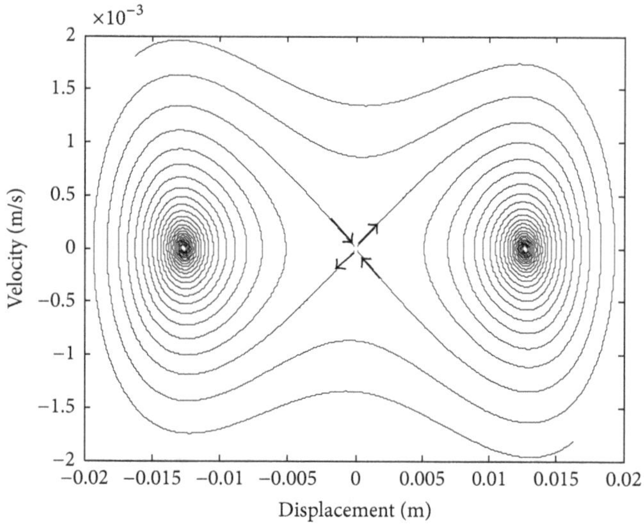

FIGURE 5: Numerical simulation for region (II), for $(K_1, K_2) = (0.02, -0.01)$.

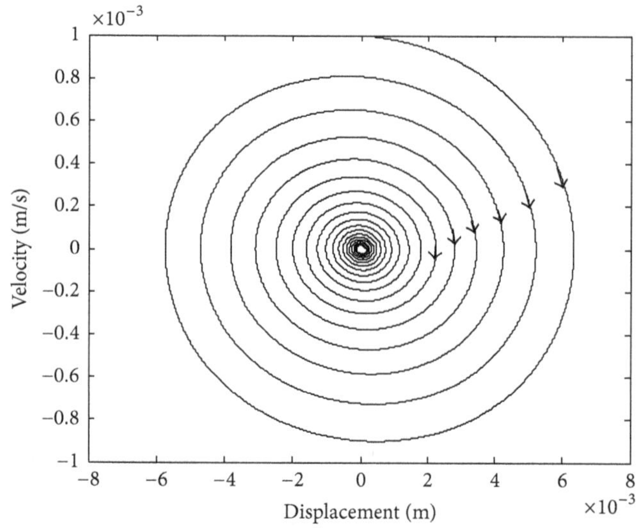

FIGURE 6: Numerical simulation for region (III), for $(K_1, K_2) = (-0.02, -0.01)$.

(III) (i.e., in bifurcation set P^-), the two sinks move closer to the saddle as the parameters approach bifurcation set, P^-, and the stability region shrinks to zero. In region (III), the two sinks disappear and the saddle (0, 0) becomes a sink. This is a supercritical pitchfork bifurcation with an attracting sink region (III) (Figure 6). From regions (III) to (V), the eigenvalues for the Jacobian matrix cross the imaginary axis; however, there is no Hopf bifurcation. This means that the dynamic behavior is linear, which is a similar situation to that when a stable equilibrium point becomes an unstable equilibrium point (Figure 7). Between region (V) and region (I), boundary P^+ must be crossed. This is called a subcritical pitchfork bifurcation.

In terms of the equilibria and Figure 2, K_1 determines the number of equilibrium points and K_2 determines the

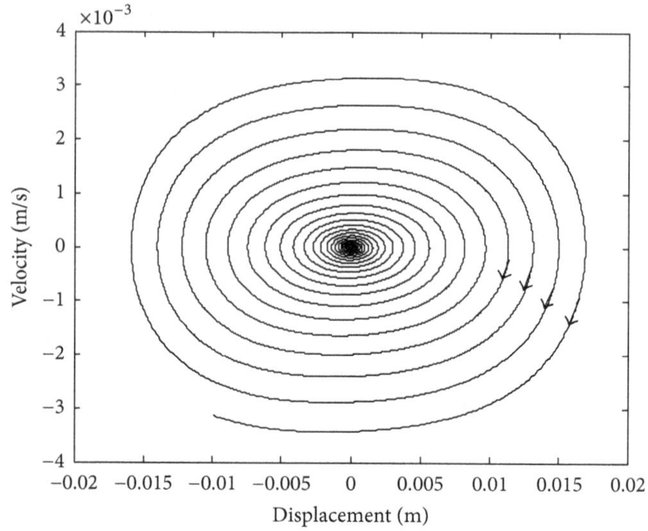

FIGURE 7: Numerical simulation for region (V), for $(K_1, K_2) = (-0.02, 0.01)$.

local behavior (e.g., stability or instability), and whether there are homoclinic orbits. Clearly, only a positive damping coefficient results in a stable system.

4. Analytical Tools for Observing Nonlinear Dynamics in a Vehicle Suspension System

In (3), the amplitude of the road excitation, $\tilde{a}$, equals 0.05 m. We performed a series of numerical simulations using (3) to determine the characteristics of this system. Dynamic behaviors can be more clearly observed over a range of parameter values in bifurcation diagrams, which are widely used to describe transitions from periodic motion to chaotic motion in dynamic systems. The resulting bifurcation diagram is presented in Figure 8. This figure clearly shows chaotic motion in region **A**. Region **C** presents period-1 motion and there is a stable equilibrium point. In this region, each response was characterized using a phase portrait, a Poincaré map (velocity versus phase angle), and a frequency spectrum. Figure 9 shows that the equilibrium point of the system is stable if parameter, Ω, exists in region **C**. It is well-known that road excitation covers a wide range frequencies, and higher vehicle velocities increase excitation from the road surface. The range of human sensitivity to frequency is from 4 to 8 Hz (about 25.12 rad/s to 50.24 rad/s); therefore, we limited this investigation to the case of $\Omega = 40.0$ rad/s. Figure 10 shows the periodic motion in the system in cases where the parameter exists in region **C** (i.e., $\Omega = 8.2$ rad/s and $\Omega = 40$ rad/s). The amplitude of road excitation is defined as 0.05 m, while the response of the system (even in periodic motion) has an amplitude of near 0.35 m (Figure 9). Clearly, this response is not acceptable. When parameter Ω exists in region **B**, quasiperiodic motion occurs in this region. As shown in Figures 11–13, the quasiperiodic trajectory is restricted to an annular-like region of the state space and differs considerably from periodic trajectories. When

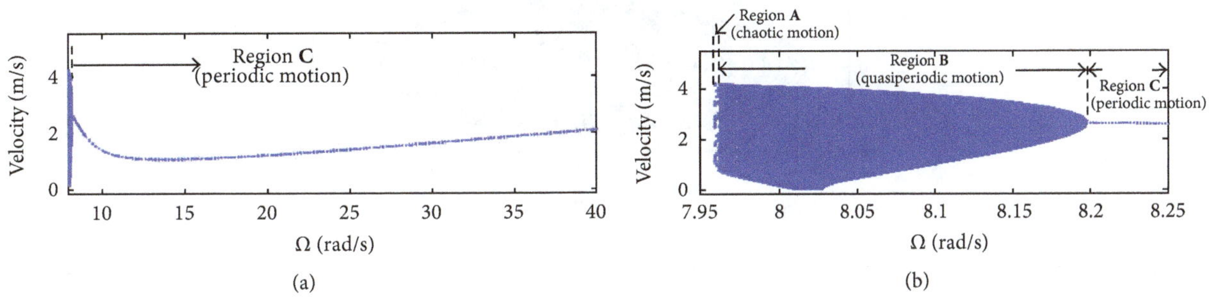

FIGURE 8: Bifurcation diagram for the system when the amplitude of the road excitation, $\tilde{a}$, is 0.05 m: (a) $\Omega = 7.96$ rad/s to 40 rad/s, (b) $\Omega = 7.96$ rad/s to 8.25 rad/s.

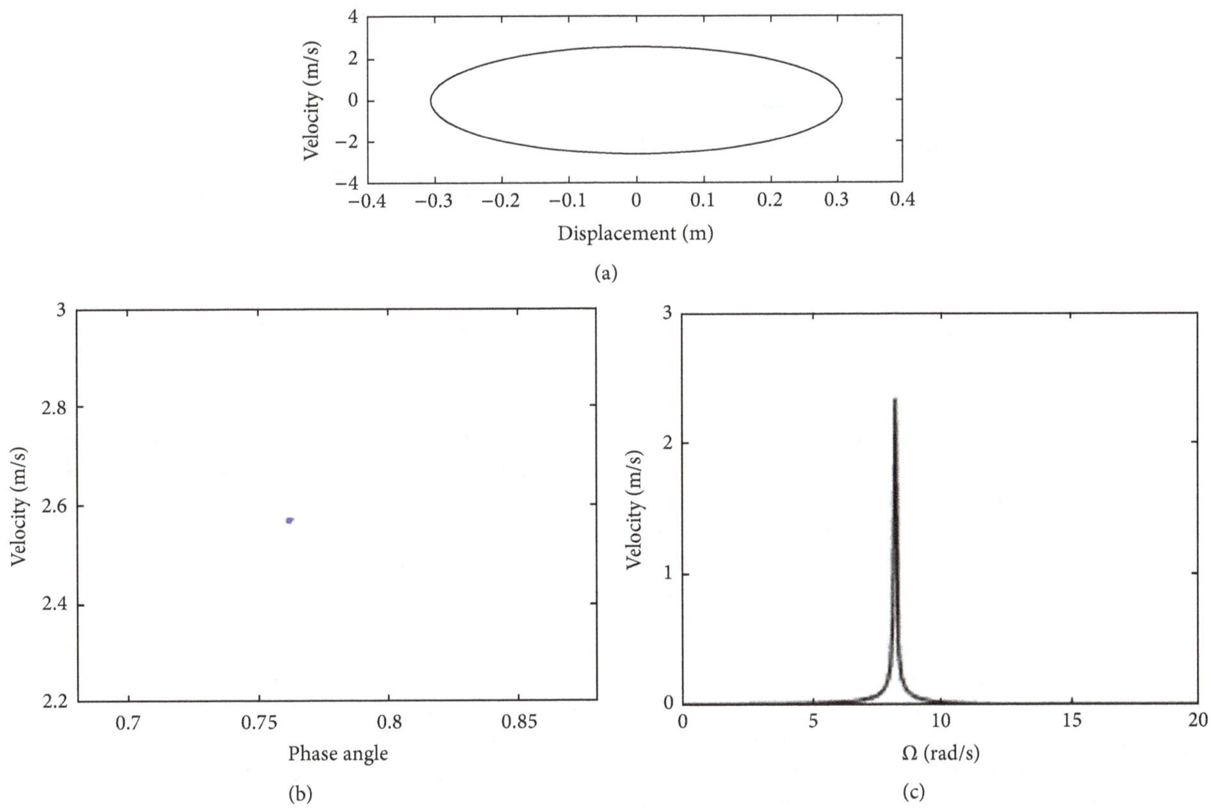

FIGURE 9: The period-one orbit at $\Omega = 8.2$ rad/s: (a) the phase portrait, (b) the Poincaré map, and (c) the frequency spectrum.

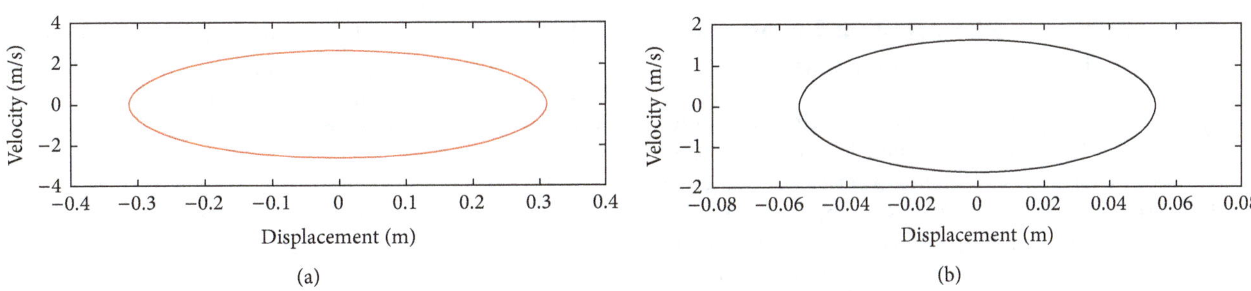

FIGURE 10: Phase portrait: (a) $\Omega = 8.2$ rad/s, (b) $\Omega = 40$ rad/s.

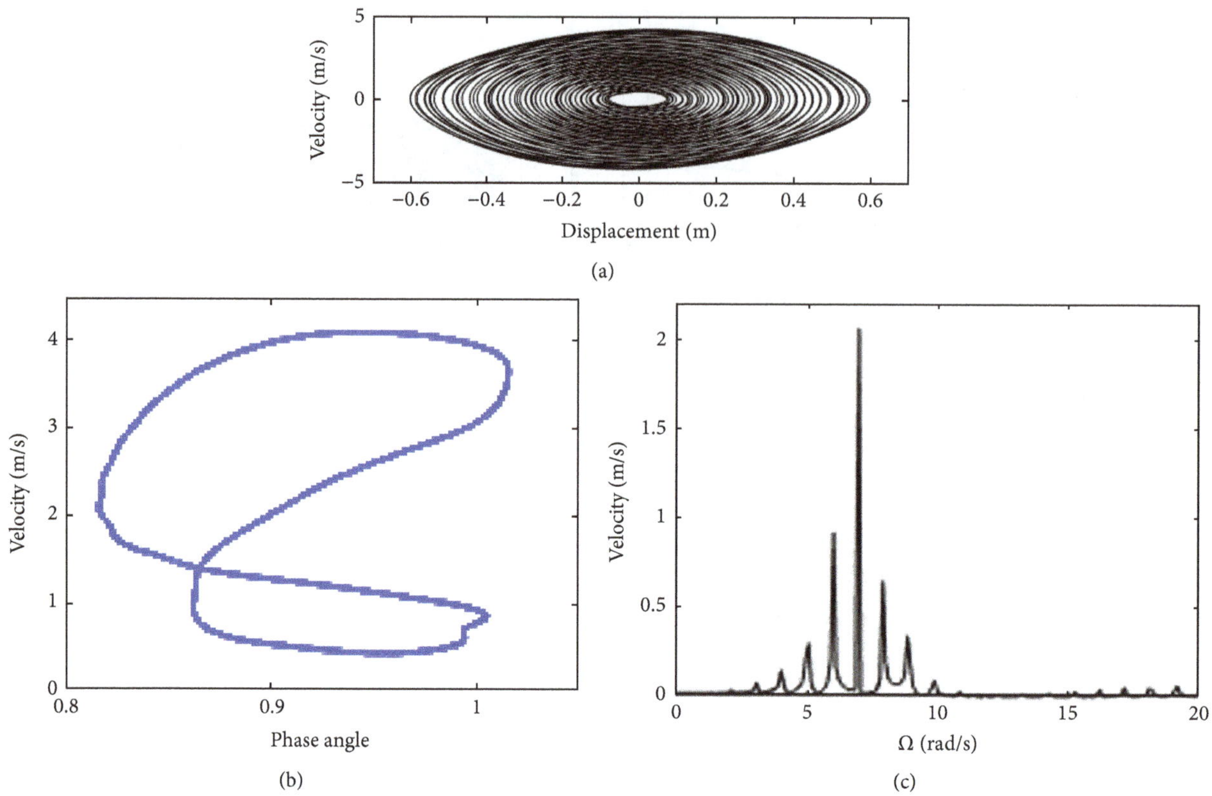

FIGURE 11: Quasiperiodic orbit at $\Omega = 8.0$ rad/s: (a) the phase portrait, (b) the Poincaré map, and (c) the frequency spectrum.

$\Omega = 8.0$, 8.05, or 8.15, the system exhibits a quasiperiodic motion and the map forms a continuous closed orbit in the Poincaré section, as shown in Figures 11–13. As the forcing frequency (Ω) continues decreasing into region **A** (Figure 14), chaotic motion produces chatter vibration. The transition sequence for the system in this paper is from periodic motion to quasiperiodic motion and finally to chaotic motion. The particular features of these two descriptors characterize the chaotic behavior. The Poincaré map contains an infinite set of points, which are referred to as a strange attractor. The frequency spectrum of the chaotic motion is a continuous broad spectrum. The strange attractor and continuous Fourier spectrum are strong indicators of chaos. Figure 14 shows clear indications of chaotic behavior in region **A**. Figures 14(a)–14(c), respectively, show the phase portraits, the Poincaré maps, and frequency spectra.

5. Analysis of Chaotic Phenomena: Lyapunov Exponent and Lyapunov Dimension

The analysis presented in Section 4 is insufficient to confidently identify chaotic motion in a suspension system. In this section, we adopt Lyapunov exponents to verify the existence of chaotic motion. In every dynamic system, the spectrum of Lyapunov exponents (λ) [20] indicates variations in the lengths, areas, and volumes of the phase space. Verifying the existence of chaos requires calculation of only the largest exponent to indicate whether nearby trajectories diverge ($\lambda > 0$) or converge ($\lambda < 0$), on average. Any bounded motion within a system that has at least one positive Lyapunov exponent is defined as chaotic. The Lyapunov exponents for periodic motion are not positive.

Figure 15 plots the evolution of the largest Lyapunov exponent in a suspension system. This value is computed using the algorithm presented by Wolf et al. [20]. Figure 14 shows that chaotic motion began at approximately $\Omega = 7.965$ rad/s, because at point (P_1), the sign of the largest Lyapunov exponent changed from negative to positive as the parameter, Ω, slowly decreased. Between points P_1 and P_2, the largest Lyapunov exponent approached zero. The existence of a forcing frequency, Ω, in these regions denotes the motion in the system as quasiperiodic. When the value of Ω is greater than P_1 (such as when $\Omega = 8.2$ rad/s), the Lyapunov exponents obtained using (3) are $\lambda_1 = -0.0010375$ and $\lambda_2 = -0.0013232$. The sum of these values is $\lambda_1 + \lambda_2 = -0.0023607$; this negative value demonstrates that the motion of mass at these values eventually converges to a stable limit cycle. Using the Lyapunov exponents for a dynamical system where $\lambda_1 \geq \cdots \geq \lambda_n$, Kaplan and Yorke [32] provided an estimation of the Lyapunov dimension, d_L, as follows:

$$d_L = j + \frac{1}{\left|\lambda_{j+1}\right|}\sum_{i=1}^{j}\lambda_i, \tag{19}$$

where j is the largest integer that satisfies $\sum_{i=1}^{j}\lambda_i > 0$. This technique yields a Lyapunov dimension of $d_L = 1$ for

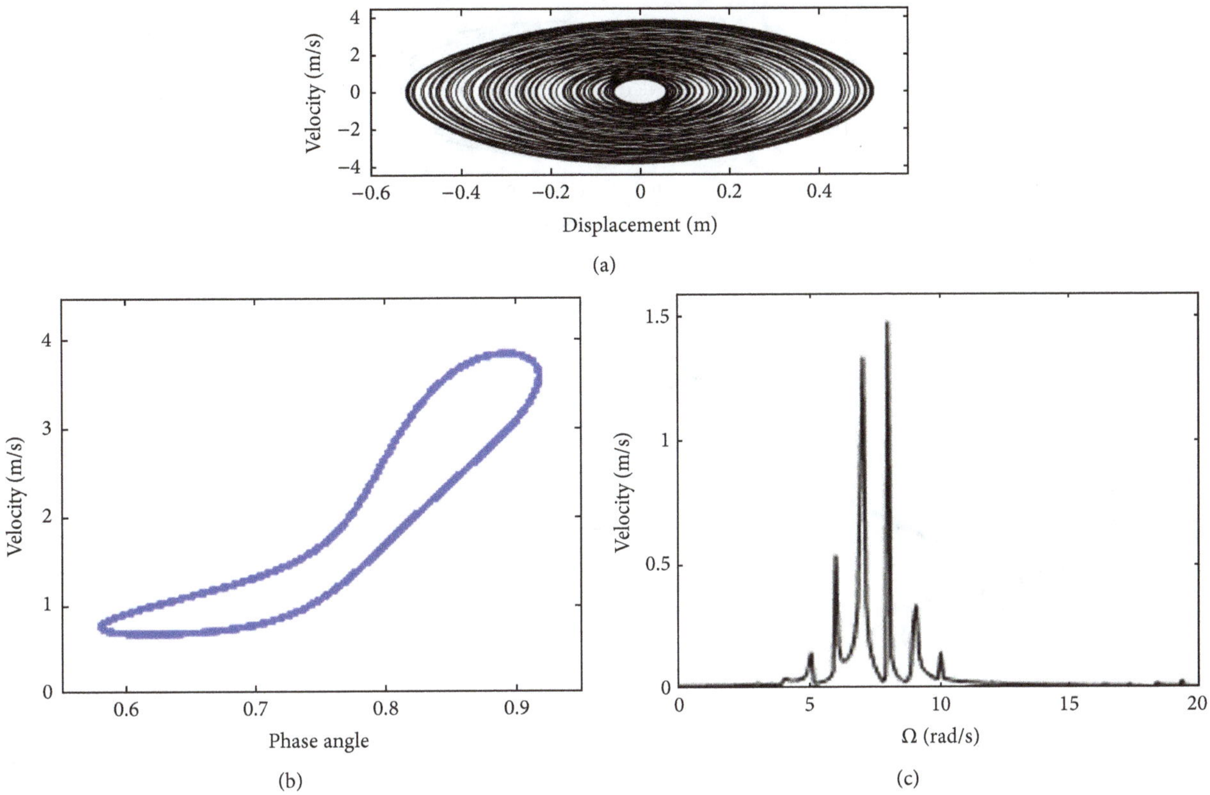

FIGURE 12: Quasiperiodic orbit at $\Omega = 8.05$ rad/s: (a) the phase portrait, (b) the Poincaré map, and (c) the frequency spectrum.

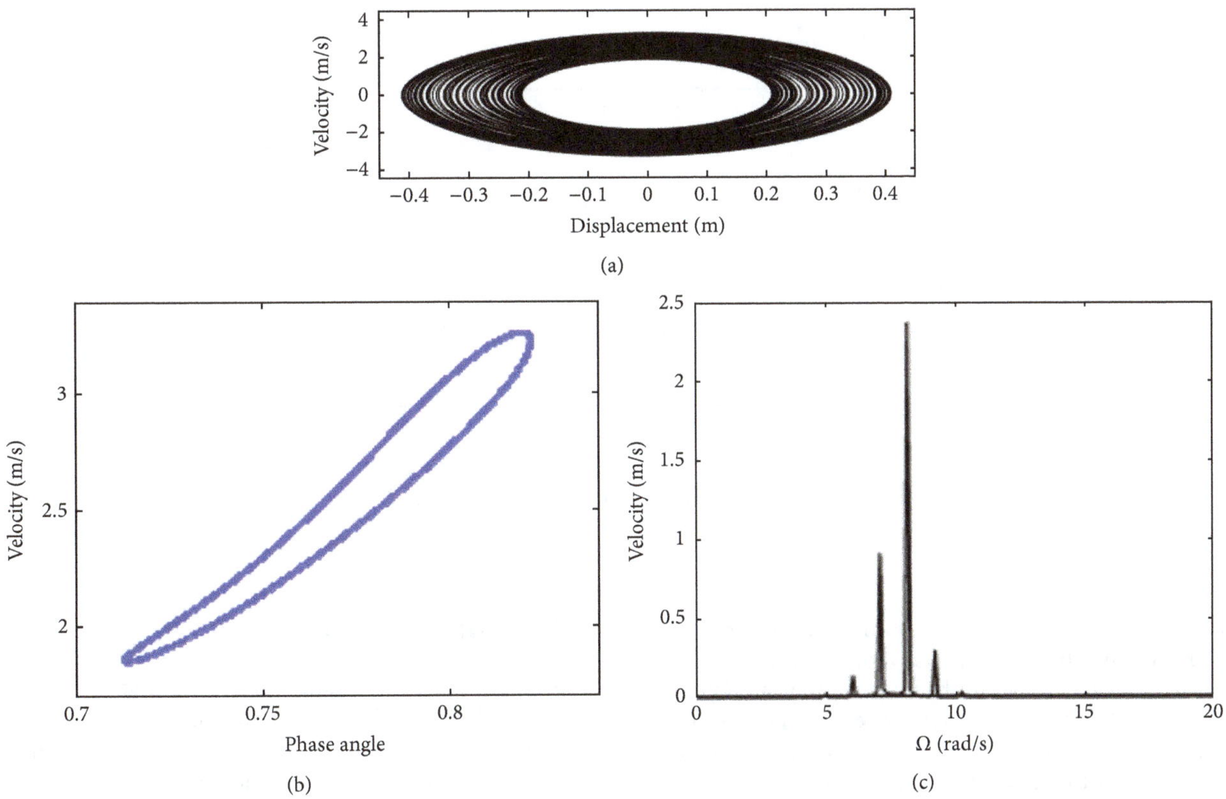

FIGURE 13: Quasiperiodic orbit at $\Omega = 8.15$ rad/s: (a) the phase portrait, (b) the Poincaré map, and (c) frequency spectrum.

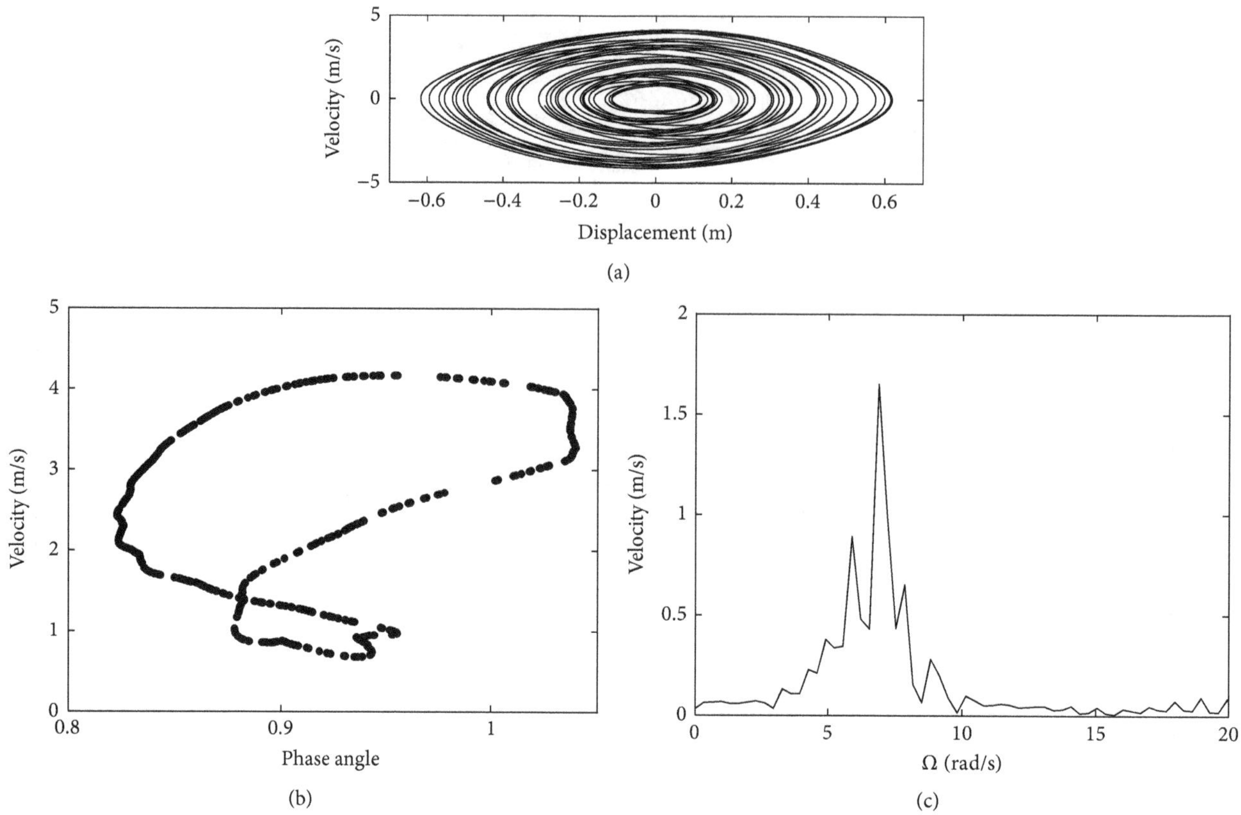

FIGURE 14: Chaotic motion at $\Omega = 7.964$ rad/s: (a) the phase portrait, (b) Poincaré map, and (c) the frequency spectrum.

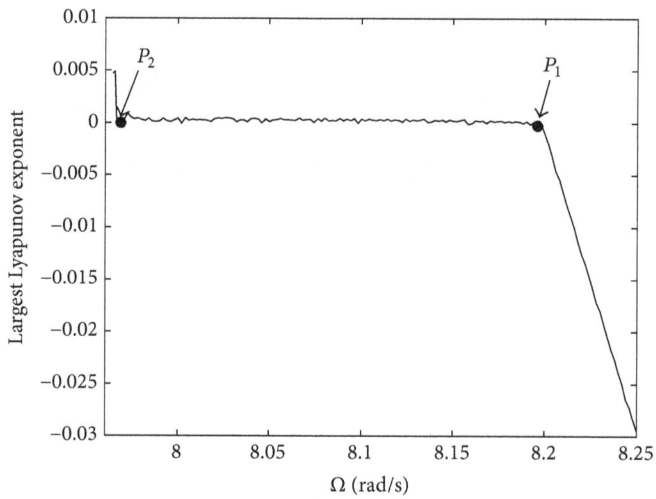

FIGURE 15: Evolution of the largest Lyapunov exponent.

$\Omega = 8.2$ rad/s. The fact that the Lyapunov dimension is an integer means that the system has periodic motion. When parameter Ω increases beyond the bifurcation point (such as when $\Omega = 7.964$ rad/s), the Lyapunov exponents are $\lambda_1 = 0.0054396$ and $\lambda_2 = -0.092153$, and the Lyapunov dimension is $d_L = 1.059$. In this case, the Lyapunov dimension is not an integer, indicating that the system exhibits chaotic motion.

6. Controlling Chaos

Improving the performance of a dynamic system and/or avoiding chaotic behavior require periodic motion, which is more important when working under specific conditions. Cai et al. [26] suggested a simple, effective control method that converts chaotic motions into periodic motion using the linear state feedback as a system variable. This approach

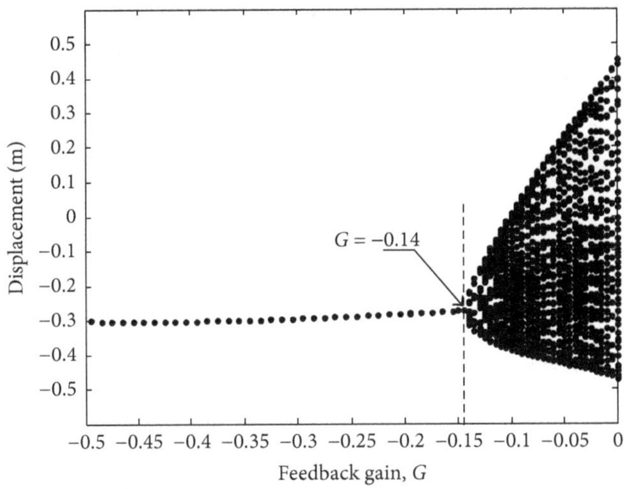

Figure 16: The bifurcation diagram for a system with state feedback control, where G denotes the feedback gain.

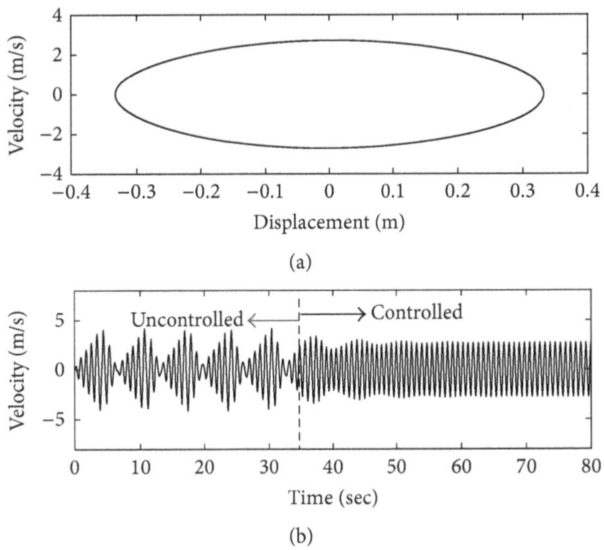

Figure 17: Transforming chaotic motion to a desired period-one orbit for $G = -0.3$ and $\Omega = 7.964$ rad/s: (a) the phase portrait for the controlled system when the state feedback control signal is introduced after 35 s and (b) the time responses of the velocity.

to achieving an n-dimensional dynamic system is explained briefly as follows:

$$\dot{x} = f(x,t), \tag{20}$$

where $x(t) \in R^n$ is the state vector and $f = (f_1, \ldots, f_i, \ldots, f_n)$, where f_i is a linear or a nonlinear function and f includes at least one nonlinear function. If $f_k(x,t)$ is a nonlinear function that gives chaotic motion in (20), then only one term of the state feedback for the available system variable, x_m, is added to the equation that includes f_k, as follows:

$$\dot{x}_k = f_k(x,t) + Gx_m, \quad k, m \in \{1, 2, \ldots, n\}, \tag{21}$$

where G is the feedback gain. Other functions remain in their original forms.

System (3) with state feedback control can be written as follows:

$$\begin{aligned} \frac{dy_1}{dt} &= y_2 \\ \frac{dy_2}{dt} &= -\omega^2 y_1 - B_1 y_1{}^3 - B_2 y_2 - B_3 y_2{}^3 \\ &\quad + \tilde{a}\Omega^2 \sin(\Omega t) + Gy_2. \end{aligned} \tag{22}$$

In the absence of state feedback control, $G = 0$ and (3) describes chaotic motion when $\Omega = 7.964$ rad/s. Figure 16 shows the resulting bifurcation diagram for a system with state feedback control. This figure clearly shows that chaotic motion occurs at approximately $G > -0.14$ and disappears at approximately $G \le -0.14$. When feedback gain, G, falls below -0.14, stable periodic motion occurs. As shown in Figure 17, a control signal is applied to the system at 35 s to govern the chaotic oscillation. In this manner, state feedback control can be used to suppress chaotic motion and improve the performance of a vehicle's suspension system. The form of the control gain in the original system equation (1) is as follows:

$$\widetilde{G} = G \cdot m. \tag{23}$$

The physical meaning of gain, $\widetilde{G}$, is control gain and the unit of the gain is N·s/kg·m.

In many general systems that involve chaotic dynamics, this state feedback technique does not require preliminary or on-line analysis of the system dynamics and can be implemented by adding the feedback of a suitable variable to the original system with sufficient control gain to suppress the development of chaos. In this regard, the state feedback control technique is simple and effective approach to chaos suppression.

7. Conclusions

This study examined the nonlinear dynamic behavior in a vehicle suspension system with hysteretic damping. Our objective was to develop a method to control chaotic vibration. Dynamic behavior was observed over the entire range of parameter values of the bifurcation sets and bifurcation diagram. Using bifurcation analysis, we identified a codimension-two bifurcation structure with double-zero degeneracy in the system. The bifurcation diagram indicates the existence of quasiperiodic orbit and chaotic motion, thereby demonstrating the effectiveness of using Lyapunov exponents as a tool for identifying chaotic motion. Finally, we proposed a state feedback control method to suppress chaotic motions and improve the ride provided by an automotive suspension system. Simulation results confirm the feasibility of the proposed method. Our results provide insight into the operating conditions under which undesirable dynamic motion takes place in vehicle suspension systems. Our results also provide a useful reference for engineers involved in the design of control systems for vehicle suspensions. It should be noted that in practical engineering, phenomena such as actuator faults, time-varying delays, unmodeled dynamics, and dynamics disturbances often occur simultaneously, and these problems are beyond the scope of the current study. In future work, we will seek to control nonlinear systems with time-varying delays.

Conflicts of Interest

The authors declare that there are no conflicts of interest regarding the publication of this paper.

Acknowledgments

The authors would like to thank the Ministry of Science and Technology of the Republic of China, Taiwan, for financially supporting this research under Contract no. MOST 105-2221-E-212-016.

References

[1] H. Chen, Z.-Y. Liu, and P.-Y. Sun, "Application of constrained script H_∞ control to active suspension systems on half-car models," *Journal of Dynamic Systems, Measurement, and Control*, vol. 127, no. 3, pp. 345–354, 2005.

[2] X. Song, M. Ahmadian, S. Southward, and L. Miller, "Parametric study of nonlinear adaptive control algorithm with magneto-rheological suspension systems," *Communications in Nonlinear Science and Numerical Simulation*, vol. 12, no. 4, pp. 584–607, 2007.

[3] L.-M. Gui, W.-K. Shi, and W. Liu, "A semi-active suspension design for off-road vehicle base on Magneto-rheological technology," in *Proceedings of the 2012 9th International Conference on Fuzzy Systems and Knowledge Discovery, FSKD 2012*, pp. 2565–2568, China, May 2012.

[4] F. Tu, Q. Yang, C. He, and L. Wang, "Experimental Study and Design on Automobile Suspension Made of Magneto-Rheological Damper," in *Proceedings of the International Conference on Future Energy, Environment, and Materials, FEEM 2012*, pp. 417–425, Hong Kong, April 2012.

[5] P. W. Nugroho, H. Du, W. Li, and G. Alici, "Implementation of Adaptive Neuro Fuzzy Inference System controller on magneto rheological damper suspension," in *Proceedings of the 2013 IEEE/ASME International Conference on Advanced Intelligent Mechatronics: Mechatronics for Human Wellbeing, AIM 2013*, pp. 1399–1403, Australia, July 2013.

[6] Arnold VI. Geometric Methods in the Theory of Ordinary Differential Equations. New York: Springer-Verlag. 1988.

[7] P. J. Holmes, "Bifurcations to divergence and flutter in flow-induced oscillations: a finite dimensional analysis," *Journal of Sound and Vibration*, vol. 53, no. 4, pp. 471–503, 1977.

[8] J. Guckenheimer and P. J. Holmes, *Nonlinear Oscillations, Dynamical System and Bifurcation of Vector Fields*, Springer-Verlag, New York, 2013.

[9] C.-Y. Tseng and P.-C. Tung, "Dynamics of nonlinear structure with magnetic actuator," *Japanese Journal of Applied Physics*, vol. 34, no. 1, pp. 374–382, 1995.

[10] K. Yagasaki, "Codimension-two bifurcations in a pendulum with feedback control," *International Journal of Non-Linear Mechanics*, vol. 34, no. 6, pp. 983–1002, 1999.

[11] R. Tian, Q. Cao, and S. Yang, "The codimension-two bifurcation for the recent proposed SD oscillator," *Nonlinear Dynamics*, vol. 59, no. 1-2, pp. 19–27, 2010.

[12] S. Yin, P. Shi, and H. Yang, "Adaptive Fuzzy Control of Strict-Feedback Nonlinear Time-Delay Systems With Unmodeled Dynamics," *IEEE Transactions on Cybernetics*, vol. 46, no. 8, pp. 1926–1938, 2016.

[13] S. Yin, H. Yang, H. Gao, J. Qiu, and O. Kaynak, "An adaptive NN-based approach for fault-tolerant control of nonlinear time-varying delay systems with unmodeled dynamics," *IEEE Transactions on Neural Networks and Learning Systems*, vol. 28, no. 8, pp. 1902–1913, 2017.

[14] S. Li, S. Yang, and W. Guo, "Investigation on chaotic motion in hysteretic non-linear suspension system with multi-frequency excitations," *Mechanics Research Communications*, vol. 31, no. 2, pp. 229–236, 2004.

[15] R. D. Naik and P. M. Singru, "Resonance, stability and chaotic vibration of a quarter-car vehicle model with time-delay feedback," *Communications in Nonlinear Science and Numerical Simulation*, vol. 16, no. 8, pp. 3397–3410, 2011.

[16] M. Borowiec, G. Litak, and M. I. Friswell, "Nonlinear response of an oscillator with a magneto-rheological damper subjected to external forcing," *Applied Mechanics and Materials*, vol. 5-6, pp. 277–284, 2006.

[17] G. Litak, M. Borowiec, M. I. Friswell, and K. Szabelski, "Chaotic vibration of a quarter-car model excited by the road surface profile," *Communications in Nonlinear Science and Numerical Simulation*, vol. 13, no. 7, pp. 1373–1383, 2008.

[18] J. Marzbanrad and A. Keshavarzi, "Chaotic vibrations of a nonlinear air suspension system under consecutive half sine speed bump," *Indian Journal of Science and Technology*, vol. 8, pp. 72–84, 2015.

[19] J. Marzbanrad and A. Keshavarzi, "A Chaotic viewpoint on a quarter-car air suspension model with numerical and experimental analysis[J]," *Journal of Cahiers Des Sciences*, vol. 23, 2014.

[20] A. Wolf, J. B. Swift, H. L. Swinney, and J. A. Vastano, "Determining Lyapunov exponents from a time series," *Physica D: Nonlinear Phenomena*, vol. 16, no. 3, pp. 285–317, 1985.

[21] E. Ott, C. Grebogi, and J. A. Yorke, "Controlling chaos," *Physical Review Letters*, vol. 64, no. 11, pp. 1196–1199, 1990.

[22] E. R. Hunt, "Stabilizing high-period orbits in a chaotic system: the diode resonator," *Physical Review Letters*, vol. 67, no. 15, pp. 1953–1955, 1991.

[23] C. Cai, Z. Xu, W. Xu, and B. Feng, "Notch filter feedback control in a class of chaotic systems," *Automatica*, vol. 38, no. 4, pp. 695–701, 2002.

[24] G. Litak, M. Borowiec, M. Ali, L. M. Saha, and M. I. Friswell, "Pulsive feedback control of a quarter car model forced by a road profile," *Chaos, Solitons & Fractals*, vol. 33, no. 5, pp. 1672–1676, 2007.

[25] H. Du, K. Y. Sze, and J. Lam, "Semi-active H_∞ control of vehicle suspension with magneto-rheological dampers," *Journal of Sound and Vibration*, vol. 283, no. 3–5, pp. 981–996, 2005.

[26] C. Cai, Z. Xu, and W. Xu, "Converting chaos into periodic motion by state feedback control," *Automatica*, vol. 38, no. 11, pp. 1927–1933, 2002.

[27] M. Ashkenazi and S.-N. Chow, "Normal forms near critical points for differential equations and maps.," *IEEE Transactions*

on Circuits and Systems II: Express Briefs, vol. 35, no. 7, pp. 850–862, 1988.

[28] Y. N. Bibikov, *Local theory of nonlinear analytic ordinary differential equations*, vol. 702 of *Lecture Notes in Mathematics*, Springer-Verlag, Berlin, 1979.

[29] L. O. Chu and H. Kokubu, "Normal Forms for Nonlinear Vector Fields— Part I: Theory and Algorithm," *IEEE Transactions on Circuits and Systems II: Express Briefs*, vol. 35, no. 7, pp. 863–880, 1988.

[30] P. Holmes, "dynamics of a nonlinear oscillator with feedback control. i: local analysis," *Journal of Dynamic Systems, Measurement, and Control*, vol. 107, no. 2, pp. 159–165, 1985.

[31] IMSL, Inc., Users Manual IMSL MATH/LIBRARY, 1989.

[32] J. L. Kaplan and J. A. Yorke, "Chaotic behavior of multidimensional difference equations," in *Functional Differential Equations and Approximation of Fixed Points (Proc. Summer School and Conf., Univ. Bonn, Bonn, 1978)*, vol. 730 of *Lecture Notes in Math.*, pp. 204–227, Springer, Berlin, Germany, 1979.

3

An Extended Car-Following Model at Signalised Intersections

Hongxing Zhao, Ruichun He, and Changxi Ma

School of Traffic and Transportation, Lanzhou Jiaotong University, Lanzhou 730070, China

Correspondence should be addressed to Ruichun He; tranman@163.com

Academic Editor: Antonio Comi

An extended car-following model is proposed on the basis of experimental analysis to improve the performance of the traditional car-following model and simulate a microscopic car-following behaviour at signalised intersections. The new car-following model considers vehicle gather and dissipation. Firstly, the parameters of optimal velocity, generalised force and full velocity difference models are calibrated by measured data, and the problems and causes of the three models are analysed with a realistic trajectory simulation as an evaluation criterion. Secondly, an extended car-following model based on the full optimal velocity model is proposed by considering the vehicle gather and dissipation. The parameters of the new car-following model are calibrated by the measured data, and the model is compared with comparative models on the basis of isolated point data and the entire car-following process. Simulation results show that the optimal velocity, generalised force, and full velocity difference models cannot effectively simulate a microscopic car-following behaviour at signalised intersections, whereas the new car-following model can avoid a collision and has a high fit degree for simulating the measured data of the car-following behaviour at signalised intersections.

1. Introduction

A car-following model is a mathematical description of the movement of a car in the same lane given the change in the moving state of the front car under the case of no overtaking; this model is also a link or bridge between macroscopic traffic flow theory and microscopic traffic flow model [1]. In recent years, an increasing number of mathematical models on car-following behaviour have been developed on the basis of experimental observations and theoretical analyses; the cornerstone for simulation modelling and traffic control, including early linear models proposed by Chandler et al. [2] and Herman et al. [3] and nonlinear models by Reuschel [4], Pipes [5], Gazis et al. [6], Newell [7], Bando et al. [8], Helbing and Tilch [9], and Jiang et al. [10], has been established. Chandler et al. [2] proposed the first prototype car-following model in 1958 at the General Motors Research Laboratory. This model was based on an intuitive hypothesis that the acceleration of a driver is proportional to the relative velocity of the front car. An initial calibration of this model used wire-linked vehicles to examine the responses of eight test subjects to a 'realistic' speed profile of a lead vehicle for 30 min on a test track. A rapid development of this model subsequently followed; Herman et al. [3] studied local and asymptotic stabilities of traffic flow in the fleet and discussed the influence of subadjacent leading vehicle car-following behaviour. Reuschel [4] and Pipes [5] first used analytical methods to study the problem of car-following behaviour; many subsequent models were developed from the Pipes model. Gazis et al. [6] discussed various nonlinear follow-the-leader models of traffic flow given available observational and experimental data that focus on steady-state flow equations. Newell [7] believed that the stimulus of a driver originates from the headway but not the relative speed; thus, Newell proposed a nonlinear car-following model with the headway as a stimulus. An optimal velocity (OV) model proposed by Bando et al. [8] is a favourable car-following model because it can describe many properties of actual traffic flow, such as instability of traffic flow, evolution of traffic congestion, and formation of stop-and-go waves. However, a comparison of empirical data indicates that the OV model exhibits significantly high acceleration and unrealistic deceleration; Helbing and Tilch [9] proposed a generalised force (GF) model by considering the negative velocity difference to overcome the abovementioned limitation. The simulation showed that the GF model is poor in the delay time of a car motion; owing to this problem, Jiang et al. [10] modified the model in 2001 by considering the negative and positive velocity differences and developed a full velocity difference (FVD) model. Subsequently, many similar works have been performed by several other authors [11–43].

The exploration of an urban signal intersection is important in traffic flow research because it can provide a theoretical basis for intersection signal timing. In recent years, the research on traffic flow theory of signal intersection has also gained certain favourable results. Sasaki and Nagatani [44] used the OV model to study traffic flow controlled by traffic lights on a single-lane roadway and found that the saturation of current occurs at the critical density; moreover, the critical density of a dynamical transition depends on the cycle time of the traffic light and strategy. Tang et al. [45] analysed the car-following behaviour in a road traffic system that contains signal lights and presented a traffic flow model that considers the signal light influence. However, the two models have not been calibrated or verified using the measured data; therefore, Yu et al. [46–49] proposed the car-following model which considers the influence of a signal lamp on the basis of the experimental analysis; the measured data were mined to find endogenous variables as the input variables of the car-following model by using a grey correlation analysis method; then, various extended car-following models were proposed on the basis of the FVD model; the results of the numerical simulations indicated that the extended car-following models can improve traffic flow stability. Fitting verification through the measured data remains lacking despite various car-following models that consider the influence of signal lamps in the literature [46–49]. Subsequently, Yu and Shi [50] used the OV and GF models to simulate the car-following behaviour at signalised intersections on the basis of the simulation system that analyses the problems and their causes; a new car-following model was proposed, and a fitting verification was conducted on the basis of the measured data. Similarly, Li and Yu [51] proposed a new model and improved the fit degree of the car-following model to the actual trajectory of the vehicle.

In summary, many important research results have been achieved on the car-following model. However, analysing a car-following behaviour at signalised intersections and studying the adaptability of the car-following models to vehicle trajectory simulations are limited. The measured data are used to calibrate the related car-following model and conduct the fitting analysis in the literature [50, 51], and the new car-following model at signalised intersections is proposed on the basis of the analysis results; however, only the car-following behaviours of the gather vehicles is considered, whereas those of the dissipated vehicles are disregarded. Thus, studying the car-following model at signalised intersections remains necessary, especially based on measured data; moreover, a new car-following model with a high fit degree must be proposed. In contrast to the literature [50, 51], the present study verifies the fit of the OV, GF, and FVD models for the measured car-following behaviour, analyses existing problems and their causes, and proposes a new car-following model on the basis of the measured data by considering the vehicle gather and dissipation behaviours.

The remainder of this paper is organised as follows. The data sources are introduced in Section 2. The analysis results of the OV, GF, and FVD models are presented in Section 3. An extended car-following model at signalised intersections is proposed in Section 4. A comparative analysis is conducted and then discussed in Section 5 to illustrate the effectiveness of the improved car-following model. Finally, the paper is summarised, and conclusions are drawn in Section 6.

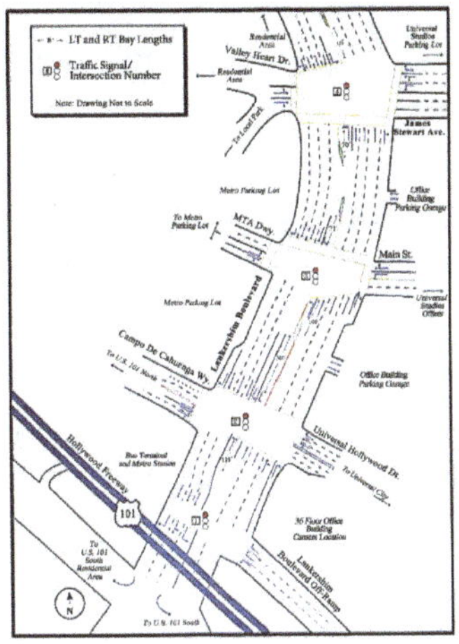

FIGURE 1: Study area: Lankershim Boulevard, where 101 is US Highway 101, and 1, 2, 3, and 4 are the intersection numbers.

2. Data Source

The Federal Highway Administration (FHWA) has been a leader in developing traffic simulation models since the 1970s. Commercial traffic simulation packages remained lacking in the marketplace before the FHWA assumed a leadership role. The Traffic Analysis Tools Programme of the FHWA launched the Next Generation SIMulation (NGSIM) programme to help achieve an extensive acceptance of using microsimulation systems and ensure that the tools provide accurate results. The data for the simulation analysis used in the present study are derived from the NGSIM programme, and the US Highway Agency provides the data free of charge [52]. Lankershim Boulevard dataset was collected under the NGSIM programme. The researchers for the NGSIM programme collected detailed vehicle trajectory data from Lankershim Boulevard in the Universal City neighbourhood of Los Angeles, CA, on 16 June 2005. The study area, as illustrated in Figure 1, consisted of bidirectional data of three- to four-lane arterial segments and a complete coverage of three signalised intersections; the study area is approximately 500 m in length. The Lankershim Boulevard dataset was selected as the data source. Data acquisition period is from 8:30 a.m. to 9:00 a.m.; at this time, the traffic state has just recovered from the morning to flat peak, the traffic density is relatively moderate, and the car-following behaviour is widespread and has robust representativeness. These data were collected using five video cameras mounted on the roof of a high building, and the vehicle trajectory data were transcribed from the video through NG-VIDEO, a customised software application that was developed for the NGSIM programme and recorded in the database.

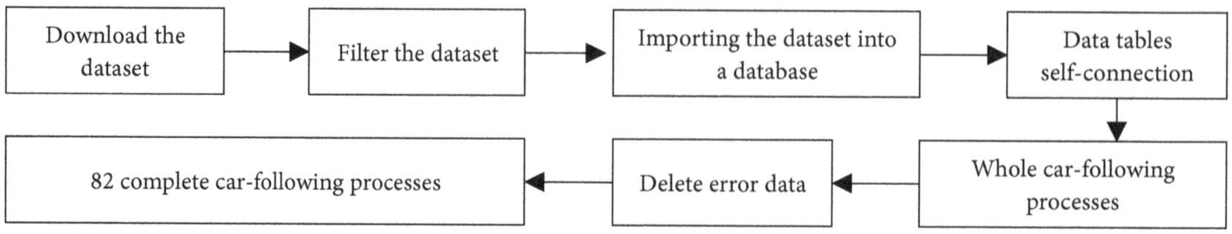

FIGURE 2: Data processing flow.

These vehicle trajectory data provided the precise location of each vehicle within the study area every one-tenth of a second. The data include solid-state information, such as vehicle number, type, width and length, and dynamic information, such as vehicle lane, intersection number, leading vehicle number, following vehicle number, acceleration, speed, distance headway, and global time. The vehicle number is the unique identification of each vehicle. We can determine the change in the position and motion state in the continuous time and then determine the relatively complete car-following process through the vehicle number and time information. The data of the car-following process can be divided into the road and intersection. The present study focuses on the car-following model at signalised intersections. The dataset must be processed to obtain the data required for this study. The data processing flow is illustrated in Figure 2. Firstly, we filtered the data that did not accumulate near the intersection on the basis of the intersection number. Then, we imported the filtered dataset into the database. Furthermore, we built a self-connection of data tables on the basis of the intersection number, leading vehicle number, following vehicle number and global time, and thereby obtaining numerous complete car-following processes. On this basis, we deleted the car-following process that had apparent errors in the dataset. Examples of these errors included the following: vehicle distance headway was less than 5 m, and the leading or following vehicle had unreasonable acceleration/deceleration and velocity. Then, we obtained 82 complete car-following processes that included 3,924 isolated point data. The data of the two car-following processes are summarised in Table 1. The data in Table 1 include only the car-following track for every 0.5 s given the limited length of this paper. In Table 1, $a_{n-1}(t)$ is the acceleration/deceleration of the $n-1$th car, that is, the leading car at time t; $v_{n-1}(t)$ is the velocity of the $n-1$th car at time t; $a_n(t)$ is the acceleration/deceleration of the *nth* car, that is, the following car at time t; $v_n(t)$ is the velocity of the nth at time t; $\Delta x_n(t)$ is the distance headway between the nth car and its leading car $n-1$ at time t. At the signal intersection, the vehicle cannot traverse the intersection smoothly, the vehicle decelerates, and the idle speed is maintained for a certain period. Then, the vehicle speeds away from the intersection when the red light is turned off and the green light is turned on. This phenomenon is called complete parking. Simultaneously, the vehicles that arrive during the green-light stage will decelerate given the influence of the queuing vehicles during the red-light stage and the queuing vehicles in front will accelerate. This phenomenon is called incomplete parking. In the two phenomena, the driver will have a different car-following behaviour. Thus, Processes 1 and 2 in Table 1 correspond to the situation of incomplete and complete parking, respectively, which are selected to represent the two types of car-following behaviour at the intersection. The two typical processes are simulated to judge the adaptability of the car-following model to analyse the car-following track.

3. Model Simulation and Analysis

The vehicle trajectory fitting simulation of the intersection is conducted on the basis of the OV, GF, and FVD models. The establishment of a traffic flow model will eventually return to the actual application. We can obtain a realistic model only by identifying the parameters of the model on the basis of numerous observations and an in-depth analysis of the actual traffic phenomenon. Thus, we randomly select 48 complete car-following processes that include 2626 isolated point data as learning samples to calibrate the model. Furthermore, the adaptive verification and analysis are conducted using the OV, GF, and FVD models which are calibrated by the actual car-following data.

3.1. OV Model. Bando et al. [8] proposed the OV model in 1995 to describe the car-following behaviour on a single-lane highway. The model was presented by introducing the OV function to optimise the OV in accordance with the distance headway. The motion formula is expressed as follows:

$$a_n(t) = \kappa \left(V\left(\Delta x_n(t)\right) - v_n(t) \right), \quad (1)$$

where $v_n(t)$ is the velocity of the nth car at time t; $\Delta x_n(t) = x_{n-1}(t) - x_n(t)$ is the distance headway between the nth car and its leading car $n-1$ at time t; $x_{n-1}(t)$ and $x_n(t)$ are the positions of the $n-1$th and nth cars, respectively; κ is the sensitivity parameter of the driver, and $V(\cdot)$ is the OV function which can be formulated as follows:

$$V\left(\Delta x_n(t)\right) = v_1 + v_2 \tanh\left(c_1\left(\Delta x_n(t) - l\right) - c_2\right), \quad (2)$$

where l is the vehicle length, and v_1, v_2, c_1, and c_2 are the parameters of the OV function that lacks physical meaning and must be calibrated.

Helbing et al. [9] calibrated the OV model, but the calibration data are obtained from the experimental site and are not the measured data at the signalised intersection. Thus, recalibrating the OV model by using the measured data of the car-following process at signalised intersections is required. The 2626 isolated point data are selected as learning samples, and the parameters of the OV model are calibrated

TABLE 1: Measured data of two typical car-following processes.

Car-following processes	Number of isolated points	$a_{n-1}(t)$ (m·s^{-2})	$v_{n-1}(t)$ (m·s^{-1})	$a_n(t)$ (m·s^{-2})	$v_n(t)$ (m·s^{-1})	$\Delta x_n(t)$ (m)
	1	-0.55	2.64	0.00	10.27	19.88
	2	-1.67	2.19	2.01	10.07	15.51
	3	0.95	1.97	-4.75	9.30	13.57
	4	-1.01	1.54	-5.65	6.98	10.45
	5	-1.19	1.16	-3.58	4.30	8.59
	6	1.52	0.94	-2.15	2.84	7.77
	7	-0.45	1.02	-1.85	1.34	7.40
1	8	-1.53	1.17	-1.03	1.23	7.32
	9	1.73	1.11	-1.29	1.16	7.33
	10	2.19	1.34	-0.52	1.44	7.24
	11	0.88	2.32	-0.74	1.06	7.54
	12	2.55	3.39	2.22	1.75	8.23
	13	2.93	4.54	2.31	2.33	9.13
	14	2.61	6.26	2.41	3.51	10.52
	15	3.62	8.01	2.62	5.23	11.91

Table 1: Continued.

Car-following processes	Number of isolated points	$a_{n-1}(t)$ (m·s^{-2})	$v_{n-1}(t)$ (m·s^{-1})	$a_n(t)$ (m·s^{-2})	$v_n(t)$ (m·s^{-1})	$\Delta x_n(t)$ (m)
	1	0.58	5.12	3.14	8.30	12.24
	2	-0.79	4.79	-3.62	5.41	9.91
	3	-2.97	3.88	-3.42	3.97	8.67
	4	-2.32	0.19	-2.26	1.33	7.87
	5	0.00	0.00	-1.99	0.45	7.66
	6	0.00	0.00	0.00	0.00	7.64
	7	0.00	0.00	0.00	0.00	7.64
	8	0.00	0.00	0.00	0.00	7.64
	9	0.00	0.00	0.00	0.00	7.64
	10	0.00	0.00	0.00	0.00	7.64
	11	0.00	0.00	0.00	0.00	7.64
	12	0.00	0.00	0.00	0.00	7.64
	13	0.00	0.00	0.00	0.00	7.64
	14	0.00	0.00	0.00	0.00	7.64
	15	0.00	0.00	0.00	0.00	7.64
	16	0.00	0.00	0.00	0.00	7.64
	17	0.00	0.00	0.00	0.00	7.64
	18	0.00	0.00	0.00	0.00	7.64
	19	0.00	0.00	0.00	0.00	7.64
	20	0.00	0.00	0.00	0.00	7.64
	21	0.00	0.00	0.00	0.00	7.64
	22	0.00	0.00	0.00	0.00	7.64
	23	0.00	0.00	0.00	0.00	7.64
	24	0.00	0.00	0.00	0.00	7.64
	25	0.00	0.00	0.00	0.00	7.64
2	26	0.00	0.00	0.00	0.00	7.64
	27	0.00	0.00	0.00	0.00	7.64
	28	0.00	0.00	0.00	0.00	7.64
	29	0.00	0.00	0.00	0.00	7.64
	30	0.00	0.00	0.00	0.00	7.64
	31	0.00	0.00	0.00	0.00	7.64
	32	0.00	0.00	0.00	0.00	7.64
	33	0.00	0.00	0.00	0.00	7.64
	34	0.00	0.00	0.00	0.00	7.64
	35	0.00	0.00	0.00	0.00	7.64
	36	0.00	0.00	0.00	0.00	7.64
	37	0.00	0.00	0.00	0.00	7.64
	38	0.00	0.00	0.00	0.00	7.64
	39	0.00	0.00	0.00	0.00	7.64
	40	0.00	0.00	0.00	0.00	7.64
	41	0.00	0.00	0.00	0.00	7.64
	42	0.00	0.00	0.00	0.00	7.64
	43	0.00	0.00	0.00	0.00	7.64
	44	0.00	0.00	0.00	0.00	7.64
	45	0.50	0.03	0.00	0.00	7.64
	46	1.84	1.36	0.00	0.00	8.10
	47	1.94	1.62	0.21	0.02	9.03
	48	3.57	4.50	-0.12	0.08	10.27
	49	1.27	5.03	3.48	1.25	11.62
	50	3.33	7.21	-0.25	2.41	13.50
	51	0.24	7.51	3.75	3.64	15.86
	52	1.42	8.58	3.19	7.56	17.00
	53	0.67	8.69	-3.39	8.29	17.97
	54	0.00	9.58	0.00	8.00	19.63

through the optimisation algorithm. In literature [50, 51, 53], particle swarm optimisation and genetic algorithms are used to calibrate the model, correspondingly. In contrast to literature [50, 51, 53], we select the standard artificial bee colony (ABC) algorithm which performs better than the other heuristic algorithms [54] in terms of calibrating the model.

The error between the actual and the simulated velocities can be used as the criterion of the consistency degree of the vehicle state in simulating the car-following process because the velocity can describe the running state of the vehicle in real time. The distance headway is the key parameter of the transition from the microscopic behaviour of a vehicle to the macroscopic phenomenon of the traffic flow. The error of the distance headway can directly reflect the consistency degree of the car-following phenomenon which is obtained by the individual vehicle on the basis of applying the OV model. Thus, the distance headway error should also be used as a criterion to evaluate the effect of the car-following model. That is, the mean absolute relative error of the velocity and the distance headway are considered, simultaneously, the optimisation criteria which can be formulated as follows:

$$EC = w_1 MARE(\Delta x) + w_2 MARE(v), \quad (3)$$

where w_1 and w_2 are the weight parameters. Simulating the microscopic behaviour of vehicles and analysing the macroscopic traffic phenomena are equally important because of the application of the car-following model for the analysis of the car-following behaviour at signalised intersection. That is, the distance headway and the velocity errors must be accorded with the same attention. Thus, we set w_1 and w_2 to 0.5 and 0.5, correspondingly. $MARE(\Delta x)$ and $MARE(v)$ correspond to the mean absolute relative error of the distance headway and the velocity and can be formulated as follows:

$$MARE(\Delta x) = \frac{1}{n}\sum_{i=1}^{n}\left|\frac{\Delta x - \Delta\widehat{x}}{\Delta\widehat{x}}\right|, \quad (4)$$

$$MARE(v) = \frac{1}{n}\sum_{i=1}^{n}\left|\frac{v - \widehat{v}}{\widehat{v}}\right|, \quad (5)$$

where n is the number of learning sample, Δx is the simulated distance headway of the OV model, $\Delta\widehat{x}$ is the empirical distance headway, v is the simulated velocity of the OV model, and $\widehat{v}$ is the empirical velocity.

The ABC algorithm is selected as the optimisation algorithm, and (3) can be minimised by optimising the parameters of the OV model. The number of employed, onlooker, and scout bees is set to 100, 100, and 1, respectively, and the maximum number of stagnation and the number of iteration in the ABC algorithm are set to 1600 and 3000, correspondingly. The same resulting parameters are obtained consistently through repeated analyses of the experiment. Moreover, the algorithm converges to these resulting parameters, and the number of first ABC iteration of the result is at most 2000 times. Thus, the algorithm has converged to the global optimal solution. The resulting parameters of the OV model are $\kappa = 0.70$, $v_1 = 2.04$, $v_2 = 1.99$, $c_1 = 18.07$, and $c_2 = 99.93$.

We selected two complete car-following processes, as presented in Table 1, to verify the adaptability of the calibrated OV model to the car-following trajectory at the signal intersection and test the OV model. In Table 1, the first point data of the two car-following process data are used as the initial state of the simulation. The complete simulation of the car-following process can then be obtained on the basis of the evolution of the model. The simulation results are depicted in Figures 3 and 4.

Figure 3(a) demonstrates the simulation analysis of the acceleration/deceleration of the OV model for car-following Process 1. This figure displays that the deceleration is faster than the measured data at the early deceleration stage, but the subsequent deceleration stage is slower than the measured data. Moreover, the simulation results of the OV model at the acceleration stage indicate a slow and insufficient acceleration over the measured data. Figure 3(b) exhibits the simulation analysis of the velocity of the OV model in car-following Process 1. This figure presents that the velocity of the early deceleration stage is less than the measured data, but the velocity of the subsequent deceleration stage is greater than the measured data. Moreover, the velocity of the OV model at the acceleration stage is less than the measured data. Figure 3(c) illustrates the simulation analysis of the distance headway of the OV model in car-following Process 1. This figure depicts that the distance headway of the entire deceleration stage is greater than the measured data given the significant deceleration at the early deceleration stage. The distance headway of the subsequent deceleration and early acceleration stages is greater than the measured data given the minimal deceleration at the succeeding deceleration stage. Furthermore, the simulation value of the distance headway for a period is close to 5 m given the minimal deceleration. Thus, the following car has the risk of a rear-end collision. The comprehensive analysis of Figures 3(a), 3(b), and 3(c) indicate the same results as those obtained through the simulation analysis of acceleration/deceleration, velocity, and distance headway. The OV model simulation of car-following Process 1 has the problem of significant deceleration at the early deceleration stage and minimal deceleration at the subsequent deceleration stage. Thus, the distance headway of the following car is approximately 5 m for a period, thereby resulting in the risk of rear-end collision for the following car. Furthermore, the OV model demonstrates a slow and insufficient acceleration over the measured data at the acceleration stage. Therefore, the fit degree is relatively low between the simulation and the measured trajectories when the OV model simulates car-following Process 1.

Figure 4(a) presents the simulation analysis of the acceleration/deceleration of the OV model in car-following Process 2. This figure shows that the deceleration is slower than the measured data at the deceleration stage, and the problem of the OV model simulation of car-following Process 2 at the acceleration stage is consistent with that of car-following Process 1. The simulation results of the OV model show slow and insufficient acceleration over the measured data. Figure 4(b) illustrates the simulation analysis of the velocity of the OV model in car-following Process 2. This figure depicts that the velocity is greater in the deceleration stage than

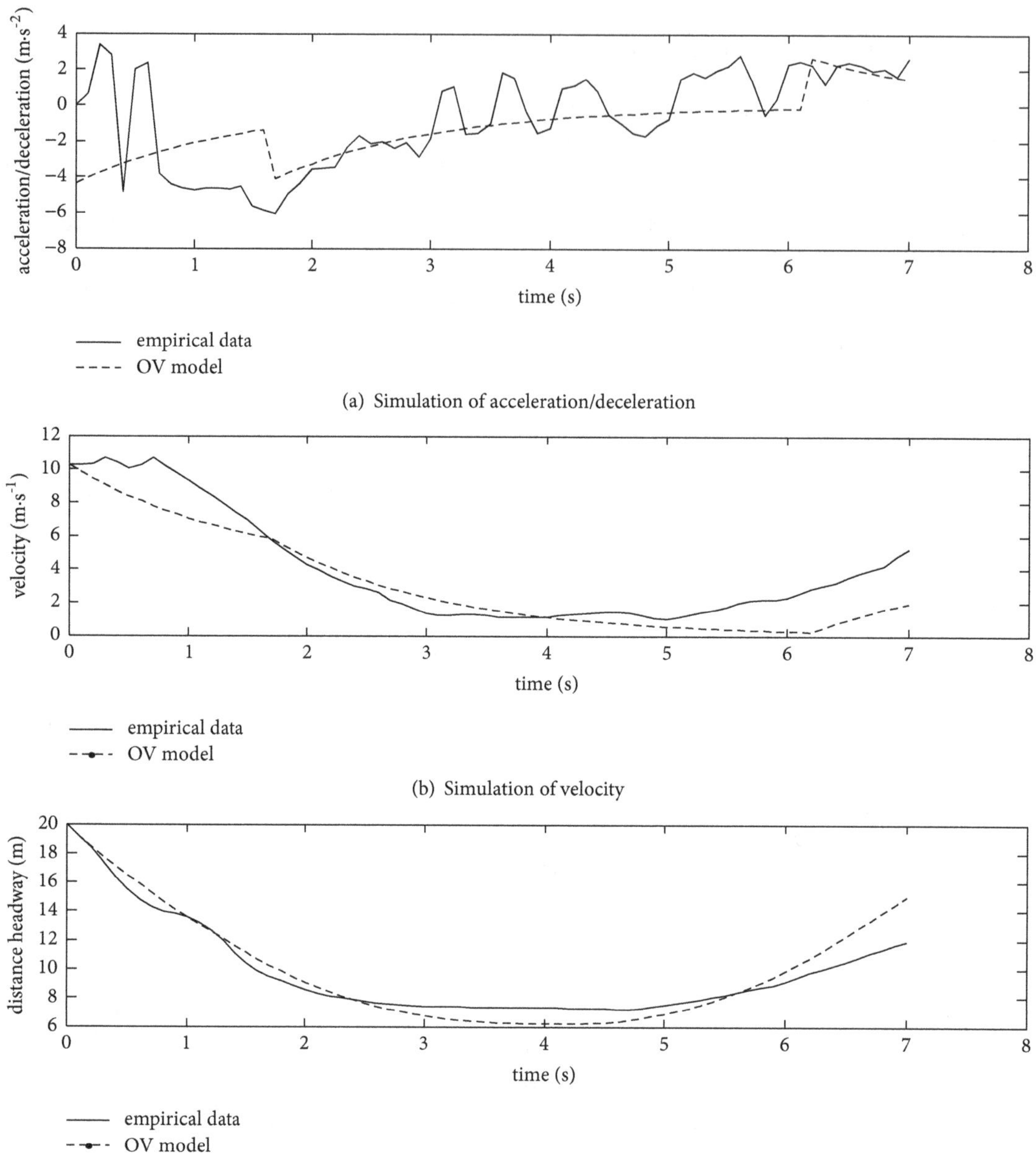

(a) Simulation of acceleration/deceleration

(b) Simulation of velocity

(c) Simulation of distance headway

FIGURE 3: OV model simulations of car-following Process 1.

in the measured data, and the problem of the OV model simulation of car-following Process 2 at the acceleration stage is consistent with that of car-following Process 1. Moreover, the velocity is less in the OV model than in the measured data. Figure 4(c) depicts the simulation analysis of the distance headway of the OV model in car-following Process 2. This figure illustrates that the distance headway of the following car is less than 5 m in the entire idling stage given the minimal deceleration at the deceleration stage, thereby causing a risk of a rear-end collision for the following car. Furthermore, the distance headway at the acceleration stage is greater in the OV model than in the measured data given the minimal acceleration. The comprehensive analyses of Figures 4(a), 4(b), and 4(c) depict the same results as those obtained through the simulation analysis of acceleration/deceleration, velocity, and distance headway. The OV model simulation of car-following Process 2 shows a minimal deceleration at the deceleration stage. Thus, the distance headway of the following car is less than 5 m in the entire idling stage, thereby resulting in a rear-end collision of the following car. The OV model shows a slow and insufficient acceleration over the measured data at the acceleration stage. Therefore, the fit degree is relatively low between the simulation and the measured trajectories when the OV model simulates car-following Process 2.

The results of the simulation analysis of car-following Processes 1 and 2 using the OV model show that the OV model simulation of the gather vehicle has a minimal

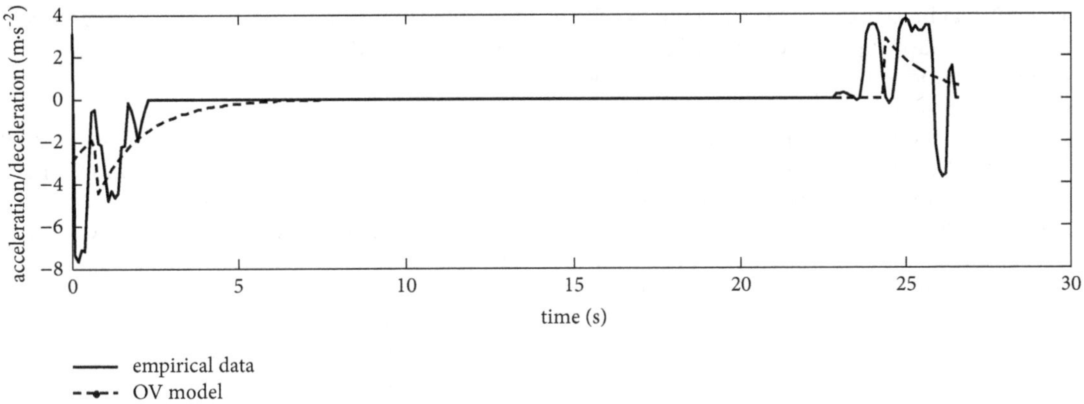

(a) Simulation of acceleration/deceleration

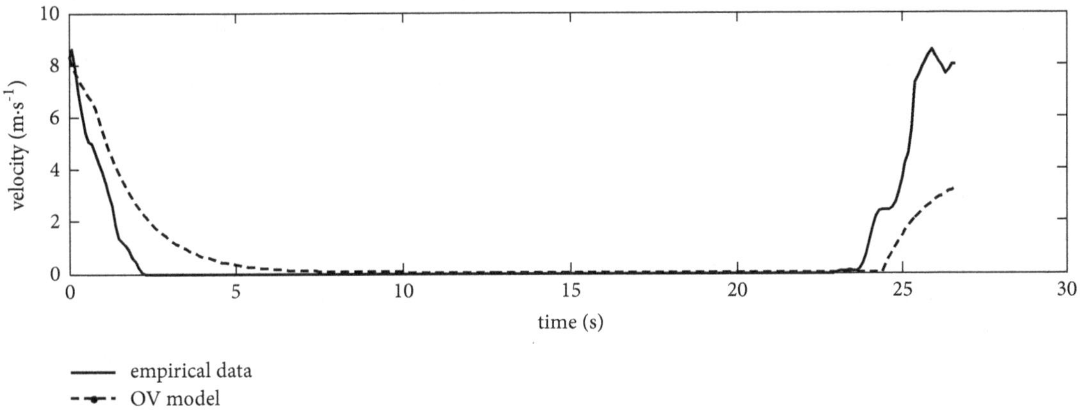

(b) Simulation of velocity

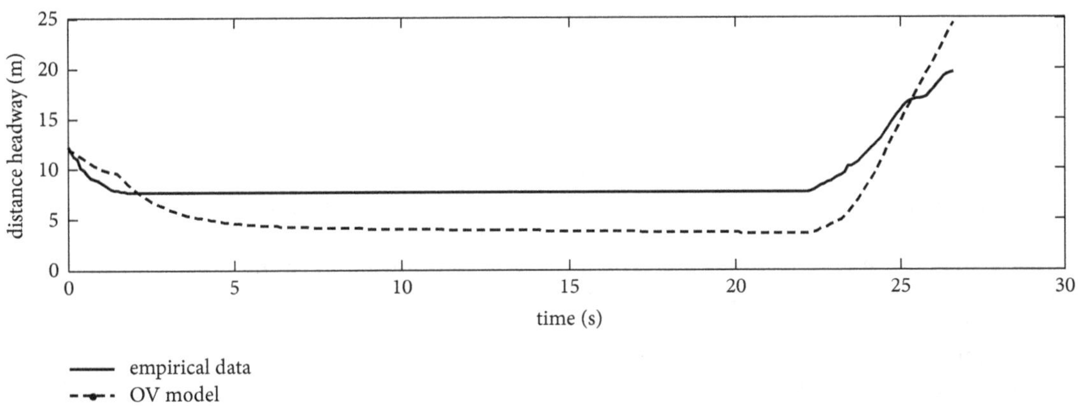

(c) Simulation of distance headway

FIGURE 4: OV model simulations of car-following Process 2.

deceleration at the deceleration stage, thereby causing the following car to be subjected to the risk of a rear-end collision. The OV model simulation of the dissipation vehicle has a slow acceleration at the acceleration stage. Comprehensively, the fit degree is relatively low using the OV model simulation of car-following Processes 1 and 2.

3.2. GF Model. Helbing and Tilch [9] calibrated the OV model with the empirical data and found that unrealistically high acceleration and deceleration appear in the OV model. To obtain improved results, Helbing and Tilch developed a GF model which is formulated as follows:

$$a_n(t) = \kappa \left(V\left(\Delta x_n(t)\right) - v_n(t)\right) + \lambda H\left(-\Delta v_n(t)\right)\Delta v_n(t), \tag{6}$$

where λ is the parameter of the GF model, $\Delta v_n(t) = v_{n-1}(t) - v_n(t)$ is the velocity difference between the nth car and its leading car $n-1$ at time t, and $v_{n-1}(t)$ and $v_n(t)$ correspond to the velocities of the $n-1$th and nth cars, and $H(\cdot)$ is the Heaviside function.

Helbing et al. [9] calibrated the parameters of the GF model by using the car-following data of the experimental site. Similar to the OV model, the calibration data are not the measured data at the signalised intersection. Therefore, the GF model is recalibrated by using the measured car-following process data at signalised intersections. Equation (3) is selected as the optimisation criterion, and the ABC algorithm is used to calibrate the model. The number of employed, onlooker, and scout bees is set to 100, 100, and 1, respectively, and the maximum number of stagnation and the number of iterations in the ABC algorithm are set to 1900 and 3000, correspondingly. The same resulting parameters are obtained consistently through repeated analysis of the experiment. Moreover, the algorithm converges to these resulting parameters, and the number of first ABC iterations of the result is at most 2000 times. Thus, the algorithm has converged to the global optimal solution. The resulting parameters of the GF model are $\kappa = 0.11$, $v_1 = 10.51$, $v_2 = 10.38$, $c_1 = 2.13$, $c_2 = 11.01$, and $\lambda = 1.26$.

We selected two complete car-following processes, as displayed in Table 1, to verify the adaptability of the calibrated GF model to the car-following trajectory at the signal intersection. In Table 1, the first point data of the two process data are used as the initial state of the simulation. Then, the complete simulation of the car-following process can be obtained on the basis of the evolution of the model. The simulation results are presented in Figures 5 and 6.

Figure 5(a) exhibits the simulation analysis of the acceleration/deceleration of the GF model in car-following Process 1. This figure shows that the deceleration is faster than the measured data at the early deceleration stage, but the subsequent deceleration stage is slower than the measured data. The acceleration simulation of the GF model is smaller in the acceleration stage than in the measured data. Figure 5(b) displays the simulation analysis of the velocity of the GF model in car-following Process 1. This figure shows that the velocity simulation of the GF model is less than the measured data at the early deceleration and greater than the measured data at the subsequent deceleration stage, similar to the OV model. Figure 5(c) presents the simulation analysis of the distance headway of the GF model in car-following Process 1. This figure shows that the distance headway of the entire deceleration stage is greater than the measured data given the significant deceleration at the early deceleration stage, and the distance headway gradually reduces at the subsequent deceleration stage considering the slow deceleration. Moreover, the distance headway simulation of the GF model at the acceleration stage is close to the measured data. The comprehensive analyses of Figures 5(a), 5(b), and 5(c) show the same results as those obtained through the simulation analysis of acceleration/deceleration, velocity, and distance headway. The GF model simulation of car-following Process 1 has the problem of a significant deceleration in the early deceleration stage and the minimal deceleration at the subsequent deceleration stage. Moreover, the simulation results of the GF model at the acceleration stage are better than the OV model, thereby ensuring that the following car avoids the risk of a rear-end collision. The simulation of car-following Process 1 shows that the GF model is better than the OV model, but the fit degree remains relatively low between the simulation and the measured trajectories when the GF model simulates car-following Process 1.

Figure 6(a) illustrates the simulation analysis of the acceleration/deceleration of the GF model in car-following Process 2. This figure presents that the deceleration is slower than the measured data in the early deceleration stage, and the deceleration is greater than the measured data at the subsequent deceleration stage. The problem of the GF model simulation of car-following Process 2 at the acceleration stage is consistent with that of the OV model. The simulation result shows a slower and insufficient acceleration over the measured data. Figure 6(b) depicts the simulation analysis of the velocity of the GF model in car-following Process 2. This figure shows that the simulation velocity is greater than the measured data at the deceleration stage, and the simulation velocity of the GF model at the acceleration stage is smaller than the measured data. Figure 6(c) demonstrates the simulation analysis of the distance headway of the GF model in car-following Process 2. This figure shows that the distance headway of the following car is close to 5 m during the entire idling stage given the minimal deceleration at the deceleration stage, thereby resulting in the risk of a rear-end collision of the following car. Moreover, the distance headway of the GF model at the acceleration stage is greater than the measured data given the minimal acceleration. The comprehensive analyses of Figures 6(a), 6(b), and 6(c) show the same results as those obtained through the simulation analysis of acceleration/deceleration, velocity, and distance headway. The GF model simulation of car-following Process 2 has a problem similar to the OV model. The deceleration of the GF model simulation process is smaller than the measured data at the deceleration stage, thereby causing the distance headway of the following car to be close to 5 m at the idling stage. The GF model shows a slow and insufficient acceleration over the measured data at the acceleration stage. Therefore, the fit degree remains relatively low between the simulation and the measured trajectories when the OV model simulates car-following Process 2.

The results of the simulation analysis of car-following Processes 1 and 2 by the GF model show that the GF model simulation of the gather and dissipation vehicles has the same problem as the OV model considering the minimal deceleration at the deceleration stage, thereby resulting in the risk of a rear-end collision of the following car. Moreover, the GF model shows a slow and insufficient acceleration over the measured data at the acceleration stage. Comprehensively, the fit degree remains relatively low by using the GF model to simulate car-following Processes 1 and 2.

3.3. FVD Model. Jiang et al. [10] proposed the FVD model by considering the negative and positive velocity differences to avoid an unrealistic acceleration. This model can explain the instance; that is, if the leading car is fast, then the car will not stop, although its headway is smaller than the safe distance. The model can be formulated as follows:

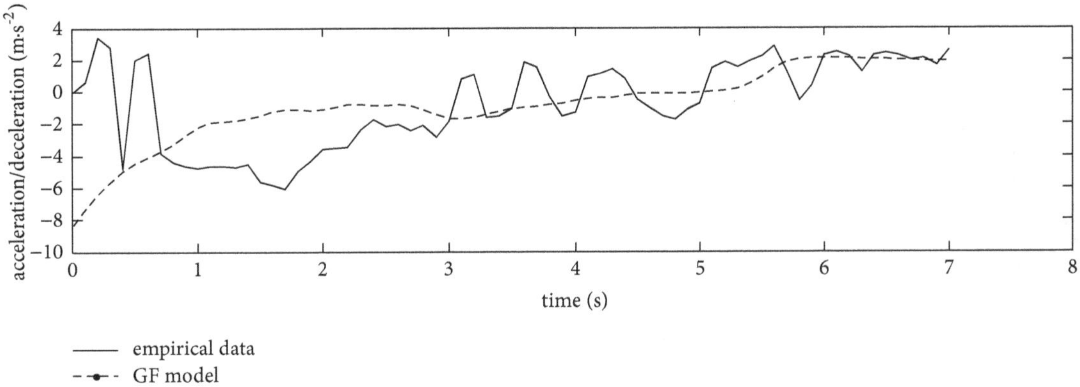

(a) Simulation of acceleration/deceleration

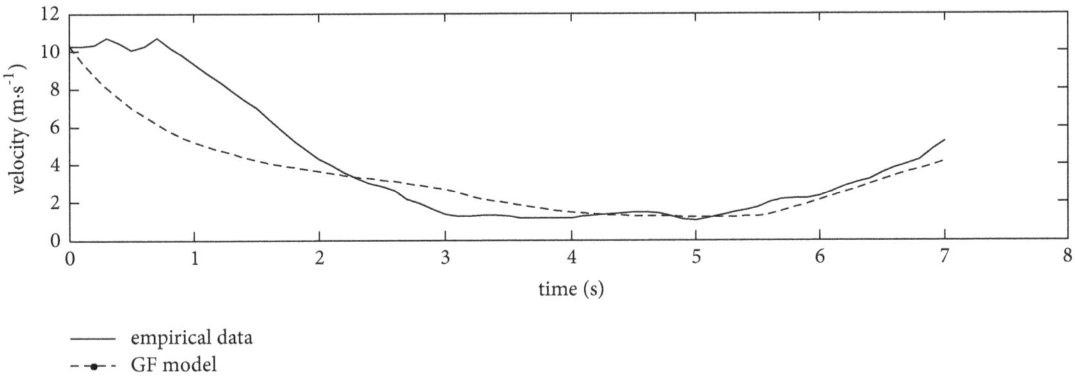

(b) Simulation of velocity

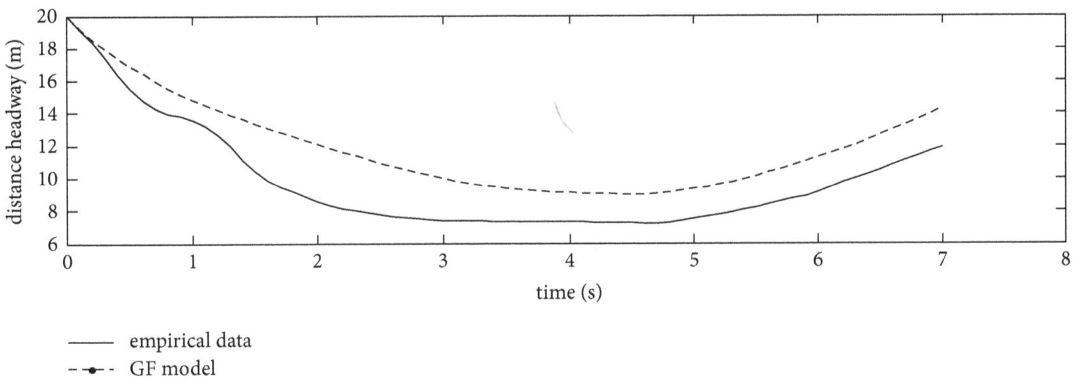

(c) Simulation of distance headway

FIGURE 5: GF model simulations of car-following Process 1.

$$a_n(t) = \kappa \left(V\left(\Delta x_n(t)\right) - v_n(t)\right) + \lambda \Delta v_n(t), \quad (7)$$

where λ is the parameter of the FVD model.

Similarly, the FVD model is recalibrated by using the measured car-following process data at signalised intersections. Equation (3) is selected as the optimisation criterion, and the ABC algorithm is used to calibrate the model. The number of employed, onlooker, and scout bees is set to 100, 100, and 1, respectively, and the maximum number of stagnation and the number of iterations in the ABC algorithm are set to 1900 and 3000, correspondingly. The same resulting parameters are obtained consistently through a repeated analysis of the experiment. Moreover, the algorithm converges to these resulting parameters, and the number of the first ABC iterations of the result is at most 2000 times. Thus, the algorithm has converged to the global optimal solution. The resulting parameters of the FVD model are $\kappa = 1.93$, $v_1 = 2.49$, $v_2 = 2.45$, $c_1 = 19.69$, $c_2 = 98.14$, and $\lambda = 0.63$.

We selected two complete car-following processes in Table 1 to test the FVD model to verify the adaptability of the calibrated FVD model to the car-following trajectory at the signal intersection. The first point data of the two process data are used as the initial state of the simulation. The complete simulation of the car-following process can then be obtained on the basis of the evolution of the model. The simulation results are demonstrated in Figures 7 and 8.

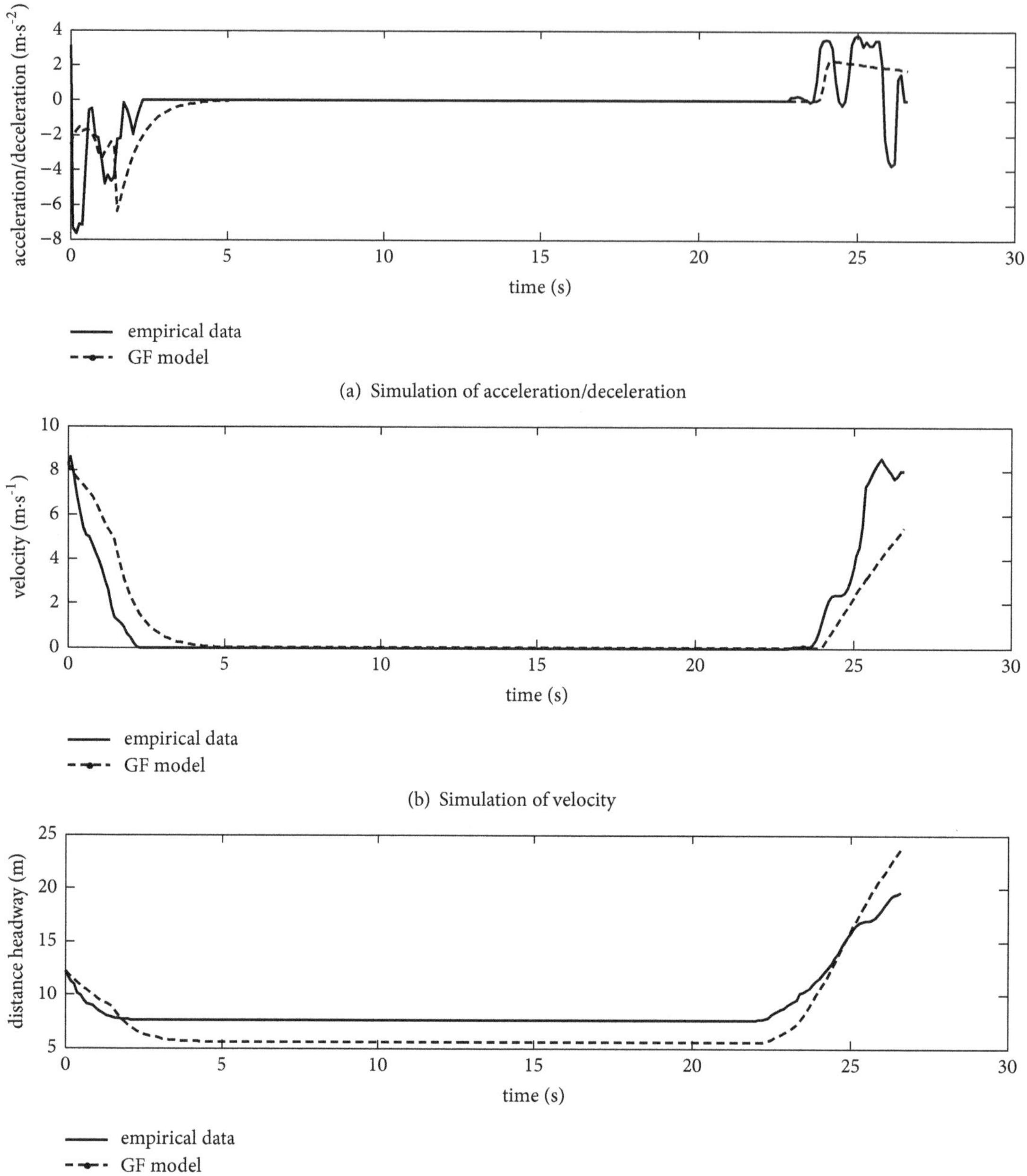

(a) Simulation of acceleration/deceleration

(b) Simulation of velocity

(c) Simulation of distance headway

FIGURE 6: GF model simulations of car-following Process 2.

Figures 7(a), 7(b), and 7(c) exhibit the simulation analyses of the acceleration/deceleration, velocity, and distance headway of the FVD model in car-following Process 1, respectively. Figure 7 displays that the FVD model simulation of car-following Process 1 has the same problem as the OV and GF models. The deceleration of the FVD model is faster than the measured data at the early deceleration stage, but the subsequent deceleration stage is slower than the measured data. Moreover, the simulation results of the FVD model indicate a significant acceleration at the early acceleration stage and avoid an acceleration delay, but the simulation of the entire acceleration process is insufficiently smooth. Overall, the simulation results of the FVD model avoid the risk of a rear-end collision which exists in the OV model. Simultaneously, the problem of acceleration insufficiency in the OV and GF models is solved, but the fit degree remains relatively low between the simulation and measured trajectories when the FVD model is used to simulate car-following Process 1.

Figures 8(a), 8(b), and 8(c) plot the simulation analyses of the acceleration/deceleration, velocity, and distance headway of the FVD model in car-following Process 2, correspondingly. Figure 8 presents that the fit degree of the FVD model simulation in car-following Process 2 at the deceleration process is unfavourable. The deceleration of the FVD model is slower than the measured data at the

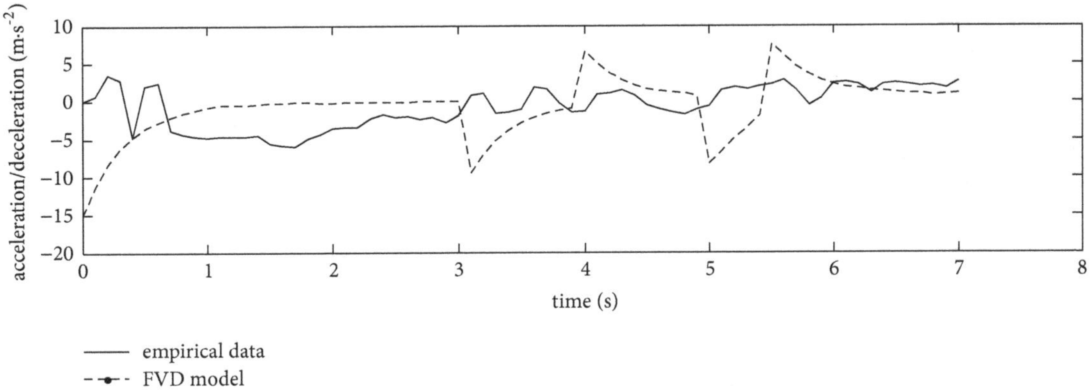

(a) Simulation of acceleration/deceleration

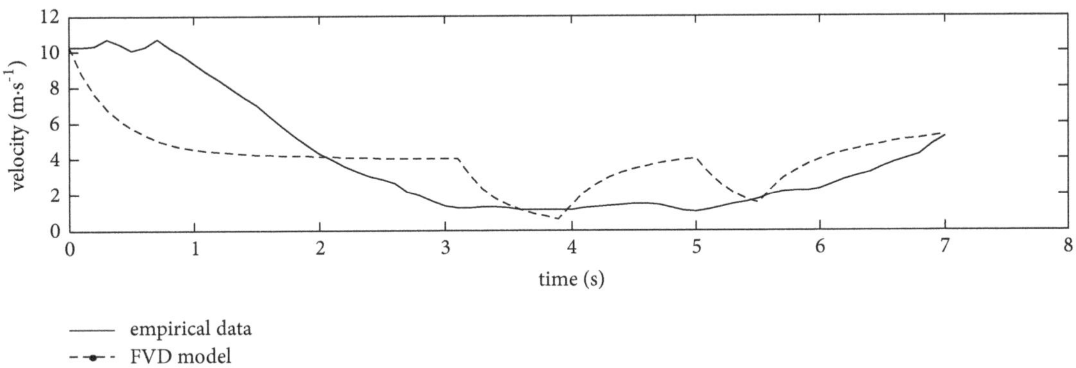

(b) Simulation of velocity

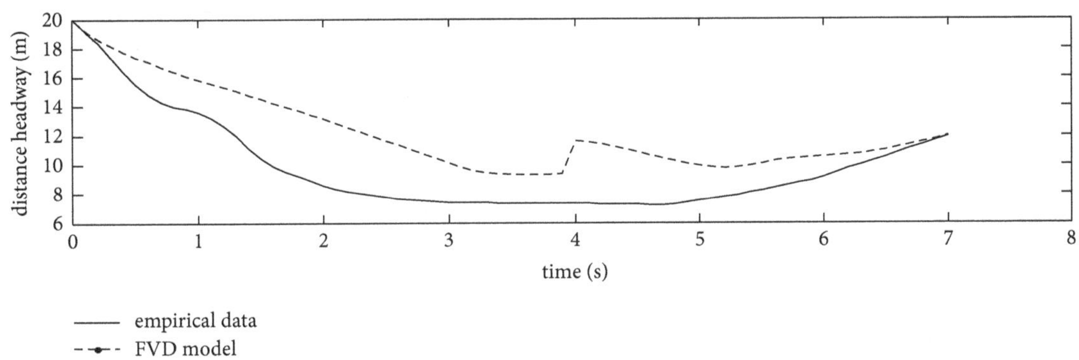

(c) Simulation of distance headway

FIGURE 7: FVD model simulation of car-following Process 1.

entire deceleration process given the phenomenon of large deceleration at the early deceleration stage, thereby causing the distance headway to become larger than the measured data at the idling stage. The simulation of the FVD model at the acceleration stage avoids the problem of acceleration delay. The acceleration is significant at the early acceleration stage, but the FVD model demonstrates a slow and insufficient acceleration over the measured data at the subsequent acceleration stage, thereby resulting in a deviation between the simulation trajectory and the measured data. Overall, the simulation results of the FVD model avoid the risk of a rear-end collision which exists in the OV model. Simultaneously, the problem of acceleration insufficiency in the OV and GF models at the early acceleration stage is solved. However, the FVD model still has insufficient acceleration at the acceleration stage, and the fit degree remains relatively low between the simulation and the measured trajectories when the FVD model is used to simulate car-following Process 2.

The simulation analysis results of car-following Processes 1 and 2 by the FVD model show that the FVD model has several advantages over the OV model and the GF model, thereby avoiding the risk of a rear-end collision. Moreover, the problem of slow and insufficient acceleration at the early acceleration is solved. However, slow deceleration and insufficient acceleration still occur at the subsequent deceleration and acceleration stage. Comprehensively, the fit degree remains relatively low when the FVD model is used to simulate car-following Processes 1 and 2.

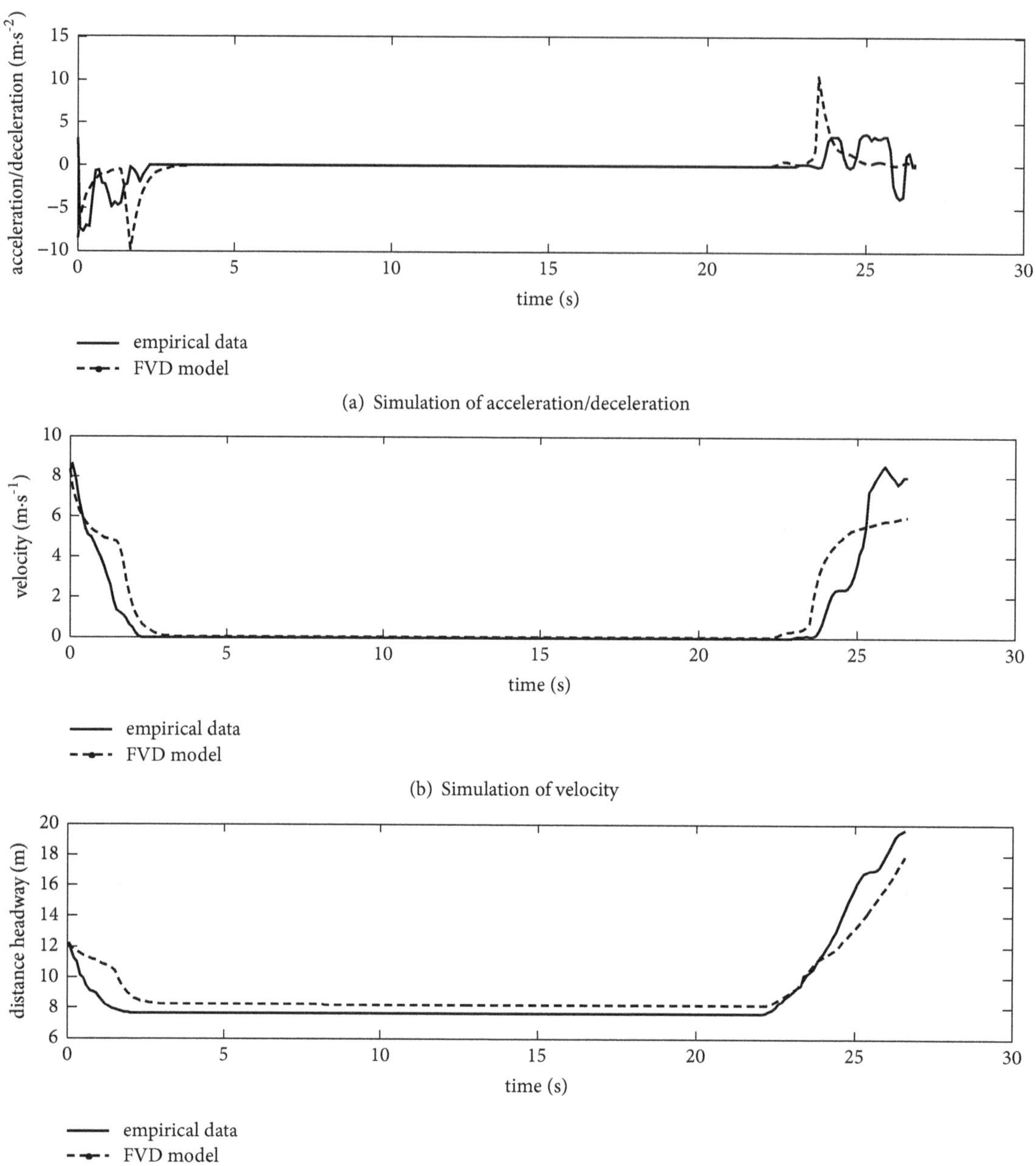

(a) Simulation of acceleration/deceleration

(b) Simulation of velocity

(c) Simulation of distance headway

FIGURE 8: FVD model simulations of car-following Process 2.

3.4. Other Models. The exploration of an urban signal intersection is important in traffic flow research because it can provide a theoretical basis for intersection signal timing. To describe car-following behaviour at signalised intersection, literature [46–49] proposed various extended car-following models on the basis of the FVD model under the analysis of measured data; literature [51] also proposed an extended car-following model based on the General Motor car-following model. Literature [46] proposed an extended car-following model considering multiple leading cars' acceleration that adapts to vehicle automation and safety early warning systems; literature [47] proposed an improved car-following model considering the velocity difference of an immediately ahead car; literature [48, 49] suggested essentially the same type of car-following model that improved the car-following model considering headway changes with memory on a single lane. Therefore, the car-following models proposed in literature [47, 49] are selected as comparative models, considering the comparability of the improved model proposed in this study. The car-following models proposed in literature [47, 49] can be expressed as follows, respectively.

$$a_n(t) = \kappa\left(V\left(\Delta x_n(t)\right) - v_n(t)\right) + \lambda \Delta v_n(t) + \gamma\left(v_{n-1}(t) - v_{n-1}(t-1)\right), \tag{8}$$

$$a_n(t) = \kappa\left(V\left(\Delta x_n(t)\right) - v_n(t)\right) + \lambda \Delta v_n(t) + \gamma\left(\Delta x_n(t) - \Delta x_n(t-1)\right), \tag{9}$$

In (8) and (9), $v_{n-1}(t) - v_{n-1}(t-1)$ is velocity difference of the immediately leading car and $\Delta x_n(t) - \Delta x_n(t-1)$ is the headway changes with memory between the nth car and its leading car $n-1$ at time t; γ denotes the sensitivity parameter and the other parameters remain constant.

In the previous analysis, the problems and causes of OV, GF, and FVD model to simulate a microscopic car-following behaviour at signalised intersections are illustrated; it is necessary to propose an extended car-following model to overcome the defects of the OV, GF, and FVD models in simulating the car-following process at signalised intersections. The car-following models proposed in literature [47, 49] are selected as comparative models to illustrate the effectiveness of the improved model. Therefore, these two models need to be recalibrated using the measured car-following data at signalised intersection. Similarly, (3) is selected as the optimisation criterion, and the ABC algorithm is used to calibrate the model. The number of employed, onlooker, and scout bees is set to 100, 100, and 1, respectively, and the maximum number of stagnation and the number of iterations in the ABC algorithm are set to 2200 and 3000, correspondingly. Thus, the algorithm has converged to the global optimal solution. The resulting parameters of the car-following model proposed in literature [47] are $\kappa = 0.48$, $v_1 = 3.21$, $v_2 = -3.15$, $c_1 = -13.58$, $c_2 = -74.57$, $\lambda = 0.58$, and $\gamma = 0.08$ and those of the car-following model proposed in literature [49] are $\kappa = 0.37$, $v_1 = 3.01$, $v_2 = 2.57$, $c_1 = 14.60$, $c_2 = 80.21$, $\lambda = 0.59$, and $\gamma = 0.14$. Notably, the results obtained in this study are not the same as those obtained in literature [47, 49] because of the different car-following data sources. In addition, these two models are selected as comparative models to illustrate the effectiveness of the improved model in the subsequent sections. Therefore, the car-following processes shown in Table 1 will not be used in analysing the two car-following models.

4. Extended Car-Following Model

In Section 3, the OV, GF, and FVD models are used to simulate car-following Processes 1 and 2. The experimental analysis results show that the three models are ineffective for simulating car-following Processes 1 and 2. In the OV, GF, and FVD models, the influence of distance headway on the driving behaviour is insufficiently considered when the vehicles gather at the signalised intersection. In literature [46, 51], actual data were collected through the grey correlation analysis method to obtain a conclusion that the leading vehicle acceleration evidently influences the following vehicle acceleration. In an actual traffic situation, drivers focus on the effect of distance headway while the vehicles gather at signalised intersections. The deceleration amplitude is minimal when the distance headway is large, but the deceleration amplitude increases when the distance headway is small. However, the drivers focus on the acceleration of the front vehicle when the vehicle is dissipated. Therefore, an extended FVD (EFVD) model is proposed to improve the fit degree of the car-following model to simulate the car-following trajectory at signalised intersections. The EFVD model can be formulated as follows:

$$\begin{aligned} a_n(t) = \kappa \left(V\left(\Delta x_n(t)\right) - v_n(t)\right) + \lambda \Delta v_n(t) \\ + \mu_1 H\left(-\Delta v_n(t)\right)\left(\Delta x_n(t) - \mu_2\right) \\ + \mu_3 H\left(\Delta v_n(t)\right) a_{n-1}(t), \end{aligned} \tag{10}$$

where $a_{n-1}(t)$ is the acceleration/deceleration of the $n-1$th car at time t, u_1 is the sensitivity parameter of the distance headway when the vehicles gather, and u_2 is the safe driving distance headway of the driver's deceleration that is different from the parking distance headway. The driver focuses on a minimal deceleration to decelerate when $\Delta x_n(t) \geq u_2$, and if $\Delta x_n(t) < u_2$, then the driver will focus on a large deceleration to decelerate. u_3 is the sensitivity parameter of the acceleration/deceleration of the $n-1$th car. u_1, u_2, and u_3 must be calibrated, and the other parameters remain constant.

The EFVD model is calibrated by using the measured car-following data at signalised intersections. Equation (3) is selected as the optimisation criterion, and the ABC algorithm is used to calibrate the model. The number of employed, onlooker, and scout bees is set to 100, 100, and 1, respectively, and the maximum number of stagnation and the number of iterations in the ABC algorithm are set to 2500 and 3000, correspondingly. The same resulting parameters are obtained every time through repeated analysis of the experiment. Moreover, the algorithm converges to these resulting parameters, and the number of the first ABC iterations of the result is at most 2000 times. Thus, the algorithm has converged to the global optimal solution. The resulting parameters of the FVD model are $\kappa = 0.31$, $v_1 = 5.71$, $v_2 = 5.65$, $c_1 = 6.76$, $c_2 = 67.01$, $\lambda = 0.68$, $\mu_1 = 0.44$, $\mu_2 = 7.27$, and $\mu_3 = 0.31$.

We selected two complete car-following processes, as listed in Table 1, to verify the adaptability of the calibrated EFVD model to the car-following trajectory at the signal intersection and test the EFVD model. The first point data of the two car-following process data are used as the initial state of the simulation. The complete simulation of the car-following process can then be obtained on the basis of the evolution of the model. The simulation results are presented in Figures 9 and 10.

Figures 9(a), 9(b), and 9(c) plot the simulation analyses of the acceleration/deceleration, velocity, and distance headway, respectively, of the EFVD model to car-following Process 1, correspondingly. Figure 9 illustrates that the deceleration of the EFVD model is faster than the measured data at the entire deceleration stage, and the simulation results of the EFVD model at the acceleration stage avoid the problem of acceleration delay and insufficient acceleration. Overall, the EFVD model solves the problem of deceleration rapidly at the early deceleration stage. Moreover, the deceleration is slow at the subsequent deceleration stage in the OV, GF, and FVD models, thereby avoiding the risk of a rear-end collision in the OV model. Simultaneously, the proposed method solves the problem of the insufficient acceleration of the OV and GF models and the unsmoothed acceleration fitting of the FVD model. The EFVD simulation results of car-following Process 1 show a certain improvement over the OV, GF, and FVD models.

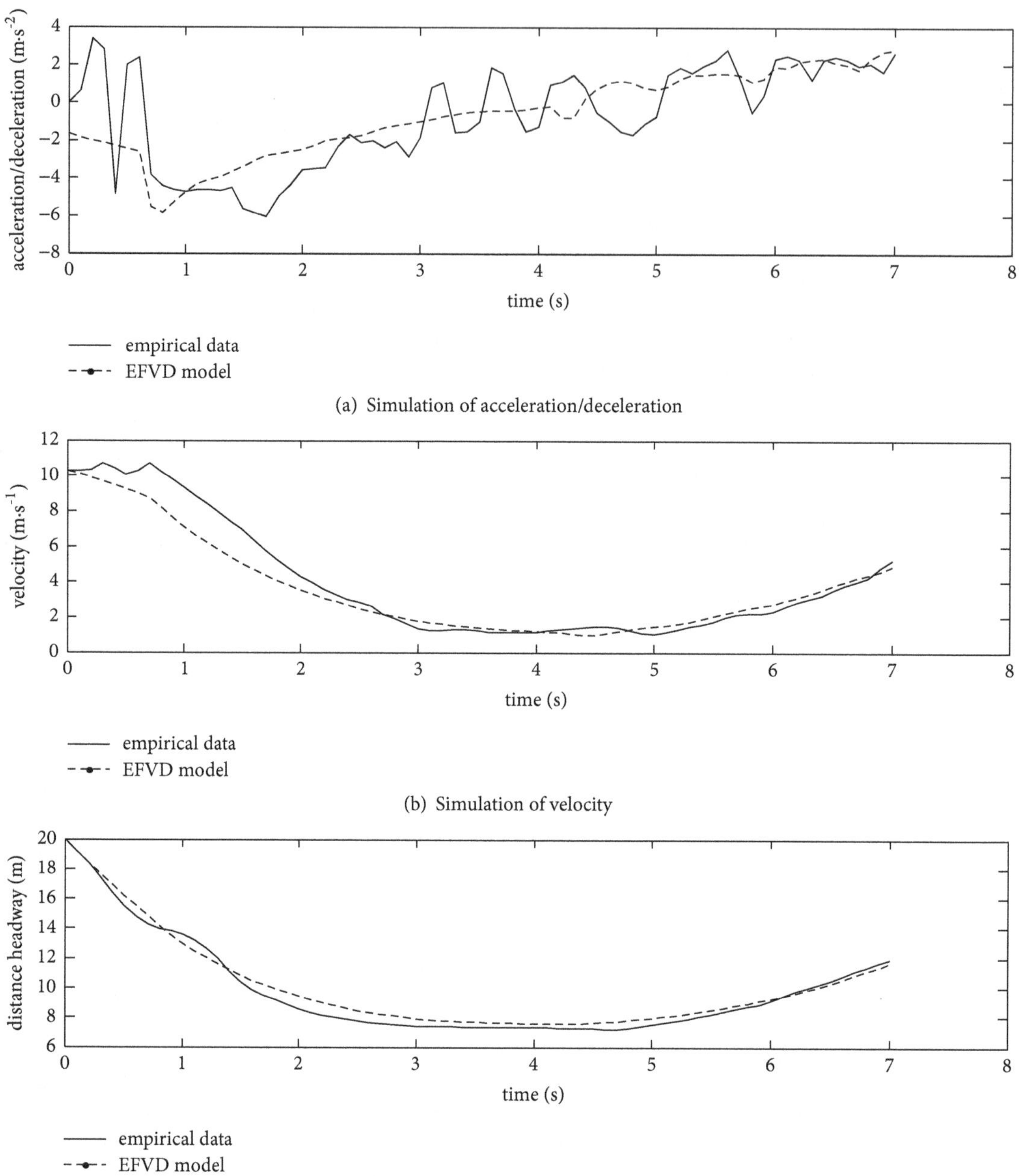

(a) Simulation of acceleration/deceleration

(b) Simulation of velocity

(c) Simulation of distance headway

FIGURE 9: EFVD model simulations of car-following Process 1.

Figures 10(a), 10(b), and 10(c) plot the simulation analyses of the acceleration/deceleration, velocity, and distance headway of the EFVD model in car-following Process 2, respectively. Figure 10 depicts that the deceleration of the EFVD model is slower than the measured data at the entire deceleration stage, but the distance headway at the idling stage is close to the measured data. The simulation of the EFVD model has a problem of advance acceleration at the acceleration stage but avoids the problem of insufficient acceleration. For car-following Process 2, the EFVD model shows slow deceleration but avoids the risk of a rear-end collision. The proposed model also solves the problem of insufficient acceleration. The EFVD simulation results of car-following Process 2 show a certain improvement over the OV, GF, and FVD models.

The simulation analysis results of car-following Processes 1 and 2 by the EFVD model show that the proposed method can eliminate the rapid deceleration at the early deceleration stage and slow deceleration at the subsequent deceleration stage. The fit degree of car-following Processes 1 and 2 are better in the proposed method than in the OV, GF, and FVD models. However, several problems, such as faster deceleration than the measured data in car-following Process 1 and slower deceleration than the measured data in car-following Process 2, remain. Moreover, the EFVD model simulation of car-following Process 2 has a problem of

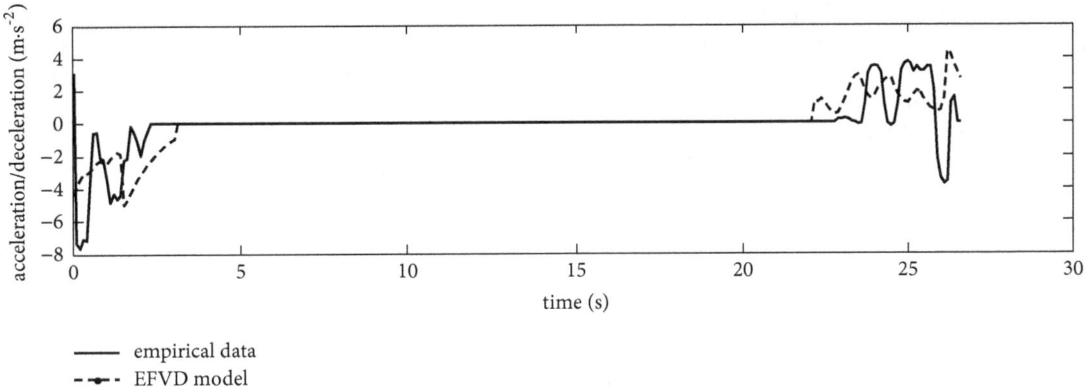

(a) Simulation of acceleration/deceleration

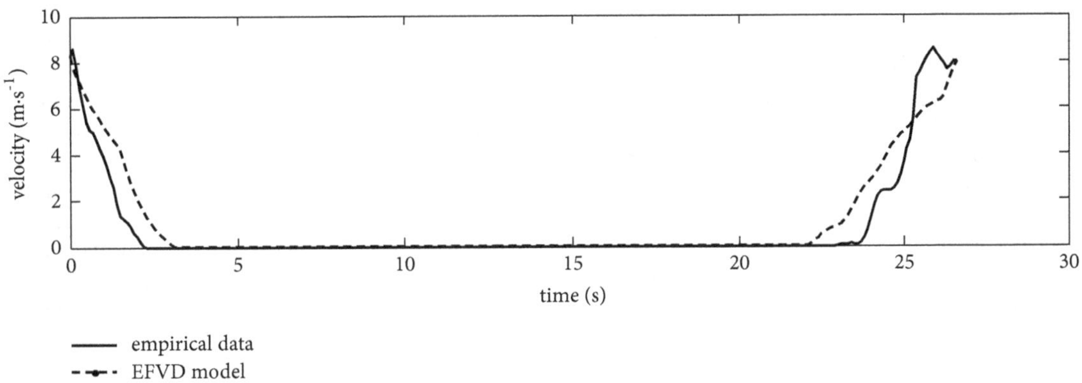

(b) Simulation of velocity

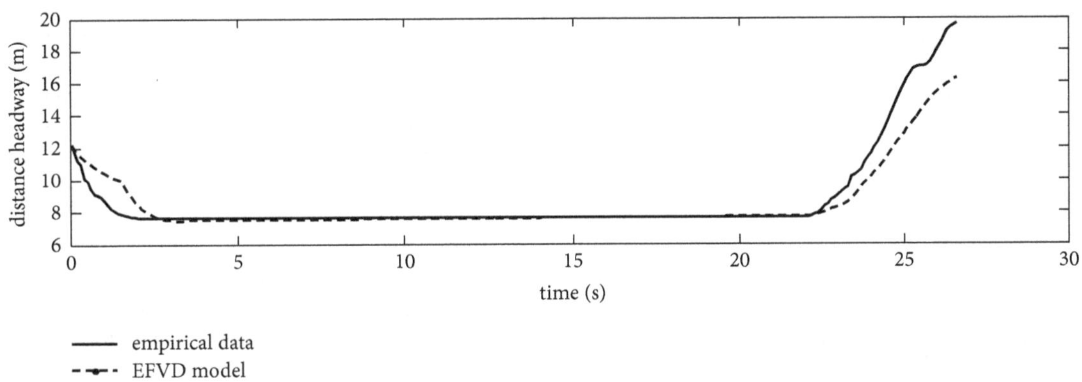

(c) Simulation of distance headway

FIGURE 10: EFVD model simulations of car-following Process 2.

advance acceleration. The EFVD model is more effective in simulating car-following Processes 1 and 2 than the OV, GF, and FOV models. Similarly, the EFVD model considers the influence of distance headway at the deceleration stage. The model can avoid the problem of early deceleration when the distance headway is large, whereas this model can induce a large deceleration and avoid the risk of a rear-end collision when the distance headway is small. Simultaneously, the problem of low fit degree at the deceleration stage in the OV, GF, and FVD models is solved. The proposed model considers the influence of the acceleration of the leading car at the acceleration stage, although the proposed model causes the problem of advance acceleration. However, the EFVD model solves the problem of acceleration insufficiency and slower acceleration at the late acceleration stage of the FVD model than the OV and GF models.

5. Comparative Analysis

In Section 3, we verified the problems of the OV, GF, and FVD models in fitting the actual car-following behaviour at the signal intersection through the two typical car-following processes. The analysis of its causes indicates that the OV, GF, and FVD models do not consider the car-following distance headway when the vehicle decelerates. Simultaneously, the OV, GF, and FVD models also do not consider the acceleration of the leading vehicle when the vehicle accelerates. The two factors lower the fitting degree of the OV, GF, and FVD models when fitting the actual car-following behaviour at signalised intersections. On the basis of this analysis, we further proposed the EFVD model in Section 4 based on the FVD model. Then, the EFVD model is verified on the basis of the two typical car-following processes. The experimental simulation shows that the EFVD model can solve the problems in the OV, GF, and FVD models. We used three models in this section for comparative analysis to fully demonstrate the superiority of the EFVD model. In this study, we firstly analysed the problems of the OV, GF, and FVD models in fitting the actual car-following behaviour at the signal intersection. Then, we proposed the EFVD model. Thus, we compared the EFVD model with the FVD model, which is better than the OV and GF models, through car-following Processes 1 and 2. The parameters of the FVD and EFVD models are constant, and the comparative results are plotted in Figures 11 and 12, respectively.

Figures 11 and 12 display that the EFVD model is better than the FVD model in simulating car-following Processes 1 and 2. To accurately explain the improvement effect of the EFVD model, (11) is selected as the evaluation criterion to measure the fitting effect of the EFVD and FVD models. Equation (9) can be formulated as follows:

$$PAD = \frac{AD^A - AD^B}{AD^B} \tag{11}$$

where PAD is the prior percentage of the fitting precision of A model to B model when $PAD < 0$. A small PAD represents an improved fitting effect of A model relative to B model while representing a degraded fitting effect of the A model relative to the B model. AD^A and AD^B denote the fitting deviation of certain car-following data using A and B models, respectively, as follows:

$$AD = \left|(0.5v + 0.5\Delta x) - \left(0.5\widehat{v} + 0.5\widehat{\Delta x}\right)\right|, \tag{12}$$

where Δx is the calculated distance headway of the car-following model, $\Delta\widehat{x}$ is the measured distance headway, v is the calculated velocity of the car-following model, and $\widehat{v}$ is the measured velocity.

The calculated results of the fitting deviation of car-following data shown in Table 1 are summarised in Table 2. Table 2 lists the three car-following data for the fitting deviation that is larger in the EFVD model than in the FVD model when simulating car-following Process 1. The 12 other types of car-following data have better fitting results than the FVD model with a minimum improvement of −65.10%. In comparison with the FVD model, the fitting effect of 80% car-following data is improved in car-following Process 1 of the EFVD model simulation. Car-following Process 2 has six car-following data at the early acceleration stage for the fitting deviation that is larger in the EFVD model than in the FVD model. The 48 other car-following data have better fitting results than the FVD model. In comparison with the FVD model, the fitting effect of 88.89% car-following data is improved in car-following Process 2 of the EFVD model simulation. However, the verification analysis is only for the two car-following processes shown in Table 1. Thus, we tested the EFVD model by using the 32 other complete car-following processes, which included 922 isolated point data. In the follow-up comparison process, we abbreviated the FVD model as Model 1. Simultaneously, the car-following models proposed in literature [47, 49] were selected as comparative models and abbreviated as Models 2 and 3, respectively, considering the comparability of the EFVD model proposed in this study. In the testing process, each group of test samples was placed in the model, and the deviation between the output of the model and the measured data was calculated using (12). Then, (11) was selected as the evaluation criterion for the EFVD model compared with Models 1, 2, and 3.

Models 1, 2, and 3 were used to compare and analyse the results, as presented in Figure 13. The graph shows only the comparison results of all data points with starting data of 1 and a gap of 4 to clarify Figure 13. The isolated point used in the EFVD model compared with Models 1, 2, and 3 is only 602 because of the influence of $v_n(t-1)$ and $\Delta x_n(t-1)$ in Models 2 and 3, respectively. Figure 13 illustrates that the PAD value is less than 0 for most of the isolated points when comparing the EFVD with Models 1, 2, and 3. The results show that the fitting degree of the EFVD model for the 602 isolated point data is better than that of Models 1, 2, and 3 for most of the isolated points. However, the fitting degree of partially isolated points in the EFVD model is inferior to Models 1, 2, and 3. According to the statistics of the simulation results, the fitting degree of the EFVD model has 60, 140, and 134 isolated points, which indicates that this model is worse than Models 1, 2, and 3, respectively. Overall, the fitting effect of the EFVD model remains better than that of Models 1, 2, and 3. Fitting effects of 90.03%, 76.74%, and 77.74% isolated points are improved when the EFVD model is compared with Models 1, 2, and 3, respectively. In literature [53], a significant shortage in the comparative testing of the car-following models based on isolated point data was emphasised in calibrating the parameters of the car-following model, and a comprehensive testing of the entire car-following process was lacking, thereby resulting in inaccurate test results. In literature [53], the authors also highlighted that the test of the car-following model should be based on the entire car-following process. Thus, the EFVD model was retested on the basis of the entire car-following process; that is, the first set of data for each car-following process and all subsequent model inputs were based on the computational evolution of the model itself. Furthermore, (12) was used to calculate the deviation of each set of data for each process. The fitting deviation of each car-following process is then calculated using the following:

$$ED = \frac{1}{n}\sum_{i=1}^{n} AD_i, \tag{13}$$

where n is the number of isolated points that are included in each of the car-following processes and AD_i is the deviation for each isolated point data. The EFVD model was tested on the basis of the 32 complete car-following processes. Models 1, 2, and 3 were adopted as the comparative model. The ED

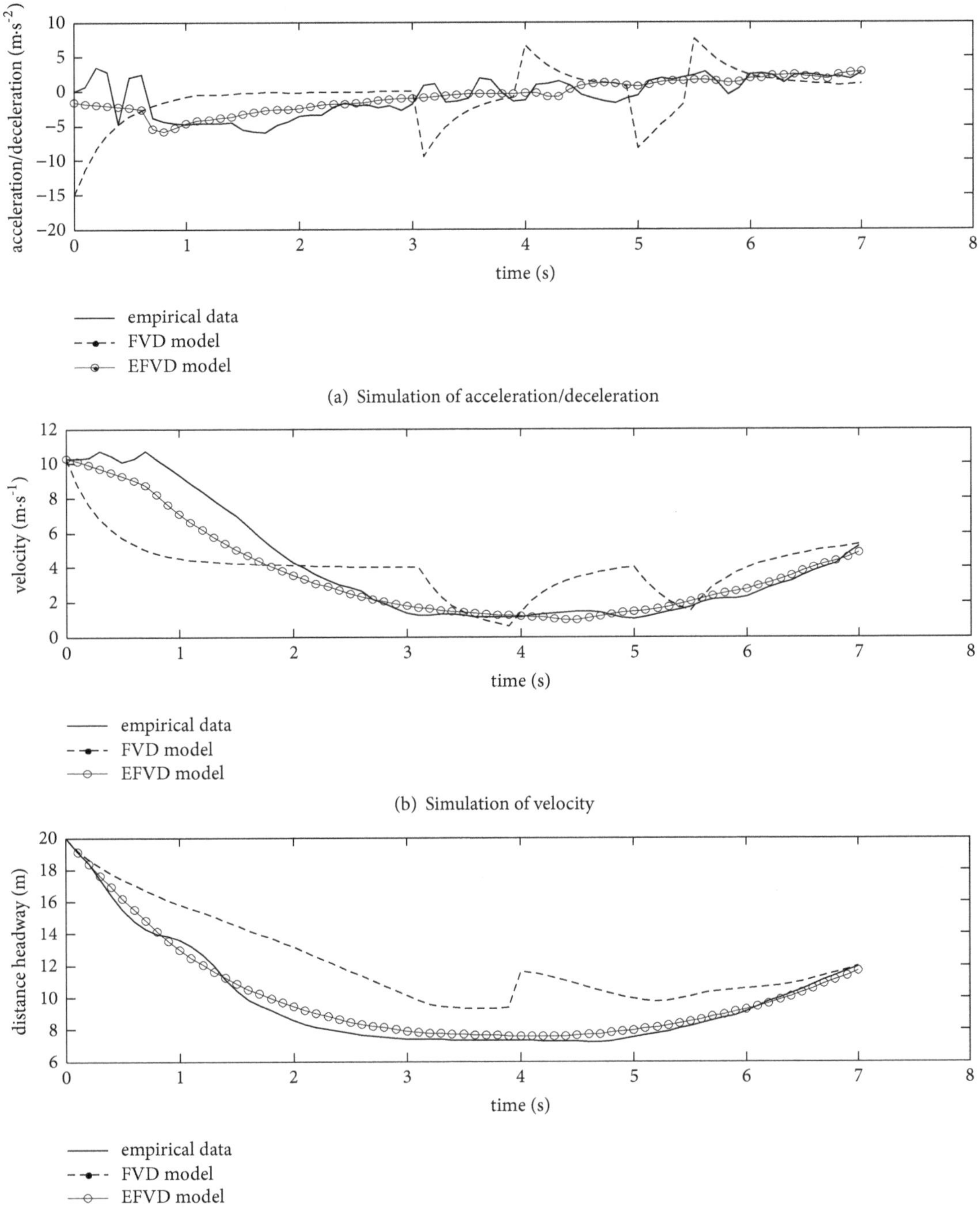

Figure 11: Simulation comparison of car-following Process 1.

of the different models are presented in Table 3, and the comparative results are summarised in Table 4.

Tables 3 and 4 list six, eight, and nine test car-following processes for the simulation deviation that is larger in the EFVD model than in Models 1, 2, and 3, respectively. The remaining car-following processes indicate that the simulation deviation is better in the EFVD model than in Models 1, 2, and 3, and the improvement of the simulation is evident. According to the statistics of the simulation results, the fitting effect of 90.03%, 76.74%, and 77.74% test car-following processes is improved when the EFVD model is compared with Models 1, 2, and 3, correspondingly. Therefore, the EFVD model from the overall perspective is better than Models 1, 2, and 3 in simulating the car-following process at signalised intersections.

The comprehensive analysis results indicate the EFVD model several advantages over Models 1, 2, and 3, whether the proposed model is tested on the basis of the isolated

Table 2: Fitting deviation of car-following data in Table 1.

Car-following processes	Number of isolated points	*AD* of the FVD model	*AD* of the EFVD model	*PAD* (EFVD model compared with FVD model)
	1	0.00	0.00	0.00%
	2	1.26	0.06	-95.58%
	3	1.26	1.40	11.64%
	4	0.66	0.78	18.20%
	5	2.18	0.05	-97.59%
	6	2.52	0.18	-93.00%
	7	2.69	0.48	-82.13%
1	8	1.08	0.27	-74.81%
	9	0.76	0.18	-76.21%
	10	2.79	0.03	-99.00%
	11	2.66	0.42	-84.24%
	12	0.81	0.28	-65.10%
	13	1.53	0.27	-82.44%
	14	0.90	0.04	-95.19%
	15	0.07	0.10	40.19%

Table 2: Continued.

Car-following processes	Number of isolated points	*AD* of the FVD model	*AD* of the EFVD model	*PAD* (EFVD model compared with FVD model)
2	1	0.00	0.00	0.00%
	2	0.92	1.28	-27.90%
	3	1.71	1.62	-5.28%
	4	3.00	2.48	-17.31%
	5	1.14	1.05	-8.10%
	6	0.55	0.15	-72.29%
	7	0.72	0.45	-38.44%
	8	0.64	0.48	-24.07%
	9	0.62	0.50	-18.14%
	10	0.61	0.50	-17.29%
	11	0.61	0.50	-16.89%
	12	0.60	0.51	-15.57%
	13	0.60	0.50	-16.87%
	14	0.60	0.50	-17.39%
	15	0.60	0.49	-17.51%
	16	0.60	0.49	-17.43%
	17	0.59	0.49	-17.25%
	18	0.59	0.49	-17.02%
	19	0.59	0.49	-16.77%
	20	0.59	0.49	-16.50%
	21	0.59	0.49	-16.22%
	22	0.58	0.50	-14.92%
	23	0.58	0.49	-16.22%
	24	0.58	0.48	-16.73%
	25	0.58	0.48	-16.83%
	26	0.58	0.48	-16.74%
	27	0.57	0.48	-16.55%
	28	0.57	0.48	-16.31%
	29	0.57	0.48	-16.04%
	30	0.57	0.48	-15.76%
	31	0.57	0.48	-15.46%
	32	0.56	0.48	-14.15%
	33	0.56	0.48	-15.44%
	34	0.56	0.47	-15.94%
	35	0.56	0.47	-16.03%
	36	0.56	0.47	-15.93%
	37	0.55	0.47	-15.72%
	38	0.55	0.47	-15.47%
	39	0.55	0.47	-15.18%
	40	0.55	0.47	-14.89%
	41	0.55	0.47	-14.58%
	42	0.54	0.47	-13.66%
	43	0.54	0.46	-14.74%
	44	0.54	0.46	-15.11%
	45	0.54	0.88	63.61%
	46	0.49	0.96	94.77%
	47	0.32	1.00	211.04%
	48	0.64	1.17	83.57%
	49	0.57	0.87	53.46%
	50	0.42	1.53	261.01%
	51	1.53	1.42	-6.86%
	52	2.99	2.23	-25.32%
	53	2.99	2.27	-24.06%
	54	2.25	1.84	-18.30%

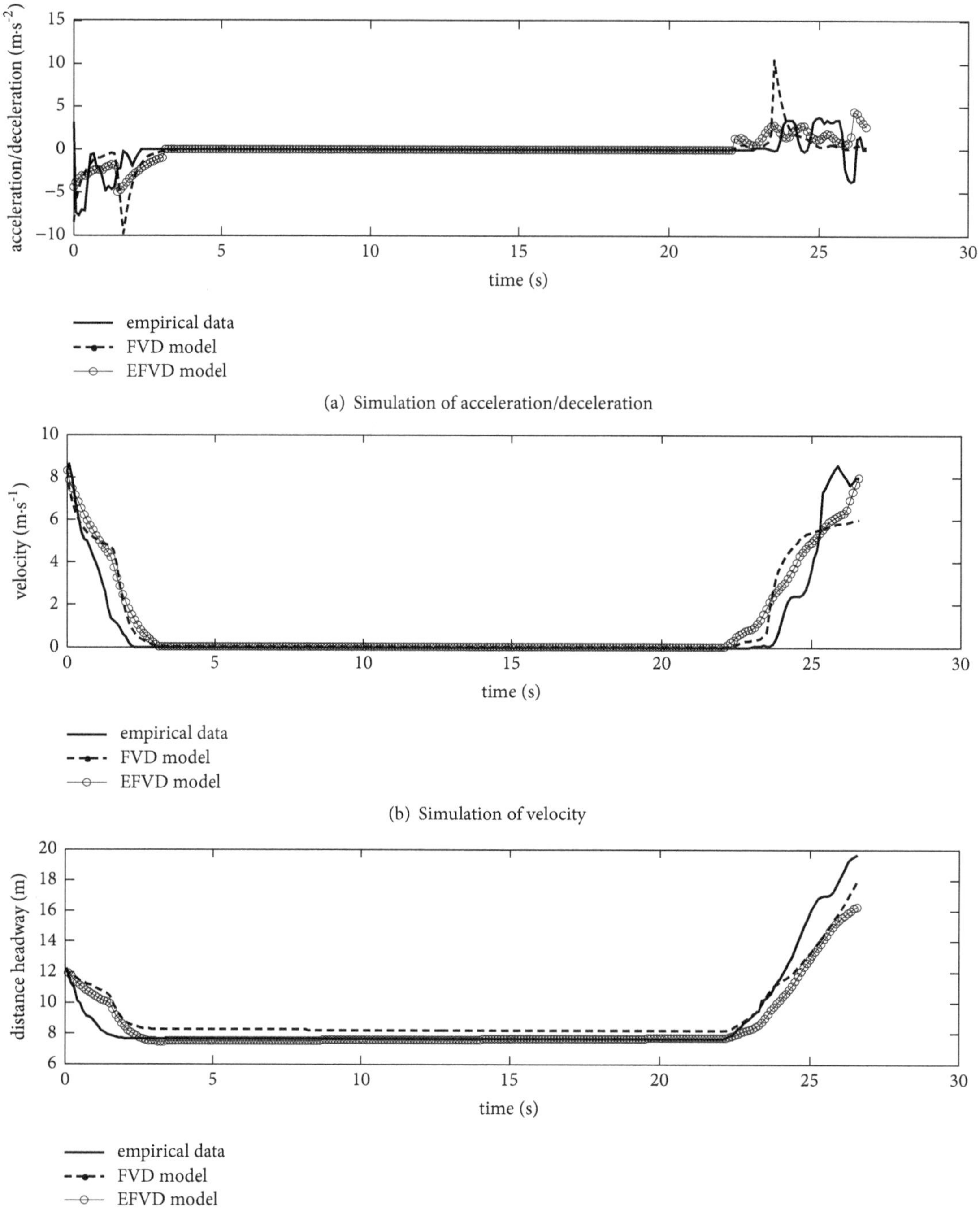

(a) Simulation of acceleration/deceleration

(b) Simulation of velocity

(c) Simulation of distance headway

FIGURE 12: Simulation comparison of car-following Process 2.

point data or the entire car-following process. In combination with the previous analysis, Model 1 ignores the influence of distance headway on the driving behaviour when the vehicles gathered at the signalised intersections and insufficient acceleration occurs at the acceleration stage. Furthermore, we conducted an in-depth analysis of Models 2 and 3. Model 2 was proposed considering the velocity difference of the leading car, and Model 3 was proposed considering the distance headway changes with memory. In comparison with the EFVD model, although Model 2 was proposed based on Model 1 considering the velocity difference of the leading car, the influence of the distance headway on the deceleration behaviour of the drivers was neglected. Model 3 was proposed based on Model 1 considering the distance headway change with memory, but the distance headway at different deceleration stages had a different effect on the deceleration behaviour of the drivers, and the velocity difference of the leading car was also neglected in Model

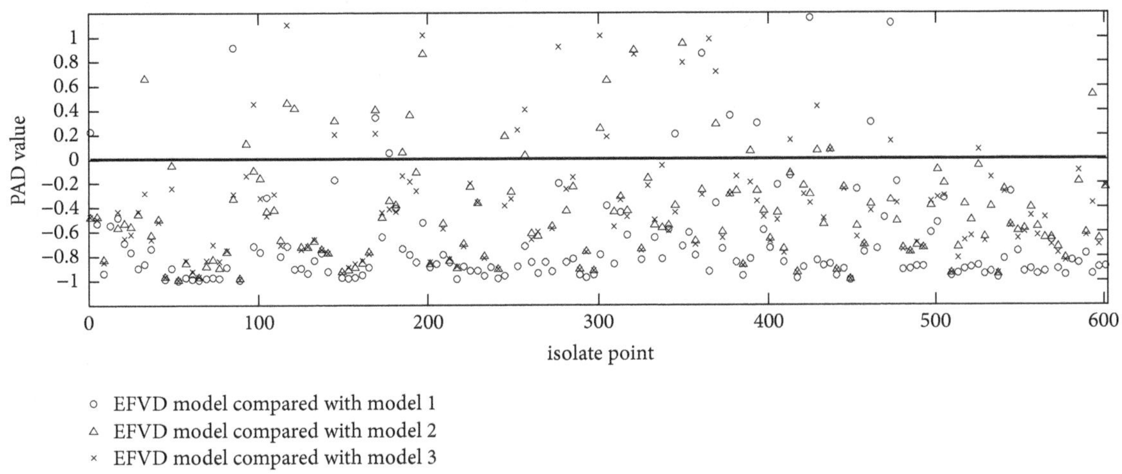

FIGURE 13: Comparison of the EFVD with Models 1, 2, and 3.

TABLE 3: *ED* of the different models.

Car-following process	*ED* of the EFVD model	*ED* of Model 1	*ED* of Model 2	*ED* of Model 3
1	0.18	0.74	0.30	0.29
2	0.52	0.45	0.48	0.51
3	0.27	1.32	0.52	0.32
4	0.81	1.81	0.91	0.93
5	0.38	1.15	0.41	0.43
6	0.24	0.71	0.31	0.36
7	0.42	0.50	0.31	0.31
8	0.52	0.92	0.54	0.52
9	0.30	1.54	0.52	0.35
10	0.22	0.42	0.31	0.32
11	0.32	0.57	0.23	0.23
12	0.20	1.06	0.22	0.21
13	0.55	0.45	0.66	0.66
14	0.36	0.61	0.18	0.17
15	0.54	1.06	0.76	0.82
16	0.37	0.29	0.31	0.31
17	0.24	0.40	0.33	0.32
18	0.70	0.83	0.78	0.76
19	0.31	0.76	0.25	0.28
20	0.04	0.64	0.14	0.13
21	0.32	0.98	0.50	0.41
22	0.14	0.24	0.16	0.17
23	0.17	0.44	0.22	0.19
24	0.93	1.70	1.28	1.36
25	0.49	2.02	1.55	1.61
26	0.81	1.68	1.18	1.29
27	0.61	1.44	0.65	0.67
28	0.49	0.27	0.54	0.56
29	0.42	0.40	0.25	0.24
30	0.11	0.94	0.40	0.39
31	0.31	0.66	0.22	0.23
32	0.50	0.36	0.63	0.63

TABLE 4: Analysis of the EFVD compared with Models 1, 2, and 3 on the basis of the car-following process.

Car-following process	(*ED* of the EFVD model– *ED* of Model 1)/*ED* of Model 1	(*ED* of the EFVD model– *ED* of Model 2)/*ED* of Model 2	(*ED* of the EFVD model– *ED* of Model 3)/*ED* of Model 3
1	-74.95%	-37.67%	-36.29%
2	13.82%	6.64%	2.01%
3	-79.81%	-48.52%	-17.10%
4	-55.29%	-10.90%	-13.30%
5	-66.72%	-7.87%	-11.25%
6	-65.69%	-20.66%	-32.48%
7	-16.34%	35.44%	36.70%
8	-43.03%	-2.70%	1.82%
9	-80.28%	-41.18%	-13.43%
10	-48.32%	-29.23%	-32.13%
11	-43.65%	39.08%	41.40%
12	-81.29%	-9.63%	-6.49%
13	20.70%	-16.95%	-17.23%
14	-40.81%	105.85%	115.86%
15	-48.66%	-28.35%	-34.07%
16	29.12%	21.87%	21.34%
17	-39.12%	-27.23%	-25.46%
18	-15.54%	-9.91%	-7.78%
19	-58.82%	23.47%	12.56%
20	-93.75%	-70.83%	-70.07%
21	-67.72%	-37.08%	-23.61%
22	-40.38%	-12.78%	-15.55%
23	-61.74%	-22.40%	-12.21%
24	-45.58%	-27.83%	-31.74%
25	-75.79%	-68.41%	-69.61%
26	-51.94%	-31.81%	-37.16%
27	-57.77%	-7.37%	-9.47%
28	79.26%	-9.79%	-13.19%
29	3.55%	67.01%	73.15%
30	-88.85%	-73.68%	-73.28%
31	-52.98%	42.94%	35.14%
32	39.30%	-20.69%	-20.00%

3. In this study, the improvement of the EFVD model is different from that of Models 2 and 3. We distinguished the vehicles that gather and dissipation based on the velocity of following and leading vehicles. In comparison with Model 2, the EFVD model emphasised the effect of acceleration on car-following behaviour only at the acceleration stage that was different from that of Model 2. In comparison with Model 3, the EFVD model highlighted the influence of headway distance on vehicle deceleration behaviour at different deceleration stages. Moreover, we determined the average safe driving distance headway of the driver's deceleration based on parameter u_2. When the distance headway is greater than u_2, the vehicle will decelerate with a small deceleration. By contrast, when the distance headway is smaller than u_2, the vehicle will decelerate with a large deceleration. This improvement method avoids the risk of a rear-end collision.

Comprehensively, the EFVD model distinguishes the vehicles that gather and dissipation based on the velocity of following and leading vehicles. The effects of different factors on the car-following behaviour were considered at different stages. The EFVD model can improve the deficiencies in the OV, GF, and FVD models. However, we still need to explain the worse fitting degree of the EFVD model than Models 1, 2, and 3 in the partially isolated point and car-following process. We analysed the EFVD model and the simulation of the car-following behaviour. The main reasons for the insufficient fitting of the EFVD model were caused by parameters u_2 and $a_{n-1}(t)$ in the EFVD model, and u_2 was the average safe driving distance headway of the driver's deceleration which is decided by numerous car-following processes. In the actual car-following environment, the safe driving distance headway of the driver's deceleration is inconsistent with u_2, given the influence of the driver's characteristics. When the driver's driving behaviour is aggressive and unsteady, the driver's deceleration safety distance is often less than the parameter u_2. When the driver's driving behaviour is conservative and careful, the driver's deceleration safety distance is often greater than parameter u_2. In these two cases, the EFVD model was insufficient to simulate deceleration behaviour because of the influence of parameter u_2. The EFVD model also considers the effect of $a_{n-1}(t)$ in solving the problem of insufficient acceleration in the FVD model. However, this effect results in the problem of advanced acceleration at the early acceleration stage. The analysis of the experimental simulation indicates that when the car-following process contains more data points, the acceleration data points in the early stage are relatively less, and the fitting degree improvement of the entire car-following process brought by parameter $a_{n-1}(t)$ is better than the fitting degree error caused by the advanced acceleration. Conversely, when the car-following process contains less data points, the acceleration data points in the early stage are relatively more, and the advanced acceleration caused by parameter $a_{n-1}(t)$ has a relatively worse effect on the fitting degree improvement of the entire car-following process. The two factors combined cause the EFVD model to have several defects. Overall, the improvement effect of the EFVD model in most of the test data is evident, although the EFVD model is worse than Models 1, 2, and 3 in the partially isolated point and car-following process.

6. Conclusion

The car-following behaviour at signalised intersections is different from that on the road. The driving behaviour when the vehicle is gathered is also different from the driving behaviour when the vehicle is dissipated. The OV, GF, and FVD models are tested by using two types of car-following processes, namely, incomplete and complete parking, to address the abovementioned problem. The results show that the deceleration of the three models is larger than the measured data at the early stage of deceleration. However, the deceleration is smaller than the measured data at the subsequent deceleration stage. Moreover, the simulation of the car-following behaviour at signalised intersections using the OV and GF models demonstrates the risk of a rear-end collision. The OV and GF models at the acceleration stage have the problem of acceleration insufficiency in the entire acceleration stage, although the FVD model has a significant acceleration in the early acceleration stage. However, the acceleration remains insufficient at the subsequent acceleration stage. We consider the effect of distance headway on the following behaviour when the vehicle is gathered and the effect of the acceleration of the front vehicle on the following behaviour when the vehicle is dissipated to solve the deficiencies of the three models because the driver has different emphases when the vehicle is gathered and when dissipated. Furthermore, we propose the EFVD model on the basis of the FVD model. The experimental analysis confirms that the EFVD model can improve the deficiencies in the OV, GF, and FVD models. Furthermore, the 622 isolated point data of the 32 complete car-following processes are used to test the EFVD model, FVD model, and two other models which are adopted as the comparative model. The results show that the simulation fit is better in the EFVD model than in the FVD model for most of the test data. Thus, the extended model is effective. The EFVD model is more reasonable than the OV, GF, FVD, and two other models for analysing the car-following behaviour at signalised intersections.

However, a certain deficiency is observed in the EFVD model considering the influence of parameters u_2 and $a_{n-1}(t)$, similar to the previous analysis. Thus, improving this model should firstly focus on driver character analysis and dynamic adjustment of the parameter u_2 value in combination with driver's character, and the determination of other parameters should also be combined with the driver's character. Secondly, we should dynamically adjust the coefficient of $a_{n-1}(t)$ in accordance with the velocity of the following vehicle considering the problem of advance acceleration at the early acceleration stage in this model. We can then further improve the fitting degree of the EFVD model for simulating the measured data of the car-following behaviour at signalised intersections. The car-following model summarises the characteristics of many drivers and describes these characteristics mathematically. The simulation results show the average driving behaviour. In the actual driving behaviour, the car-following model cannot fully fit the trajectory of the vehicle given the influence of various conditions. Therefore, we propose various new car-following models that adapt to the automatic driving environment with the rapid development of information and vehicle technologies. In the next stage, we will focus on analysing the influencing factors of the car-following model at signal intersections under the automatic driving environment. We will further improve the traditional car-following model and develop an enhanced car-following model at signalised intersections to adapt to the automatic driving environment.

Conflicts of Interest

The authors declare that they have no conflicts of interest.

Acknowledgments

The work described in this paper was fully supported by the National Nature Science Foundation of China (no. 61364026). Any opinions, findings, conclusions, or recommendations expressed in this paper do not reflect the views of the Government of the China for the Innovation and Technology Support Program of the Innovation and Technology Fund.

References

[1] M. Brackstone and M. McDonald, "Car-following: a historical review," *Transportation Research Part F: Traffic Psychology and Behaviour*, vol. 2, no. 4, pp. 181–196, 1999.

[2] R. E. Chandler, R. Herman, and E. Montroll, "Traffic dynamics: studies in car following," *Operations Research*, vol. 6, no. 2, pp. 165–184, 1958.

[3] R. Herman, E. W. Montroll, R. B. Potts, and R. W. Rothery, "Traffic dynamics: analysis of stability in car following," *Operations Research*, vol. 7, no. 1, pp. 86–106, 1959.

[4] A. Reuschel, "Fahrzeugbewegungen in der kolonne," vol. 4, no. 3, 1950.

[5] L. A. Pipes, "An operational analysis of traffic dynamics," *Journal of Applied Physics*, vol. 24, pp. 274–281, 1953.

[6] D. C. Gazis, R. Herman, and R. W. Rothery, "Nonlinear follow-the-leader models of traffic flow," *Operations Research*, vol. 9, no. 4, pp. 545–567, 1961.

[7] G. F. Newell, "Nonlinear effects in the dynamics of car following," *Operations Research*, vol. 9, no. 2, pp. 209–229, 1961.

[8] M. Bando, K. Hasebe, A. Nakayama, A. Shibata, and Y. Sugiyama, "Dynamical model of traffic congestion and numerical simulation," *Physical Review E Statistical Physics Plasmas Fluids & Related Interdisciplinary Topics*, vol. 51, no. 2, 1995.

[9] D. Helbing and B. Tilch, "Generalized force model of traffic dynamics," *Physical Review E: Statistical, Nonlinear, and Soft Matter Physics*, vol. 58, no. 1, pp. 133–138, 1998.

[10] R. Jiang, Q. Wu, and Z. Zhu, "Full velocity difference model for a car-following theory," *Physical Review E: Statistical, Nonlinear, and Soft Matter Physics*, vol. 64, part 2, no. 1, 2001.

[11] X. Zhao and Z. Gao, "A new car-following model: full velocity and acceleration difference model," *The European Physical Journal B*, vol. 47, no. 1, pp. 145–150, 2005.

[12] A. K. Gupta and V. K. Katiyar, "Phase transition of traffic states with on-ramp," *Physica A: Statistical Mechanics and Its Applications*, vol. 371, no. 2, pp. 674–682, 2006.

[13] A. K. Gupta and V. K. Katiyar, "A new anisotropic continuum model for traffic flow," *Physica A: Statistical Mechanics and Its Applications*, vol. 368, no. 2, pp. 551–559, 2006.

[14] T. Wang, Z.-Y. Gao, and X.-M. Zhao, "Multiple velocity difference model and its stability analysis," *Acta Physica Sinica*, vol. 55, no. 2, pp. 634–640, 2006.

[15] A. K. Gupta and V. K. Katiyar, "A new multi-class continuum model for traffic flow," *Transportmetrica*, vol. 3, no. 1, pp. 73–85, 2007.

[16] H. X. Ge, R. J. Cheng, and Z. P. Li, "Two velocity difference model for a car following theory," *Physica A Statistical Mechanics & Its Applications*, vol. 387, no. 21, pp. 5239–5245, 2008.

[17] Z. Wen-Xing and J. Lei, "Nonlinear analysis of a synthesized optimal velocity model for traffic flow," *Communications in Theoretical Physics*, vol. 50, no. 2, pp. 505–510, 2008.

[18] H. X. Gong, H. C. Liu, and B.-H. Wang, "An asymmetric full velocity difference car-following model," *Physica A: Statistical Mechanics and Its Applications*, vol. 387, no. 11, pp. 2595–2602, 2008.

[19] G. H. Peng and D. H. Sun, "A dynamical model of car-following with the consideration of the multiple information of preceding cars," *Physics Letters A*, vol. 374, no. 15-16, pp. 1694–1698, 2010.

[20] W. X. Zhu and J. Lei, "Stability and kinku2013antikink soliton solution for total generalized optimal velocity model," *International Journal of Modern Physics C*, vol. 19, no. 9, 2008.

[21] T. Q. Tang, Y. H. Wu, L. Caccetta, and H. J. Huang, "A new car-following model with consideration of roadside memorial," *Physics Letters A*, vol. 375, no. 44, pp. 3845–3850, 2011.

[22] Y. F. Jin, M. Xu, and Z. Y. Gao, "KdV and kink-antikink solitons in an extended car-following model," *Journal of Computational and Nonlinear Dynamics*, vol. 6, no. 1, pp. 52–63, 2011.

[23] T. Q. Tang, C. Y. Li, Y. H. Wu, and H. J. Huang, "Impact of the honk effect on the stability of traffic flow," *Physica A Statistical Mechanics & Its Applications*, vol. 390, no. 20, pp. 3362–3368, 2011.

[24] Y. Li, D. Sun, W. Liu et al., "Modeling and simulation for microscopic traffic flow based on multiple headway, velocity and acceleration difference," *Nonlinear Dynamics*, vol. 66, no. 1-2, pp. 15–28, 2011.

[25] S. Jin, D.-H. Wang, and X.-R. Yang, "Non-lane-based car-following model with visual angle information," *Transportation Research Record*, no. 2249, pp. 7–14, 2011.

[26] Y. F. Jin and M. Xu, "Bifurcation analysis of the full velocity difference model," *Chinese Physics Letters*, vol. 27, no. 4, pp. 53–55, 2010.

[27] S. Jin, D. H. Wang, Z. Y. Huang, and P. F. Tao, "Visual angle model for car-following theory," *Physica A: Statistical Mechanics & Its Applications*, vol. 390, no. 11, pp. 1931–1940, 2011.

[28] A. K. Gupta and V. K. Katiyar, "Analyses of shock waves and jams in traffic flow," *Journal of Physics A: Mathematical and Theoretical*, vol. 38, no. 19, pp. 4069–4083, 2005.

[29] S. Jin, D. H. Wang, C. Xu, and Z. Y. Huang, "Staggered car-following induced by lateral separation effects in traffic flow," *Physics Letters A*, vol. 376, no. 3, pp. 153–157, 2012.

[30] T. Tang, H. Huang, and H. Shang, "A new pedestrian-following model for aircraft boarding and numerical tests," *Nonlinear Dynamics*, vol. 67, no. 1, pp. 437–443, 2012.

[31] T. Q. Tang, Y. P. Wang, X. B. Yang, and Y. H. Wu, "A new car-following model accounting for varying road condition," *Nonlinear Dynamics*, vol. 70, no. 2, pp. 1397–1405, 2012.

[32] T. Tang, Y. Shi, Y. Wang, and G. Yu, "A bus-following model with an on-line bus station," *Nonlinear Dynamics*, vol. 70, no. 1, pp. 209–215, 2012.

[33] G.-H. Peng and R.-J. Cheng, "A new car-following model with the consideration of anticipation optimal velocity," *Physica A: Statistical Mechanics and Its Applications*, vol. 392, no. 17, pp. 3563–3569, 2013.

[34] A. K. Gupta, "A section approach to a traffic flow model on networks," *International Journal of Modern Physics C*, vol. 24, no. 5, 2013.

[35] A. K. Gupta, S. Sharma, and P. Redhu, "Analyses of lattice traffic flow model on a gradient highway," *Communications in Theoretical Physics*, vol. 62, no. 3, pp. 393–404, 2014.

[36] A. K. Gupta and I. Dhiman, "Analyses of a continuum traffic flow model for a nonlane-based system," *International Journal of Modern Physics C*, vol. 25, no. 10, p. 4, 2014.

[37] A. K. Gupta and P. Redhu, "Analyses of the driver's anticipation effect in a new lattice hydrodynamic traffic flow model with passing," *Nonlinear Dynamics*, vol. 76, no. 2, pp. 1001–1011, 2014.

[38] J. I. Ge and G. Orosz, "Dynamics of connected vehicle systems with delayed acceleration feedback," *Transportation Research Part C: Emerging Technologies*, vol. 46, pp. 46–64, 2014.

[39] T. Tang, W. Shi, H. Shang, and Y. Wang, "A new car-following model with consideration of inter-vehicle communication," *Nonlinear Dynamics*, vol. 76, no. 4, pp. 2017–2023, 2014.

[40] W.-X. Zhu and L.-D. Zhang, "A speed feedback control strategy for car-following model," *Physica A: Statistical Mechanics and Its Applications*, vol. 413, pp. 343–351, 2014.

[41] T. Zhou, D. Sun, Y. Kang, H. Li, and C. Tian, "A new car-following model with consideration of the prevision driving behavior," *Communications in Nonlinear Science and Numerical Simulation*, vol. 19, no. 10, pp. 3820–3826, 2014.

[42] A. K. Gupta and I. Dhiman, "Phase diagram of a continuum traffic flow model with a static bottleneck," *Nonlinear Dynamics*, vol. 79, no. 1, pp. 663–671, 2015.

[43] S. Sharma, "Lattice hydrodynamic modeling of two-lane traffic flow with timid and aggressive driving behavior," *Physica A: Statistical Mechanics and Its Applications*, vol. 421, pp. 401–411, 2015.

[44] M. Sasaki and T. Nagatani, "Transition and saturation of traffic flow controlled by traffic lights," *Physica A: Statistical Mechanics and Its Applications*, vol. 325, no. 3-4, pp. 531–546, 2003.

[45] T. Tang, G. Xu, H.-J. Huang, and Y. Xue, "A traffic flow model considering signal light influence and its numerical simulation," *Acta Physica Sinica*, vol. 57, no. 1, pp. 56–60, 2008.

[46] S. Yu and Z. Shi, "An extended car-following model at signalized intersections," *Physica A: Statistical Mechanics and Its Applications*, vol. 407, pp. 152–159, 2014.

[47] S. Yu, X. Zhao, Z. Xu, and Z. Shi, "An improved car-following model considering the immediately ahead cars velocity difference," *Physica A Statistical Mechanics & Its Applications*, vol. 461, pp. 446–455, 2016.

[48] S. Yu and Z. Shi, "An improved car-following model considering headway changes with memory," *Physica A: Statistical Mechanics and its Applications*, vol. 421, pp. 1–14, 2015.

[49] S. Yu and Z. Shi, "An extended car-following model considering vehicular gap fluctuation," *Measurement*, vol. 70, pp. 137–147, 2015.

[50] S. Yu and Z. Shi, "Car-following model on vehicles arrival during the red phase," *Journal of Highway and Transport*, vol. 27, no. 11, pp. 93–100, 2014.

[51] X. H. Li and S. Yu, "Analysis and modeling for car-following behaviors on arrival traffic flow of signalized intersection," *Journal of Transportation Systems Engineering and Information Technology*, vol. 15, no. 2, pp. 102–108, 2015.

[52] https://www.fhwa.dot.gov/publications/research/operations/its/06135/index.cfm.

[53] D. H. Wang, P. F. Tao, S. Jin, and D. F. Ma, "Method of calibrating and validating car-following model," *Journal of Jilin University (Engineering and Technology Edition)*, vol. s1, pp. 59–65, 2011.

[54] G. Zhu and S. Kwong, "Gbest-guided artificial bee colony algorithm for numerical function optimization," *Applied Mathematics and Computation*, vol. 217, no. 7, pp. 3166–3173, 2010.

4

The Roles of Car Following and Lane Changing Drivers' Anticipations during Vehicle Inserting Process: A Structural Equation Model Approach

Li Li ,[1] Dong Zhang ,[2] Zhi-gang Xu ,[3] Ping Wang,[1] and Gui-ping Wang[1]

[1]*School of Electronics & Control Engineering, Chang'an University, Middle Section of Nan Er Huan, Xi'an, Shaanxi 710064, China*
[2]*School of Transportation & Logistics, Dalian University of Technology, No. 2 Linggong Rd., Gan Jingzi District, Dalian, Liaoning 116024, China*
[3]*School of Information Engineering, Chang'an University, Middle Section of Nan Er Huan, Xi'an, Shaanxi 710064, China*

Correspondence should be addressed to Dong Zhang; zhangdong@dlut.edu.cn and Zhi-gang Xu; xuzhigang@chd.edu.cn

Academic Editor: Stefano de Luca

Anticipating ability is a skill that drivers count on to handle risky tasks in the traffic. This paper explores how the drivers of lane changing vehicle and its immediately car follower anticipate surrounding vehicles' movements and adjust their manoeuvers during vehicle inserting process. The drivers' anticipating mechanisms are modelled in the framework of structural equation model and estimated from field data. Results show that the change of lane changing type or traffic signal affects the drivers' anticipation. Increased vehicle speed impels subject driver to anticipate driving condition in further future, but the stimulus is lower than the one coming from the kinematic comparisons of subject vehicle and other vehicles. The drivers care more about the vehicles' interactions with which they are personally involved than the one to which they are only onlookers. The drivers' responses to the counterpart vehicle's movements depend on the progress of vehicle insertion and their roles in vehicle interactions.

1. Introduction

A lane changing (LC) manoeuver is often accompanied by high collision risk, especially in congested traffic [1, 2]. The mutual misunderstanding between the drivers of LC vehicle (LCV) and surrounding car following vehicles (CFV) is a major source of the risk. However, the cases of successful LC mostly happen in real traffic, when the drivers seem to be able to predict the risk and make adjustments in advance to avoid it. Such driving skill is usually named as *anticipating ability* or *anticipation*.

Driver's anticipation can reduce vehicle speed fluctuation and fuel consumption [3], but how this ability works and how to model it is still under study. Hofmann et al. [4] found that the anticipation gradually forms with driving experience, and some hints, such as turning light of the vehicle ahead, let the LCV driver know the counterpart's intention. Chan et al. [5] reported the weaker hazard anticipation of the novice drivers from the experienced drivers. Many existing LC decision models assume that the LCV's manoeuvre is impelled by a onetime and irrevocable decision of the vehicle driver, so the anticipation is often modelled as some predictive vehicle kinematic variables, such as vehicle gap or delay of travel time, measured at the moment of making LC decision [6, 7]. However, the assumption has been challenged by the pieces of evidence obtained from recent empirical studies [8, 9]. They found that the LCV's inserting process could last over ten seconds, and the preparation of such insertion is even longer. Some researchers have realized the risk of ignoring the continuity of LCV manoeuvre in traffic simulation, so they modelled the LCV's preparation movements before its insertion as a series of consecutive decisions followed by longitudinal manoeuvre adjustments [10–12]. In the models, the LCV driver's anticipation works continuously during the process of seeking an acceptable inserting gap in LC target lane. The LCV's longitudinal acceleration or deceleration was abstracted as the trade-off of driver's LC desire and his/her anticipated collision risk with surrounding vehicle if conducting LC [13, 14]. It is noteworthy that most related studies are concerned with the LC preparing process on the original lane, but the anticipation should also work when the driver executes the inserting manoeuvre from the original lane to the target lane. Unfortunately, few efforts have been devoted to analyse anticipative inserting manoeuvre of the LCV driver [15].

More importantly, the anticipation is the behavioural basis of formulating LC model and CF model into a unified framework. Such integrated model can reduce the simulating error of many existing microscopic traffic flow models in the scenario when the LCV transits from the state of inserting preparation to the one of inserting execution [16, 17]. The modelling effort of CFV driver's anticipation in CF scenario has lasted for decades (see Section 2), but there rarely exist studies of how the skill works in LC scenario. In fact, the absence of CFV's anticipation could make potential conflict brought by LC manoeuver evolve to be a real collision.

To fill the existing research gaps, this study models and analyses the anticipating ability of the drivers of the LCV and its surrounding CFVs during the LCV's inserting process. The process is divided into two consecutive periods. The working mechanisms of the drivers' anticipation in two periods are formulated into two structural equation models (SEMs) correspondingly. A driver's anticipation to another vehicle is abstracted as a latent variable of the SEM. Vehicle trajectories and the indicators reflecting driving environment are utilized as the observable variables of the model. The transmission path of the drivers' anticipation on their behaviours and the influence degrees are identified based on the estimations of the SEMs.

The rest content is arranged as follows. Section 2 introduces the characteristics of driver's anticipation. Section 3 describes the modelling effort to figure out the working mechanism of the anticipation during the LCV's inserting process. Section 4 depicts data collection work. The model estimations are reported and discussed in Section 5. Section 6 concludes the main findings and points out the directions for future studies.

2. Characteristics of Driver's Anticipation

Figure 1(a) shows the minimum platoon involving a LC scenario, including a LCV, its following vehicle (FV), and leading vehicle (LV) on LC target lane. The LV's movement is assumed to affect the drivers of the LCV and the FV, but the LV driver does not suffer backward influences from the two following vehicles. The study period of this paper is the LCV's inserting process, and its detailed definition refers to the content in Appendix. The process can be divided into two stages by the entrance moment (EM, see Figure 1(a)), when body centre of the LCV crosses lane marking line. The period before the EM is *anticipating period*, while the one after the EM is *relaxing period*. The anticipations of the LCV driver and the FV drivers begin to work at the start of the anticipating period. Three driving relationships exist in each period. The FV and the LCV involve an LC relationship in anticipating period, and the CF relationships existing in this period include the following: (1) The FV follows the LV longitudinally on the target lane, and (2) the LCV follows the LV laterally on the original lane. In the relaxing period, the FV turns to follow the LCV laterally on target lane, and the other two CF relationships continue.

Anticipating ability of the drivers in the LCV and the FV assumes working in both longitudinal and lateral directions of lane traffic. This assumption has been verified by previous studies of CF behaviour. The longitudinal anticipation is modelled as a CF driver's estimation of one or more leaders' movements in one or more future moments [18–20]. Lateral anticipation is presented as that a staggered leader in the same lane or the neighbouring lane of the CFV brings the driver's lateral disturbance and triggers his or her deceleration [21, 22]. In some LC cases, the FV driver, who originally follows the LV in target lane, could anticipate adjacent LCV driver's inserting intention and decelerate voluntarily to create a larger inserting gap for the LCV [21]. Meanwhile, the LCV driver has a great possibility of anticipating the FV driver's decelerating intention and take advantage of the opportunity to finish the insertion. Under similar anticipation of the two drivers working simultaneously, they can coordinate their manoeuvers to ensure the vehicle safety during the LCV's inserting process. After the EM, the two drivers' anticipation continues working in the relaxing period. The LCV's relationship with the FV changes from the LC couple to a lateral CF couple, although its inserting manoeuvres continue after the EM. In this period, the two vehicles begin recovering their desired spacing with the LCV and the LV ahead, respectively [23–25]. Based on the previous findings mentioned above, the assumption of this study that the two drivers' anticipation works during the whole LCV's inserting process is reasonable.

Besides the nonexclusiveness and temporal continuity, other characteristics of the drivers' anticipation should be taken into account when modelling manoeuvres of the LCV and FV during the LCV's inserting process.

(1) Universality: driver's anticipations exist in every LC scenario. Sometimes it can be indirectly inferred from vehicle manoeuvers, while other times it only exists in the driver's mind. For instance, the FV driver's conservative anticipation to the adjacent LCV in a cooperative LC scenario can be inferred from observable predeceleration of the FV before the EM [26]. On the contrary, the FV driver who involves in a forced LC scenario has aggressive anticipation to the LCV. It induces the FV driver postponing the deceleration till or after the EM when the LCV forcibly but successfully is inserted into the target lane.

(2) Gradualness: each driver's anticipation is gradually formed during the LCV's inserting process as he or she continuously observes the other vehicles' movements. It is hard to imagine that a driver can accurately recognize the other driver's intentions by counting on the first impression only.

3. Model

3.1. Structural Equation Model. The characteristics of the driver's anticipation mentioned above belongs to the domain of human psychology. How to model and analyse the psychological influences on the manoeuvres of the LCV and the FV is the critical issue of this study.

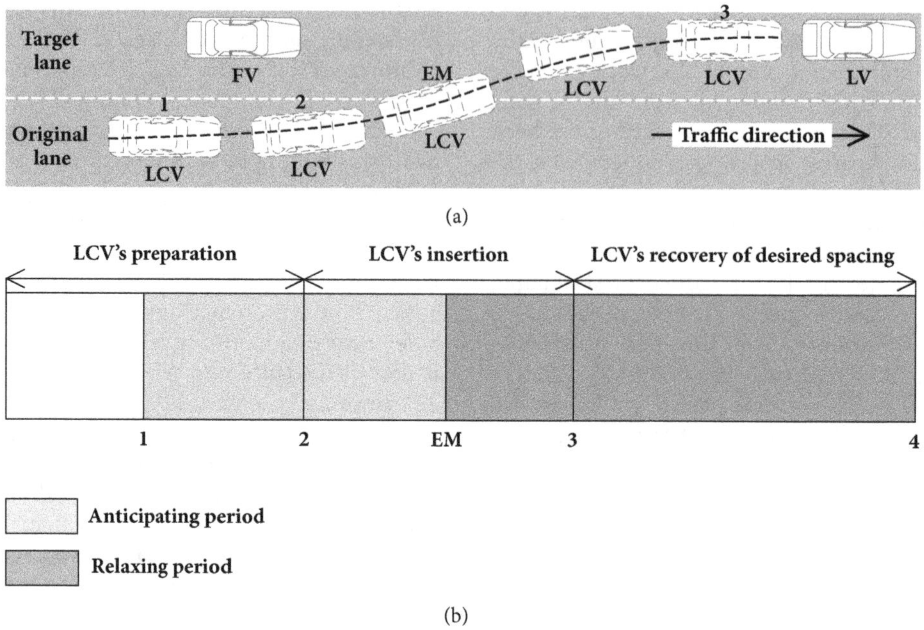

FIGURE 1: (a) A kind of LCV's insertion. (b) Timeline of the inserting process.

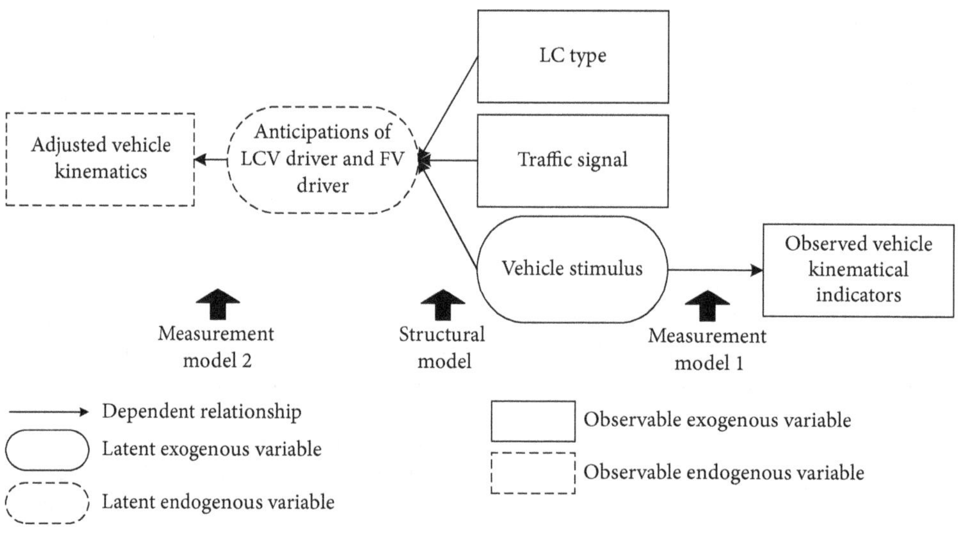

FIGURE 2: Conceptual model.

A conceptual model is proposed under the framework of SEM to depict how the psychological traits affect the drivers' behaviours. SEM provides an empirical approach to identify the cause-effect relationship that is formulated based on solid theories or empirical results [27]. It can identify the effects of personal risk perception and task loads on drivers' attitudes of driving safety [28]. The conceptual model consists of one structural model and two measurement models. The structural model hypothesizes the influential paths of the external stimuli to the drivers' anticipations, which could come from traffic signals, LC type, and other vehicles' movements. The vehicle stimuli can be inferred from some observed variables that reflect vehicle kinematics at the start of a study period. This inferring system is formulated as the "measurement model 1" in Figure 2. The "measurement model 2" in the figure is another system to infer how the drivers adjust vehicle manoeuvers at the end of the period based on their anticipations.

The structural and measurement models are made of latent variables, observed variables, and their dependent relationships. They are illustrated as the normal rectangles, the rounded rectangles, and the directional lines in Figure 2. The dependent relationship presents as a directional line pointing from a causal variable to an affected variable. The anticipations of the LCV driver and the FV driver to future movements of surrounding vehicles are abstracted as the latent

TABLE 1: Latent variable list.

Endogenous latent variable		Exogenous latent variable	
Name	Meaning	Name	Meaning
LA_{F2P}	Anticipation of the FV driver to the LV	LS_{L4F}	The LV stimulus for the FV driver
LA_{F2LC}	Anticipation of the FV driver to the LCV	LS_{LC4F}	The LCV stimulus for the FV driver
LA_{LC2F}	Anticipation of the LCV driver to the FV	$LS_{L\&LC4F}$	The LV&LCV stimulus for the FV driver
LA_{LC2L}	Anticipation of the LCV driver to the LV	LS_{L4LC}	The LV stimulus for the LCV driver
		LS_{F4LC}	The FV stimulus for the LCV driver
		$LS_{L\&F4LC}$	The LV&FV stimulus for the LCV driver

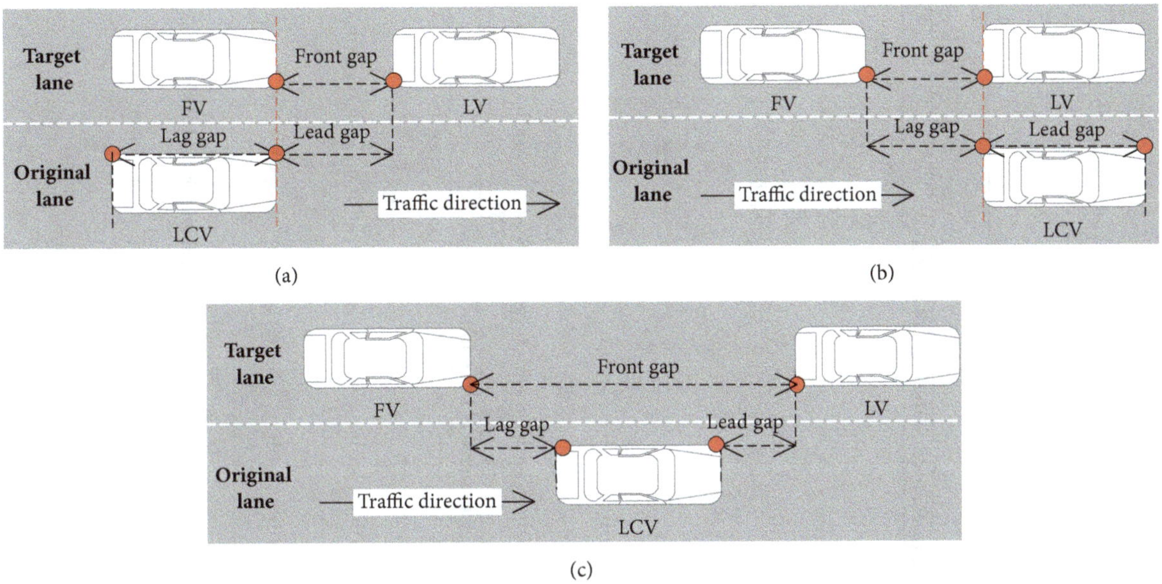

FIGURE 3: (a) Forward insertion. (b) Backward insertion. (c) Median insertion.

variables listed in Table 1, as they cannot always be observed from vehicle trajectories. The observed variables listed in Table 2 are defined from the view points of the LCV driver and the FV driver in specific scenarios. Their definitions will be detailed in Section 3.2.

3.2. Variables

3.2.1. Scenario-Dependent Variable. The LCV could be inserted into target lane in different ways [29], so three inserting scenarios illustrated in Figure 3 are defined in this study. In a forward inserting scenario, the LCV overtakes the FV to find opportunity of conducting the insertion. In a backward inserting scenario, the LCV lets the LV pass it and cuts in the backward gap. In a median inserting scenario, the LCV has run aside the gap between the LV and the FV for a while and finds a proper time to fill in the gap. The observed variables are measured at the start and the end of the two study periods. Their settings are defined in Table 2.

The settings of some variables at the start of the anticipating period depend on the observing point of each driver in the scenarios defined above. Here take the lag gap observed by the FV driver (G_{lag4F}), for instance. In the forward inserting scenario (see Figure 3(a)), lag gap appears in the negative direction of the FV driver while in the positive direction of the LCV driver, it sets G^1_{lag4F} <0 and G^1_{lag4LC} >0. Note that the LCV driver needs to estimate the FV driver's reaction to the LV's movements, so the two drivers are assumed to stay in the same line when observing the front gap (G^1_{front} in Table 2) at the start of the anticipating period. Similarly, the FV driver is assumed to observe the lead gap (G^1_{lead} in the table) as the LCV driver does.

3.2.2. Scenario-Independent Variable. Besides the scenario-dependent variable, the values of some variables do not relate to the scenarios defined above. Front gap always appears in the positive direction of the FV in each scenario (see Figure 2), so it holds the positive value as an input or output of drivers' anticipations ($G^1_{front}/G^{EM}_{front}/G^n_{front}$ >0 in Table 2). Speed difference is often used to measure the degree of driving safety. A driver can evaluate the collision risk to another vehicle from varying trends of their speed differences. For example, if the FV is slower than the LV ($SD^1_{F\&L}/SD^{EM}_{F\&L}/SD^n_{F\&L}$ >0 in the table), its head-end collision risk to the LV is lower, and *vice versa*. Hence, a positive speed difference between the two vehicles ($SD_{F\&L}/SD_{F\&LC}/SD_{LC\&L}$ in the table) indicates a safer driving condition than the condition

TABLE 2: Observed variables and their settings.

Anticipating period		Relaxing period	
Name (Unit)	Measurement	Name (Unit)	Measurement
Exogenous observed variable (measured at the start of a period)			
S_F^1 (m/s)	--	S_F^{EM} (m/s)	--
S_{LC}^1 (m/s)	--	S_{LC}^{EM} (m/s)	--
G_{front}^1 (m)	$G_{front}^1 >0$	G_{front}^{EM} (m)	$G_{front}^{EM} >0$
G_{lag4F}^1/G_{lag4LC}^1 (m)	Forward LC: $G_{lag4F}^1 < 0,\ G_{lag4LC}^1 > 0$ Backward LC: $G_{lag4F}^1 > 0,\ G_{lag4LC}^1 < 0$ Normal LC: $G_{lag4F}^1 > 0,\ G_{lag4LC}^1 < 0$	G_{lag4F}^{EM} (m) G_{lag4LC}^{EM} (m)	$G_{lag4F}^{EM} >0$ Do not consider
G_{lead}^1 (m)	Forward LC: $G_{lead}^1 > 0$ Backward LC: $G_{lead}^1 < 0$ Normal LC: $G_{lead}^1 > 0$	G_{lead}^{EM} (m)	$G_{lead}^{EM} >0$
$SD_{F\&L}^1$ (m/s)	S_L^1-S_F^1	$SD_{F\&L}^{EM}$ (m/s)	S_L^{EM}-S_F^{EM}
$SD_{F\&LC}^1$ (m/s)	S_{LC}^1-S_F^1	$SD_{F\&LC}^{EM}$ (m/s)	S_{LC}^{EM}-S_F^{EM}
$SD_{LC\&L}^1$ (m/s)	S_L^1-S_{LC}^1	$SD_{LC\&L}^{EM}$ (m/s)	S_L^{EM}-S_{LC}^{EM}
LCT (binary)	0=Discretionary LC; 1=Mandatory LC	LCT (binary)	Do not consider
TS (binary)	0=Red; 1=Green	TS (binary)	0=Red; 1=Green
Endogenous observed variable (measured at the end of a period)			
G_{front}^{EM} (m)	$G_{front}^{EM} >0$	G_{front}^{n} (m)	$G_{front}^{n} >0$
G_{lag4F}^{EM} (m)	$G_{lag4F}^{EM} >0$	G_{lag4F}^{n} (m)	$G_{lag4F}^{n} >0$
G_{lag4LC}^{EM} (m)	$G_{lag4LC}^{EM} <0$	G_{lag4LC}^{n} (m)	Do not consider
G_{lead}^{EM} (m)	$G_{lead}^{EM} >0$	G_{lead}^{n} (m)	$G_{lead}^{n} >0$
$SD_{F\&L}^{EM}$ (m/s)	S_L^{EM}-S_F^{EM}	$SD_{F\&L}^{n}$ (m/s)	S_L^{n}-S_F^{n}
$SD_{F\&LC}^{EM}$ (m/s)	S_{LC}^{EM}-S_F^{EM}	$SD_{F\&LC}^{n}$ (m/s)	S_{LC}^{n}-S_F^{n}
$SD_{LC\&L}^{EM}$ (m/s)	S_L^{EM}-S_{LC}^{EM}	$SD_{LC\&L}^{n}$ (m/s)	S_L^{n}-S_{LC}^{n}

Note: (1) S_F: the FV's speed; S_{LC}: the LCV's speed; S_L: the LV's speed; G_{front}: front gap for the FV or LCV driver; G_{lag4F}: lag gap for the FV driver; G_{lag4LC}: lag gap for the LCV driver; G_{lead}: lead gap for the FV or LCV driver; $SD_{F\&L}$: speed difference between the FV and the LV; $SD_{F\&LC}$: speed difference between the FV and the LCV; $SD_{LC\&L}$: speed difference between the LCV and the LV; LCT: LC type; TS: the state of traffic signal.

(2) The variable measured at the first second, the EM, or the last instant of the LCV's inserting process is marked by the superscript "1", "EM", or "n", respectively.

the negative difference happens. At the end of the anticipating period or in the relaxing period, the LCV enters target lane, so each vehicle's position is fixed and the meanings of the gaps do not change for each driver ($G_{lag4F}^{EM}/G_{lag4F}^{n}$ >0, $G_{lead}^{EM}/G_{lead}^{n}$ >0 and G_{lag4LC}^{EM} <0 in the table).

3.3. Model Definition. Two SEMs are formulated to specify the conceptual model developed in Section 3.1, and each of them consists of one structural model and two measurement models. Their definitions will be introduced in this section.

3.3.1. Definition of Structural Model. The structural model is to infer whether and how much an external stimuli could affect a driver's anticipations of other vehicles' future movements. The path of transmitting the effect is illustrated in Figure 4 as a directional line pointing from an exogenous observed variable to an endogenous latent variable. The former one models a stimulus while the latter one models the anticipation. Both of the LCV driver and the FV driver are involved in the driving relationships defined in Section 2 and observe the external stimuli in different ways. The two facts are the basis of structural model development.

The FV longitudinally follows the LV on LC target lane, and it also maintains LC relationship in the anticipating period or lateral CF relationship in the relaxing period with the LCV. The anticipations of the FV driver to the developing trends of the two relationships are abstracted as the LS_{L4F} and LS_{LC4F}. The driver also predicts the relative moving trend of the LCV and the LV ($LS_{L\&LC4F}$). The influences of these three vehicle stimuli on the FV driver are modelled by "$LS_{L4F} \longrightarrow LA_{F2L}$" and "$LS_{LC4F}/LS_{L\&LC4F} \longrightarrow LA_{F2LC}$" in Figures 4(a) and 4(b). In a similar way, the LCV driver's anticipations of future moving trends of the FV and the LV can be modelled as "$LS_{F4LC}/LS_{L\&F4LC} \longrightarrow LA_{LC2F}$" and "$LS_{L4LC} \longrightarrow LA_{LC2L}$" in Figure 4(a) and $LS_{L4LC} \longrightarrow LA_{LC2L}$ in Figure 4(b). Note that the LCV driver is assumed to be immune to the backward influences of the FV any longer in the relaxing period (LS_{F4LC} and $LS_{L\&F4LC}$), as the driver has encroached into the target lane after the EM and occupied a leading position with respect to the FV. Hence, LS_{F4LC} and $LS_{L\&F4LC}$ are excluded from Figure 4(b). Besides adjacent vehicles' movements, a drivers' anticipation of collision risks is also related to subject vehicle speeds. A higher vehicle speed makes the driver more sensitive to an external stimulus. So speed of the LCV or the PF is assumed to affect its driver's anticipations to other vehicles' movements. These are modelled as "$S_F^1 \longrightarrow LA_{F2L}/LA_{F2LC}$" and "$S_{LC}^1 \longrightarrow LA_{LC2F}/LA_{LC2L}$" in Figure 4(a), "$S_F^{EM} \longrightarrow LA_{F2L}/LA_{F2LC}$" and "$S_{LC}^{EM} \longrightarrow LA_{LC2L}$" in Figure 4(b).

Moreover, LC type (LCT) can reflect the influence of LC intention degree on a driver's behaviour. With respect to a discretionary LC, a mandatory LC usually arouses the LCV driver's stronger LC intention and inserting manoeuvers of the LCV could be more aggressive. Its impact on the FV could be harder too. Such influence is modelled as "$LCT \longrightarrow LA_{F2LC}/LA_{LC2F}$" in Figure 4(a). LCT is excluded from Figure 4(b), as the LCV has entered the target lane after the EM and the variance of LC type should be ineffective for the drivers. In addition, the influence of traffic signals (TS) on a driver's decision at urban arterial cannot be ignored. This factor is assumed to affect the drivers' anticipations at the anticipating and relaxing periods. It is modelled as "$TS \longrightarrow LA_{F2L}/LA_{F2LC}/LA_{LC2F}/LA_{LC2L}$" in Figure 4(a) and "$TS \longrightarrow LA_{F2L}/LA_{F2LC}/LA_{LC2L}$" in Figure 4(b).

3.3.2. Definition of Measurement Model. There exist two measurement models in Figures 4(a) and 4(b). One measurement model is used to infer the stimulus of surrounding vehicle perceived by a driver at the start of the anticipating or relaxing period. The stimulus is assumed to derive from vehicle spacing and speed difference, which is modelled as "$LS_{L4F}/LS_{L\&F4LC} \longrightarrow G_{front}^1/SD_{F\&L}^1$", "$LS_{LC4F} \longrightarrow G_{lag4F}^1/SD_{F\&LC}^1$", "$LS_{L\&LC4F}/LS_{L4LC} \longrightarrow G_{lead}^1/SD_{LC\&L}^1$", and "$LS_{F4LC} \longrightarrow G_{lag4LC}^1/\ SD_{F\&LC}^1$" in Figure 4(a) and "$LS_{L4F} \longrightarrow G_{front}^{EM}/SD_{F\&L}^{EM}$", "$LS_{LC4F} \longrightarrow G_{lag4F}^{EM}/SD_{F\&LC}^{EM}$", and "$LS_{L\&LC4F}/LS_{L4LC} \longrightarrow G_{lead}^{EM}/SD_{LC\&L}^{EM}$" in Figure 4(b). Another measurement model is to identify how the two drivers adjust vehicle spacing and speed difference at the end of a period based on their anticipations of other vehicles' future movements. The FV driver can control its spacing and speed difference to the LV, while the LCV driver can control the ones to the LV. These are modelled as "$LA_{F2L} \longrightarrow G_{front}^{EM}/SD_{F\&L}^{EM}$" and "$LA_{LC2L} \longrightarrow G_{lead}^{EM}/\ SD_{LC\&L}^{EM}$" in Figure 4(a) and "$LA_{F2L} \longrightarrow G_{front}^{n}/SD_{F\&L}^{n}$" and "$LA_{LC2L} \longrightarrow G_{lead}^{n}/SD_{LC\&L}^{n}$" in Figure 4(b). The drivers of the LCV and the FV jointly control lag gap and the speed difference between them in the anticipating period, so it can be modelled as "$LA_{F2LC} \longrightarrow G_{lag4F}^{EM}$", "$LA_{LC2F} \longrightarrow G_{lag4LC}^{EM}$", and "$LA_{F2LC}/LA_{LC2F} \longrightarrow SD_{F\&LC}^{EM}$" in Figure 4(a). In the relaxing period, only the FV driver hold the controls, while the LCV driver is assumed to not care the FV's movements any more after the EM. It can be reflected as "$LA_{F2LC} \longrightarrow G_{lag4F}^{n}/\ SD_{F\&LC}^{n}$" in Figure 4(b).

3.3.3. Correlated Variables. Correlation test is made to the observed variables, and test results are listed in Tables 3 and 4. The bolded coefficients indicate the pairs of the variable whose correlations are modelled in the SEMs, which are illustrated in Figure 4 as the normal rectangles connected by curved lines. The correlations are set based on the logic of driving behaviour. Since the anticipations of the LCV driver and the FV driver are assumed to work simultaneously after the start of anticipating period, the exogenous variables in Figure 4(a) are set to be uncorrelated. At the end of the anticipating and relaxing period, it assumes that the kinematical indicators whose values are determined by the same driver are set to be correlated.

4. Data Collection

An arterial link in Shanghai with 200 m length is selected as the study site. Traffic video was captured at a roadside building in the height of 70 m to cover the whole link (see Figure 5). The trajectories of 250 group vehicles were extracted from the video by self-developed software [30] in 10 hz and

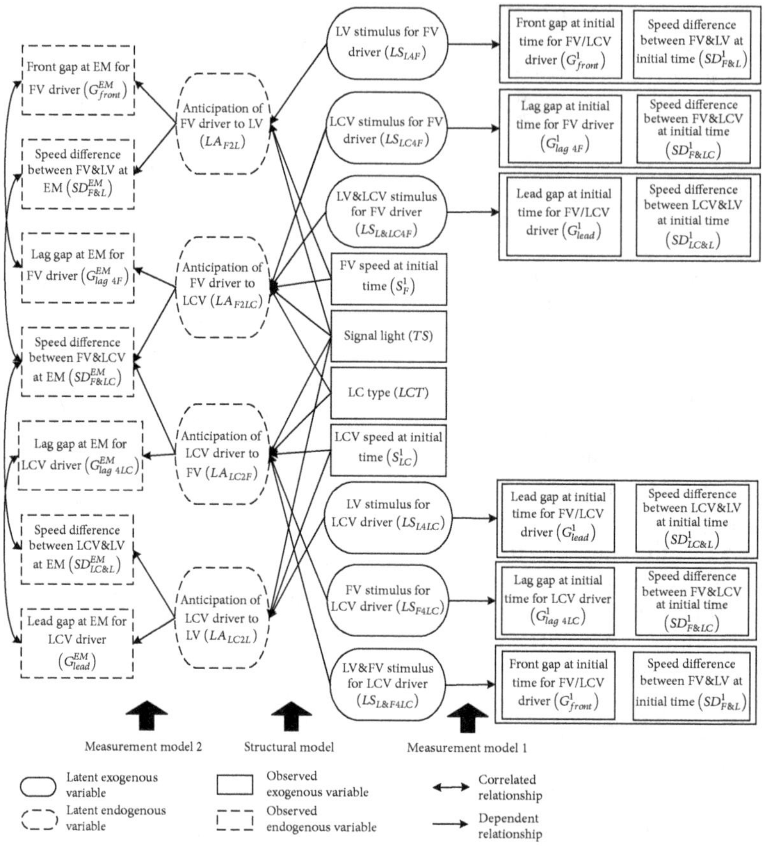

(a) The SEM in the anticipating period

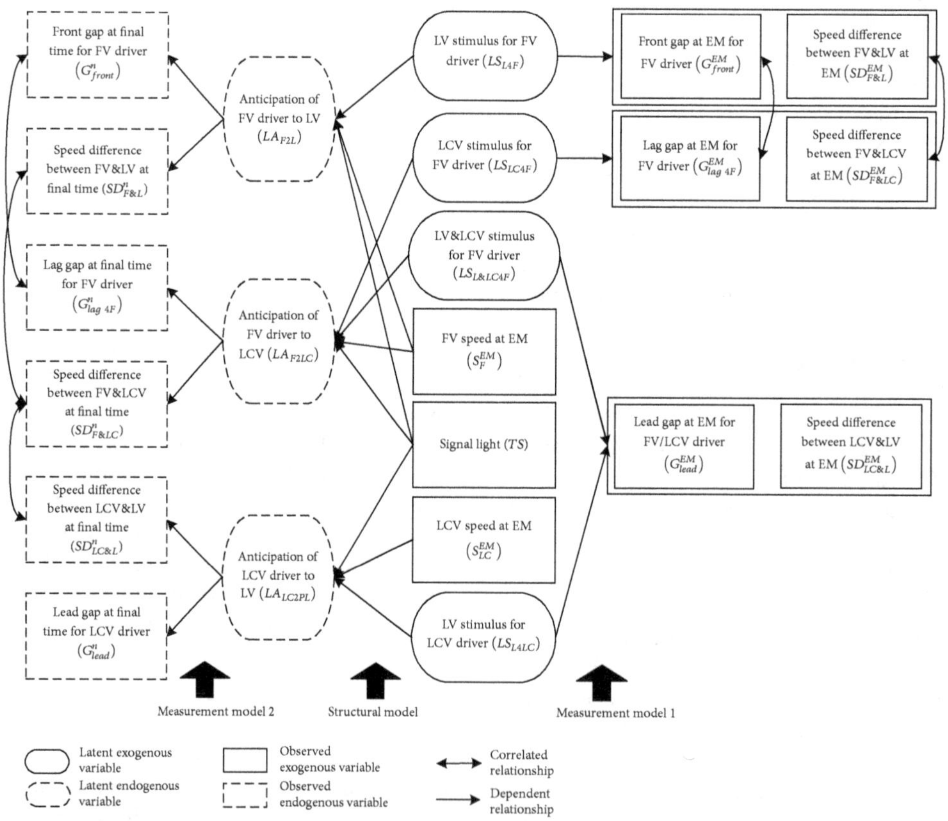

(b) The SEM in the relaxing period

FIGURE 4

TABLE 3: Correlation of the exogenous observed variables in the anticipating/relaxing period.

	S_F^1/S_F^{EM}	S_{LC}^1/S_{LC}^{EM}	$G_{front}^1/G_{front}^{EM}$	$G_{lag4F}^1/G_{lag4F}^{EM}$	$G_{lag4LC}^1/G_{lag4LC}^{EM}$	G_{lead}^1/G_{lead}^{EM}	$SD_{F\&L}^1/SD_{F\&L}^{EM}$	$SD_{F\&LC}^1/SD_{F\&LC}^{EM}$	$SD_{LC\&L}^1/SD_{LC\&L}^{EM}$
S_F^1/S_F^{EM}	1.000/1.000	--	--	--	--	--	--	--	--
S_{LC}^1/S_{LC}^{EM}	0.234*/0.443*	1.000/1.000	--	--	--	--	--	--	--
$G_{front}^1/G_{front}^{EM}$	0.411*/0.374*	0.197*/0.436*	1.000/1.000	--	--	--	--	--	--
$G_{lag4F}^1/G_{lag4F}^{EM}$	0.422*/0.308*	0.148*/-0.187*	0.253*/**0.294***	1.000/1.000	--	--	--	--	--
$G_{lag4LC}^1/G_{lag4LC}^{EM}$	-0.422*/-0.308	-0.148*/0.187*	-0.253*/-0.294*	-1.000*/-1.000*	1.000/--	--	--	--	--
G_{lead}^1/G_{lead}^{EM}	0.220*/0.241*	0.229*/0.544*	0.894*/0.879*	-0.143*/-0.087	0.143*/--	1.000/1.000	--	--	--
$SD_{F\&L}^1/SD_{F\&L}^{EM}$	-0.474*/-0.423*	-0.039/0.001	-0.605*/-0.369*	-0.311*/-0.174*	0.311*/--	-0.460*/-0.283	1.000/1.000	--	--
$SD_{F\&LC}^1/SD_{F\&LC}^{EM}$	-0.063/-0.291*	0.956*/0.692*	0.078/0.168*	0.024/-0.463*	-0.024/--	0.169*/0.383*	0.103/**0.331***	1.000/1.000	--
$SD_{LC\&L}^1/SD_{LC\&L}^{EM}$	-0.093/-0.168*	-0.176*/-0.412*	-0.594*/-0.555*	0.265*/0.079	-0.265*/--	-0.703*/-0.626*	0.557*/0.470*	-0.153*/-0.305*	1.000/1.000

Note: (1) a/b: a is coefficient of variable correlation in the anticipating period, and b is the one in the relaxing period. (2) *: the coefficient is significant at 95% confidence level.

TABLE 4: Correlation of the endogenous observed variables in the anticipating/relaxing period.

	$G^{EM}_{front}/G^{n}_{front}$	$G^{EM}_{lag4F}/G^{n}_{lag4F}$	$G^{EM}_{lag4LC}/G^{n}_{lag4LC}$	$G^{EM}_{lead}/G^{n}_{lead}$	$SD^{EM}_{F\&L}/SD^{n}_{F\&L}$	$SD^{EM}_{F\&LC}/SD^{n}_{F\&LC}$	$SD^{EM}_{LC\&L}/SD^{n}_{LC\&L}$
$G^{EM}_{front}/G^{n}_{front}$	1.000/1.000	--	--	--	--	--	--
$G^{EM}_{lag4F}/G^{n}_{lag4F}$	**0.284*/0.572***	1.000/1.000	--	--	--	--	--
$G^{EM}_{lag4LC}/G^{n}_{lag4LC}$	-0.284*/-0.572*	-1.000*/-1.000	1.000/--	--	--	--	--
$G^{EM}_{lead}/G^{n}_{lead}$	0.943*/0.896*	-0.047/0.239*	**0.047/--**	1.000/1.000	--	--	--
$SD^{EM}_{F\&L}/SD^{n}_{F\&L}$	-0.499*/-0.106	-0.211*/-0.062	0.211*/--	-0.448*/-0.070	1.000/1.000	--	--
$SD^{EM}_{F\&LC}/SD^{n}_{F\&LC}$	0.076/0.223*	-0.409*/0.231*	0.409*/--	0.220*/0.187*	**0.306*/0.349***	1.000/1.000	--
$SD^{EM}_{LC\&L}/SD^{n}_{LC\&L}$	-0.514*/-0.345*	0.125*/-0.276*	-0.125*/--	-0.579*/-0.280*	0.675*/0.685*	**-0.496*/-0.352***	1.000/1.000

Note: (1) "a/b": "a" is coefficient of variable correlation in the anticipating period, and "b" is the one in the relaxing period. (2) *: the coefficient is significant at 95% confidence level.

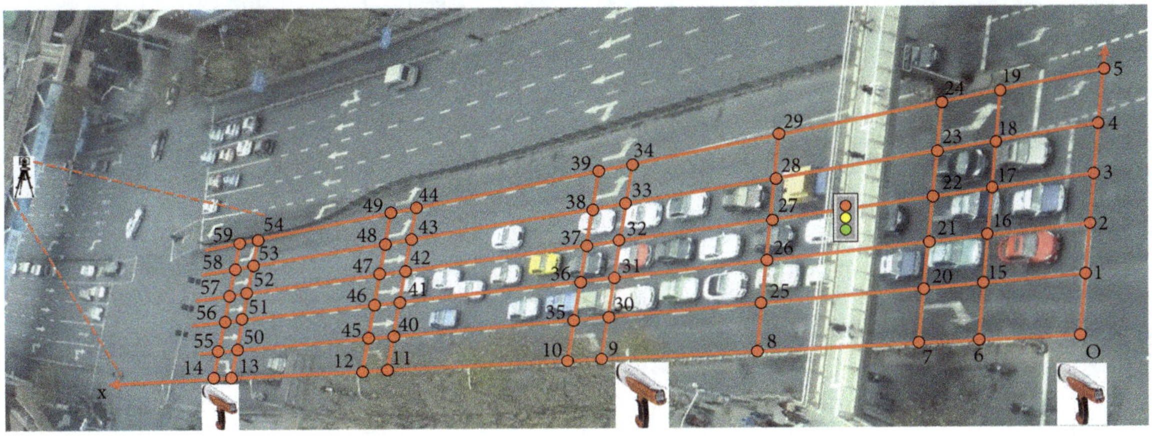

FIGURE 5: Study site.

calibrated to real-world coordinate using the method in [31]. The trajectory recording points locate at the head or rear bumper of a vehicle, where the subject vehicle collides with other vehicles most possibly. The noise in the raw trajectories was filtered referring to the method in [32]. The speed of some vehicles was calculated from the calibrated trajectories and compared with the values measured by three radar guns located at the cross-sections marked in the figure. The error of the speed calculated from the trajectories was found to be limited in 0.1 m/s. The vehicle kinematical indicators in Table 2 were calculated from the trajectories and used to estimate the SEMs proposed before. The information of traffic signal and LC type was recorded from a pedestrian overpass at upstream intersection. The LCV's inserting manoeuvre is a discretionary LC if LC original and target lanes serve the same direction traffic; otherwise, it is a mandatory LC.

5. Results Interpretation and Discussion

The statistical software *Stata* is applied to formulate and estimated the two SEMs. The models are estimated at a 95% confidence level, and the goodness-of-fit measures show they fit the data well: Cronbach alpha value = 0.81/0.83; root-mean-square-error of approximation (RMSEA) = 0.08/0.04; comparative fit index (CFI) = 0.90/0.93; standardized root mean squared residual (SRMR) = 0.07/0.05. Tables 5 and 6 list the estimated unstandardized and standardized coefficients. The former ones will be interpreted in Section 5.1, while the latter ones are used to calculate the indirect path effects listed in Table 7, which will be interpreted in Section 5.2.

5.1. Model Coefficient

5.1.1. Structural Model. The estimated coefficients of structural model measure how much the external stimuli appeared at the start of the anticipating and relaxing periods impacts anticipations of the drivers in the FV and the LCV. Results show that an increase of vehicle speed could enhance the drivers' anticipations in both period ("$S_F^1 \longrightarrow LA_{F2L}/LA_{F2LC}$"=0.189/0.164>0 for the FV driver, "$S_{LC}^1 \longrightarrow LA_{LC2F}/LA_{LC2L}$"=0.146/ 0.149>0 for the LCV driver in Table 5; "$S_F^{EM} \longrightarrow LA_{F2L}/LA_{F2LC}$"=0.386/0.192>0 for the FV driver, and "$S_{LC}^{EM} \longrightarrow LA_{LC2L}$"=0.348>0 for the LCV driver in Table 6). Such a result verifies the authors' speculation that a driver's anticipation of other vehicles' moving trends is related to the subject vehicle's speed. A faster vehicle could increase its driver's sensitivity to external stimuli and impel him or the driver to predict driving condition in the further future. But compared with the speed variance of subject vehicle, other vehicles' stimuli impact the drivers' anticipations heavier ("$LS_{L4F} \longrightarrow LA_{F2L}$"=0.967>0.189="$S_F^1 \longrightarrow LA_{F2L}$" and "$LS_{LC4F} \longrightarrow LA_{F2LC}$"=0.897>0.164= "$S_F^1 \longrightarrow LA_{F2LC}$" for the FV driver, "$LS_{F4LC} \longrightarrow LA_{LC2F}$"=0.623>0.146="$S_{LC}^1 \longrightarrow LA_{LC2F}$" and "$LS_{L4LC} \longrightarrow LA_{LC2L}$" = |0.985| > 0.149 = "$S_{LC}^1 \longrightarrow LA_{LC2L}$" for the LCV driver in Table 5; "$LS_{L4F} \longrightarrow LA_{F2L}$"=0.753>0.386="$S_F^{EM} \longrightarrow LA_{F2L}$" and "$LS_{LC4F} \longrightarrow LA_{F2LC}$"=0.913>0.192= "$S_F^{EM} \longrightarrow LA_{F2LC}$" for the FV driver, "$LS_{L4LC} \longrightarrow LA_{LC2L}$"=0.858>0.348="$S_{LC}^{EM} \longrightarrow LA_{LC2L}$" for the LCV driver in Table 6). The stimuli come from each driver's perception of vehicle gap or speed difference that exists between (1) subject vehicle and another vehicle and (2) two other vehicles. The stimulus coming from the former vehicle couple is found to impact the drivers' anticipations more than the one of the latter couple at each period ("$LS_{L\&LC4F} \longrightarrow LA_{F2LC}$"=0.232<0.897="$LS_{LC4F} \longrightarrow LA_{F2LC}$" for the FV driver, "$LS_{L\&F4LC} \longrightarrow LA_{LC2F}$" = | − 0.320| < 0.623 = "$LS_{F4LC} \longrightarrow LA_{LC2F}$" for the LCV driver in Table 5; "$LS_{L\&LC4F} \longrightarrow LA_{F2LC}$" = | − 0.330| < 0.913 = "$LS_{LC4F} \longrightarrow LA_{F2LC}$" for the FV driver in Table 6).

Besides the stimuli of vehicle kinematics, the change of LC type is also found to affect the drivers' anticipations. The LCV driver produces a stronger anticipation in a mandatory LC scenario than in a discretionary one ("$LCT \longrightarrow LA_{LC2F}$"=0.136 in Table 5), as the driving task is riskier in the mandatory LC scenario. What is interesting is that the FV driver has a weaker anticipation in a mandatory LC scenario than in a discretionary one ("$LCT \longrightarrow LA_{F2LC}$"=-0.164<0.136="$LCT \longrightarrow LA_{LC2F}$" in Table 5). It could be attributed to the fact that the FV driver can identify the LCV driver's intention more easily in the mandatory LC scenario, but it becomes harder for the driver involved in discretionary

TABLE 5: Estimated SEM in the anticipating period.

Structural model				
Latent variable	**(⟵) Exogenous variable**	**C.**	**S.C.**	**S.E.**
LA_{F2L}	S_F^1	2.585*	0.189*	0.025
	TS	-9.006*	-0.137*	0.032
	LS_{L4F}	8.908*	0.967*	0.010
LA_{F2LC}	S_F^1	0.829*	0.164*	0.030
	LCT	-3.822*	-0.164*	0.059
	TS	-4.526*	-0.186*	0.066
	LS_{LC4F}	3.022*	0.897*	0.040
	$LS_{L\&LC4F}$	0.780*	0.232*	0.114
LA_{LC2F}	S_{LC}^1	0.076*	0.146*	0.057
	LCT	1.107*	0.136*	0.051
	TS	-0.314	-0.037	0.067
	LS_{F4LC}	0.848*	0.623*	0.244
	$LS_{L\&F4LC}$	-0.415*	-0.320*	0.213
LA_{LC2L}	S_{LC}^1	0.596*	0.149*	0.065
	TS	0.447*	0.006*	0.001
	LS_{L4LC}	-13.668*	-0.985*	0.009
Measurement model				
Exogenous observed variable	**(⟵) Latent variable**	**C.**	**S.C.**	**S.E.**
G_{front}^1	LS_{L4F}	11.873*	0.964*	0.003
$SD_{F\&L}^1$		1	0.475*	0.040
G_{lag4F}^1	LS_{LC4F}	1	0.221*	0.022
$SD_{F\&LC}^1$		0.803*	0.312*	0.081
G_{lead}^1	$LS_{L\&LC4F}$	-10.168*	-0.957*	0.003
$SD_{LC\&L}^1$		1	0.766*	0.040
G_{lead}^1	LS_{L4LC}	1	0.061*	0.007
$SD_{LC\&L}^1$		0.575*	0.286*	0.106
G_{lag4LC}^1	LS_{F4LC}	4.176*	0.765*	0.023
$SD_{F\&LC}^1$		1	0.337*	0.045
G_{front}^1	$LS_{L\&F4LC}$	1	0.074*	0.004
$SD_{F\&L}^1$		1.646*	0.723*	0.027
Measurement model				
Endogenous observed variable	**(⟵) Latent variable**	**C.**	**S.C.**	**S.E.**
G_{front}^{EM}	LA_{F2L}	1	0.956*	0.011
$SD_{F\&L}^{EM}$		-0.058*	-0.520*	0.069
G_{lag4F}^{EM}	LA_{F2LC}	1	0.916*	0.012
$SD_{F\&LC}^{EM}$		0.140*	0.321*	0.085
G_{lag4LC}^{EM}	LA_{LC2F}	2.539*	0.805*	0.128
$SD_{F\&LC}^{EM}$		1	0.801*	0.083
G_{lag4LC}^{EM}	LA_{LC2L}	1	0.819*	0.039
$SD_{LC\&L}^{EM}$		-0.122*	-0.900*	0.015

Note: (1) C.: coefficient, S.C.: standard coefficient, S.E.: standard error.
(2) "∗": the coefficient is significant at 95% confidence level.
(3) "⟵": the path direction shown in Figure 4(a).

TABLE 6: Estimated SEM in the relaxing period.

Structural model				
Latent variable	**(⟵) Exogenous variable**	**C.**	**S.C.**	**S.E.**
LA_{F2L}	S_F^{EM}	2.012*	0.386*	0.050
	TS	-8.289*	-0.355*	0.048
	LS_{L4F}	2.039*	0.753*	0.025
LA_{F2LC}	S_F^{EM}	0.821*	0.192*	0.075
	TS	-2.554*	-0.134*	0.066
	LS_{LC4F}	5.394*	0.913*	0.045
	$LS_{L\&LC4F}$	-0.965*	-0.330*	0.167
LA_{LC2L}	S_{LC}^{EM}	1.850*	0.348*	0.063
	TS	-11.404*	-0.358*	0.058
	LS_{L4LC}	3.465*	0.858*	0.028
Measurement model				
Exogenous observed variable	**(⟵) Latent variable**	**C.**	**S.C.**	**S.E.**
G_{front}^{EM}	LS_{L4F}	1	0.162*	0.008
$SD_{F\&L}^{EM}$		-0.403*	-0.428*	0.057
G_{lag4F}^{EM}	LS_{LC4F}	1	0.163*	0.016
$SD_{F\&LC}^{EM}$		0.727*	0.386*	0.123
G_{lead}^{EM}	$LS_{L\&LC4F}$	-5.491*	-0.793*	0.031
$SD_{LC\&L}^{EM}$		1	0.895*	0.060
G_{lead}^{EM}	LS_{L4LC}	1	0.174*	0.011
$SD_{LC\&L}^{EM}$		-0.319*	-0.404*	0.142
Endogenous observed variable	**(⟵) Latent variable**	**C.**	**S.C.**	**S.E.**
G_{front}^{n}	LA_{F2L}	1	0.912*	0.005
$SD_{F\&L}^{n}$		-0.024*	-0.080*	0.006
G_{lag4F}^{n}	LA_{F2LC}	1	0.979*	0.005
$SD_{F\&LC}^{n}$		0.022*	0.108*	0.023
G_{lead}^{n}	LA_{LC2L}	1	0.883*	0.016
$SD_{LC\&L}^{n}$		-0.044*	-0.212*	0.017

Note: (1) C.: coefficient; S.C.: standard coefficient; S.E.: standard error.
(2) "∗": the coefficient is significant at 95% confidence level.
(3) "⟵": the path direction shown in Figure 4(b).

Table 7: Indirect path effects.

Anticipating period								
Exogenous variable (⟶)	Via endogenous latent variable (⟶)	Endogenous observed variable						
		G^{EM}_{front}	G^{EM}_{lag4F}	G^{EM}_{lag4LC}	G^{EM}_{lead}	$SD^{EM}_{F\&L}$	$SD^{EM}_{F\&LC}$	$SD^{EM}_{LC\&L}$
S^1_F	LA_{PF2PL}/LA_{PF2LC}	**2.585**	0.829	--	--	-0.149	0.116	--
S^1_{LC}	LA_{LC2F}/LA_{LC2L}	--	--	0.192	0.596	--	0.076	-0.072
LCT	LA_{F2LC}/LA_{LC2F}	--	-3.822	2.810	--	--	0.571	--
TS	$LA_{F2L}/LA_{F2LC}/LA_{LC2L}$	-9.006	-4.526	--	0.447	0.522	-0.633	-0.054
LS_{L4F}	LA_{F2L}	8.908	--	--	--	-0.516	--	--
LS_{LC4F}	LA_{F2LC}	--	3.022	--	--	--	0.423	--
$LS_{L\&LC4F}$	LA_{F2LC}	--	0.780	--	--	--	0.109	--
LS_{L4LC}	LA_{LC2L}	--	--	--	-13.668	--	--	1.667
LS_{F4LC}	LA_{LC2F}	--	--	2.153	--	--	0.848	--
$LS_{L\&F4LC}$	LA_{LC2F}	--	--	-1.053	--	--	-0.415	--
Relaxing period								
Exogenous variable (⟶)	Via endogenous latent variable (⟶)	Endogenous observed variable						
		G^{n}_{front}	G^{n}_{lag4F}	G^{n}_{lag4LC}	G^{n}_{lead}	$SD^{n}_{F\&L}$	$SD^{n}_{F\&LC}$	$SD^{n}_{LC\&L}$
S^{EM}_F	LA_{F2L}/LA_{F2LC}	2.012	0.821	--	--	-0.048	0.018	--
S^{EM}_{LC}	LA_{LC2L}	--	--	--	1.850	--	--	-0.081
TS	$LA_{F2L}/LA_{F2LC}/LA_{LC2L}$	-8.289	-2.554	--	-11.404	0.198	-0.056	0.501
LS_{L4F}	LA_{F2L}	2.039	--	--	--	-0.048	--	--
LS_{LC4F}	LA_{F2LC}	--	5.394	--	--	--	0.118	--
$LS_{L\&LC4F}$	LA_{F2LC}	--	-0.965	--	--	--	-0.021	--
LS_{L4LC}	LA_{LC2L}	--	--	--	3.465	--	--	-0.152

Note: (1) "--" means that no indirect path exists between the exogenous and endogenous observed variables, or the path effect is insignificant.
(2) ⟶: the path direction shown in Figure 4.

LC scenario. After all, in the latter case the LCV driver can head to the destination without conducting a LC.

We find that the FV driver's anticipation decreases when traffic signal switches from red to green in the two periods ("$TS \longrightarrow LA_{F2L}/LA_{F2LC}$"=-0.137/-0.186<0 in Table 5; "$TS \longrightarrow LA_{F2L}/LA_{F2LC}$"=-0.355/-0.134<0 in Table 6). A red signal can help the FV driver build a reliable expectation that all vehicles ahead have to decelerate in the near future, so it is easier for the driver to anticipate other vehicles' movements. Obviously, this trend becomes harder to predict in the green phase. In contrast, the LCV driver's anticipation of the FV's movements is hardly affected by the switch of the traffic signal in the anticipating period ("$TS \longrightarrow LA_{LC2F}$" is insignificant in Table 5). Its influence on the LCV driver's anticipation of the LV is much slighter than that of other factors ("$TS \longrightarrow LA_{LC2L}$"=0.006< "$S^1_{LC} \longrightarrow LA_{LC2L}$"=0.149< "$LS_{L4LC} \longrightarrow LA_{LC2L}$" = | − 0.985| in Table 5). The insensitiveness of the LCV driver to the traffic signal in the anticipating period is reasonable, because no matter in the red or green phase, the driver needs to keep safe gap and speed difference with the FV before subject vehicle inserts into target lane. So the LCV driver focuses on predicting the FV's future movements, which is hardly affected by the switch of the traffic signal. When the LCV enters target lane in the relaxing period, traffic signal turns to be influential for the driver ("$TS \longrightarrow LA_{LC2L}$"=0.358 in Table 6), as the driver's attention switches from conducting the insertion to following the LV. The future moving trend of the LV becomes a critical issue that needs the LCV driver cares.

5.1.2. Measurement Model. The estimations of two measurement models of the SEMs can tell us (1) which kinematical indicator is major source of vehicle stimuli for a driver at start of a period; (2) which kinematical indicator is mainly adjusted by a driver at end of a period based on his or her anticipation of counterpart's future movements. The results will be interpreted in a temporal sequence from start of the anticipating period to end of the relaxing period.

At the start of the anticipating period, a larger aspect of the LV's stimulus perceived by the FV driver comes from front gap, rather than the speed difference with the LV ("$LS_{L4F} \longrightarrow G^1_{front}$"=0.964>0.475="$LS_{L4F} \longrightarrow SD^1_{F\&L}$" and "$LS_{L\&LC4F} \longrightarrow G^1_{lead}$"=| − 0.957| >0.766="$LS_{L\&LC4F} \longrightarrow SD^1_{LC\&L}$" in Table 5). It impels the FV driver to shorten the spacing to the LV. But this adjustment could be weakened by the FV driver's anticipation to the relationship between the LV and the LCV ("$LS_{L\&LC4F} \longrightarrow G^1_{lead}$"=-0.957<0 and "$LS_{L4F} \longrightarrow G^1_{front}$"=0.964>0 in the table). On the contrary, a larger aspect of the LV's stimulus perceived by the LCV driver derives from vehicle speed difference ("$LS_{L4LC} \longrightarrow SD^1_{LC\&L}$" =0.286>0.061="$LS_{L4LC} \longrightarrow G^1_{lead}$" and "$LS_{L\&F4LC} \longrightarrow SD^1_{F\&L}$"=0.732> 0.074="$LS_{L\&F4LC} \longrightarrow G^1_{front}$" in the table). As the LC couple, the FV driver is concerned more about the speed difference to the LCV ("$LS_{LC4F} \longrightarrow SD^1_{F\&LC}$"=0.312>0.221="$LS_{LC4F} \longrightarrow G^1_{lag4F}$" in the table), while the LCV driver pays more attention to his or her lag gap with the FV ("$LS_{F4LC} \longrightarrow G^1_{lag4LC}$"=0.765>0.337="$LS_{F4LC} \longrightarrow SD^1_{F\&LC}$" in the table). Then at the end of the anticipating period, based on their anticipations to other vehicles, the drivers of the FV and the LCV adjust their vehicle manoeuvres to reduce collision risk with other vehicles. In addition, their main adjusted kinematical indicators related to the LV are the ones they were concerned of at start of the period. The FV driver cares more about front gap ("$LA_{F2L} \longrightarrow G^{EM}_{front}$"=0.956> | − 0.520|="$LA_{F2L} \longrightarrow SD^{EM}_{F\&L}$" in Table 5), while the LCV driver cares more about the speed difference with the LV ("$LA_{LC2L} \longrightarrow SD^{EM}_{LC\&L}$"=| − 0.900| >0.819="$LA_{LC2L} \longrightarrow G^{EM}_{lag4LC}$" in the table). As for their mutual driving relationship, the LCV driver still focuses on adjusting lag gap with the FV ("$LA_{F2LC} \longrightarrow G^{EM}_{lag4F}$"=0.916>0.321="$LA_{F2LC} \longrightarrow SD^{EM}_{F\&LC}$" in the table), which is also concerned by the driver at the start of this period. However, the FV driver's interest has been changed to lag gap ("$LA_{LC2F} \longrightarrow G^{EM}_{lag4LC}$"=0.805>0.801="$LA_{LC2F} \longrightarrow SD^{EM}_{F\&LC}$" in the table), rather than the speed difference with the LCV which is the FV driver's main concern at the start of the anticipating period.

An interesting finding is that, in the anticipating period, the drivers' anticipations of their counterparts' movements are affected by the factors differed with the ones they observed at the start of the period. This result could be caused by asymmetric information received by the drivers. Since the LCV driver has a clear target when conducting the insertion, the driver is naturally concerned of the acceptable inserting gap in target lane from the start of the anticipating period. In contrast, the FV driver does not know the LCV driver's intention at the first view in most cases, so the most compelling stimulus for the FV driver is the observable speed difference between the two vehicles. After continuous observations of each other during the anticipating period, both drivers care more about the spacing between them at end of this period. This is because of the fact that lag gap usually achieves the minimum value at the EM, which is the smallest spacing of the two vehicles. Moreover, we find that the drivers' attitudes to the LV are consistent during the anticipating period. It could result of their complete control of the relationships with the LV. They can balance the collision risk to the LV and the one to each other and make the best tactic of adjusting vehicle manoeuvers.

At the start of the relaxing period, both of the FV driver and the LCV driver confirm the LCV's successful insertion, so they follow a similar behavioural logic in this period. The drivers are concerned about the same type of vehicle kinematical indicators at the start of the period, and they adjust the same type of indicators at the end of the period. At the start of the period, speed difference matters more for the drivers than vehicle gap ("$LS_{L4F} \longrightarrow SD^{EM}_{F\&L}$"=|−0.428| >0.162="$LS_{L4F} \longrightarrow G^{EM}_{front}$", "$LS_{LC4F} \longrightarrow SD^{EM}_{F\&LC}$"=0.386>0.163="$LS_{LC4F} \longrightarrow G^{EM}_{lag4F}$", and "$LS_{L\&LC4F} \longrightarrow SD^{EM}_{LC\&L}$"=0.895> | − 0.793|= "$LS_{L\&LC4F} \longrightarrow G^{EM}_{lead}$" for the FV driver and "$LS_{L4LC} \longrightarrow SD^{EM}_{LC\&L}$"=| − 0.404| >0.174="$LS_{L4LC} \longrightarrow G^{EM}_{lead}$" for the LCV driver in Table 6). At end of this period, they adjust vehicle gap more than speed difference ("$LA_{F2PL} \longrightarrow G^n_{front}$" = 0.912> | − 0.080|="$LA_{F2L} \longrightarrow SD^n_{F\&L}$", and "$LA_{F2LC} \longrightarrow G^n_{lag4F}$"=0.979>0.108="$LA_{F2LC} \longrightarrow SD^n_{F\&LC}$"

for the FV driver and "$LA_{LC2L} \longrightarrow G^{n}_{lead}$"=0.883> | – 0.212|= "$LA_{LC2L} \longrightarrow SD^{n}_{LC\&L}$" for the LCV driver in the table). In the authors' view, such behavioural similarity of the drivers should be attributed to the following facts. When the drivers confirm the fact that the LCV has entered target lane after the EM, recovering to the desired gap with the leading vehicle becomes their main targets. So vehicle gap is the foremost adjusting indicator. Some studies used this logic to explain the relaxation phenomenon in traffic flow [24, 25]. To achieve the targets, the vehicle manoeuvre adjustments start from the EM, when the speed difference of the two vehicles is relatively high. So speed difference is the drivers' main stimulus at the EM, and recovery of the desired gaps could be postponed after speed difference decreases.

5.2. Indirect Path Effect. In the SEMs illustrated in Figure 4, some indirect paths connect observed or latent exogenous variables with observed endogenous variables via latent endogenous variables. A path transmits external stimuli each driver perceived at the start of a period via his or her anticipations to the adjusted manoeuvers at the end of the period. The cause-effect paths have physical meanings, so their values are calculated from the unstandardized coefficients. The indirect path effects are interpreted as in Table 7.

Take the effect of the path between initial speed of the FV and final front gap in the anticipating period ("$S^{1}_{F} \longrightarrow LA_{F2L} \longrightarrow G^{EM}_{front}$"=2.585, bolded in Table 7) as an example. A 1 m/s increase of the FV speed at the start of the period could induce its driver leaving an additional 2.585 m from the LV at the end of the period. This result is easy to understand. In order to offset the higher collision risk caused by the increased vehicle speed, the FV driver needs to enlarge the front gap to maintain enough emergency braking distance to the LV. The physical meanings of other path effects can be interpreted in this way. To give a big picture about cause-effect relationships between input and output variables of an indirect path, we focus on comparing the estimated effects from a general perspective to obtain knowledge, rather than interpreting the effects one by one.

5.2.1. Different Vehicle Stimuli. We find that when the drivers of the LCV and the FV perceive other vehicles' stimuli, their response could be more severe when adjusting the driving relationships they personally involved (LS_{LC4F} and LS_{PF4LC}), than the condition they are only observers ($LS_{L\&LC4F}$ and $LS_{L\&F4LC}$) ("$LS_{LC4F} \longrightarrow LA_{F2LC} \longrightarrow G^{EM}_{lag4F}/SD^{EM}_{F\&LC}$"=3.022/0.423> 0.780/0.109="$LS_{L\&LC4F} \longrightarrow LA_{F2LC} \longrightarrow G^{EM}_{lag4F}/SD^{EM}_{F\&LC}$" for the FV driver, "$LS_{F4LC} \longrightarrow LA_{LC2F} \longrightarrow G^{EM}_{lag4LC}/SD^{EM}_{F\&LC}$"=2.153/0.848> | – 1.053|/| – 0.415|="$LS_{L\&F4LC} \longrightarrow LA_{LC2F} \longrightarrow G^{EM}_{lag4LC}/SD^{EM}_{F\&LC}$" for the LCV driver in the anticipating period in Table 7; "$LS_{LC4F} \longrightarrow LA_{F2LC} \longrightarrow G^{n}_{lag4F}/SD^{n}_{F\&LC}$"=5.394/0.118> | – 0.965|/| – 0.021|="$LS_{L\&LC4F} \longrightarrow LA_{F2LC} \longrightarrow G^{n}_{lag4F}/SD^{n}_{F\&LC}$" for the FV driver in the relaxing period in the table). In other words, each driver is more sensitive to the direct vehicle stimuli than to the indirect ones.

For example, when the FV driver perceives a one-unit increase in direct stimulus from the LCV at the start of the anticipating period, the driver's adjustment of lag gap at end of the period is more than the adjustment of lag gap responding to indirect stimulus from the LCV and the LV ("$LS_{LC4F} \longrightarrow LA_{F2LC} \longrightarrow G^{EM}_{lag4F}$"=3.022>0.780="$LS_{L\&LC4F} \longrightarrow LA_{F2LC} \longrightarrow G^{EM}_{lag4F}$" in the upper part of Table 7). From the estimation of the measurement model, we know that a one-unit increase of the LCV stimulus for the FV driver derives from a 1m increase of initial lag gap ("$LS_{LC4F} \longrightarrow G^{1}_{lag4F}$"=1 in Table 5), or a 0.803 m/s increase of their initial speed difference ("$LS_{LC4F} \longrightarrow SD^{1}_{F\&LC}$"=0.803 in Table 5). In contrast, for the indirect stimulus from the interaction between LCV and LV perceived by the FV driver, one unit of its increase is caused by a 10.168m decrease of initial lead gap ("$LS_{L\&LC4F} \longrightarrow G^{1}_{lead}$"=-10.168 in Table 5) or a 1m/s increase of initial speed difference between the LCV and the LV ("$LS_{L\&LC4F} \longrightarrow SD^{1}_{LC\&L}$"=1 in Table 5). Since it needs more variants to arouse indirect stimulus ($LS_{L\&LC4F}$) than direct stimulus (LS_{LC4F}) (| – 10.168| > 1 in gap, 1>0.803 in speed difference), $LS_{L\&LC4F}$ has a relatively smaller effect on the FV driver than LS_{LC4F}.

In addition, the LCV's direct stimulus for the FV driver could be enhanced in the anticipating period by the indirect stimulus coming from the interaction between the LCV and the LV ("$LS_{LC4F} \longrightarrow LA_{F2LC} \longrightarrow G^{EM}_{lag4F}/SD^{EM}_{F\&LC}$"=3.022/0.423>0, "$LS_{L\&LC4F} \longrightarrow LA_{F2LC} \longrightarrow G^{EM}_{lag4F}/SD^{EM}_{F\&LC}$"= 0.780/0.109>0 in the upper part of Table 7). This result indicates that, as a passive acceptor of the LCV's insertion, the FV driver is more sensitive to external stimuli when driving condition is uncertain before the EM. As active executor of the LC movement, the LCV driver is less sensitive than the FV driver before the EM, and direct stimulus of the FV perceived by the LCV driver could be weakened by the indirect stimulus coming from the FV and the LV ("$LS_{F4LC} \longrightarrow LA_{LC2F} \longrightarrow G^{EM}_{lag4LC}/SD^{EM}_{F\&LC}$"=2.153/0.848>0, "$LS_{F4LC} \longrightarrow LA_{LC2F} \longrightarrow G^{EM}_{lag4LC}/SD^{EM}_{F\&LC}$"=-1.053/-0.415<0 in the upper part of the table). Different performances of the two drivers at the anticipating period are determined by asymmetric information received by them. The LCV driver always knows more than the FV driver does about when and where to insert into the target lane. However, such asymmetry disappears after the EM. When the insertion is actually successful, the FV driver performs less sensitively to the external stimuli in the relaxing period than the LCV driver does in the anticipating period. The direct stimulus of the LCV for the FV driver could be weakened by the indirect one from the interaction of the LCV and the LV ("$LS_{LC4F} \longrightarrow LA_{F2LC} \longrightarrow G^{n}_{lag4F}/SD^{n}_{F\&LC}$"=5.394/0.118>0, "$LS_{L\&LC4F} \longrightarrow LA_{F2LC} \longrightarrow G^{n}_{lag4F}/SD^{n}_{F\&LC}$"=-0.965/-0.021<0 in the lower part of the table).

There is another interesting finding. In the anticipating period, the responses of the FV driver and the LCV driver to direct stimulus of the counterpart vehicle could be different. The FV driver prefers to keep a larger lag gap when perceiving the LCV stimulus ("$LS_{LC4F} \longrightarrow LA_{F2LC} \longrightarrow G^{EM}_{lag4F}$"=3.022>2.153="$LS_{F4LC} \longrightarrow LA_{LC2F} \longrightarrow G^{EM}_{lag4LC}$" in the upper part of Table 7), while the LCV driver is more likely to increase speed difference with the FV ("$LS_{F4LC} \longrightarrow LA_{LC2F} \longrightarrow SD^{EM}_{F\&LC}$"=0.848> 0.423="$LS_{LC4F} \longrightarrow LA_{F2LC} \longrightarrow SD^{EM}_{F\&LC}$" in the upper part of the table). With respect to the LCV driver, the FV driver is more conservative and would always play a risk-averse role in their interactions, e.g., actively decelerating before the EM.

5.2.2. Different Periods. We find that the drivers of the FV and the LCV use different strategies to adjust their vehicle kinematics during the two periods. The FV manoeuver turns to be gentle from the anticipating period to the relaxing period gradually, but the LCV driver's manoeuver performs in an opposite way.

A 1 m/s increase of the FV speed (S^1_F/S^{EM}_F) or the traffic signal switching from red to green (TS), or a one-unit increase of the LV stimulus (LS_{L4F}) could arouse less adjustment of the FV manoeuvre to the front gap at end of the relaxing period than at the end of the anticipating period ("$S^{EM}_F/TS/LS_{L4F} \longrightarrow LA_{F2L} \longrightarrow G^n_{front}$"=2.012/| − 8.289|/2.039<2.585/| − 9.006|/ 8.908="$S^1_F/TS/LS_{L4F} \longrightarrow LA_{F2L} \longrightarrow G^{EM}_{front}$" in Table 7). Besides, the FV driver also reduces less speed difference with the LV ($SD^n_{F\&L}$) or with the LCV ($SD^n_{F\&LC}$) in the relaxing period than in the anticipation period ("$S^{EM}_F/TS/LS_{L4F} \longrightarrow LA_{F2L} \longrightarrow SD^n_{F\&L}$"=| − 0.048|/0.198/|− 0.048| < | − 0.149|/0.522/| − 0.516|="$S^1_F/TS/LS_{L4F} \longrightarrow LA_{F2L} \longrightarrow SD^{EM}_{FL}$" and "$S^{EM}_F/TS/LS_{L4F} \longrightarrow LA_{F2L} \longrightarrow SD^n_{F\&LC}$"=0.018/| − 0.056|/0.118/-0.021<0.116/| − 0.633|/0.432/ 0.109="$S^1_F/TS/LS_{L4F} \longrightarrow LA_{F2L} \longrightarrow SD^{EM}_{F\&LC}$" in the table). The only exception happens when the FV driver adjusts the lag gap with the LCV. With increased vehicle speed or change of traffic signal, the FV driver conducts more adjustment of the lag gap in the anticipating period than in the relaxing period ("$S^1_F/TS \longrightarrow LA_{F2LC} \longrightarrow G^{EM}_{lag4F}$"=0.829/| − 4.526| >0.821/| − 2.554|="$S^{EM}_F/TS \longrightarrow LA_{F2LC} \longrightarrow G^n_{lag4F}$" in the table). However, with a one-unit increase of LCV-involved stimulus, the FV driver could execute less adjustment of the lag gap in the anticipating period ("$LS_{LC4F}/LS_{L\&LC4F} \longrightarrow LA_{F2LC} \longrightarrow G^n_{lag4F}$"=5.394/| − 0.965| >3.022/0.780="$LS_{LC4F}/LS_{L\&LC4F} \longrightarrow LA_{F2LC} \longrightarrow G^{EM}_{lag4F}$"). This exception reflects a feature of the FV driver's behaviour: the driver can tolerate a small gap with the LCV temporally when the LCV tries to insert into target lane in the anticipating period and recovers to the desired following spacing with the LCV after it enters target lane in the relaxing period. This feature is also found by previous studies [21, 26]. It means that, in the relaxing period, the FV could recover desired lag gap to the LCV in a gentler way than it does in the anticipating period by maintaining less speed differences with the LV and the LCV.

In addition, the LCV driver is found to have a stronger intention to narrow speed difference with the LV in the relaxing period than in the anticipating period ("$S^{EM}_{LC}/LS_{L4LC} \longrightarrow LA_{LC2L} \longrightarrow SD^n_{LC\&L}$"=-0.081/-0.152< -0.072/1.667="$S^1_{LC}/LS_{L4LC} \longrightarrow LA_{LC2L} \longrightarrow SD^{EM}_{LC\&L}$" in Table 7). However, the LCV driver has to maintain a larger lead gap in the relaxing period than in the anticipating period ("$S^{EM}_{LC}/LS_{L4LC} \longrightarrow LA_{LC2L} \longrightarrow G^n_{lead}$"=1.850/3.465>0.596/ -13.668="$S^1_{LC}/LS_{L4LC} \longrightarrow LA_{LC2L} \longrightarrow G^{EM}_{lead}$" in the table). This result could be caused by speed advantage of the LV with respect to the LCV. Since the LV is hardly affected by the vehicles behind it, its speed could be higher than the LCV and the FV in most part of the LCV's inserting process, especially heavy interactions existing between the LCV and the FV in the anticipating period. In the limited time of the relaxing period, the LCV driver is difficult to shorten lead gap with the LV. Even if the gap is shortened, the speed difference between the two vehicles is enlarged ("$TS \longrightarrow LA_{LC2L} \longrightarrow G^n_{lead}/SD^n_{LC\&L}$" =-11.404/0.501 in the table). Recovering to the desired gap with the LV could take the LCV driver a relatively long time.

5.2.3. Traffic Signal. The influence of traffic signal on a driver's behaviour is verified by study results. We find that the manoeuvre adjustments of the FV and the LCV at the end of a period could be affected by the switch of traffic signal. In some cases, this influence is even stronger than that of vehicle stimulus. For example, when traffic signal switches from red to green, the FV leaves a shorter front gap at end of the anticipating period or relaxing period ("$TS \longrightarrow LA_{F2L} \longrightarrow G^{EM}_{front}$"=-9.006<0 and "$TS \longrightarrow LA_{F2L} \longrightarrow G^n_{front}$"=-8.289<0 in Table 7). The shortened spacing outweighs the increased one of the front gap caused by the LV's stimulus ("$TS \longrightarrow LA_{F2L} \longrightarrow G^{EM}_{front}$"=| − 9.006| >8.908="$LS_{L4F} \longrightarrow LA_{F2L} \longrightarrow G^{EM}_{front}$" and "$TS \longrightarrow LA_{F2L} \longrightarrow G^n_{front}$"=| − 8.289| >2.039="$LS_{L4F} \longrightarrow LA_{F2L} \longrightarrow G^n_{front}$" in the table). Similar influences can also be found in the variance of other vehicle kinematical indicator ("$TS \longrightarrow LA_{F2LC} \longrightarrow G^{EM}_{lag4F}$"=| − 4.526| >3.022="$LS_{LC4F} \longrightarrow LA_{F2LC} \longrightarrow G^{EM}_{lag4F}$", "$TS \longrightarrow LA_{LC2L} \longrightarrow G^n_{lead}$"=| − 11.404| >3.465="$LS_{L4LC} \longrightarrow LA_{LC2L} \longrightarrow G^n_{lead}$", "$TS \longrightarrow LA_{F2L} \longrightarrow SD^n_{F\&L}$"= 0.198> | − 0.048|="$LS_{L4F} \longrightarrow LA_{F2L} \longrightarrow SD^n_{F\&L}$", and "$TS \longrightarrow LA_{LC2L} \longrightarrow SD^n_{LC\&L}$"=0.501> | − 0.152|= "$LS_{L4LC} \longrightarrow LA_{LC2L} \longrightarrow SD^n_{LC\&L}$" in the table). This finding reminds us of the necessity of considering the role of traffic signals in the analysis of drivers' psychology and behaviour, though this issue is rarely addressed in existing studies [33]. In real traffic, the driver receives much dynamic information, such as the instructions provided by traffic control devices, vehicle or roadside lighting, etc. For instance, rear braking light and turning light of the vehicle could deliver a clear message about subject driver's intention to the surrounding drivers, while traffic signals indicate the right of way in specific traffic movement. All of such information could affect the driver's anticipations of other vehicles' movements, so they should be considered in the models simulating the LC or CF behaviour during the LCV's inserting process.

6. Conclusions

This study analyses the working mechanism of anticipating ability of the drivers in the LCV's inserting process in virtue of SEM. The developed two SEMs are calibrated with data extracted from vehicle trajectories and manual observations.

We find that (1) increasing the speed of the FV or the LCV could impel subject driver to anticipate driving condition in the further future, but the stimuli are lower than the ones the driver perceives by comparing the kinematics of subject vehicle with other vehicles; (2) the variance of vehicle gap or speed difference could affect the anticipations of the LCV driver and the FV driver, but which factor is the main source for a driver depends on the driver's role in the anticipating or relaxing period; (3) the drivers care more about the stimulus coming from the vehicle interaction he or she is involved in than the one he or she only observes; (4) the FV driver makes gentler adjustment of vehicle manoeuvre in the relaxing period than in the anticipating period, while the LCV driver acts in the opposite way; (5) the drivers' anticipations of other vehicles' movements could be affected by the change of LC type and traffic signal, and they adjust vehicle manoeuvres in different ways.

The findings of this study can be used to improve the simulation accuracy of LC traffic. For instance, the parameters of LC model or CF model should be adjusted in a time-dependent way, as this study finds that the main kinematical indicators each driver cares about could vary with his/her role in a specific LC stage. Another promising application is to consider the drivers' anticipating ability into motion planning algorithm of autonomous vehicle [34–36]. One of the most challenges in this field is to predict a human driver's response to the movements of autonomous vehicle, and this task could demand a lot of sensing and computing resources of the vehicle. If the human anticipating ability in complex driving circumstance, such as the LCV executing insertion in urban traffic, is considered in the algorithm of autonomous vehicle, it could help release the sensing and computing burden.

Nonetheless, this study leaves ample room for future improvements. First, since vehicle kinematical indicators are recorded only at the start or end of a period, the influence of driver's anticipation on his or her behaviours is estimated on a temporal average level. In fact, such ability keeps working during the whole LCV's inserting process. Future studies should take dynamics of the drivers' anticipations into account and produce a more accurate estimation of this influence. In addition, only the cases of successful LC are analysed in this study, so the length of the inserting process is limited. Sometimes the insertion is time-consuming or even unsuccessful [8, 9]. These cases should be considered in future study. Finally, only three vehicles directly involved in the inserting process are analysed here. However, more surrounding vehicles could affect or be affected by the LCV, such as its follower in the original lane. Their responses to the inserting movement are related to the LCV's impact on traffic flow, so the analysis of their manoeuvres could improve our understanding of traffic flow stability in LC scenario.

Appendix

Study Period

The start of the anticipating period shown in Figure 1(b) (instant 1) is defined according to relative positions of the LCV and the FV in the three scenarios shown in Figure 3.

(1) In the forward inserting scenario (see Figure 3(a)), the LCV intends to overtake the FV and find opportunity to conduct an LC. Instant 1 refers to the moment that head bumpers of the LCV and the FV align, because from this moment the FV driver definitely notices the LCV and begins to anticipate its future movements.

(2) In the backward inserting scenario (see Figure 3(b)), instant 1 is defined as the moment when the vehicles' back bumpers align.

(3) In the median inserting scenario (see Figure 3(c)), the LCV has run aside the gap between the LV and the FV for a while, and the variance of its acceleration could be too slight to catch the FV driver's attention. So instant 1 is defined as the moment when the LCV starts its lateral movement, or its turn signal is activated that can be seen by the FV driver.

It is noteworthy that the definitions of instant 1 above are set to ensure the LCV and the FV drivers' anticipation being working at that moment. In most cases, the LCV driver predicts the FV's movements in advance. To analyse the two drivers' interactive anticipations, the start of anticipating period should be defined as the moment the FV driver definitely anticipates the LCV's movement as well. After instant 1, the LCV driver usually synchronizes his or her vehicle speed along with that of the FV [16, 17] until finding an acceptable gap to cut in.

The anticipating period ends at the EM and the relaxing period starts then, even though the LCV's insertion has not ended. Such a definition which derives from a fact that both of the LCV and the FV drivers realize the LCV has entered target lane successfully after the EM. The relaxing period is assumed to end when the LCV's body is back to parallel to lane marker line of the target lane (instant 3 in Figures 1(a) and 1(b)). The relaxing period defined here could be shorter than the one defined in other studies [21], in which the period does not end until the FV recovers to its desired space with the LCV (instant 4 in Figure 1(b)). This paper does not follow the previous definition, as the study scenario here is urban arterial, rather than the freeway in those studies. Due to the disturbance of traffic signals and complex driving circumstances in urban road, driver's desired space with his or her leading vehicle should be redefined. For instance, drivers in a discharging flow at a green phase or the ones in a queuing flow at a red phase could find it difficult to keep a steady desire space. That is why the relaxing period assumes to end when the LCV's inserting movement is finished.

Conflicts of Interest

The authors declare that they have no conflicts of interest.

Acknowledgments

This work was supported by the National Key Research and Development Program of China (Grant no. 2018YFB0105104), the National Natural Science Foundation of China (Grants nos. 51505037 and 71701031), the Key Research and Development Program of Shaanxi Province (Grants nos. 2018ZDCXL-GY-05-04, 2018ZDCXL-GY-05-07-02, and 2018ZDXM-GY-013), and the Fundamental Research Funds for the Central Universities (Grants nos. 300102328101, 300102328401, 300102328205, and 300102328501).

References

[1] J. Weng, G. Du, D. Li, and Y. Yu, "Time-varying mixed logit model for vehicle merging behavior in work zone merging areas," *Accident Analysis & Prevention*, vol. 117, pp. 328–339, 2018.

[2] H. Park, C. Oh, J. Moon, and S. Kim, "Development of a lane change risk index using vehicle trajectory data," *Accident Analysis & Prevention*, vol. 110, pp. 1–8, 2018.

[3] R. Thijssen, T. Hofman, and J. Ham, "Ecodriving acceptance: An experimental study on anticipation behavior of truck drivers," *Transportation Research Part F: Traffic Psychology and Behaviour*, vol. 22, pp. 249–260, 2014.

[4] P. Hofmann, G. Rinkenauer, and D. Gude, "Preparing lane changes while driving in a fixed-base simulator: Effects of advance information about direction and amplitude on reaction time and steering kinematics," *Transportation Research Part F: Traffic Psychology and Behaviour*, vol. 13, no. 4, pp. 255–268, 2010.

[5] E. Chan, A. K. Pradhan, A. Pollatsek, M. A. Knodler, and D. L. Fisher, "Are driving simulators effective tools for evaluating novice drivers' hazard anticipation, speed management, and attention maintenance skills?" *Transportation Research Part F: Traffic Psychology and Behaviour*, vol. 13, no. 5, pp. 343–353, 2010.

[6] Z. Zheng, "Recent developments and research needs in modeling lane changing," *Transportation Research Part B: Methodological*, vol. 60, pp. 16–32, 2014.

[7] Y. Xi and M. Crisler, "A review of lane change definitions and identification methods," in *Proceedings of the 92nd Transportation Research Board Annual Meeting, Transportation Research Board of the National Academies*, Washington, D.C, 2013.

[8] J. Sun, J. Ouyang, and J. Yang, "Modeling and analysis of merging behavior at expressway on-ramp bottlenecks," *Transportation Research Record*, vol. 2421, pp. 74–81, 2014.

[9] L. Zhao, J. Sun, and H. Zhang, "Observations and analysis of multistep-approach lane changes at expressway merge bottlenecks in shanghai, China," *Transportation Research Record*, no. 2395, pp. 73–82, 2013.

[10] A. Kesting, M. Treiber, and D. Helbing, "General lane-changing model MOBIL for car-following models," *Transportation Research Record: Journal of the Transportation Research Board*, vol. 1999, no. 1, pp. 86–94, 2007.

[11] C. F. Choudhury, V. Ramanujam, and M. E. Ben-Akiva, "Modeling acceleration decisions for freeway merges," *Transportation Research Record: Journal of the Transportation Research Board*, no. 2124, pp. 45–57, 2009.

[12] C. F. Choudhury and M. E. Ben-Akiva, "Modelling driving decisions: A latent plan approach," *Transportmetrica A: Transport Science*, vol. 9, no. 6, pp. 546–566, 2013.

[13] W. Schakel, V. Knoop, and B. Van Arem, "Integrated lane change model with relaxation and synchronization," *Transportation Research Record*, vol. 2316, pp. 47–57, 2012.

[14] A. Kesting, M. Treiber, M. Schönhof, and D. Helbing, "Extending adaptive cruise control to adaptive driving strategies," *Transportation Research Record*, no. 2000, pp. 16–24, 2007.

[15] D. Yang, L. Zhu, B. Ran, Y. Pu, and P. Hui, "Modeling and Analysis of the Lane-Changing Execution in Longitudinal Direction," *IEEE Transactions on Intelligent Transportation Systems*, vol. 17, no. 10, pp. 2984–2992, 2016.

[16] X. Wan, P. Jin, L. Zheng, Y. Cheng, and B. Ran, "Speed synchronization process of merging vehicles from the entrance ramp," *Transportation Research Record*, no. 2391, pp. 11–21, 2013.

[17] X. Wan, P. J. Jin, F. Yang, and B. Ran, "Merging preparation behavior of drivers: How they choose and approach their merge positions at a congested weaving area," *Journal of Transportation Engineering*, vol. 142, no. 9, Article ID 05016005, 2016.

[18] D. Ngoduy, "Macroscopic effects of multianticipative driving behavior on traffic flow characteristics," *Transportation Research Record: Journal of the Transportation Research Board*, no. 2124, pp. 103–112, 2009.

[19] N. Farhi, H. Haj-Salem, and J.-P. Lebacque, "Multianticipative piecewise-linear car-following model," *Transportation Research Record*, vol. 2315, no. 1, pp. 100–109, 2012.

[20] B. Gunay, "Car following theory with lateral discomfort," *Transportation Research Part B: Methodological*, vol. 41, no. 7, pp. 722–735, 2007.

[21] Z. Zheng, S. Ahn, D. Chen, and J. Laval, "The effects of lane-changing on the immediate follower: anticipation, relaxation, and change in driver characteristics," *Transportation Research Part C: Emerging Technologies*, vol. 26, pp. 367–379, 2013.

[22] C. F. Choudhury and M. M. Islam, "Modelling acceleration decisions in traffic streams with weak lane discipline: A latent leader approach," *Transportation Research Part C: Emerging Technologies*, vol. 67, pp. 214–226, 2016.

[23] A. Duret, S. Ahn, and C. Buisson, "Passing rates to measure relaxation and impact of lane-changing in congestion," *Computer-Aided Civil and Infrastructure Engineering*, vol. 26, no. 4, pp. 285–297, 2011.

[24] S. Kim and B. Coifman, "Driver relaxation impacts on bottleneck activation, capacity, and the fundamental relationship," *Transportation Research Part C: Emerging Technologies*, vol. 36, pp. 564–580, 2013.

[25] C. Appert-Rolland and J. Du Boisberranger, "Macroscopic relaxation after on-ramps in real data and in cellular automata simulations," *Transportation Research Part C: Emerging Technologies*, vol. 34, pp. 162–175, 2013.

[26] P. Hidas, "Modelling vehicle interactions in microscopic simulation of merging and weaving," *Transportation Research Part C: Emerging Technologies*, vol. 13, no. 1, pp. 37–62, 2005.

[27] R. H. Hoyle, *Handbook of structural equation modeling*, The Guilford Press, New York, NY, USA, 2012.

[28] T. Ram and K. Chand, "Effect of drivers' risk perception and perception of driving tasks on road safety attitude," *Transportation Research Part F: Traffic Psychology and Behaviour*, vol. 42, pp. 162–176, 2016.

[29] V. L. Knoop, M. Keyvan-Ekbatani, M. de Baat, H. Taale, and S. P. Hoogendoorn, "Lane Change Behavior on Freeways: An Online Survey Using Video Clips," *Journal of Advanced Transportation*, vol. 2018, Article ID 9236028, 11 pages, 2018.

[30] L. Zhang, H. Wang, and L. Li, "Vehicle detection and tracking in video from moving airborne platform," *Journal of Computational Information Systems*, vol. 10, no. 1, pp. 1–8, 2014.

[31] K. Ismail, T. Sayed, and N. Saunier, "A methodology for precise camera calibration for data collection applications in urban traffic scenes," *Canadian Journal of Civil Engineering*, vol. 40, no. 1, pp. 57–67, 2013.

[32] M. Montanino and V. Punzo, "Trajectory data reconstruction and simulation-based validation against macroscopic traffic patterns," *Transportation Research Part B: Methodological*, vol. 80, pp. 82–106, 2015.

[33] J. Barceló, M. Kuwahara, and M. Miska, "Traffic Data Collection and Its Standardization," in *Traffic Data Collection and its Standardization*, vol. 144 of *International Series in Operations Research & Management Science*, pp. 1–10, Springer New York, New York, NY, 2010.

[34] Z. Xu, K. Zhang, H. Min, Z. Wang, X. Zhao, and P. Liu, "What drives people to accept automated vehicles? Findings from a field experiment," *Transportation Research Part C: Emerging Technologies*, vol. 95, no. 10, pp. 320–334, 2018.

[35] Z. Xu, M. Wang, F. Zhang, S. Jin, J. Zhang, and X. Zhao, "PaTAVTT: A Hardware-in-the-Loop Scaled Platform for Testing Autonomous Vehicle Trajectory Tracking," *Journal of Advanced Transportation*, vol. 2017, Article ID 9203251, 2017.

[36] Y. Liu, J. Guo, J. Taplin, and Y. Wang, "Characteristic analysis of mixed traffic flow of regular and autonomous vehicles using cellular automata," *Journal of Advanced Transportation*, vol. 2017, Article ID 8142074, 2017.

5

Exploring the Relationships between Subjective Evaluations and Objective Metrics of Vehicle Dynamic Performance

Jianyou Zhao,[1] Jing Liu,[1,2] Liping Yang,[3] and Ping He[2]

[1]*School of Automobile, Chang'an University, Xi' an 710064, China*
[2]*School of Mechanical and Electrical Engineering, Anhui Jianzhu University, Hefei 230601, China*
[3]*School of Automobile and Traffic Engineering, Hefei University of Technology, Hefei 230009, China*

Correspondence should be addressed to Liping Yang; yangliping3696@163.com

Academic Editor: Juan C. Cano

This study explored the relationships between subjective evaluations and objective metrics of vehicle dynamic performance. First, a real vehicle test was performed to measure the acceleration performance under different conditions, and participants' subjective evaluations of the acceleration performance were investigated. Second, correlation analysis was conducted to explore relationships between each subjective evaluation and its corresponding objective metric as well as between the overall subjective evaluation and three individual subjective evaluations. Finally, an overall subjective evaluation model related to the three objective metrics was established based on the Probabilistic Neural Network (PNN). The analysis results demonstrated that the correlation coefficients of the three groups of data were greater than 0.5 and that each subjective evaluation was significantly correlated with its corresponding objective metric. The individual subjective evaluation of the climbing acceleration performance had the largest effect on the overall subjective evaluation, with a correlation coefficient of 0.47. The established overall subjective evaluation model was relatively reliable, with a prediction accuracy of 90%. This study furthered the existing knowledge of the methods for evaluating vehicle dynamic performance. The proposed overall subjective evaluation model improves the reliability of vehicle dynamic evaluations and offers a theoretical basis for vehicle manufacturers to improve automobile performance.

1. Introduction

Before prototype vehicles are put into production, many performance indicators are tested to assess whether a vehicle achieves the design goals and to identify differences between the prototype vehicle and competing vehicles. Among the vehicle performance indicators, the dynamic performance is the most basic and important as it indicates that vehicles can reach the average travel speed under longitudinal force when driving straight on good roads. The dynamic performance is often tested using three indices: maximum speed, acceleration time, and maximum gradient [1]. The dynamic performance directly affects vehicle efficiency and is therefore one of the most important performance indicators for users as well as a key area in which vehicle manufacturers can improve their competitiveness.

To improve the validity of vehicle dynamic performance evaluations, certain domestic and foreign scholars have continued to explore evaluation methods for this performance indicator. According to the evaluation process, these methods can be divided into subjective evaluation methods and objective evaluation methods. In subjective evaluation methods, trained evaluation personnel or experienced drivers evaluate the vehicle performance based on their experiences in road tests. In objective evaluation methods, researchers analyze vehicle operation data collected via computer simulations or road tests to assess the vehicle performance. These two evaluation methods differ in application range and evaluation effectiveness. In practical applications, researchers select the appropriate method based on actual demands. For example, vehicle manufacturers in Germany primarily adopt the subjective evaluation system presented by Bernd Heißing et al. This evaluation system includes acceleration performance, and certain poor performance issues (such as acceleration jitter and acceleration pitching) are taken as the evaluation indices for acceleration performance during vehicle startup [2]. The subjective evaluation system of UK Lotus is also relatively complete and includes dynamic performance feel as an evaluation item. The related indices mainly focus on the dynamic performance feel under different acceleration conditions, including full throttle, partial throttle, starting acceleration, and overtaking acceleration [3]. A company in Japan uses an evaluation system that examines the vehicle dynamic performance more comprehensively based on evaluation indices related to driving force and drivability [4]. Vehicles from different manufacturers have different characteristics and styles; therefore, each manufacturer adopts different subjective evaluation indices. However, manufacturers in China typically use foreign evaluation systems to evaluate vehicle dynamic performance, as the subjective evaluation systems that have been developed by Chinese manufacturers and researchers are not sufficient. Based on existing research and applications, Li Bing presented a subjective evaluation system for dynamic performance and braking performance; compared to other evaluation systems [5], this proposed system is relatively complete. Based on the existing subjective evaluation index system, Wang et al. proposed a method to determine the weight of the subjective evaluation index of vehicle dynamic performance based on entropy method [6]. Combining the subjective evaluation system of vehicle dynamic performance with the weight coefficient is helpful to improve the reliability of the evaluation results. In summary, the subjective evaluation systems developed in foreign countries are more comprehensive than those developed in China.

The subjective evaluation method is widely used to evaluate vehicle performance because it is easy to perform. This method requires items to be described with demotic language and requires the use of experienced drivers, but certain problems such as strong subjectivity and poor reproducibility cannot be avoided completely. With the development of science and technology, the objective evaluation method for vehicle performance is gradually overcoming the shortcomings of the subjective evaluation method. The objective evaluation method can evaluate vehicle performance based on vehicle operation data collected from computer simulations or road tests. These objective evaluation methods for vehicle dynamic performance can be further divided into two categories. The first category includes evaluating the performance of a vehicle through a bench test, a simulation test, or a driving simulation experiment [7, 8] or presenting evaluation methods of dynamic performance based on simulation platforms [9]. The second category evaluates the dynamic performance through road tests; for instance, Wei proposed an objective evaluation system with evaluation indices related to dynamic performance and drivability as well as the test conditions of all indices [10]. Zhou put forward a new method to evaluate the dynamic performance of the vehicle; this method evaluates the vehicle dynamic performance through the driving performance feeling index [11]. The results of the PFI evaluation method and the conventional vehicle dynamic performance evaluation method showed that the PFI is a more objective and comprehensive dynamic performance evaluation method. In addition, with the development of electric vehicles, some scholars have devoted themselves to study the dynamic and economy evaluation methods of electric vehicles or evaluate their dynamic performance [12–15]. These studies laid the foundation for objective evaluations of vehicle performance.

Compared with the subjective evaluation method, the objective evaluation method has a smaller error and more intuitive results. However, when the objective evaluation method is used for certain types of performance indicators, such as comfort and drivability, the evaluation performance is poor. In contrast, such performance indicators can be effectively tested using the subjective evaluation method; thus, the two types of evaluation methods complement each other. To further explore the relationships between these two evaluation methods, certain scholars have explored the relationships between subjective evaluation items and objective metrics of vehicle handling stability. The related research can be divided into two categories: the first category explored the linear relationship between subjective evaluation items and objective metrics through regression analysis [16, 17] and the second category used neural networks to explore the nonlinear relationship between subjective evaluation items and objective metrics (King et al., 2003) [18]. These studies provided theoretical guidance for exploring the relationships between subjective and objective evaluation methods in terms of other vehicle performance metrics. Dynamic performance is an important index for measuring vehicle performance; however, certain real vehicle tests related to dynamic performance require higher costs and longer cycles, so no relevant research has explored the relationships between the two evaluation methods for vehicle dynamic performance based on real vehicle test data. As a result, vehicle manufacturers suffer from uncertainty and a lack of knowledge when selecting a dynamic performance evaluation method.

To improve the existing knowledge of the evaluation methods for vehicle dynamic performance, this paper explores the relationships between the subjective evaluation items and objective indices for dynamic performance. This work was conducted as follows. First, the evaluation indices were selected, and road tests were performed to collect the subjective evaluation data and objective measurement data for all indices. Second, Pearson correlation analysis was applied to explore the relationships between each subjective evaluation and its corresponding objective metric as well as between the overall subjective evaluation and the three individual subjective evaluations. Finally, an overall subjective evaluation model related to the objective metrics was established based on the Probabilistic Neural Networks (PNN). This model combines the subjective evaluation method with the objective evaluation method to improve the reliability of dynamic performance evaluation.

2. Materials and Methods

2.1. Real Vehicle Test

2.1.1. Test Instrumentation and Participants. The test vehicle, with a displacement of 1.6 L, an age of six months, and mileage of 7,359 km, passed a quality review performed by senior

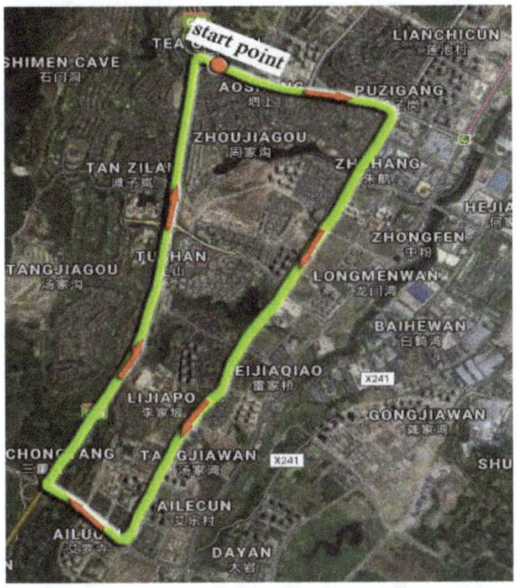

FIGURE 1: Diagram of the test route.

FIGURE 2: Diagram of the field test.

technicians. During the test, researchers used opaque cards to cover all of the logos on the test vehicle in order to prevent the vehicle brand from impacting participant evaluations. An electronic control unit with an output frequency of 30 Hz was installed on the test vehicle and was used to record the automotive operation data, including speed and engine speed. In addition, a car recorder was installed on the test vehicle to record environmental information, which helps to extract target data during data processing.

To improve the reliability of the test results, 105 experienced drivers participated in the test, including 89 male drivers and 16 female drivers. All participants provided written informed consent. The average age of the participants was 41, and their average mileage was 390,000 km. The participants were not involved in any serious traffic accidents or traffic violations in the past 3 years. The statistics indicated that these participants had excellent driving skills, extensive driving experience, and strong safety awareness. In the course of recruiting participants, the drivers' vehicles were investigated, and some applicants who drove the same vehicle or a similar vehicle as the test car did not participate in the test to avoid the effects of driver familiarity with the test vehicle.

2.1.2. Test Route. To test vehicle acceleration performance under different conditions, a 10-km-long closed loop route with both an urban road and a highway was selected as the test route shown in Figure 1. The urban road includes several signal-controlled intersections, which were used to test the starting acceleration, and a sloped section, which was used to test the climbing acceleration. The highway was used to test the driving acceleration. The road surface of the test route is smooth, clean, and dry asphalt pavement. The test vehicle was able to maintain a free driving condition because of the low traffic volume on this route, reducing the impacts of the external environment on the test vehicle. Figure 2 is the test process diagram.

2.1.3. Test Procedure. During the vehicle test, the participants were required to drive along the selected route and to perform assigned tasks on corresponding sections. Before the formal test, each participant was informed of the driving tasks they would conduct during the test and was asked to freely drive on the road near the starting point for 5 min to become familiar with the test vehicle. In addition, the participants were asked to notice the acceleration performance during the assigned tasks so that they could complete the subsequent questionnaire.

During the formal test, participants drove through an intersection, a sloped section (gradient of approximately 2.5°), and a highway section in turn, and they were required to complete the following three tasks: starting acceleration (0-30 km/h), climbing acceleration (from the bottom to the top of the slope), and highway driving acceleration (30-100 km/h). Although no interventions were imposed on the test vehicle or participant during the tasks, certain details still required the participants' attention. First, the participants were directed to stop in front of the stop lines on the premise of safety when arriving at an intersection with a red signal; otherwise, they were instructed to stop behind the preceding car, leaving a distance of approximately one car. This requirement was intended to keep the preceding car from impacting the test car during the starting acceleration performance test. To ensure consistent test distance, the participants were directed to stop the car when they arrived at the starting point of the climbing test; then, they were directed to start the car and to naturally accelerate to the top of the slope. The participants were instructed to start the driving acceleration test when the speed slowed to 30 km/h in the merging area of the highway and were again reminded to freely drive when the speed reached 100 km/h for the first time.

The test time lasted from November 2 to November 5, 2016. Throughout the test, the weather was clear, and the ambient relative humidity was no greater than 75%. The

Table 1: Subjective evaluation items for acceleration performance.

Num	Index	Content
1	Starting acceleration performance	What did you think of the acceleration performance when the vehicle started accelerating at the intersection?
2		What did you think of the ride comfort when the vehicle started accelerating at the intersection?
3	Climbing acceleration performance	What did you think of the climbing acceleration performance?
4	Driving acceleration performance	What did you think of the acceleration performance during the highway driving task?
5		What did you think of the riding comfort during the highway driving task?
6	Overall acceleration performance	In general, what do you think of the overall acceleration performance of the vehicle?

maximum temperature was approximately 283.15 K, and the temperature fluctuations were minimal.

2.2. Questionnaire Survey. After the vehicle test was completed, each participant filled out the subjective evaluation questionnaire based on the driving experience. This questionnaire asked questions related to the participants' subjective feelings regarding the starting acceleration, climbing acceleration, and highway driving acceleration. The subjective evaluation questionnaire included three components. The first component covered basic information about the participant, including gender, age, driving experience, driving type, and total mileage. The second component consisted of individual subjective evaluation items intended to investigate the participants' feelings toward each type of acceleration performance. The third component addressed the overall subjective evaluation item to investigate the participants' feelings toward the overall acceleration performance. The second and third components contained 6 items in total, and the specific contents of each item are shown in Table 1.

2.3. Test Indices. Li Bing proposed a set of dynamic performance evaluation indices that include acceleration jitter, acceleration pitch, starting acceleration performance, and climbing performance [5]. However, it is difficult to collect objective data for certain indices, such as acceleration jitter and acceleration pitch, in real vehicle tests. In contrast, it is easy to obtain objective data for the starting acceleration performance, climbing performance, and driving acceleration performance in real vehicle tests. Therefore, these three indices were selected for evaluating the automotive dynamic performance.

The acceleration time of each test was obtained according to the collected data. As a result, this study used acceleration time for the objective evaluation data. The objective evaluation data for the starting acceleration test, climbing acceleration test, and driving acceleration test corresponded to the acceleration time from 0 to 30 km/h at the start of the intersection, the acceleration time from the bottom to the top of the slope, and the acceleration time from 30 to 100 km/h on the highway, respectively. The three groups of objective evaluation data were denoted as T_1, T_2, and T_3, respectively.

The questionnaire was used to obtain the subjective evaluation data for the dynamic performance. As shown in Table 1, the starting acceleration performance and climbing acceleration performance each included two items. The mean of item 1 and item 2 was taken as the subjective evaluation for the starting acceleration performance and was denoted as S_1. Similarly, the mean of item 4 and item 5 was taken as the subjective evaluation for the driving acceleration performance and was denoted as S_3. Item 3 was considered the subjective evaluation for the climbing acceleration performance. In addition, the subjective evaluation of the overall acceleration performance was denoted as S.

3. Results

3.1. Descriptive Statistics. Before analysis, outliers were eliminated based on Pauta criterion which is used to eliminate outliers in a large sample that follows a normal distribution [19]. After processing, 95 sets of effective data remained.

3.1.1. Individual Subjective Evaluations. The subjective evaluation results of the selected indices are shown in Figure 3, where the X-axis represents the five evaluation ratings of each item and the Y-axis represents the number of participants corresponding to each evaluation rating. Figure 3 shows that more than half of the participants thought that the driving acceleration of the test vehicle was moderately fast or fast, approximately 43% of the participants deemed that the starting acceleration of the test vehicle was moderately fast or fast, and only 21% of the participants stated that the climbing acceleration of the test vehicle was moderately fast or fast. Overall, the driving acceleration performance of the test vehicle was better than the starting acceleration performance, whereas the climbing acceleration performance was worse than both. Figure 3 illustrates that the subjective evaluation results of the three indices have certain differences, which need to be verified.

3.1.2. Acceleration Time. As shown in Table 2, in the starting acceleration test, the maximum and minimum durations for the test vehicle to accelerate from 0 to 30 km/h were 15.5 s and 6.5 s, respectively, and the average time was 10.2 s. The maximum and minimum durations for the test vehicle

TABLE 2: Acceleration time statistics of the three indices (units: s).

Statistics	Maximum	Minimum	Mean	Standard deviation
Acceleration time of starting	15.5	6.5	10.2	2.0
Acceleration time of climbing	29.5	18.5	23.5	2.7
Acceleration time of driving	70.5	20.5	38.1	5.4

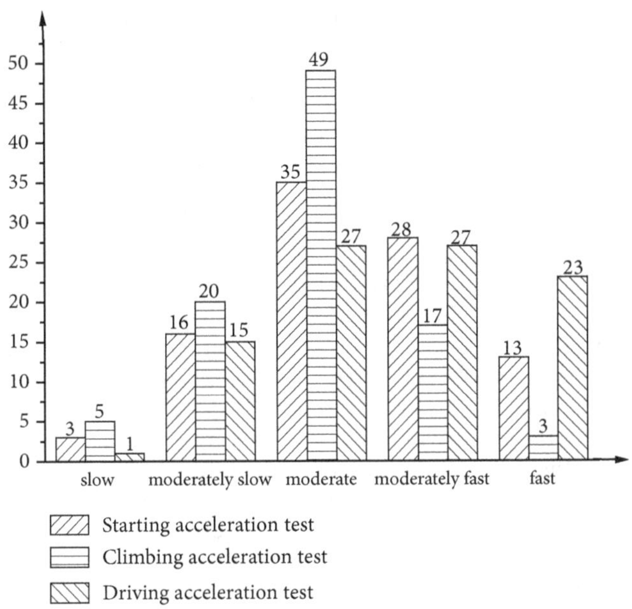

FIGURE 3: Subjective evaluation results of the three different conditions.

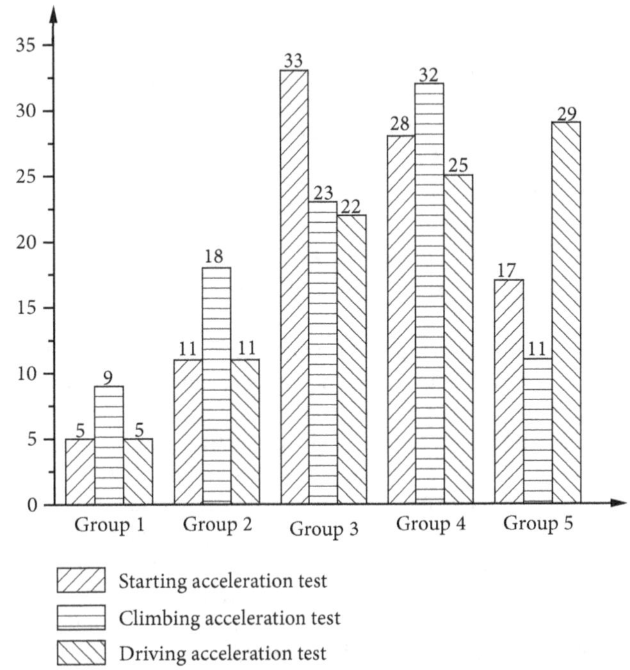

FIGURE 4: Acceleration time results of the three different conditions.

to climb from the bottom of the slope to the top were 29.5 s and 18.5 s, respectively, and the average time was 23.5 s. The maximum and minimum durations for the test vehicle to accelerate from 30 to 100 km/h were 70.5 s and 20.5 s, respectively, and the average time was 38.1 s.

To compare the subjective evaluation results with the objective acceleration time data, the continuous acceleration time data were divided into five groups because the subjective evaluation results had five ratings. The grouping method is described as follows. Assuming that $a_i = \max\{T_{ij}\}$, $bi = \min\{T_{ij}\}$ (T_{ij} ($i = 1, 2, 3$; $j = 1, 2, \cdots, 95$) denote the acceleration time of index i of participant j) and that $t_i = (a_i - b_i)/5$, then the acceleration time data of all participants for index i can be divided into five groups according to t_i, i.e., $[b_i, b_i + t_i)$, $[b_i + t_i, b_i + 2t_i)$, $[b_i + 2t_i, b_i + 3t_i)$, $[b_i + 3t_i, b_i + 4t_i)$, and $[b_i + 4t_i, a_i)$. These groups were labeled group 5, group 4, group 3, group 2, and group 1, respectively. The group with larger sequence numbers had shorter acceleration time. In other words, better dynamic performance is indicated by a larger sequence number. The number of participants in each group is shown in Figure 4.

As shown in Figure 4, in the starting acceleration test, the acceleration time of relatively few participants fell within the ranges of 13.7-15.5s (corresponding to Group 1) and 11.9-13.7s (corresponding to Group 2) and only 17% of the participants accelerated from 0 to 30 km/h with no less than 11.9s. In the climbing acceleration test, the acceleration time for the participants was primarily distributed in Group 3 (between 22.9s and 25.1s) and Group 4 (between 20.7s and 22.9s), and the proportion of participants in Group 3 and Group 4 was 58%. In the driving acceleration test, the acceleration time for the participants was mainly distributed in Group 4 (between 30.5s and 40.5s) and Group 5 (between 20.5s and 30.5s), and 57% of the participants accelerated from 30 km/h to 100 km/h in less than 40.5 s.

3.1.3. Comparison between Two Groups of Data. The subjective evaluation results used 5 ratings: slow, moderately slow, moderate, moderately fast, and fast. Therefore, the acceleration time data were divided into 5 corresponding groups: Groups 1, 2, 3, 4, and 5. Comparing the subjective evaluation results with the acceleration time data for the starting acceleration performance and driving acceleration performance shows that an approximately equal number of participants had fast or moderately fast evaluation results and had acceleration time that was distributed in Groups 4 and 5. As shown in Figures 3 and 4, a total of 41 participants considered the starting acceleration performance to be better than average (evaluation results of moderately fast or fast), and 43 participants had a starting acceleration time that was distributed in Groups 4 and 5. A total of 50 participants considered the driving acceleration performance to be better

Table 3: Analysis results of the subjective evaluations and objective metrics.

Correlation coefficients	S_1	S_2	S_3
T_1	-0.561**		
T_2	-	-0.543**	
T_3	-	-	-0.678**

Note. $*p < 0.05$; $**\ p < 0.01$.

than average, and 54 participants had a driving acceleration time that was distributed in Groups 4 and 5. However, for the climbing acceleration performance, the number of participants showed larger differences in certain evaluation ratings and corresponding groupings. The above analysis indicated that certain relationships might exist between the subjective evaluation results and the acceleration time data; thus, further research should be conducted.

3.2. Correlation Analysis. To further explore the relationships between the subjective and objective evaluation methods, the correlation analysis method was adopted to determine the relationships between each subjective evaluation and its corresponding acceleration time as well as between the overall subjective evaluation and three individual subjective evaluations; this method has also been adopted to explore the relation between the objective and subjective workload assessment measures [20]. The correlation analysis method takes the Pearson Product Moment Correlation Coefficient (PPMCC) as statistical index and this index has values between -1 and 1 and is often denoted as r. The formula for calculating the PPMCC is as follows:

$$r = \frac{\sum_{j=1}^{n}\left(S_j - \overline{S}\right)\left(T_j - \overline{T}\right)}{\sqrt{\sum_{j=1}^{n}\left(S_j - \overline{S}\right)^2}\sqrt{\sum_{j=1}^{n}\left(T_j - \overline{T}\right)^2}} \quad (1)$$

where S_i is subjective evaluation of the jth participant; $\overline{S}$ is average subjective evaluation of all participants; T_i is acceleration time of the jth participant; $\overline{T}$ is average acceleration time of all participants.

3.2.1. Individual Subjective Evaluations and Acceleration Time. In this study, the starting acceleration performance, climbing acceleration performance, and driving acceleration performance were selected to measure the automotive dynamic performance. Each index contains two groups of data: one group contains the acceleration time data, and the other group consists of the subjective evaluation data. Although the two groups of data used different evaluation methods, they both measured the automotive dynamic performance. To explore whether or not the two types of evaluation results have a relationship, the individual subjective evaluation and the corresponding objective metric of the three indices were used in a correlation analysis. The analysis results are shown in Table 3, where S_i and T_i $(i = 1, 2, 3)$ denote the subjective evaluation data and objective metric data, respectively.

Table 4: Analysis results of the overall evaluation and individual evaluations.

Correlation coefficients	S_1	S_2	S_3
S	0.221*	0.473**	0.319**

Note. $*p < 0.05$, $**\ p < 0.01$.

Table 3 illustrates that each individual evaluation result was significantly negatively correlated with the corresponding acceleration time data. In other words, the evaluation score was lower when more time was spent on each test. This relationship is in agreement with the actual situation; thus, the evaluation results of the subjective method and objective method are consistent with each other.

As shown in Table 3, the subjective evaluation result for the driving acceleration performance had the strongest correlation with the corresponding acceleration time data (r=-0.678), and the correlation coefficient of the starting acceleration performance was similar to that of the climbing acceleration performance (r=-0.561 and r=-0.543, respectively). The two evaluation results had the strongest correlation for the driving acceleration performance, possibly due to the test conditions. Unlike the other two indices, the driving acceleration performance was tested on the highway, which had better conditions than the other roads and few disturbances; these improved conditions might improve the validity of the test. The correlation coefficient of the two evaluation results for the climbing acceleration performance was lower than that for the other two indices, possibly due to the geometry of the sloped road section. The sloped road section had a dip angle that might hinder the participants from perceiving the acceleration performance. Therefore, a closed, flat test route should be selected for performance tests to avoid the external environment (such as road line type) affecting test participant perceptions.

3.2.2. Individual Subjective Evaluations and the Overall Evaluation. The overall subjective evaluation investigated the participants' feelings toward the overall acceleration performance, whereas the individual subjective evaluations investigated the participants' feelings toward each type of acceleration performance. To explore the relationships between the two groups of data, the corresponding data were subjected to correlation analysis to explore the impacts of the individual subjective evaluations on the overall subjective evaluation. The analysis results are shown in Table 4, where S_i $(i = 1, 2, 3)$ denotes the individual subjective evaluation data and S denotes the overall subjective evaluation data.

Table 4 shows that the individual subjective evaluation items were significantly positively correlated with the overall subjective evaluation items. In other words, the scores of the overall evaluations were associated with the scores of the individual evaluation. However, the three individual subjective evaluation items have different impacts on the overall subjective evaluation. The subjective evaluation for the climbing acceleration performance had the largest effect on the overall subjective evaluation (r=0.473), followed by that of the driving acceleration performance (r=0.319) and then that of the starting acceleration performance (r=0.221).

As noted above, the individual subjective evaluations for climbing acceleration and driving acceleration had greater effects on the overall evaluation than that of the starting acceleration. This observation could be caused by the high demands that were made on the dynamic performance during climbing acceleration and driving acceleration; thus, these two types of conditions can more directly reflect the dynamic performance. As a result, the climbing acceleration and driving acceleration had larger effects on the overall evaluation of the dynamic performance. This result showed that car manufacturers should focus on the dynamic performance of climbing acceleration and high gear acceleration to improve driver satisfaction.

3.3. Model of the Overall Subjective Evaluation. The overall subjective evaluation of vehicle dynamic performance is an important index and one that vehicle manufacturers focus on. If the acceleration time in different situations could reflect the overall subjective evaluation of dynamic performance, then the overall subjective evaluation results can be predicted through these objective metrics to avoid the effects of individual differences of drivers on the overall evaluation results.

As noted above, the relationships between the individual subjective evaluations and objective metrics showed that the acceleration time can affect the corresponding subjective evaluation. In addition, the results indicated that the individual subjective evaluations can affect the overall subjective evaluations of drivers. As a result, the acceleration time in different situations may affect the overall subjective evaluation of the dynamic performance.

To explore the relationships between the three acceleration times and the overall subjective evaluation, a four-dimensional scatter plot was drawn using the test data. In Figure 5, the X-, Y-, and Z-axes represent the acceleration times of the starting test, climbing test, and driving test, respectively, and the five colors correspond to the five ratings of the overall subjective evaluation. This figure shows that the overall subjective evaluation nonmonotonically decreased with increasing climbing acceleration time and that the acceleration times of the other two tests also affected the overall subjective evaluation. The overall subjective evaluation decreased with increasing driving acceleration time when the acceleration time of the climbing test was approached, and this trend was more significant when the starting acceleration time was longer. In summary, a nonlinear relationship existed between the overall subjective evaluation and the acceleration times under the three different conditions. Thus, further research must be conducted to explore this complex relationship.

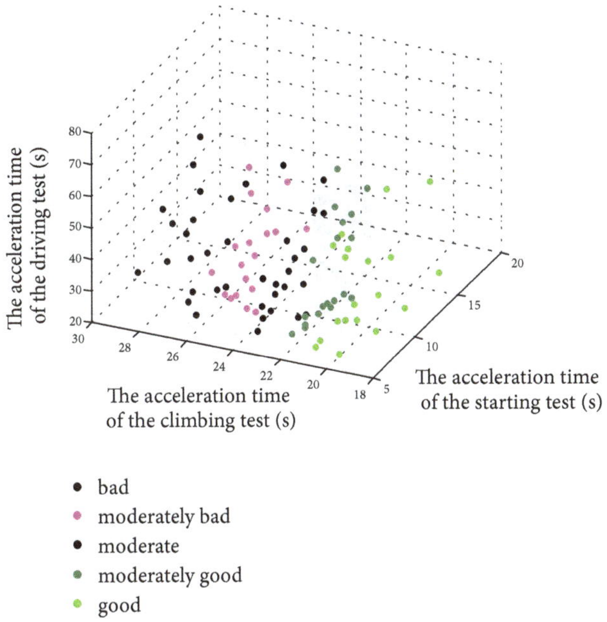

FIGURE 5: Scatter diagram of the overall subjective evaluation.

3.3.1. PNN Model Summary. As shown in Figure 5, the relationships between the overall subjective evaluation and the acceleration times under different situations were nonlinear; thus, methods used for exploring nonlinear relationships between variables should be adopted. A previous study that explored the relationships between subjective and objective evaluations of vehicle handling showed that artificial neural networks (ANNs) are a powerful tool for exploring nonlinear relationships between subjective evaluations and objective metrics [21]. Therefore, this paper adopted ANNs to explore the relationships between subjective evaluations and objective metrics of dynamic performance.

ANNs include a variety of algorithms, and each algorithm has different features and application ranges. Because the output variable is a categorical variable and the values of the input and output variables in the training data set were both obtained, the Probabilistic Neural Networks (PNN) were adopted. The structure of PNN is as shown in Figure 6.

PNN is a forward neural network that was proposed by Specht in 1989 [22]. Parzen proposed treating the Gauss function as the basis function to form the estimation method of joint probability density and Bayesian optimization rule. Based on this proposal, Specht constructed a neural network of the classification estimation of probability density and parallel processing. As a result, the PNN has not only the characteristics of general neural networks but also excellent generalization ability and fast learning ability. As shown in Figure 6, the PNN structure consists of an input layer, a hidden layer, and an output layer. The input layer is responsible for introducing characteristic vectors into the networks, and the number of neurons in this layer is equal to the dimensions of the input eigenvector. The hidden layer is connected to the input layer through connection weights; this layer calculated the matching degree between the input characteristic vectors and each pattern in the training set and delivered the distance of these two groups of data to the Gauss function to obtain the outputs of this layer. The number of neurons in the hidden layer is equal to the number of samples. The output layer is responsible for connecting hidden layer units of each category and outputting the class with the highest score; the number of neurons in this layer is equal to the number of sample categories.

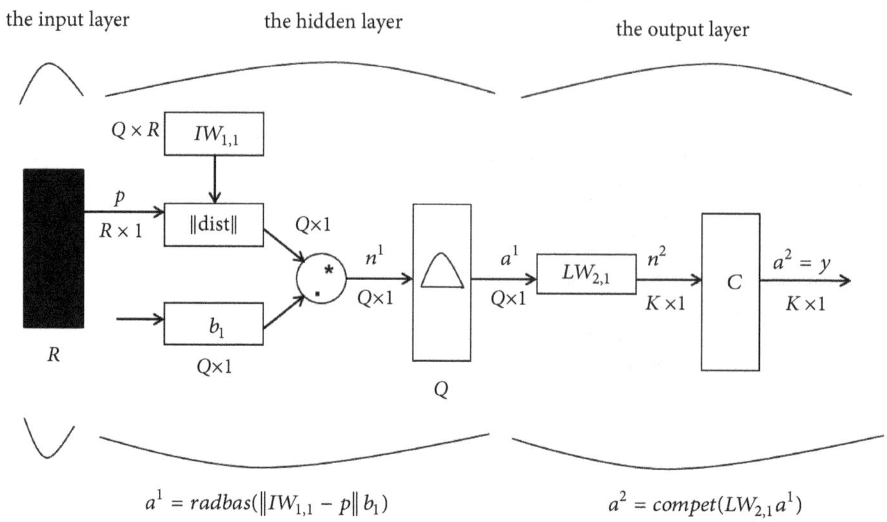

FIGURE 6: Structure of the PNN.

3.3.2. Model Establishment. In this paper, the PNN was used to establish an overall subjective evaluation model, and the following steps were taken.

First, the input variables and output variables were determined.

In this paper, the input variables for the neural network were the acceleration time data from the three tests, and the output variable was the participants' overall subjective evaluation, with values of 1-5 corresponding to bad, moderately bad, moderate, moderately good, and good, respectively.

Second, the PNN model was trained. The training process of PNN is as follows:

(1) Determine the center of radial basis function of neurons in hidden layer.

Supposing P and T denote the input matrix and the output matrix of the training set, P and T are as follows:

$$P = \begin{bmatrix} p_{11} & p_{12} & \cdots & p_{1Q} \\ p_{21} & p_{22} & \cdots & p_{2Q} \\ \vdots & \vdots & & \vdots \\ p_{R1} & p_{R2} & \cdots & p_{RQ} \end{bmatrix}$$
$$T = \begin{bmatrix} t_{11} & t_{12} & \cdots & t_{1Q} \\ t_{21} & t_{22} & \cdots & t_{2Q} \\ \vdots & \vdots & & \vdots \\ t_{R1} & t_{R2} & \cdots & t_{RQ} \end{bmatrix} \qquad (2)$$

In the matrix P and T, the variable p_{ij} represents the ith input variable of the jth training sample; the variable t_{ij} represents the ith output variable of the jth training sample; R represents dimensions of input variables and the K represents dimensions of output variables corresponding to the number of the classification; Q represents the number of sample size of training set.

In the PNN algorithm, each neuron in the hidden layer corresponds to a training sample. In other words, the radial basis function center of Q neurons in hidden layer is as follows:

$$C = P' \qquad (3)$$

(2) Determine the threshold of the hidden layer neurons.

Assume that the threshold of neurons in hidden layer is $b_1 = [b_{11}, b_{12}, \cdots, b_{1Q}]$, and $b_{11} = b_{12} = \cdots = b_{1Q} = 0.8326/spread$, *spread* denotes the expansion speed of radial basis function.

(3) Determine the weights between the hidden layer and the output layer.

When the center and the threshold of radial basis function of the hidden layer neurons are determined, the output of the hidden layer neurons can be calculated by the following formula:

$$a^i = \exp\left(-\|C - p_i\|^2 b_1\right), \quad i = 1, 2, \cdots, Q \qquad (4)$$

where $p_i = [p_{i1}, p_{i2}, \cdots, p_{iR}]'$ is the ith training sample vector.

The connection weight W between the hidden layer and the output layer in PNN algorithm is taken as the output matrix of the training set:

$$W = \mathrm{t} \qquad (5)$$

(4) Calculate the output of output layer neurons.

When the connection weights between the hidden layer and the output layer neurons are determined, the output of the output layer neurons can be calculated as follows:

$$n^i = LW_{2,1}a^i, \quad i = 1, 2, \cdots, Q$$
$$y^i = compet\left(n^i\right), \quad i = 1, 2, 3 \cdots, Q \qquad (6)$$

The training and optimization programs were developed using MATLAB. All the effective data were randomly divided into two groups: one group included 75 samples that were used to train the neural network and the other group included 20 samples that were used to verify the prediction effect of the model.

Finally, the validity of the model was verified.

The prediction error rate was determined by comparing the true values with the values predicted by the established model.

3.3.3. Model Verification. The prediction values and actual values for 20 groups of test samples are shown in Figure 7.

As shown in Figure 7, among all the testing samples, only two prediction values (Nos. 4 and 6) were not in agreement with the actual values, and the remaining values were in absolutely agreement with the actual values, meaning that the prediction accuracy was 90%. In addition, for Nos. 4 and 6, the prediction values were only one rating higher than the actual values. In other words, little difference was noted between the actual values and the incorrect prediction values. In summary, the prediction accuracy of the overall evaluation model was relatively high. Although the correct prediction rate failed to reach 100%, the incorrect prediction values were relatively close to the actual values. Therefore, according to the acceleration time of the three indices, the participants' overall subjective evaluations can be predicted using the established PNN model, and the subjective evaluation results can be used to improve the dynamic performance of vehicles to meet the diverse demands of users.

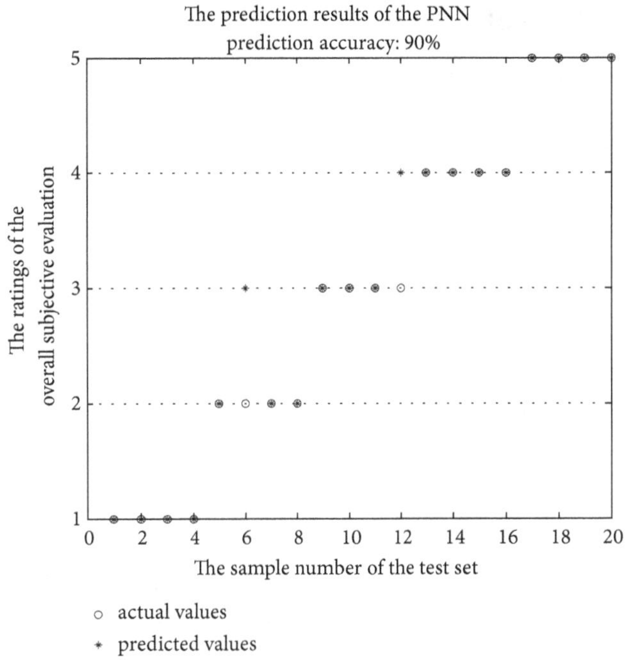

FIGURE 7: Prediction results of the overall subjective evaluations.

4. Conclusions

Evaluating the dynamic performance is an important task in vehicle production, and subjective evaluation and objective evaluation are two commonly used methods for this task. However, no relevant research has explored the relationships between the subjective evaluation and objective evaluation so that the validity of two evaluation methods cannot be assessed.

In this study, first, a real vehicle test was conducted to collect vehicle acceleration data under different conditions, and the participants' subjective feelings of the dynamic performance were investigated using a questionnaire. Second, based on the collected data, the correlation analysis method was adopted to explore the relationships between each individual subjective evaluation item and the corresponding objective metric as well as between the overall subjective evaluation item and the three individual subjective evaluation items. The results showed that individual subjective evaluations were significantly correlated with the corresponding objective metrics, and the correlation coefficients were greater than 0.5 (the correlation coefficients between the subjective evaluations and objective metrics for the starting acceleration performance, climbing acceleration performance, and driving acceleration performance were 0.561, 0.543, and 0.678, respectively). The climbing acceleration performance had the largest effect on the overall subjective evaluation, and the correlation coefficient of the two groups of data was 0.473. Finally, an overall subjective evaluation model related to the three objective metrics was established based on the PNN, and the verification results showed that the prediction accuracy of the proposed model was relatively high.

This study explored the relationships between subjective evaluations and objective evaluations of dynamic performance based on selected indices. This research furthered the theoretical knowledge of the methods for evaluating vehicle dynamic performance and offered methodological guidance for exploring the relationships between subjective evaluations and objective indices. In addition, an overall subjective evaluation model of dynamic performance was established based on real vehicle test data; this model used the acceleration time data obtained under different situations as input variables. Furthermore, because this model predicted the overall subjective evaluation through objective test data, it can avoid the effects of subjectivity on the overall evaluation of dynamic performance. As a result, the proposed model had the advantages of high practicability and reliability and can be used by vehicle manufacturers to optimize vehicle designs. However, in the current study, the acceleration times under different conditions were identified as the objective indices. Although these indices can strongly reflect the acceleration performance of the vehicle, they cannot completely reflect the drivability of the car. Therefore, in future research, the evaluation indices, particularly the objective indices that correspond to the subjective evaluations, should be further perfected so that the relationships between the two methods could be explored in depth to improve the reliability of the evaluation results and to further improve the competitiveness of vehicles in the market.

Conflicts of Interest

The authors declare that they have no conflicts of interest.

Acknowledgments

This paper is supported by the "National Natural Science Foundation of Anhui Province" (no. 1708085ME125) and the "Scientific Research Projects Foundation in Higher Education of Anhui Educational Department" (no. KJ2017A485).

References

[1] Z. S. Yu, *Automobile Theory*, The Mechanical Industry Press, Beijing, China, 5th edition, 2009.

[2] B. Heißing and H. J. Brandl, *The subjective evaluations of vehicle dynamic performance*, Federal Highway Research Institute, Cologne, 2002.

[3] J. A. D. A. Pereira, "New fiesta: brake pedal feeling development to improve customer satisfaction," in *Proceedings of the SAE Brasil 2003 Congress & Exhibit*, Detroit, MI, USA, 2003.

[4] X. X. Tian, *The Evaluation Method Research about Automotive Subjective Performance [Master Thesis]*, Chang'an University, 2014.

[5] B. Li, *The Subjective Evaluation Method Research of Vehicle Dynamic Performance And Brake Performance [Master Thesis]*, Jilin University, 2008.

[6] S. C. Wang, D. Fu, J. J. Chen, F. T. Cai, and X. W. Zhang, "Determination of weights of subjective evaluation indexes of automobile dynamic performance based on entropy method," *Journal of Highway and Transportation Research and Development*, vol. 32, no. 7, pp. 153–158, 2015.

[7] J. Zehetner, P. Schoeggl, M. Dank, and K. Meitz, "Simulation of drivability in real-time," in *Proceedings of the SAE 2009 World Congress & Exhibition*, Detroit, MI, USA, 2009.

[8] O. Hayat, M. Lebrun, and E. Domingues, "Powertrain drivability evaluation analysis and simplification of dynamic models," in *Proceedings of the SAE 2003 World Congress & Exhibition*, Detroit, MI, USA, 2003.

[9] G.-L. Luo, F.-J. Zhang, and H.-J. Yuan, "Analysis of vehicle powertrain dynamic performance," *Journal of Beijing Institute of Technology*, vol. 22, no. 2, pp. 171–178, 2013.

[10] X. Wei and G. Rizzoni, "Objective metrics of fuel economy, performance and drivability - a review," in *Proceedings of the SAE 2004 World Congress & Exhibition*, Detroit, MI, USA, 2004.

[11] J. M. Zhou, "Application of PFI in vehicle dynamic evaluation," *Automobile & Parts*, vol. 41, pp. 82–85, 2016.

[12] M. Chen and L. Guo, "The synthetic evaluation method of the dynamic performance and economic performance of battery electric vehicle based on principal component analysis," *Applied Mechanics and Materials*, vol. 215-216, pp. 1259–1262, 2012.

[13] M. Chen and L. X. Guo, "Comprehensive evaluation of power performance and energy consumption economy of battery electric vehicle based on AHP-entropy-TOPSIS," *Journal of Machine Design*, vol. 30, no. 4, pp. 15–19, 2013.

[14] H. Xu, H. B. Fu, W. L. Dong, C. M. Yang, D. M. Wu, and H. Zhu, "Research on evaluation method of electric vehicle's dynamic performance based on battery output characteristics," *Journal of Beijing University of Chemical Technology (Natural Science)*, vol. 45, no. 3, pp. 61–66, 2018.

[15] Z. Zhang, M. R. Guo, and P. Zheng, "The pure electric vehicle dynamic performance simulation," *Applied Mechanics Materials*, vol. 260-261, pp. 353–356, 2012.

[16] X. Zhao, H. Chen, B. Gao, L. Lou, and N. Shirou, "Correlations between subjective and objective evaluations of on-center steering feel," in *Proceedings of the FISITA 2012 World Automotive Congress*, vol. 198, pp. 313–327, 2013.

[17] D. Chen, *Subjective and Objective Vehicle Handling Behavior [Ph.D. thesis]*, University of Leeds, 1997.

[18] J. Pelikan, P. Steinbauer, M. Valasek, and J. Ulehla, "Correlation of objective and subjective evaluation of vehicle handling by neural networks," *Bulletin of Applied Mechnics*, vol. 8, no. 29, pp. 1–4, 2011.

[19] Z. M. Liu, *Error and Data Processing*, Atomic Energy Press, Beijing, China, 1st edition, 1981.

[20] A. Didomenico and M. A. Nussbaum, "Interactive effects of physical and mental workload on subjective workload assessment," *International Journal of Industrial Ergonomics*, vol. 38, no. 11-12, pp. 977–983, 2008.

[21] R. P. King, D. A. Crolla, H. A. Ash, and J. Whitehead, "Identification of subjective-objective vehicle handling links using neural networks for the foresight vehicle," in *Proceedings of the SAE 2002 World Congress & Exhibition*, 2002.

[22] F. Shi, H. Wang, F. Hu, and L. Yu, *The Matlab Intelligent Algorithm Analysis about 30 Cases*, Beihang University Press, Beijing, China, 1st edition, 2011.

6

Modeling Microscopic Car-Following Strategy of Mixed Traffic to Identify Optimal Platoon Configurations for Multiobjective Decision-Making

Mudasser Seraj,[1] **Jiangchen Li**,[1] **and Zhijun Qiu**[1,2]

[1]*Department of Civil and Environmental Engineering, University of Alberta, Edmonton, Alberta, Canada*
[2]*Intelligent Transportation System Research Center, Wuhan University of Technology, Wuhan, Hubei, China*

Correspondence should be addressed to Zhijun Qiu; zhijunqiu@ualberta.ca

Guest Editor: Guoyuan Wu

Microscopic detail of complex vehicle interactions in mixed traffic, involving manual driving system (MDS) and automated driving system (ADS), is imperative in determining the extent of response by ADS vehicles in the connected automated vehicle (CAV) environment. In this context, this paper proposes a naïve microscopic car-following strategy for a mixed traffic stream in CAV settings and specified shifts in traffic mobility, safety, and environmental features. Additionally, this study explores the influences of platoon properties (i.e., intra-platoon headway, inter-platoon headway, and maximum platoon length) on traffic stream characteristics. Different combinations of MDS and ADS vehicles are simulated in order to understand the variations of improvements induced by ADS vehicles in a traffic stream. Simulation results reveal that grouping ADS vehicles at the front of traffic stream to apply Cooperative Adaptive Cruise Control (CACC) based car-following model will generate maximum mobility benefits for upstream vehicles. Both mobility and environmental improvements can be realized by forming long, closely spaced ADS vehicles at the cost of reduced safety. To achieve balanced mobility, safety, and environmental advantages from mixed traffic environment, dynamically optimized platoon configurations should be determined at varying traffic conditions and ADS market penetrations.

1. Introduction

Vehicles with diverse levels of integrated connectivity and automated control systems are considered to be pushing a technological leap towards diminished trip delay, fuel efficiency, reduced emission, and enhanced safety of road traffic. Although a purely automated vehicle-based traffic stream could take decades to become a reality, introducing and gradually increasing market shares of automated driving system-based vehicles in traffic streams would enable us to perceive and harness the potential gains from these technologies. Varied perceptions of mixed traffic streams and their collaborative motion dynamics hindered both researchers and practitioners from progressing with these technologies. Furthermore, the ideal compositions of automated vehicles in mixed traffic conditions remain unfamiliar to most. In response to these problems, this study proposes a simple yet effective car-following strategy for mixed traffic stream and measures the resulting impact on mobility, safety, and the environment. Additionally, the car-following strategy involved platoon development in a connected automated vehicle (CAV) environment and the study explores various platoon configurations to determine platoon parameters at different traffic states to obtain utmost benefits.

Numerous studies have been conducted by acclaimed researchers and practitioners to interpret the complex dynamics of combined traffic movements [1–6]. While these studies transcended in conceiving the levels of impact of automated driving technologies through simplified to complex macroscopic and mesoscopic modeling, the motivation of the present was shaped by the need of modeling microscopic car-following behavior in heterogeneous traffic in order to study macroscopic consequences from mobility, safety, and environmental perspectives. With that intention, this study gives insights into a wide variation of distinct forms of impact while simulating automated-control-enabled vehicles on ideal locations and distributions along traffic stream. These insights into mixed traffic movements and platoon characteristics will motivate researchers to consider other unattended aspects of mixed traffic dynamics (e.g., lane-changing, gap acceptance, and merging) in order to rectify perceived benefits. Similarly, traffic operational authorities can take these lessons into account to impose different control strategies (e.g., dynamic aggregated controls for manually driven vehicles, dynamic personalized controls on connected vehicles) on traffic to attain maximum improvements with regard to reduced travel time, collision rates, greenhouse gas emissions, etc.

The rest of the paper is organized as follows: The next section summarizes the existing literature on car-following models and strategies for mixed traffic and also touches on studies that identify the different form of impact that automated vehicles have on traffic. The proposed car-following strategy is described in the following section. The description of simulation procedures and the discussion on obtained results are covered, respectively, in two subsequent sections. The following section proposed an approach to obtain dynamic optimal platoon configuration for specific traffic state. The last section provides the synopsis of findings of the study and also gives recommendations for future research.

2. Literature Review

As the primary aim of this study relates to car-following strategy for mixed traffic environment, the literature related to car-following models for both forms (i.e., manual, automated) of driving system is explored here. Numerous microscopic car-following models have been proposed to imitate driving pattern of manual driving system [7–12]. Among the proposed stimulus-response-based car-following models, the intelligent driver model (IDM) is widely used in literature to depict manual driving dynamics. The ability of this model to define numerous microscopic (e.g., desired velocity and acceleration/deceleration limits) and macroscopic (e.g., capacity, capacity drop, and fundamental diagram) phenomena made it the prevalent model. On the other hand, due to rapid growth of CAV technology, the longitudinal control models for automated vehicles were also examined by researchers [13–18]. These studies provide us with structures to work on car-following strategy in mixed traffic environment and identify the extents of potential paradigm shifts.

A clear distinction of the driving system is dictated by the operational authority. While the manual driving system (MDS) represents driving systems controlled by humans, the motion dynamics of vehicles with the automated driving system (ADS) are mandated by distinct levels of automation. ADS vehicles' longitudinal movements are commonly portrayed with adaptive cruise control (ACC) and cooperative adaptive cruise control (CACC). Many studies have analyzed the contributions of longitudinal control system of ADS vehicles on traffic mobility [19–28]. While mobility was the main focus of these studies, the impact of traffic movement from safety and environmental perspective was often ignored. Yeo et al. [29] proposed an integrated car-following and lane changing model to perform microsimulation of oversaturated freeway traffic. The proposed algorithm considered complex dynamic interactions at a microscopic level to replicate vehicle movements. However, the aptitude of this model to capture possible consequences was not tested. Wang et al. [30] proposed a car-following control for autonomous vehicle and identified the impact, focusing mainly on traffic flow characteristics. Liberis et al. [31] took a macroscopic approach to identify traffic mobility parameters in a heterogeneous traffic environment. The authors used the market penetration rate of connected vehicles to estimate traffic states. Moreover, other researchers studied the impact of introducing ADS based vehicles with conventional vehicles [32–36] on flow and mobility. Reviews of these studies provide us with the opportunity to constructively examine the contributions of earlier studies, identify the necessities to improve current knowledge, and uncover the latent insights to progress promptly with CA technology.

While the mobility attributes of traffic flow were widely discussed in many studies, the safety and environmental aspects, which are equally if not more important, were relatively unexplored by a majority of the studies. The impact of automated vehicles on both safety and mobility was discussed by Fernandes and Nunes [37]. They studied platooning of ADS vehicles with different communication schemes at various flow rates to improve roadway capacity. Several studies assessed the safety aspects of CAV based traffic. According to the National Highway Traffic Safety Administration (NHTSA), a complete adaptation of CAV based traffic movements would annually prevent 439,000–615,000 crashes [38]. Li et al. [39] evaluated the impact of CACC control on reducing rear-end collisions on freeways. The study shows a reduction in safety improvements with increasing market share of ADS vehicles. Rahman and Abdel-Aty [40] compared potential improvement in longitudinal safety due to varying market penetration of connected vehicles. According to the analysis presented, the managed-lane CAC control outperformed multilane control with regard to traffic safety. The report of Zabat et al. [41] stated that the presence of boundary layer along closely spaced vehicles would reduce aerodynamic drag, resulting in reduced fuel consumption and less emission. Platoon-wide environment-friendly CACC system was studied by Wang et al. [42] and their objective assessment attained 2% fuel saving with 17% emission reductions. Mamouei et al [43] argued that fuel-economy based ACC control model would not lead to highly conservative driving dynamics of traffic.

Although the reviewed studies had remarkable contributions that helped to clarify the roles and influences of ADS vehicles in traffic, the inadequacy of multiobjective decision-making approach to address ADS vehicles' potential has influenced this research further to investigate the complex interdependencies of mixed traffic. This research seeks to contribute on three research gaps identified from the literature. These gaps are (i) the significance of ADS vehicles' position and distributions along traffic stream, (ii) the variations of traffic flow attributes (i.e., mobility, safety, and environmental) resulting from structural changes of

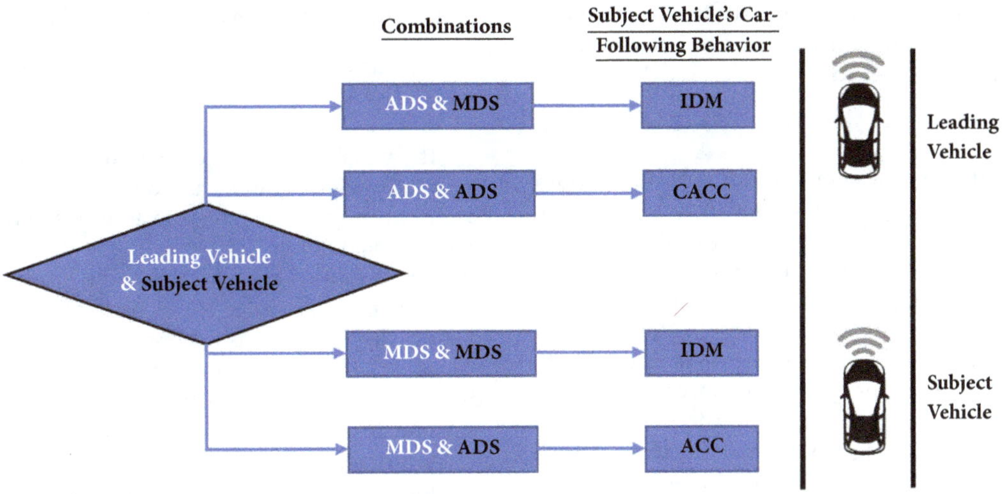

FIGURE 1: Proposed car-following strategy for mixed traffic.

CACC platoons, and (iii) adjusting platoon configurations dynamically to obtain balanced benefits from considered traffic attributes.

3. Proposed Car-Following Strategy for Individual Vehicle

Interactions and behaviors of vehicles at microscopic levels have macroscopic implications. Factors like maximum accelerations, comfortable decelerations, preferred time headways, etc. are directly linked to traffic mobility, safety, and environmental aspects. Car-following models provide individual vehicles' acceleration from dynamic interactions with adjacent vehicles, control constraints to generate velocity, and position to determine vehicle trajectory. The car-following models of vehicles in mixed traffic were schemed here to simulate real-traffic movements. As mentioned earlier, the existence of two types of vehicle driving system was considered for combined traffic. The proposed driving strategy identified all potential combinations of leading vehicle and subject vehicle based on driving systems to determine suitable car-following models.

The proposed car-following mechanism presumed that MDS vehicles would maintain conventional car-following behavior irrespective of the leading vehicle's driving system [Figure 1]. In this regard, intelligent driver model (IDM) [10] was chosen to represent manual drivers' car-following behavior. Extensive applications of this model across different studies developed this model as a perfect example to simulate MDS vehicles' car-following behavior. An enhanced version of traditional IDM was used to determine a realistic longitudinal control decision of MDS vehicles (see (1)). Discretized kinematic equations were used for all vehicles irrespective of the driving system to determine vehicle's velocity and position (see (2) and (3)).

$$\dot{v}(t+\Delta t) = a\left[1-\left(\frac{v(t)}{v_0}\right)^4 - \left(\frac{s_0+\max\left[0, v(t)\times T+(v(t)\times \Delta v(t))/2\sqrt{ab}\right]}{s}\right)^2\right] \tag{1}$$

$$v(t+\Delta t) = v(t) + a(t)\times \Delta t \tag{2}$$

$$p(t+\Delta t) = p(t) + v(t)\times \Delta t + \frac{1}{2}a(t)\times \Delta t^2 \tag{3}$$

where $\dot{v}$ is acceleration of vehicle (m/s^2), v is velocity (m/s), p is vehicle position (m), a is maximum acceleration (set as 3 m/s^2), v_0 is desired velocity (35 m/s), s_0 is leading gap at jam density (5 m), b is desirable deceleration (-3 m/s^2), Δv is velocity difference with leading vehicle, and T is preferred time headway (2.5 sec).

To demonstrate the car-following mechanism of ADS vehicles, both ACC and CACC based car-following were implemented. A MDS based leading vehicle would prompt ADS vehicle to follow ACC with relatively high preferred time headway. Provided that the leading vehicle had ADS, the subject vehicle would choose CACC based car-following. Whether the subject vehicle would join the CACC platoon depends on the leading vehicle's platoon ID. Platoon ID is an identification number assigned to an ADS vehicle that represents its order of position in the platoon. If an ADS vehicle is a part of a platoon, it will have a fixed platoon ID; otherwise its platoon ID will contain a platoon ID = 0 (zero). While travelling through roads, the built-in communication technology of ADS vehicles would enable them to identify the leading vehicles driving system as well as platoon ID. If the platoon ID of the leading vehicle was equal to the maximum platoon length, then the subject vehicle would form a new platoon by maintaining inter-platoon headway and as a leader of the new platoon. In addition, if the leading vehicle's platoon ID was lower than maximum platoon length, the subject vehicle would join the platoon by maintaining intra-platoon headway. We adopted the ACC and CACC car-following models developed in [17]. The accelerations of the subject vehicle were determined with respect to relative position and velocity. The following equation was used to determine the acceleration of the subject vehicle:

$$\dot{v}(t + \Delta t) = k_1 (\Delta p(t) - v(t) \times T - s_0) + k_2 \Delta v(t) \quad (4)$$

where k_1, k_2 are control constants for relative distance and speed, respectively ($k_1, k_2 > 0$) and $\Delta p(t)$ is position difference with leading vehicle. The stability of the proposed ACC system was proved in [17]. Suitable k_1, k_2 values were chosen according to [17] to implement realistic simulation accounting for the sensitivity of these factors. Similar approach of dual consensus was taken by Wang et al. [18] where both position and velocity consensus were considered to determine acceleration/deceleration decision. While both ACC and CACC car-following models used (4) to determine acceleration values for ADS vehicles, higher preferred time headways ($T = 1.5$ sec) distinguish ACC mode with CACC mode ($T \leq 1.0$ sec).

4. Simulation Procedures

A microscopic simulation structure was built on MATLAB to replicate vehicles' motion on a two-lane directional highway. The simulation environment was grounded on numerical analysis-based car-following behavior. All previously mentioned motion dynamic equations were coded to follow proposed car-following strategy. A stream of 20 vehicles following a controlled leading vehicle was simulated for numerous scenarios. The time headways between the vehicles in traffic stream were manipulated to simulate distinct traffic flow rates. In the simulation environment, the acceleration of the first vehicle was controlled consciously to generate multiple shockwaves and to observe the reaction of the vehicles behind it. Each simulation ran for 1000 time steps and 20 times for each scenario. The preferred time headway (T) for MDS vehicles was considered as a normally distributed variable with mean value of 2.5 sec and standard deviation of 0.5 sec. Multiple runs for each scenario were executed to ensure that the obtained outcome was free from anomaly. The average values of 20 runs were listed for analysis. In the beginning of the simulation, the first vehicle was travelling at 25 m/s for 210 time steps and then accelerated at 0.167 m/s^3 rate for 60 time steps followed by steady state (acceleration/deceleration rate = 0 m/s^3, velocity = 35 m/s) for 120 time steps. Finally, the controlled vehicle at front decelerated again at 0.167m/s^3 rate for 60 time steps to regain 25m/s velocity and moved with constant velocity for the remaining time steps. The maximum velocity was set to 35 m/s. The combinations generated from the following variables sets were simulated to represent various traffic states encountered in roadways as well as to identify the variations on improvements obtained by introducing the ADS vehicles in the connected automated vehicle (CAV) environment:

(a) Initial flow rate (veh/hr): (i) 1400, (ii) 1800, (iii) 2400

(b) ADS market share (%): (i) 25, (ii) 50, (iii) 75

(c) Maximum platoon length (vehicle): (i) 3, (ii) 4, (iii) 5, (iv) 6

(d) Inter-platoon headway (sec): (i) 2, (ii) 4, (iii) 6, (iv) 8

(e) Intra-platoon headway (sec): (i) 0.5, (ii) 0.75, (iii) 1.0, (iv) 1.25

The variables sets were restricted by the above values to limit the analysis and discussions within manageable ranges while covering a wide range of variations in traffic conditions. Platoon parameters (i.e., maximum platoon length, inter-platoon headway, and intra-platoon headway) were varied within reasonable ranges to identify observable trends. Two distinct driving systems were simulated by assigning specific values of driving system (0 for MDS, 1 for ADS). The driving system values assigned for vehicles were used to implement the proposed car-following strategy on the CAV environment. Assigned driving system values were also useful to adopt proper sets of motion dynamic equations.

5. Analysis, Results, and Findings

5.1. Impact of ADS Vehicles Location and Distribution. Before analyzing the mobility, safety, and environmental aspects of ADS vehicles on traffic, the influences of ADS vehicles location and distribution in traffic stream were explored. It was hypothesized that the positions of ADS vehicles in traffic stream dictated their impacts on the remaining vehicles. To prove this hypothesis, the proposed car-following strategy was simulated by allotting ADS vehicles at diverse combinations of positions with gradually increasing the initial flow rate and ADS market share. To clearly comprehend the significance of vehicle position more clearly and to reduce the intricacy of comprehension, only two features were analyzed: acceleration fluctuation of MDS vehicle in the vehicle group and variations of maximum traffic flow at varying traffic state. Since numerous combinations of ADS vehicles' distribution are viable at different penetration rates of ADS vehicles, only a handful of combinations were selected to cover most possible variations.

Initially, these distributions were generated by placing ADS vehicles as far apart as possible (--% Comb-1) in the vehicle stream while maintaining target ADS market share. Gradually, ADS vehicles were grouped together in different combinations. The purpose of placing ADS vehicles in such an order was to visualize and measure the impact of ADS vehicles location and distribution along the vehicle stream. The combinations are listed in Table 1. The first column of the table shows percentages of ADS vehicles in the traffic stream. The numbers in second column of Table 1 identify the position ID of ADS vehicles in the traffic stream. Other vehicles, except the positions mentioned in the table, were MDS vehicles. The last column of the table provides distinct combination name of each distribution of ADS vehicles. These combinations were simulated on developed simulation environment by virtually placing ADS vehicles in the mentioned position IDs of the vehicle stream and by following proposed car-following strategy for mixed traffic. The listed combinations in Table 1 were assumed to represent varying ranges of ADS vehicles distribution on vehicle group. Analyzing these sets of vehicle location and distribution

TABLE 1: List of ADS vehicle combinations simulated for different market penetrations.

ADS Market Share	Distribution of ADS vehicles (position)	Combination Name
25%	4, 8, 12, 16, 20	25% Comb-1
	4, 5, 10,11, 16	25% Comb-2
	5, 6, 7, 13, 14	25% Comb-3
	9, 10, 11,12 17	25% Comb-4
	2, 3, 4, 5, 6	25% Comb-5
	16, 17, 18, 19, 20	25% Comb-6
50%	2, 4, 6, 8, 10, 12, 14, 16, 18, 20	50% Comb -1
	2, 3, 6, 7, 10, 11, 14, 15, 18, 19	50% Comb -2
	2, 3, 4, 8, 9, 10, 14, 15, 16, 20	50% Comb -3
	2, 3, 4, 5, 10, 11, 12, 13, 18, 19	50% Comb -4
	2, 3, 4, 5, 6, 7, 8, 9, 10,11	50% Comb -5
75%	2, 3, 4, 6, 7, 8, 10, 11, 12, 14, 15, 16, 18, 19, 20	75% Comb-1
	2, 3, 4, 5, 6, 9,10, 11, 12, 13, 16, 17, 18, 19, 20	75% Comb-2
	2, 3, 4, 5, 6, 9, 10, 11, 12, 13, 16, 17, 18, 19, 20	75% Comb-3

provided the opportunity to shed light on resulting impacts due to ADS vehicles' position on traffic stream.

From the analysis, the simulation outcomes of the initial flow rate of 1800 veh/hr with different ADS market penetration are provided in Figure 2 to demonstrate the influences of ADS vehicles position and distribution along the stream from both microscopic and macroscopic perspective. Figure 2(a) represents the variations of maximum flow rates resulting from the proposed car-following strategy at listed combinations. Figure 2(b) shows the average coefficient of variations (CoV) of acceleration of MDS vehicles in the simulated vehicle stream. Boxplots for a specific combination were plotted from the maximum flow rate and average CoV of acceleration data of simulated scenarios with varying platoon parameters, as listed before. Macroscopic analysis on maximum flow rates at different ADS vehicle shares (Figure 2(a)) identified the pattern of gradual increment with increasing ADS shares in the traffic. Observations of different combinations revealed that combinations with scattered ADS vehicles lead to lower maximum flow rates in comparison to combinations with grouped ADS vehicles. Additionally, grouping ADS vehicles at the front of the vehicle stream (i.e., 25% Comb-6, 50% Comb-5, and 75% Comb-3) resulted in 6.7-11.5% higher maximum flow rates in comparison to the scattered distribution of ADS vehicles (i.e., 25% Comb-1, 50% Comb-1, and 75% Comb-1). Analysis on microscopic characteristics of MDS vehicles was undertaken by measuring the average CoV of acceleration at different market shares and combinations of ADS vehicles. The resulting analysis showed a gradual decreasing CoV of acceleration with increasing shares of ADS vehicles. Similar to macroscopic analysis, the maximum amount of decrease in CoV (1.69–6.63%) was observed from combinations with ADS vehicles at the front of the traffic stream grouped together. Specific analysis on maximum platoon length's influence on acceleration fluctuations of MDS vehicles revealed that increasing maximum platoon length reduced the average coefficient of variation of acceleration for ADS vehicles. Similar analysis on the other two platoon parameters (i.e., inter-platoon headway and intra-platoon headway) demonstrated a reciprocal relation with acceleration fluctuations (increasing inter- and intra-platoon headway increased the average CoV of acceleration).

The analysis of the remaining initial flow rates and ADS market share revealed that creating platoons of ADS vehicles by positioning them at the front of traffic stream would be beneficial to the rest of vehicles in the traffic stream. Furthermore, increasing market shares of ADS vehicles could gradually reduce the acceleration fluctuation of MDS vehicles. Finally, increasing the flow rates could inversely influence traffic flow improvements with a specific ADS location and distribution combination. The notion of traffic flow improvements guided the authors in this study to explore mobility improvement potentials of the proposed car-following strategy by placing ADS vehicles at ideal positions along the traffic stream.

5.2. Impact on Mobility. Since creating platoons of ADS vehicles was found to be the most effective way of acquiring associated benefits, influences of ADS vehicles on traffic mobility were examined with respect to three key variables of platooning: intra-platoon headway, inter-platoon headway, and maximum platoon length. Combinations of these three variables within listed sets were utilized to generate various platoon scenarios for simulation and analysis. The impact of these platoon structures on mobility was measured and compared with the help of two parameters: Average Travel Time (ATT) (see (5)) and Average Travel Distance (ATD) (see (6)). Later, case scores were computed by providing equal weights to ATT and ATD (see (7)). Different cases of platoon configurations were simulated and evaluated through case scores. Higher dispersion from base-case (0% ADS share) scores indicated higher mobility improvements. The objective of this analysis was to identify the optimal platoon configuration to improve mobility by increasing ATD and reducing ATT. The following equations were used to identify the mobility gains.

$$ATT = \frac{\sum_{j=1}^{J} ATT_j}{J} = \sum_{i=1}^{I} \frac{\left(p_{i,j} - p_{i,j-1}\right)}{v_{i,j}} \tag{5}$$

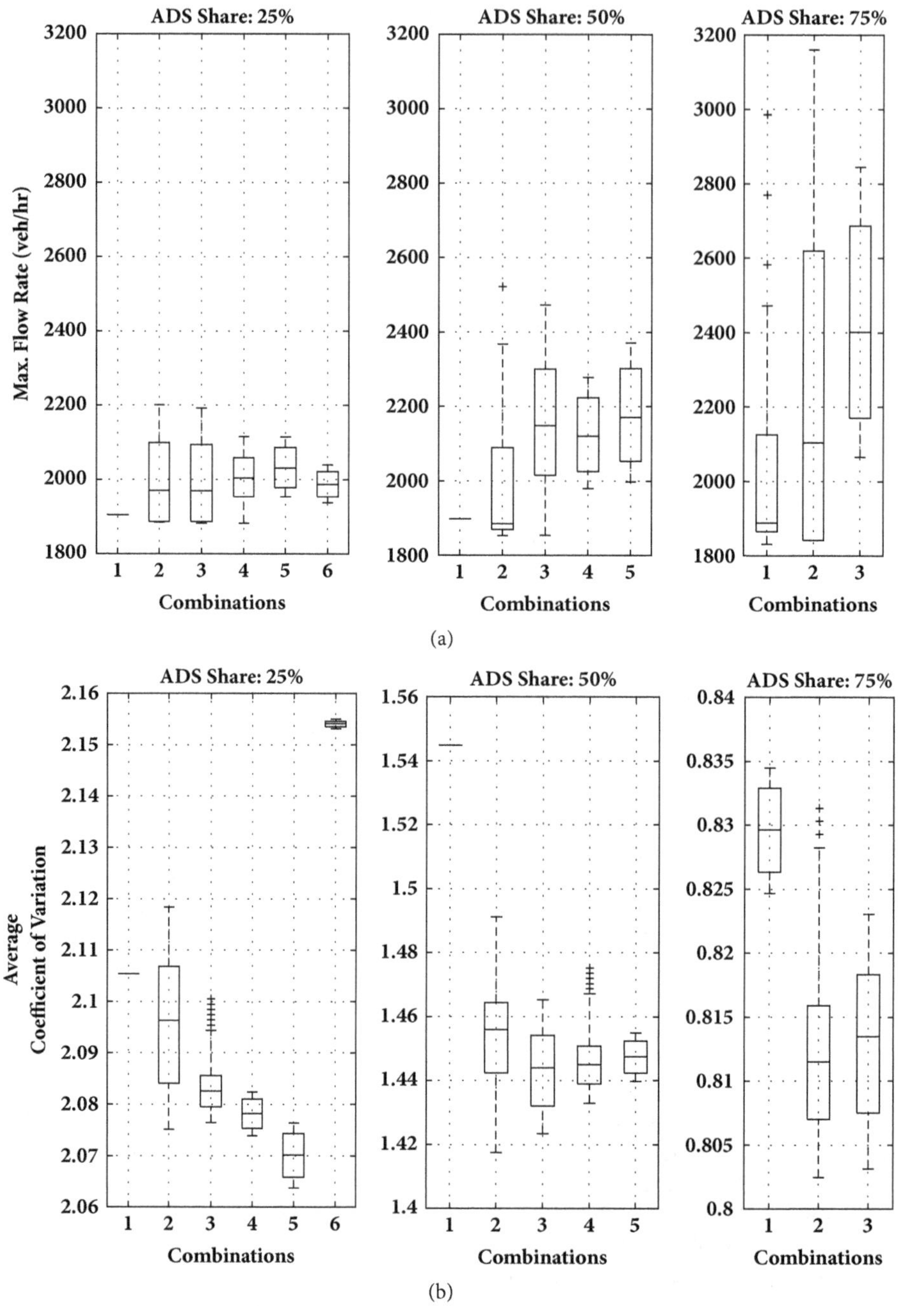

(a)

(b)

FIGURE 2: Influences of ADS vehicles' position on (a) maximum flow rate and (b) average coefficient of variation of accelerations.

$$ATD = \frac{\sum_{J=1}^{J} ATD_j}{J},$$

$$ATD_j = \frac{\left(p_{I,j} - p_{1,j}\right)}{I} \tag{6}$$

$$Score_{Case\,k} = \frac{\sum_{J=1}^{J} v_{1,j}\left(ATD_{j,Case\,k}\right)}{J} - \left(ATT_{Case\,k}\right) \tag{7}$$

Here, i is vehicle index (I = 21), j is time index (J = 1000), $v_{i,j}$ is velocity of vehicle i at time step j, $p_{i,j}$ is position of vehicle i at time step j, and $Score_{SCase\,k}$ is score of case k. Aforementioned (Section 4) platoon variables (i.e., maximum platoon length, inter-platoon headway, and intra-platoon headway) were explored to generate distinct platoon scenarios. The combinations of these parameter sets produced 64 distinct platoon configuration cases that were simulated for different traffic flows and ADS market shares to detect the capability of mobility improvements. Moreover, the limits of mobility improvements due to variation of platoon configurations were also revealed in this analysis. Obtained mobility improvements from base cases at different traffic states are presented in Figure 3(a). The three-quarter circles showed comparative mobility progresses at different flow

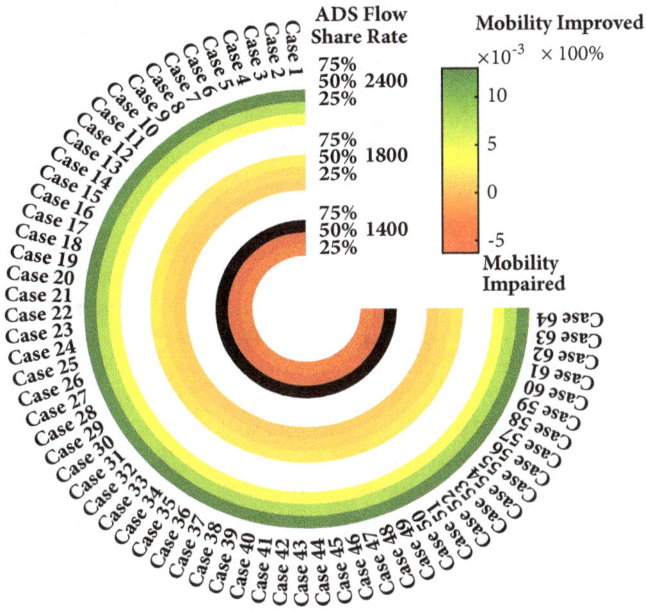

(a)

Case	Intra Platoon Headway	Inter Platoon Headway	Max. Platoon Length	Mobility Improvement	Case	Intra Platoon Headway	Inter Platoon Headway	Max. Platoon Length	Mobility Improvement
1	0.50	2	3	0.200%	33	0.50	2	5	0.204%
2	0.75	2	3	0.196%	34	0.75	2	5	0.200%
3	1.00	2	3	0.193%	35	1.00	2	5	0.196%
4	1.25	2	3	0.190%	36	1.25	2	5	0.192%
5	0.50	4	3	0.189%	37	0.50	4	5	0.198%
6	0.75	4	3	0.186%	38	0.75	4	5	0.194%
7	1.00	4	3	0.182%	39	1.00	4	5	0.190%
8	1.25	4	3	0.179%	40	1.25	4	5	0.186%
9	0.50	6	3	0.178%	41	0.50	6	5	0.193%
10	0.75	6	3	0.175%	42	0.75	6	5	0.189%
11	1.00	6	3	0.172%	43	1.00	6	5	0.185%
12	1.25	6	3	0.168%	44	1.25	6	5	0.181%
13	0.50	8	3	0.168%	45	0.50	8	5	0.188%
14	0.75	8	3	0.164%	46	0.75	8	5	0.184%
15	1.00	8	3	0.161%	47	1.00	8	5	0.180%
16	1.25	8	3	0.158%	48	1.25	8	5	0.176%
17	0.50	2	4	0.202%	49	0.50	2	6	0.204%
18	0.75	2	4	0.198%	50	0.75	2	6	0.200%
19	1.00	2	4	0.194%	51	1.00	2	6	0.196%
20	1.25	2	4	0.191%	52	1.25	2	6	0.192%
21	0.50	4	4	0.194%	53	0.50	4	6	0.198%
22	0.75	4	4	0.190%	54	0.75	4	6	0.194%
23	1.00	4	4	0.186%	55	1.00	4	6	0.190%
24	1.25	4	4	0.183%	56	1.25	4	6	0.186%
25	0.50	6	4	0.186%	57	0.50	6	6	0.193%
26	0.75	6	4	0.182%	58	0.75	6	6	0.189%
27	1.00	6	4	0.178%	59	1.00	6	6	0.185%
28	1.25	6	4	0.175%	60	1.25	6	6	0.181%
29	0.50	8	4	0.178%	61	0.50	8	6	0.188%
30	0.75	8	4	0.174%	62	0.75	8	6	0.184%
31	1.00	8	4	0.170%	63	1.00	8	6	0.180%
32	1.25	8	4	0.167%	64	1.25	8	6	0.176%

(b)

FIGURE 3: Variations of mobility benefits due to varying platoon configurations at (a) different flow rates and ADS shares and (b) specific flow rate (1800 veh/hr) and ADS share (75%).

rates and ADS market shares simulated for the analysis. The color bar in Figure 3(a) indicated the extent of generated mobility score improvements. Figure 3(b) reveals detailed analysis for a specific flow rate and ADS share. For clear understanding of the impact of platoon configurations at a specific traffic state, mobility improvements at initial flow rate of 1800 veh/hr and 75% ADS share are provided in Figure 3(b) as an example. As observed in Figure 3(b), sixty-four (64) separate platoon configurations are generated from listed parameter set (Section 4). Parameters for each case are listed in the table in Figure 3(b). The mobility improvement column was calculated by comparing the base case (flow rate = 1800 veh/hr, ADS share = 0%) with the corresponding cases and transforming the value into a percentage. Negative percentages indicate impaired mobility and positive percentages denote improved mobility resulting from a specific platoon configuration. When inspecting Figure 3(b), it was found that maximum mobility benefits (0.204% improvement on case score) could be obtained from Case 33 (platoon configuration: intra-platoon headway = 0.50 sec, inter-platoon headway = 2 sec, and max. platoon length = 5) and Case 49 (platoon configuration: intra-platoon headway = 0.50 sec, inter-platoon headway = 2 sec, and max. platoon length = 6) for that specific traffic state. A declining trend of mobility gains was captured with increasing inter and intra-platoon headway. Additionally, increasing maximum platoon length parameter showed expansion with regard to mobility which came to a halt at maximum platoon length = 5.

An exploration of Figure 3(a) revealed that increasing ADS market could bring broader mobility enhancement at higher flow rates (yellow to green bands on 2400 veh/hr flow rate). Increased ADS share at low flow rates had a diminishing effect on mobility (light red to deep red bands on 1400 veh/hr flow rate). Another finding of this analysis was that the closely spaced ADS vehicles with long platoons would generate more mobility improvements. Hence, the maximum mobility benefit was experienced in Case 33 and Case 49. Although the analysis concluded that closely spaced, long ADS platoons could attain higher mobility benefits, close proximity of ADS platoons and long chain of ADS vehicles in these platoon configurations would severely restrict merging vehicles from neighboring lanes, on-ramps, side roads, etc.

Analysis on platoon parameters at different traffic state revealed that, with other parameters being constant, increasing platoon length resulted in improved mobility gains. Similar investigation on inter-platoon headway presented that increase in inter-platoon headway would reduce traffic mobility if other two parameters remain constant at a specific traffic state. Analysis of intra-platoon headways coincides with the insights of inter-platoon headway analysis. Therefore, compactness of ADS vehicles would bring more mobility benefits in roadway sections with minimal conflict points (e.g., spans between on/off-ramps on freeways and sections between intersections in arterial). The notion of conflict points led to the next section of this study, examining the impact of ADS vehicles on traffic safety.

5.3. Impact on Safety. Although Cases 33 and 49 were found to be an obvious choice among 64 tested platoon configuration cases with respect to mobility enhancements, all aforementioned cases were examined again to identify the potential impact on traffic safety. Findings from ADS vehicles location and distribution influenced the simulation of safety improvements by placing a series of ADS vehicles at the front of traffic stream to obtain optimal benefits. Since no merging traffic was considered, the safety enhancements were examined as a measure of potentials to reduce rear-end collision risks. Three safety surrogate measures were considered in this regard: time-to-collision (TTC), time exposed time-to-collision (TET), and time integrated time-to-collision (TIT).

TTC, TET, and TIT, introduced by Hayward, Minderhoud, and Bovy [44, 45], were widely used by traffic safety researchers. The time required for two successive vehicles in the same lane to hit if they maintain their current velocity is represented by TTC. Higher TTC would indicate safer traffic condition. TTC can be used to evaluate safety of a traffic environment, since lower TTC is indicative of potential dangerous situation [46]. Both TET and TIT are derived from TTC to measure safety improvements from macroscopic standpoint. Since TET is the summation of instances when TTC are lower than threshold value, the lower TET value is expected at safer traffic conditions. TET value was measured by (9) where TTC values for each vehicle at each time stamp ($TTC_{i,j}$) were compared with the threshold TTC (TTC^*) value to calculate TET value for each scenario. TIT measures the value of TTC lower than the threshold TTC. Similar to TET, a higher TIT value indicates higher safety concerns. The values of these parameters were measured using the following equations:

$$TTC_{i,j} = \begin{cases} \dfrac{p_{i-1,j} - p_{i,j} - L}{v_{i,j} - v_{i-1,j}} & if\ v_{i,j} > v_{i-1,j} \\ Inf & if\ v_{i,j} \le v_{i-1,j} \end{cases} \tag{8}$$

$$TET = \sum_{j=1}^{J} TET_j,$$
$$TET_j = \sum_{i=1}^{I} \delta_j \Delta j, \quad \delta_j = \begin{cases} 1 & \forall 0 < TTC_{i,j} < TTC^* \\ 0 & else \end{cases} \tag{9}$$

$$TIT = \sum_{j=1}^{J} TIT_j,$$
$$TIT_j = \sum_{i=1}^{I} \left[\frac{1}{TTC_{i,j}} - \frac{1}{TTC^*} \right] .\Delta j \tag{10}$$
$$\forall 0 < TTC_{i,j} < TTC^*$$

The threshold TTC values to measure TET and TIT were set as 2.5 sec, similar to standard perception reaction time. Resulting changes with regard to safety are displayed in Figure 4. Figure 4(a) presents total TET and average TIT values over the simulation period on base cases which

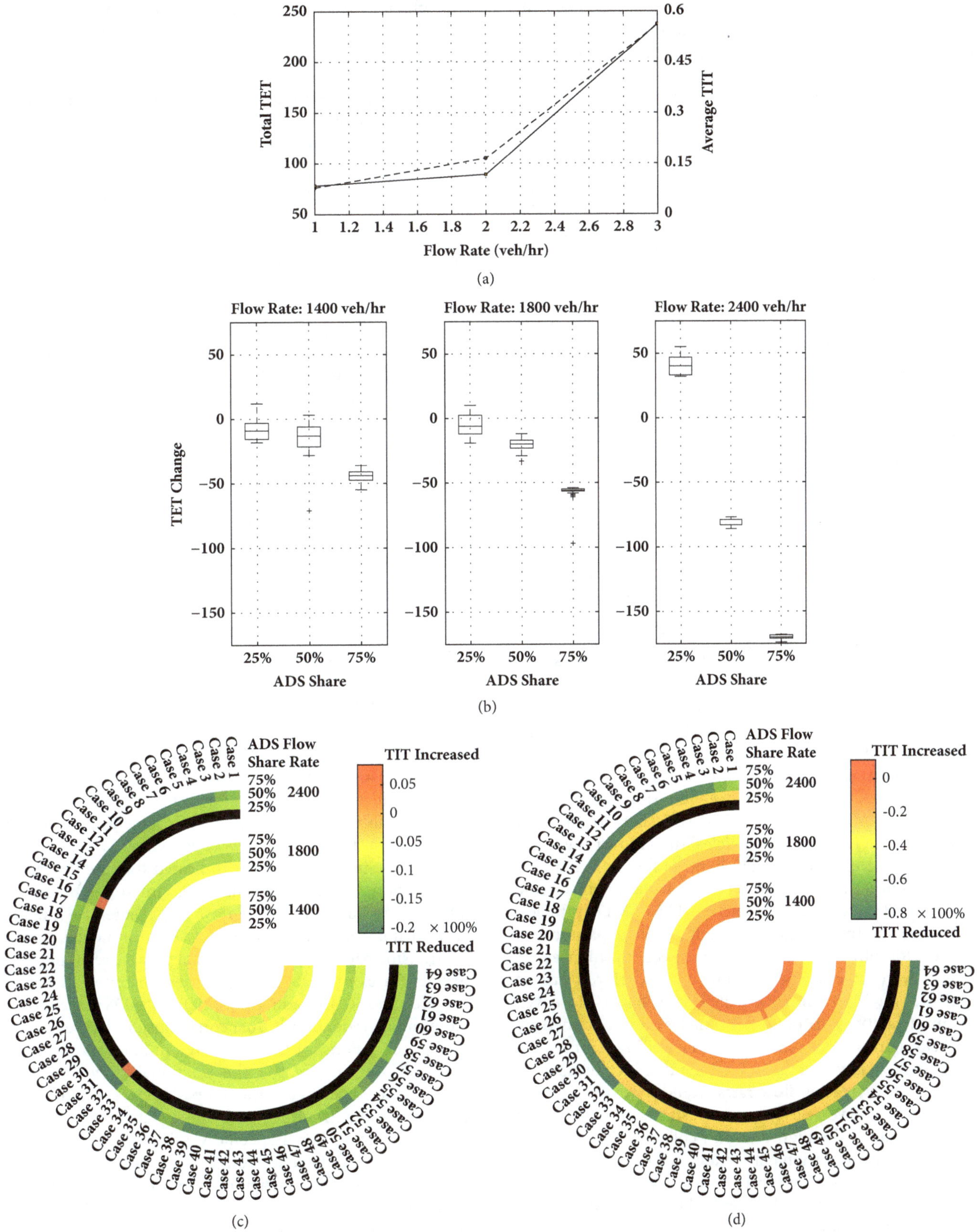

FIGURE 4: (a) Base-case safety parameter values at varying flow rates. (b) Changes in total TET, (c) variations of average TIT values considering MDS vehicles only, and (d) variations of average TIT values considering all vehicles, due to varying platoon configurations at different flow rates and ADS shares.

were utilized to measure safety improvements gained with the introduction of ADS vehicles. Figure 4(b) displays the range of changes on total TET values at different traffic states with varying platoon structures. Increasing ADS shares showed a gradual decline of total TET values. The extent of declination was much higher in higher flow rates. However, an exception was observed at high flow rates and lower ADS shares (flow rate = 2400 veh/hr, ADS share = 25%) where total TET value increased from base traffic states. Therefore, it can be stated that higher ADS share is required to bring noticeable safety improvements with increasing flow rates. Figures 4(c) and 4(d) show analysis results of average TIT changes. As shown in earlier figure (Figure 3(a)), both factional circles revealed resulting improvements on average TIT parameters. Figure 4(c) shows resulting safety improvements of the vehicle stream for different platoon configurations, ADS shares, and flow rates by comparing with base average TIT values. This analysis considered average TIT values of MDS vehicles only in the traffic stream. Average TIT values of MDS vehicles in the CAV environment were compared with corresponding vehicles on base case for this analysis. The average TIT reduction of MDS vehicles was found to be within the range of -20.76%–8.55%. Additionally, higher safety gains were achieved with shorter platoons including ADS vehicles sparsely spaced.

On the other hand, Figure 4(d) shows the analysis by comparing average TIT values of all vehicles with base case. For this analysis, it was assumed that there was no collision risk for ADS vehicles (average TIT values = 0 for ADS vehicles), irrespective of platoon configurations. Comparison between Figures 4(c) and 4(d) shows significantly higher improvements on average TIT values for all vehicles over MDS vehicles. The range of average reduction is much higher in Figure 4(d). Detailed analysis of safety enhancement for a specific traffic state provided further insights on the impact of platoon configurations. For instance, simulation results of 1800 veh/hr flow rate with 75% ADS share traffic state revealed that increasing ADS vehicles' stretch over the traffic stream resulted in greater safety benefits for the remaining vehicles. Hence, the maximum safety gain was attained from Case 16 (-10.23% reduction on average TIT of MDS vehicles) for this specific traffic state. Although a similar pattern was observed for other traffic states, unexpectedly high safety concerns were experienced for some cases (dark red strip in Figure 4(c) for 2400 veh/hr with 25% ADS share). Moreover, maximum safety gains on MDS vehicles were obtained on 50% ADS share at 1800 veh/hr flow. The findings from safety impact analysis have led us to conclude that increasing ADS vehicles with increasing flow rates would improve safety of all vehicles if ADS vehicles form short, sparse platoon in start of traffic stream. Although rear-end collision risk for MDS vehicles would proportionately reduce with increasing ADS share at comparatively high and low flow rate, this correlation between safety gain and ADS share did not hold true for flow rates near capacity level.

Exploring the evolution pattern of platoon parameters provided important insights into safety feature. While other parameters (i.e., inter-platoon headway and max. platoon length) remain the same, continuous increment of intra-platoon headway showed reduction on rear-end collision expectation. Inter-platoon headway followed similar pattern to intra-platoon headway. However, range of safety improvement in both parameters depends on maximum platoon length. Magnitude of safety gains was much higher at small platoons (i.e., max. platoon length = 3) than big platoons (i.e., max. platoon length).

5.4. Impact on Environment. Environmental implications of proposed car-following mechanism were measured with respect to fuel consumption and emission reduction. While numerous models were available and utilized in the literature [47–49], the integrational simplicity of the VT-micro model [50–52] with car-following model persuaded us to implement this model. Output from car-following models can be directly used on the VT-micro model as input to estimate environmental impact due to vehicle dynamics which makes this model a perfect candidate for this analysis. According to the VT-micro model, the fuel consumption of or emission rate of ith vehicle at time step j can be measured using the following equations:

$$\ln\left(MOE_{i,j}\right) = \begin{cases} \sum_{l=0}^{3}\sum_{m=0}^{3} K_{l,m} \times v_{i,j}^{l} \times a_{i,j}^{m} & if\ a \geq 0 \\ \sum_{l=0}^{3}\sum_{m=0}^{3} K'_{l,m} \times v_{i,j}^{l} \times a_{i,j}^{m} & if\ a < 0 \end{cases} \tag{11}$$

where $MOE_{i,j}$ is measure of effectiveness with respect to fuel consumptions, CO_2 emissions, and NO_x emissions for vehicle i at time j and $K_{l,m}$ are regression coefficients for MOEs at powers l and m. The values of regression coefficients are obtained from [50]. $v_{i,j}^{l}$ is velocity of vehicle i at time j with power l. $a_{i,j}^{m}$ is acceleration of vehicle i at time j with power m.

Analysis using the VT-micro model for the base case (0% ADS share) measured average fuel consumptions, CO_2 emissions, and NO_x emissions of the vehicles in simulated traffic stream at different flow rates, which is presented in Figure 5(a). Simulation results indicated that the lowest fuel consumption, CO_2 emission, and NO_x emission at base traffic state occurred at 1800 veh/hr flow rate. Therefore, low flow rates do not necessarily ensure low environmental impact. The transformation in environmental impact resulting from varying shares of ADS vehicles is demonstrated in Figures 5(b), 5(c), and 5(d). Gradual increments of ADS share showed a continuous reduction in fuel consumption. However, CO_2 and NO_x emissions for the traffic stream followed a different trend. As previous figures, the fractional circles displayed changes in average environmental parameters resulting from varying platoon structures and traffic states (i.e., flow rates and ADS shares). For a specific traffic state (i.e., flow rate and ADS share), the effect on environment demonstrated similar patterns to the impact on mobility. As an example, we can examine the platoon structures for 1800 veh/hr flow rate with 75% ADS share. The observations of this specific traffic state revealed that environmental benefits kept increasing with the gradual compaction of ADS vehicle in traffic stream. For instance, maximum reduction on fuel consumption was obtained for Case 33 and Case 49 (platoon configuration:

Base Case (0% ADS Share)				
Parameters	Unit	Flow Rate (veh/hr)		
		1400	1800	2400
Fuel Consumption	Litre/ 100km	8.57	8.52	8.66
CO_2 Emission	Kg/ 100km	19.71	19.64	19.9
NO_x Emission	g/ 100km	28.03	27.96	27.33

(a)

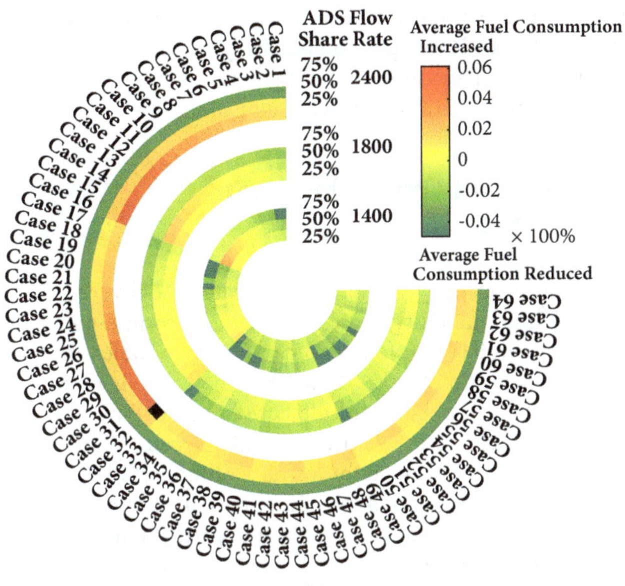

(b)

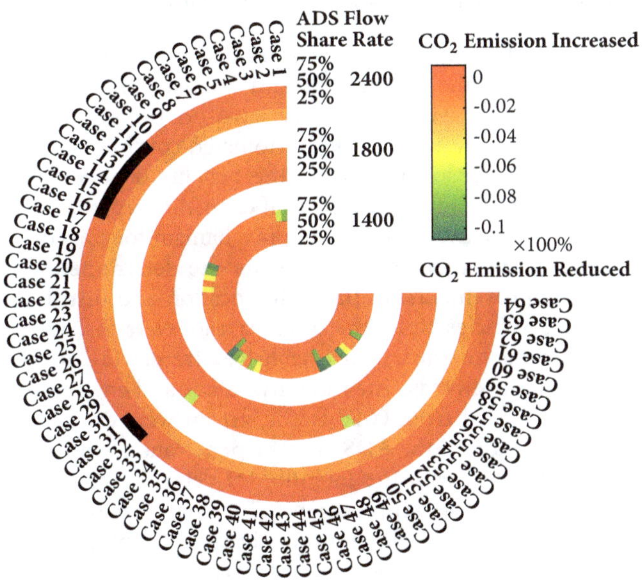

(c)

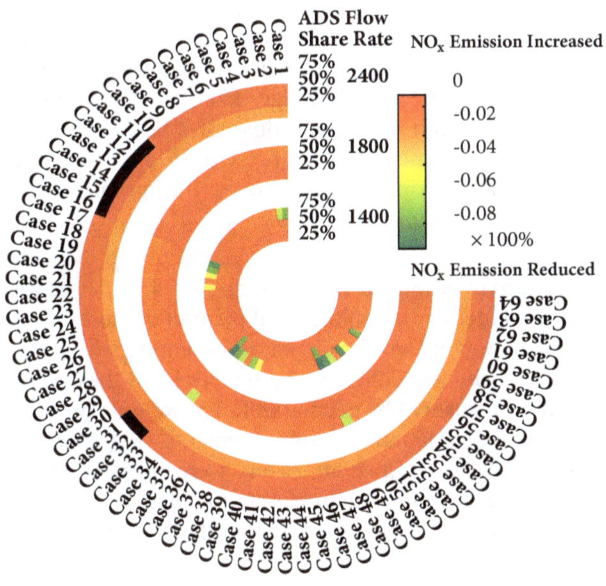

(d)

FIGURE 5: Observed variations of (a) environmental parameters at base case, (b) fuel consumption, (c) CO_2 emission, and (d) NO_x emission for varying platoon structures at different traffic state.

intra-platoon headway = 0.50 sec, inter-platoon headway = 2 sec, and max. platoon length = 5 and 6, respectively) which reduced average fuel consumption by 4.87% from the base case, whereas Cases 48 and 64 (platoon configuration: intra-platoon headway = 1.25 sec, inter-platoon headway = 8 sec, and max. platoon length = 5 and 6] reduced fuel consumption by 1.42% and 1.44%, respectively. A similar pattern was observed for the other two parameters (i.e., CO_2 emission and NO_x emission) for this traffic state. Detailed analysis of the environmental impact identified a proportional relation between fuel consumption reduction and ADS share at all simulated traffic flow rates. However, the extent of improvements varied at different flow rates. As for CO_2 and NO_x emission, the relation between emission reduction and ADS share was found to be proportionate at lower flow rates (i.e., 1400, 1800 veh/hr). At high flow rate (i.e., 2400 veh/hr), higher reduction was obtained at low ADS share. Out analysis of the environmental impact of ADS vehicles and formed platoons provided us with the insights of fuel consumption, CO_2 emission, and NO_x emission characteristics in order to make informed decision regarding platoon structures with an aim to attain optimal environmental benefits.

Close inspection of platoon parameters revealed similar inclinations to mobility. Unlike mobility gains, the degree of environmental gains was significantly higher at larger platoon sizes (i.e., max. platoon length = 5,6) in comparison to smaller platoons (i.e., max. platoon length = 3,4). Furthermore, the increments of intra and inter-platoon headway values showed steady declination of environmental benefits at a specific traffic state with other parameters being constant.

6. Identification of Optimal Platoon Parameter Set

An analysis of proposed car-following strategy delivered insights regarding mobility, safety, and environmental improvement potentials due to presence of ADS vehicles at mixed traffic conditions. One key finding of the analysis was that the expectation to obtain multiobjective improvements (i.e., mobility, safety, and environmental) from single platoon configuration was impractical. Since mobility and environmental developments maintained a reciprocal relationship with safety enhancements, a suboptimal platoon configuration could be determined to procure maximum gains from these three features. Another compelling outcome of prior analysis involved recognizing the fact that both traffic flow rates and ADS market shares impacted obtained benefits. Hence, achieving maximum mobility, safety, and environmental advantages from fixed suboptimal platoon configuration at different flow rates was unrealistic. To this end, it was necessary to present an approach that identified dynamic suboptimal platoon configurations for multiobjective decision-making purposes.

Influenced by Khondaker and Kattan [53], an analysis was performed to identify the suboptimal platoon configurations to maximize mobility, safety, and environmental gains generated by ADS vehicles. Collective influences from these three features were measured by placing different weights on them to get resulting variations on improvements (Figure 6(b)). Four sets of multiobjective functions were investigated to obtain suitable platoon structure. Sets for platoon variables were chosen from earlier analyses to identify suboptimal configurations.

The optimization of platoon variables for different multiobjective function identified each features' (i.e., mobility, safety, and environmental) individual and collective inclinations. To obtain clear and precise insights of these trends, the group of vehicles with 1800 veh/hr flow rate and 75% ADS market penetration is demonstrated in Figure 6. The improvements obtained due to ADS vehicles were scaled within [0 1] using extreme values from prior analysis of all the features (Figure 6(a)). For mobility improvements, the scenario scores were scaled within the above-mentioned range. Extreme average TIT values measured in safety impact analysis were applied to measure safety scores of different platoon configurations. Similarly, environmental score was calculated by assigning equal weights to three components of environmental impact (i.e., fuel consumption, CO_2 emission, and NO_x emission) while scaling them within the range of 0 and 1. This action was performed due to variations of units in measures of effectiveness and to bring them in the same scale for optimization. Reviews of individual features identified a gradual reduction of mobility and environmental improvements with an increase of intra and inter-platoon headway. However, safety improvements showed opposite pattern. Figure 6(b) shows the results of a set of objective functions with predefined weight put on mobility, safety, and environmental aspect. The goal of this analysis is to obtain suboptimal platoon configurations for predefined objective sets and also to identify the objective function with maximum benefits from the assorted weight sets. Analysis of combined impacts identified that maximum benefits for all objective functions were achieved with the platoon configuration of intra-platoon headway = 0.50 sec, inter-platoon headway = 2 sec, and maximum platoon length = 5/6 vehicles. The objective of this analysis was to present an approach to identify suboptimal platoon configurations suitable for specific flow rates and ADS market share with specific motivation to assist in multiobjective decision-making.

7. Conclusion and Future Extensions

The objective of the study was to obtain rationalized insight on mixed traffic movements and evaluate the impact that ADS vehicles will supposedly have on traffic. While the potential of connectivity and automated controls is astounding, the extent of harnessing the benefits depends on discerning their influences on traffic. In this regard, we have proposed a naïve car-following mechanism for mixed traffic and analyzed their motion dynamics to determine the possible improvements. Initially, the location and distributions of ADS vehicles along the traffic stream were discovered to be moving forward with established framework to obtain the highest rewards. The mobility, safety, and environmental gains obtained from CAV traffic stream were examined for varying traffic flow, ADS market penetration, and platoon configurations with the intention of determining the limits of these potential improvements. The final stage of this study was the analysis to obtain optimal platoon configurations to achieve maximum collective improvements.

The findings of the research show that, to obtain maximum mobility benefits, close and compact platoons are favorable in roadway sections without side frictions. However, segments with on-ramps, off-ramps, side roads, etc. need to be researched in the future to account for side frictions and their consequences on collective mobility, safety, and environmental gains. Identifying suboptimal platoon configurations for varying flow rates and market shares of ADS vehicles will assist traffic operation authorities to propose traffic state responsive dynamic platoon structures. Utilizing these platoon configurations will make the best use of ADS vehicles on prevailing traffic conditions to obtain maximum gains. Future research based on this study will account for vehicles with conflicting movements (i.e., lane changing, merging traffic from on-ramps, diverging traffic towards off-ramp, etc.) and propose potential improvements.

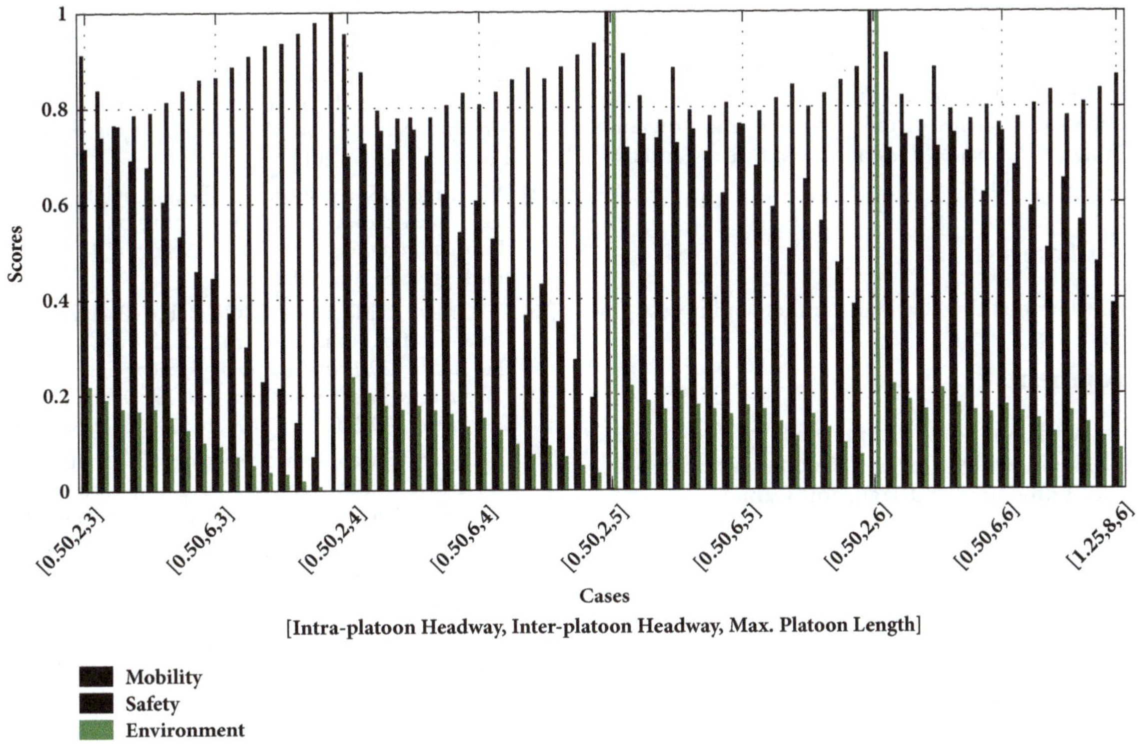

(a)

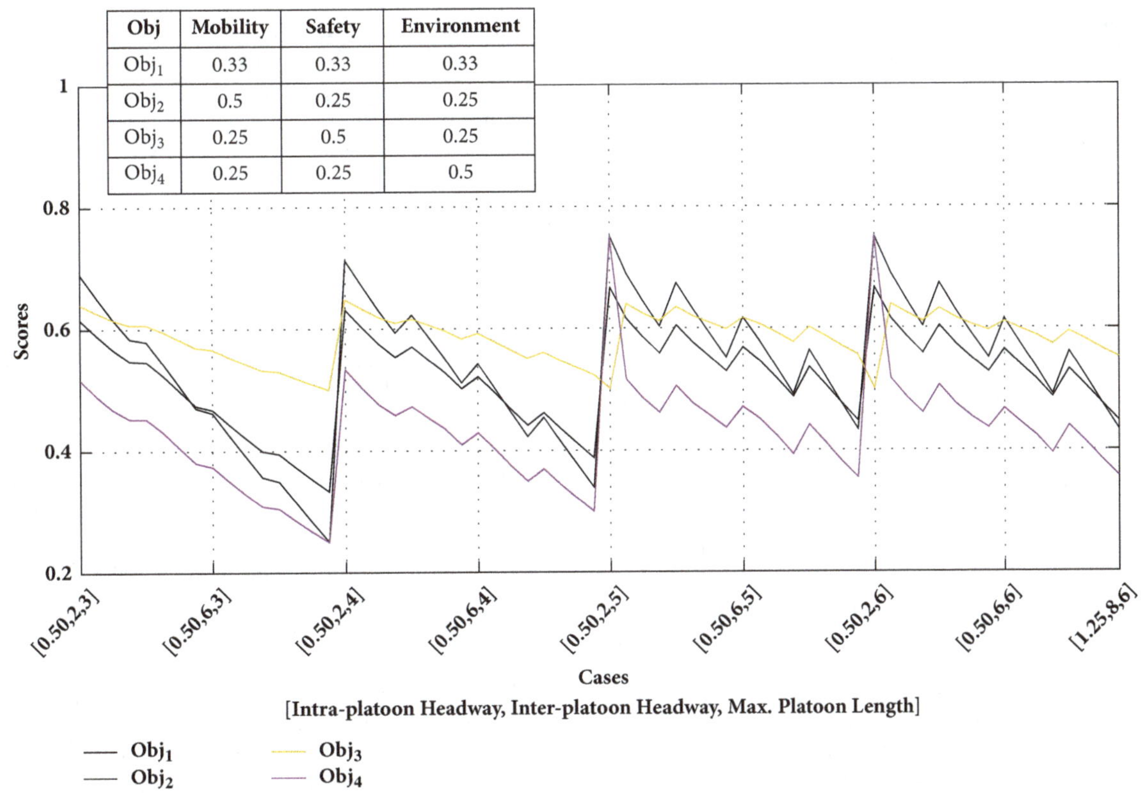

Obj	Mobility	Safety	Environment
Obj_1	0.33	0.33	0.33
Obj_2	0.5	0.25	0.25
Obj_3	0.25	0.5	0.25
Obj_4	0.25	0.25	0.5

(b)

FIGURE 6: Observed variations of (a) individual features due to diverse platoon variables listed and (b) listed multiobjective function sets resulting from changing platoon variables.

Disclosure

The contents of this paper reflect the views of the authors who are responsible for the facts and the accuracy of the data presented herein. The contents do not necessarily reflect the official views or policies of the City of Edmonton and Transport Canada. This paper does not constitute a standard, specification, or regulation.

Conflicts of Interest

The authors declare that they have no conflicts of interest.

Acknowledgments

This research work was jointly supported by the Natural Sciences and Engineering Research Council (NSERC) of Canada, City of Edmonton, and Transport Canada.

References

[1] S. Fakharian, "Evaluation of Cooperative Adaptive Cruise Control (CACC) Vehicles on Managed Lanes Utilizing Macroscopic and Mesoscopic Simulation," *Transp. Res. Rec. J. Transp. Res. Board*, vol. no, p. 16, 2016.

[2] A. Ghiasi, O. Hussain, Z. Qian, and X. P. Li, "A mixed traffic capacity analysis and lane management model for connected automated vehicles: A Maekov chain method," *Transportation Research Part B: Methodological*, vol. 106, pp. 266–292, 2017.

[3] Y. Liu, J. Guo, J. Taplin, and Y. Wang, "Characteristic Analysis of Mixed Traffic Flow of Regular and Autonomous Vehicles Using Cellular Automata," *Journal of Advanced Transportation*, vol. 2017, Article ID 8142074, 10 pages, 2017.

[4] A. Talebpour and H. S. Mahmassani, "Influence of connected and autonomous vehicles on traffic flow stability and throughput," *Transportation Research Part C: Emerging Technologies*, vol. 71, pp. 143–163, 2016.

[5] N. Bekiaris-Liberis, C. Roncoli, and M. Papageorgiou, "Highway traffic state estimation per lane in the presence of connected vehicles," *Transportation Research Part B: Methodological*, vol. 106, pp. 1–28, 2017.

[6] D. Chen, S. Ahn, M. Chitturi, and D. A. Noyce, "Towards vehicle automation: Roadway capacity formulation for traffic mixed with regular and automated vehicles," *Transportation Research Part B: Methodological*, vol. 100, pp. 196–221, 2017.

[7] R. E. Chandler, R. Herman, and E. W. Montroll, "Traffic dynamics: studies in car following," *Operations Research*, vol. 6, pp. 165–184, 1958.

[8] W. Helly, "Simulation of bottlenecks in single-lane traffic flow," in *Theory of traffic flow*, pp. 207–238, Elsevier, Amsterdam, 1961.

[9] M. Bando, K. Hasebe, A. Nakayama, A. Shibata, and Y. Sugiyama, "Dynamical model of traffic congestion and numerical simulation," *Physical Review E: Statistical, Nonlinear, and Soft Matter Physics*, vol. 51, no. 2, pp. 1035–1042, 1995.

[10] M. Treiber, A. Hennecke, and D. Helbing, "Congested traffic states in empirical observations and microscopic simulations," *Physical Review E: Statistical, Nonlinear, and Soft Matter Physics*, vol. 62, no. 2, pp. 1805–1824, 2000.

[11] P. G. Gipps, "A behavioural car-following model for computer simulation," *Transportation Research Part B: Methodological*, vol. 15, no. 2, pp. 105–111, 1981.

[12] L. Evans and R. Rothery, *Experimental measurement of perceptual thresholds in car following. Highway Research Board, Washington District of Columbia, United States*, 1973.

[13] L. C. Davis, "Optimality and oscillations near the edge of stability in the dynamics of autonomous vehicle platoons," *Physica A: Statistical Mechanics and its Applications*, vol. 392, no. 17, pp. 3755–3764, 2013.

[14] P. Y. Li and A. Shrivastava, "Traffic flow stability induced by constant time headway policy for adaptive cruise control vehicles," *Transportation Research Part C: Emerging Technologies*, vol. 10, no. 4, pp. 275–301, 2002.

[15] G. Orosz, J. Moehlis, and F. Bullo, "Delayed car-following dynamics for human and robotic drivers," in *Proceedings of the ASME 2011 International Design Engineering Technical Conferences and Computers and Information in Engineering Conference, IDETC/CIE 2011*, pp. 529–538, USA, August 2011.

[16] L. Xiao and F. Gao, "Practical string stability of platoon of adaptive cruise control vehicles," *IEEE Transactions on Intelligent Transportation Systems*, vol. 12, no. 4, pp. 1184–1194, 2011.

[17] S. G. Hu, H. Y. Wen, L. Xu, and H. Fu, "Stability of platoon of adaptive cruise control vehicles with time delay," *Transportation Letters*, pp. 1–10, 2017.

[18] Z. Wang, G. Wu, and M. Barth, "Developing a distributed consensus-based Cooperative Adaptive Cruise Control (CACC) system," *J. Adv. Transp*, vol. 2017, 2017.

[19] M. Fountoulakis, N. Bekiaris-Liberis, C. Roncoli, I. Papamichail, and M. Papageorgiou, "Highway traffic state estimation with mixed connected and conventional vehicles: Microscopic simulation-based testing," *Transportation Research Part C: Emerging Technologies*, vol. 78, pp. 13–33, 2017.

[20] Q. Xin, N. Yang, R. Fu, S. Yu, and Z. Shi, "Impacts analysis of car following models considering variable vehicular gap policies," *Physica A: Statistical Mechanics and its Applications*, vol. 501, pp. 338–355, 2018.

[21] J. Sun, Z. Zheng, and J. Sun, "Stability analysis methods and their applicability to car-following models in conventional and connected environments," *Transportation Research Part B: Methodological*, vol. 109, pp. 212–237, 2018.

[22] B. Van Arem, C. J. G. Van Driel, and R. Visser, "The impact of cooperative adaptive cruise control on traffic-flow characteristics," *IEEE Transactions on Intelligent Transportation Systems*, vol. 7, no. 4, pp. 429–436, 2006.

[23] G. N. Bifulco, L. Pariota, F. Simonelli, and R. D. Pace, "Development and testing of a fully adaptive cruise control system," *Transportation Research Part C: Emerging Technologies*, vol. 29, pp. 156–170, 2013.

[24] J. Yi and R. Horowitz, "Macroscopic traffic flow propagation stability for adaptive cruise controlled vehicles," *Transportation Research Part C: Emerging Technologies*, vol. 14, no. 2, pp. 81–95, 2006.

[25] I. A. Ntousakis, I. K. Nikolos, and M. Papageorgiou, "On Microscopic Modelling of Adaptive Cruise Control Systems," *Transportation Research Procedia*, vol. 6, pp. 111–127, 2015.

[26] A. I. Delis, I. K. Nikolos, and M. Papageorgiou, "Macroscopic traffic flow modeling with adaptive cruise control: development and numerical solution," *Computers & Mathematics with Applications*, vol. 70, no. 8, pp. 1921–1947, 2015.

[27] L. Ye and T. Yamamoto, "Modeling connected and autonomous vehicles in heterogeneous traffic flow," *Physica A: Statistical Mechanics and its Applications*, vol. 490, pp. 269–277, 2018.

[28] W.-X. Zhu and H. M. Zhang, "Analysis of mixed traffic flow with human-driving and autonomous cars based on car-following model," *Physica A: Statistical Mechanics and its Applications*, vol. 496, pp. 274–285, 2018.

[29] H. Yeo, A. Skabardonis, J. Halkias, J. Colyar, and V. Alexiadis, "Oversaturated freeway flow algorithm for use in Next Generation Simulation," *Transportation Research Record*, no. 2088, pp. 68–79, 2008.

[30] N. Chen, M. Wang, T. Alkim, and B. van Arem, "A Robust Longitudinal Control Strategy of Platoons under Model Uncertainties and Time Delays," *Journal of Advanced Transportation*, vol. 2018, Article ID 9852721, 13 pages, 2018.

[31] N. Bekiaris-Liberis, C. Roncoli, and M. Papageorgiou, "Highway traffic state estimation with mixed connected and conventional vehicles," *IEEE Transactions on Intelligent Transportation Systems*, vol. 17, no. 12, pp. 3484–3497, 2016.

[32] V. A. C. van den Berg and E. T. Verhoef, "Autonomous cars and dynamic bottleneck congestion: the effects on capacity, value of time and preference heterogeneity," *Transportation Research Part B: Methodological*, vol. 94, pp. 43–60, 2016.

[33] P. Tientrakool, Y.-C. Ho, and N. F. Maxemchuk, "Highway capacity benefits from using vehicle-to-vehicle communication and sensors for collision avoidance," in *Proceedings of the IEEE 74th Vehicular Technology Conference, VTC Fall*, The Hilton San Francisco Union Square, San Francisco, USA, September 2011.

[34] J. Vander Werf, S. Shladover, M. Miller, and N. Kourjanskaia, "Effects of Adaptive Cruise Control Systems on Highway Traffic Flow Capacity," *Transportation Research Record*, vol. 1800, pp. 78–84, 2002.

[35] D. Ni, J. Li, S. Andrews, and H. Wang, "Preliminary estimate of highway capacity benefit attainable with IntelliDrive technologies," in *Proceedings of the 13th International IEEE Conference on Intelligent Transportation Systems, ITSC 2010*, pp. 819–824, Portugal, September 2010.

[36] T.-H. Chang and I.-S. Lai, "Analysis of characteristics of mixed traffic flow of autopilot vehicles and manual vehicles," *Transportation Research Part C: Emerging Technologies*, vol. 5, no. 6, pp. 333–348, 1997.

[37] P. Fernandes and U. Nunes, "Platooning with IVC-enabled autonomous vehicles: Strategies to mitigate communication delays, improve safety and traffic flow," *IEEE Transactions on Intelligent Transportation Systems*, vol. 13, no. 1, pp. 91–106, 2012.

[38] N. H. T. S. Administration, "FMVSS No. 150 Vehicle-To-Vehicle Communication Technology For Light Vehicles," Off. Regul. Anal. Eval. Natl. Cent. Stat. Anal., no. 150, 2016.

[39] Y. Li, H. Wang, W. Wang, L. Xing, S. Liu, and X. Wei, "Evaluation of the impacts of cooperative adaptive cruise control on reducing rear-end collision risks on freeways," *Accident Analysis & Prevention*, vol. 98, pp. 87–95, 2017.

[40] M. S. Rahman and M. Abdel-Aty, "Longitudinal safety evaluation of connected vehicles' platooning on expressways," *Accident Analysis & Prevention*, vol. 117, pp. 381–391, 2018.

[41] M. Zabat, N. Stabile, S. Farascaroli, and F. Browand, "The Aerodynamic Performance Of Platoons: A Final Report," Calif. Partners Adv. Transit Highw, 1995.

[42] Z. Wang, G. Wu, P. Hao, K. Boriboonsomsin, and M. Barth, "Developing a platoon-wide Eco-Cooperative Adaptive Cruise Control (CACC) system," in *Proceedings of the 28th IEEE Intelligent Vehicles Symposium, IV 2017*, pp. 1256–1261, USA, June 2017.

[43] M. Mamouei, I. Kaparias, and G. Halikias, "A framework for user- and system-oriented optimisation of fuel efficiency and traffic flow in Adaptive Cruise Control," *Transportation Research Part C: Emerging Technologies*, vol. 92, pp. 27–41, 2018.

[44] J. C. Hayward, "Near-Miss Determination Through," *Highw. Res. Board*, pp. 24–35, 1971.

[45] M. M. Minderhoud and P. H. L. Bovy, "Extended time-to-collision measures for road traffic safety assessment," *Accident Analysis & Prevention*, vol. 33, no. 1, pp. 89–97, 2001.

[46] K. Vogel, "A comparison of headway and time to collision as safety indicators," *Accident Analysis & Prevention*, vol. 35, no. 3, pp. 427–433, 2003.

[47] M. Barth et al., "Development of a Comprehensive Modal Emissions Model," *Natl. Coop. Highw. Res. Progr*, 2000.

[48] J. Koupal, H. Michaels, M. Cumberworth, C. Bailey, and D. Brzezinski, "EPAs plan for MOVES: a comprehensive mobile source emissions model," in *Proceedings of the 12th CRC On-Road Veh. Emiss. Work*, San Diego, CA.

[49] S. Hausberger, J. Rodler, P. Sturm, and M. Rexeis, "Emission factors for heavy-duty vehicles and validation by tunnel measurements," *Atmospheric Environment*, vol. 37, no. 37, pp. 5237–5245, 2003.

[50] K. Ahn, *Microscopic Fuel Consumption and Emission Modeling*, Virginia Tech University, 1998.

[51] K. Ahn, H. Rakha, A. Trani, and M. van Aerde, "Estimating vehicle fuel consumption and emissions based on instantaneous speed and acceleration levels," *Journal of Transportation Engineering*, vol. 128, no. 2, pp. 182–190, 2002.

[52] T.-Q. Tang, Z.-Y. Yi, and Q.-F. Lin, "Effects of signal light on the fuel consumption and emissions under car-following model," *Physica A: Statistical Mechanics and its Applications*, vol. 469, pp. 200–205, 2017.

[53] B. Khondaker and L. Kattan, "Variable speed limit: a microscopic analysis in a connected vehicle environment," *Transportation Research Part C: Emerging Technologies*, vol. 58, pp. 146–159, 2015.

7

Studying the Safety Impact of Autonomous Vehicles using Simulation-Based Surrogate Safety Measures

Mark Mario Morando,[1] Qingyun Tian,[2] Long T. Truong,[1] and Hai L. Vu[1]

[1]*Monash Institute of Transport Studies, Department of Civil Engineering, Monash University, Melbourne, VIC, Australia*
[2]*School of Civil and Environmental Engineering, Nanyang Technological University, Singapore*

Correspondence should be addressed to Long T. Truong; long.truong@monash.edu

Academic Editor: Abdelaziz Bensrhair

Autonomous vehicle (AV) technology has advanced rapidly in recent years with some automated features already available in vehicles on the market. AVs are expected to reduce traffic crashes as the majority of crashes are related to driver errors, fatigue, alcohol, or drugs. However, very little research has been conducted to estimate the safety impact of AVs. This paper aims to investigate the safety impacts of AVs using a simulation-based surrogate safety measure approach. To this end, safety impacts are explored through the number of conflicts extracted from the VISSIM traffic microsimulator using the Surrogate Safety Assessment Model (SSAM). Behaviours of human-driven vehicles (HVs) and AVs (level 4 automation) are modelled within the VISSIM's car-following model. The safety investigation is conducted for two case studies, that is, a signalised intersection and a roundabout, under various AV penetration rates. Results suggest that AVs improve safety significantly with high penetration rates, even when they travel with shorter headways to improve road capacity and reduce delay. For the signalised intersection, AVs reduce the number of conflicts by 20% to 65% with the AV penetration rates of between 50% and 100% (statistically significant at $p < 0.05$). For the roundabout, the number of conflicts is reduced by 29% to 64% with the 100% AV penetration rate (statistically significant at $p < 0.05$).

1. Introduction

Autonomous vehicle (AV) technology has advanced significantly in recent years. Automakers have already provided vehicles with some automated features (e.g., self-parking) and crash avoidance features such as automated braking, forward collision warning, lane departure warning, and blind spot monitoring [1, 2]. AV testing and piloting have begun in various countries. By 2014, AV testing on roadways has been legalised in four states in the US. In Australia, AV testing has been first introduced in South Australia's roadways in 2016 [3]. The market penetration rate of AVs is estimated to be between 24% and 87% by 2045 [4, 5].

AVs have the potential to significantly improve road safety as the majority of crashes are related to driver errors, fatigue, alcohol, or drugs [6–8]. It is also expected that AVs can travel with shorter headways due to improved safety, leading to increased road and intersection capacities [9, 10]. AVs would also provide improved mobility to the disabled, those who are too young to drive, and older people [11]. Other potential benefits of AVs include enhanced productive use of travel time, fewer emissions, better fuel efficiency, and reduced parking costs [12, 13].

Implementing AVs within the road network has the potential to significantly reduce the number of crashes caused by the drivers through the gradual removal of human control [12]. Already, many vehicle manufacturers are increasing the implementation of features such as adaptive cruise control and parking assistance that enables the vehicle to park itself with minimal human intervention [14]. Many of these driver assistance features are partially automated, meaning that driver intervention is still required. Although the implementation of automated features has increased in recent years, fully AVs are yet to be legally deployed on a large scale within the road network globally. Safety benefits of fully AVs would not be maximised without a high penetration rate of AVs [12].

It should be noted that even as continuous developments are made to AV software by frontrunners such as Google and Tesla, questions arise with regard to their ability to produce a fleet of fully automated driving systems that are safe enough to negate the need for human intervention [15]. Recent research suggested that current levels of technology from frontrunners like Google and Mercedes-Benz still experience malfunctions that require the need for human intervention [16]. Further issues arise with regard to the amount of miles the cars will need to be tested for in order to obtain statistically relevant data to demonstrate the vehicles' safety. Before these vehicles are able to be deployed within road networks, it is expected that statistically relevant data must be collected to certain standards. This may be a major setback as the amount of testing required to validate a particular failure rate could be of great magnitude [15]. In addition, the safety implications in terms of network security will need to be addressed as cyberattacks on connected AVs are a major safety threat [15]. Due to the connected nature of such advanced vehicle technology, an area that needs to be addressed is the potential for such attacks causing further safety risks, counteracting the theorised safety improvement from the elimination of human control.

Several studies have attempted to examine safety benefits of AVs using different approaches. Fagnant and Kockelman [14] assumed near elimination of human errors, which is related to main factors of over 90% of crashes in the US, from AV technology. Rau et al. [17] developed a method to identify crashes, which could be addressed by AV technology, by mapping automated vehicle functions to five layers of crash information (location, precrash scenario, driving conditions, travel speed, and driver condition). Real-world data from AV testing in California has increasingly been utilised in recent research [16, 18–20]. For example, Schoettle and Sivak [18] found that AVs were not at fault in any crashes and the overall injury severity was lower for crashes involving AVs than for crashes involving human-driven vehicles (HVs).

Potential safety benefits of different connected vehicle technology and control have been evaluated in previous research. Tian et al. [21] showed that platoon-based driving can significantly reduce the chain collisions and alleviate their severity. Optimising vehicle trajectories via adjusting the platoon level reaction time and speed of the lead vehicle was shown to provide both efficiency and safety benefits [22]. It is worth noting that these studies only considered single lane traffic. Zhou et al. [23] investigated the impacts of connected AVs on the performance of freeway merging. Safety impacts were also considered, but indirectly through impacts on speed dispersion. As previous research tends to focus on freeways, there is a lack of investigations on other network settings, such as signalised intersections and roundabouts.

Using traffic microsimulation models of various network settings, Kockelman et al. [24] found that in general AVs reduce the number of potential conflicts based on surrogate safety measures and thus improve safety. In their study, a car-following model based on the Wiedemann 74 and 99 models was adopted for HVs. This model was then modified to model behaviours of AVs. For example, as AV behaviours are expected to be less stochastic, the variance of the driver random terms was set to zero. Other parameters, such as minimum acceptable gap for merging or turning, sight distance, and lane change preferences, were modified to make the AV behaviours more conservative as automakers would be very unlikely to make AVs aggressive due to their potential liability. In other words, AVs were modelled to be more cautious than human drivers and therefore had fewer potential conflicts. However, there is another possible scenario, particularly in the long run, where AV behaviours could be less conservative due to shorter headways and more aggressive acceleration as AVs are anticipated to increase road and intersection capacities. Thus, it is also important to investigate the safety impacts of AVs with such behaviours.

This paper aims to investigate safety impacts of AVs using simulation-based surrogate safety measures where AVs are modelled with anticipated behaviours, such as shorter headways. Safety performance of AVs is considered with varying penetration rates in two case studies, that is, a signalised intersection and a roundabout. The delay reduction at increasing penetration rates will also be considered.

The remainder of this paper is organized as follows. In the next section, the methodology will be outlined, followed by key results and discussion. The paper is concluded with a summary of key findings and directions for future research.

2. Methodology

To understand the safety implications of AVs, VISSIM [25] was used as the traffic microsimulation platform while Surrogate Safety Assessment Model (SSAM) [26] was employed to extract the number of potential conflicts based on surrogate safety measures from simulated data. The following subsections present the simulation platform, surrogate safety measure extraction, and then the case studies.

2.1. Simulation Platform. Behaviours of human-driven vehicles (HVs) were modelled using VISSIM's Wiedemann 99 car-following model with default parameters. This provides a reasonable base model for human drivers. This paper assumes that AVs are fully automated with level 4 automation [27]. Previous studies have shown that parameters of the VISSIM car-following model can be modified to model behaviours of AVs [2, 28]. Two sets of AV parameters adopted from Atkins [28] and PTV [29] were considered in this study. Table 1 presents a set of HV parameters and two sets of AV parameters.

Parameters for AVs reflect more assertive behaviours, such as shorter standstill distance (CC0) and shorter safety distance (lower headway CC1 and following variation CC2). Thus, shorter gaps are expected for AVs (Figure 1). Smaller values of the negative following threshold (CC4) and positive following threshold (CC5) reflect a more sensitive reaction of AVs to the acceleration or deceleration of the preceding vehicle. As AVs can strictly follow the desired speed without oscillation, CC6 is set as zero. According to Atkins [28], AVs can also have more aggressive acceleration (higher CC7 and CC8) and a higher number of observed vehicles due to connected vehicle technology. Although the exact behaviours

Table 1: Parameters for HV and AV behaviours.

Parameter	Description	HV	AV-1	AV-2
CC0 standstill distance (m)	The desired distance between stopped vehicles	1.5	0.5	0.75
CC1 headway time (s)	The gap in seconds that a vehicle keeps	0.9	0.5	0.45
CC2 following variation (m)	The distance in addition to the allowed safety distance that is permissible before the vehicle-drive unit moves closer to the proceeding vehicle	4	0	2
CC4 negative following threshold	Control speed differences during car following	−0.35	0	−0.1
CC5 positive following threshold	Control speed differences during car following	0.35	0	0.1
CC6 speed dependency of oscillation	Influence of distance on speed oscillation (the variation of speed around the desired speed)	11.44	0	0
CC7 (m/s^2)	Influence of vehicle acceleration during car-following oscillation	0.25	0.45	0.25
CC8 (m/s^2)	Desired acceleration when starting from standstill	3.5	3.9	3.5
Look-ahead distance	Number of observed vehicles the model will look ahead at	2	10	2

Note. HVs are VISSIM default parameters; AV-1 are adopted from Atkins [28]; AV-2 are adopted from PTV [29].

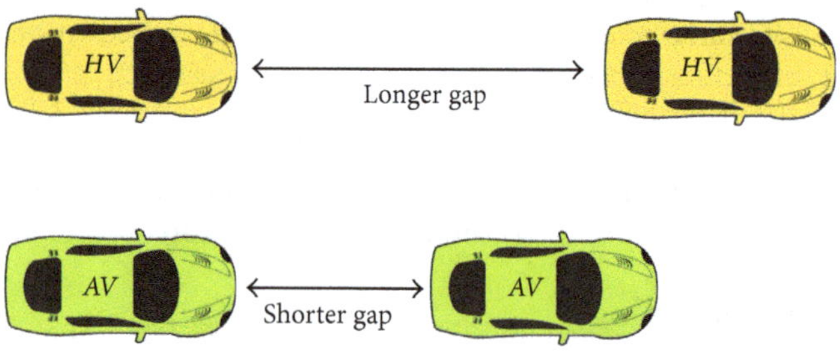

Figure 1: Shorter gaps between AVs.

of AVs are largely unknown at this stage, the modification of these parameters should be able to reflect the anticipated AV behaviours.

2.2. Surrogate Safety Measures. Surrogate safety measures are tools that can be used to assess potential conflicts of a road network. Surrogate safety measures, such as time to collision (TTC), postencroachment time (PET), deceleration rate (DR), gap time (GT), and proportion of stopping distance (PSD), are useful measurements of the safety implications [30–33]. TTC, defined as the expected time for two vehicles to collide if they remain on the same path and at the same speed, is a widely used surrogate safety measure. Figure 2 illustrates situations for TTC calculations [34]. The following equation can be used to calculate TTC:

$$\text{TTC} = \begin{cases} \dfrac{d_2}{v_2} & \text{if } \dfrac{d_1}{v_1} < \dfrac{d_2}{v_2} < \dfrac{d_1 + l_1 + w_2}{v_1} \text{ (side)} \\ \dfrac{d_1}{v_1} & \text{if } \dfrac{d_2}{v_2} < \dfrac{d_1}{v_1} < \dfrac{d_2 + l_2 + w_1}{v_2} \text{ (side)} \\ \dfrac{X_1 - X_2 - l_1}{v_2 - v_1} & \text{if } v_2 > v_1 \text{ (rear end)} \\ \dfrac{X_1 - X_2}{v_1 + v_2} & \text{(head on)}, \end{cases} \tag{1}$$

where v_1 and v_2 are vehicle speeds, l_1 and l_2 are vehicle lengths, w_1 and w_2 are vehicle widths, X_1 and X_2 are vehicle positions, and d_1 and d_2 are distances to conflict areas.

PET, defined as the time difference between when the leading vehicle occupies a location and when the trailing vehicle arrives at this location, is usually used to identify conflicts in combination with TTC [35]. Figure 3 illustrates situations for PET calculations [34].

It is generally considered that a TTC, equal to or less than 1.5 seconds, would result in an unsafe situation [26, 36, 37]. This threshold of 1.5 seconds is therefore applied for potential conflicts involving HVs (HV-HV and HV-AV). It can be argued that the TTC threshold for conflicts between AVs would be lower due to their ability to react to situations a lot faster that their human counterparts, particularly with connected vehicle technology. Therefore, a sensitivity analysis for the TTC threshold of AV-AV conflicts was conducted with two values (1 second and 0.75 seconds). A PET threshold of 5 seconds is used as the default value in SSAM [26]. Overall, a potential conflict is identified when the calculated TTC and PET are equal to or smaller than their respective threshold.

SSAM was developed for analysing simulation-based surrogate safety measures [26]. It works through the analysis of trajectory files (.trj) obtained from various traffic microsimulation programs, such as Aimsun, Paramics, and VISSIM. In this study, the trajectory files were obtained from VISSIM, which were then imported into the SSAM to identify the number of potential conflicts. Although SSAM does not have a built-in function to identify conflicts by vehicle types, it does provide the ID of vehicles involved in a conflict. Vehicle IDs and corresponding vehicle types can be extracted from VISSIM. Thus, it was possible to identify the number of potential conflicts by vehicle types (HV-HV, HV-AV, and AV-AV).

2.3. Case Studies. Using the aforementioned approach, safety impacts of autonomous vehicles were investigated in two case studies, that is, a signalised intersection and a roundabout. Details of each case study are presented in the following subsections.

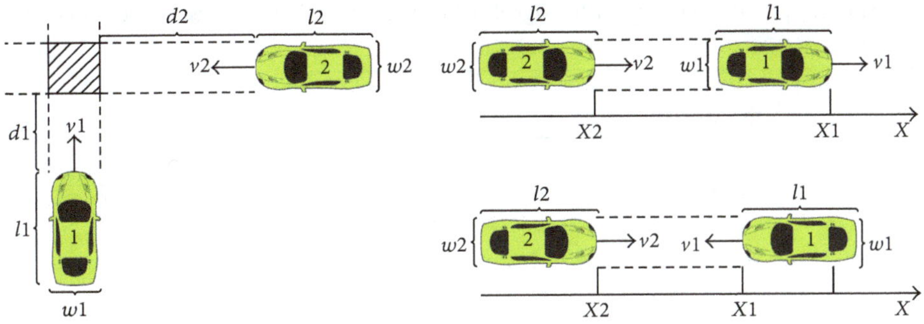

FIGURE 2: Time to collision (TTC).

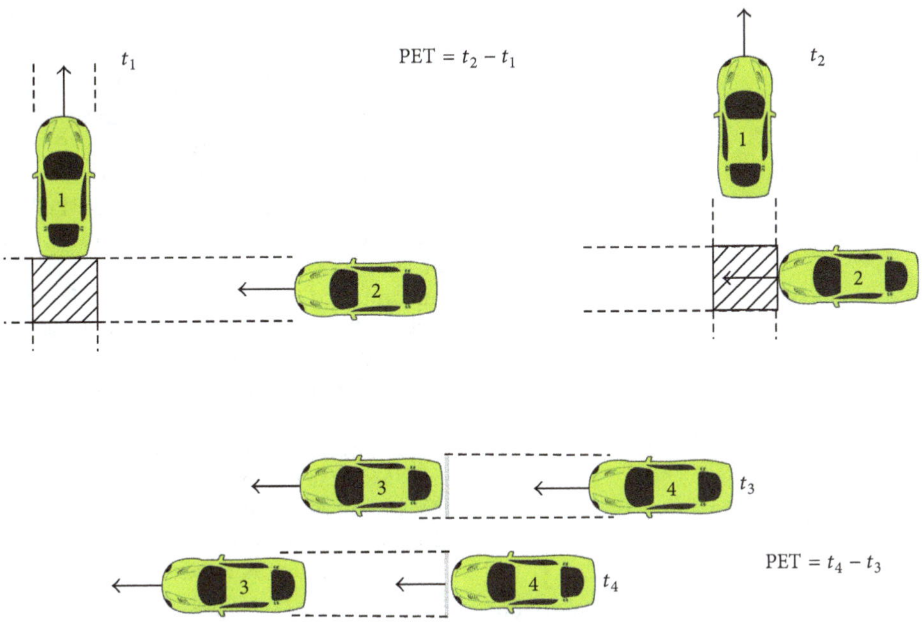

FIGURE 3: Postencroachment time (PET).

2.3.1. Signalised Intersection. The intersection of Ferntree Gully Road and Blackburn Road in Melbourne, Australia, was chosen as the first case study. Both roads are 3-lane arterial roads with several bus routes. The intersection also has a right turning lane on all four approaches and a left turning slip lane on three approaches. To ensure the model replicates the intersection as closely as possible, SCATS data (traffic volume and traffic signal phasing) for the peak period between 8 and 9 a.m. was obtained from VicRoads to calibrate the model. Traffic volumes per intersection approach range from 760 to 2260 veh/h. The traffic composition was 95% cars and 5% trucks. The bus routes with far-side and midblock bus stops were also modelled with scheduled arrival times. The intersection layout is illustrated in Figure 4.

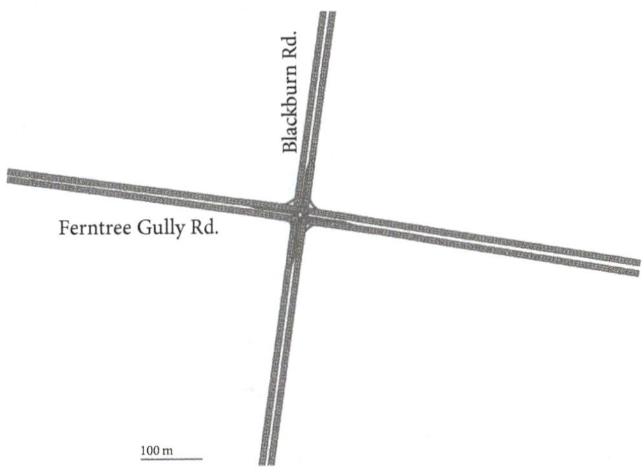

FIGURE 4: Signalised intersection case study.

The variance in desired speed distributions is another key element that varies from HVs and AVs within the network. It can be assumed that the variation in the desired speed distribution of a human driver would be much larger than that of an AV due to the nature of AVs having much more precise throttle control. Based on field conditions, the desired speed distribution for HVs ranges between 65 km/h and 75 km/h. AVs' desired speed distribution has the same mean, but a narrower range between 69 km/h and 71 km/h.

Table 2: Conflicts by types and AV penetration rates for the signalised intersection (AV-1 parameters and 0.75 s TTC threshold for AV-AV conflicts).

Penetration rate	Conflict type			Total
	AV-AV	AV-HV	HV-HV	
0% (base case)			863.3	863.3
25%	23.9	326.7	365.7	716.2
50%	85.8	389.1	148.2	623.1
75%	182.1	286.4	41.7	510.1
100%	411.9			411.9

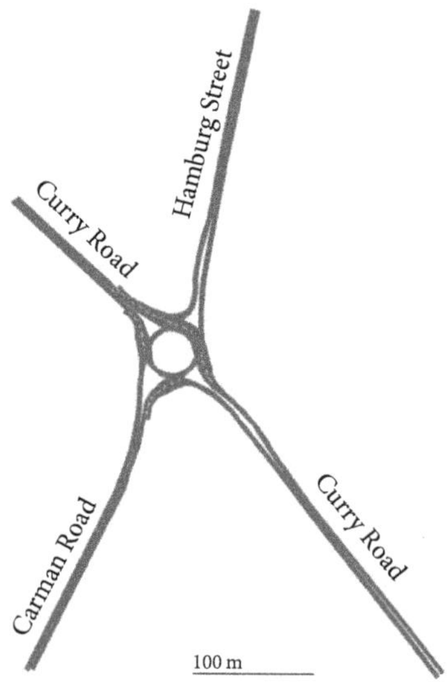

Figure 5: Roundabout case study.

2.3.2. Roundabout. The second case study was based on a roundabout model that was provided with SSAM. This is a four-legged roundabout in Schenectady, New York. Traffic volumes per lane on each approach range from about 490 to 1050 veh/h. Traffic composition was 96% cars and 4% trucks. The desired speed distribution for HVs ranges between 48 km/h and 58 km/h. Like the first case study, AVs' desired speed distribution has a narrower range between 52 km/h and 54 km/h. It is noted that desired speeds within the roundabout were modelled to be lower (between 20 and 25 km/h), reflecting reduced speed behaviours. Figure 5 depicts the layout of the roundabout.

2.3.3. Scenarios. Varying penetration rates of AVs, including 0% (base case with 100% HVs), 25%, 50%, 75%, and 100%, were analysed for each case study. For the signalised intersection case study, all buses were modelled with AV behaviours in the 100% AV scenarios, but with HV behaviours in other scenarios. In total, there were 18 scenarios. For each scenario, the simulation time was set to 1 hour, excluding a warm-up time of 10 mins. All scenarios were modelled with 10 simulation time steps per second. This maximum simulation time step was used to better accommodate AVs' shorter headways. Due to the stochastic nature of microsimulation, 20 runs with different seed numbers were performed for each scenario to obtain reliable outputs [38].

3. Results and Discussion

3.1. AV-1 Parameters

3.1.1. Signalised Intersection. Results for the first AV parameter set (AV-1) with the TTC threshold for AV-AV conflicts of 0.75 seconds are presented in Table 2. As expected, the number of conflicts between AVs and AVs increases and the number of conflicts between HVs and HVs decreases with increasing AV penetration rates. At the penetration rate of 50%, the number of conflicts between AVs and HVs is maximum. Conflicts between AVs and HVs appear more frequently with the 25% AV penetration rate than with the 75% AV penetration rate. The reason may be that, with the 75% AV penetration rate, an AV is more likely to have a conflict with another AV due to shorter headways.

Figures 6(a) and 6(b) depict the total number of conflicts and associated 95% confidence intervals (CIs) by varying AV penetration rates for both the 0.75-second and 1-second TTC thresholds. It is evident that the number of conflicts in all AV scenarios is lower compared to the base case. Safety benefits from AVs range between 15% and 52%, depending on penetration rates and TTC threshold for AV-AV conflicts. Moreover, the number of conflicts with the 50%, 75%, and 100% AV penetration rates is significantly lower compared to the base case, at $p < 0.05$. With regard to the TTC threshold of 1 second for AV-AV conflicts, the number of conflicts with the 75% penetration rate is slightly lower than that with the 50% and 100% penetration rates (Figure 6(a)). However, these differences are not statistically significant. Compared to the base case, the number of conflicts is reduced by around 20% with the 100% AV penetration rate. These results suggest that safety benefits of AVs might not be significant until a high penetration rate is achieved. However, results with the TTC threshold of 0.75 seconds for AV-AV conflicts show a consistent decline in the number of vehicle conflicts as the penetration rate is increased to 100% (Figure 6(b)). At a penetration rate of 50% and above, the decline is statistically significant at $p < 0.05$. This is further evidence that higher penetration rates of AVs are likely to provide safety benefits.

Overall, while the results demonstrate that the TTC threshold for AV-AV conflicts is a major factor in identifying potential conflicts, they consistently show the safety benefits of AVs, particularly with high AV penetration rates.

3.1.2. Roundabout. Table 3 summarises the average number of conflicts by different types and AV penetration rates for

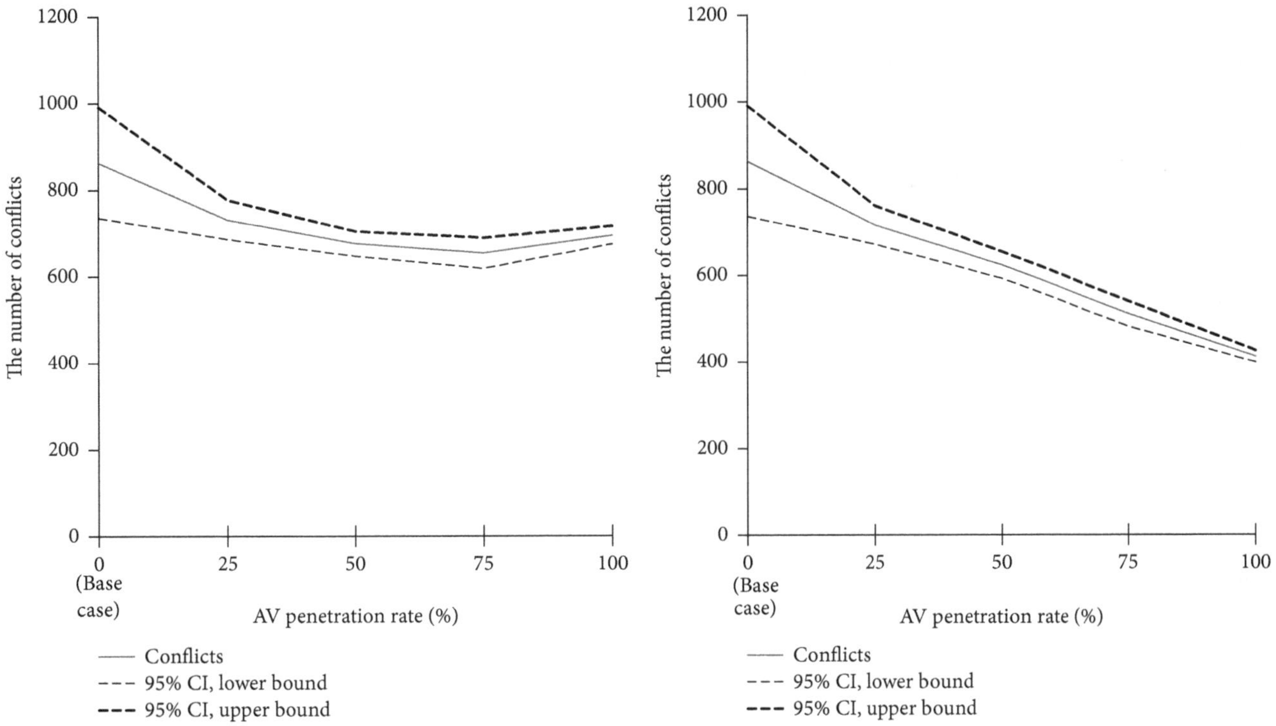

(a) 1 s TTC threshold for AV-AV conflicts

(b) 0.75 s TTC threshold for AV-AV conflicts

FIGURE 6: Total number of conflicts by AV penetration rate for the signalised intersection (AV-1 parameters).

TABLE 3: Conflicts by types and AV penetration rates for the roundabout (AV-1 parameters and 0.75 s TTC threshold for AV-AV conflicts).

Penetration rate	Conflict type			Total
	AV-AV	AV-HV	HV-HV	
0% (base case)			449.1	449.1
25%	14.8	270.1	165.0	449.9
50%	48.7	275.9	83.0	407.6
75%	108.5	204.7	18.7	331.9
100%	172.9			172.9

the roundabout case study (TTC threshold of 0.75 seconds for AV-AV conflicts). As the penetration rate increases, the number of conflicts between AVs increases whereas the number of conflicts between HVs decreases. Similar to results of the signalised intersection, the number of conflicts between AV and HV is highest with the 50% penetration rate. These results are expected. The total number of conflicts by varying AV penetration rates is further depicted in Figure 7.

Figures 7(a) and 7(b) with different TTC thresholds show that the number of conflicts increases between the base case and 25% AV penetration rate but decreases steadily between 25% and 75% AV penetration rates. In particular, the number of conflicts with 100% AV penetration rate is significantly lower at $p < 0.05$, compared to other AV penetration rates. Considering the TTC threshold of 1 second for AV-AV conflicts, AVs provide safety benefits by 6% and 29% with the penetration rate of 75% and 100%, respectively, compared to the base case (Figure 7(a)). With regard to the AV-AV TTC threshold of 0.75 seconds (Figure 7(b)), the results show increased safety benefits of between 9% and 62% for penetration rates of 50% and 100%, respectively. These results also confirm that a high AV penetration rate might be required to deliver AVs' anticipated safety benefits.

Different patterns in conflicts by penetration rates between the two case studies suggest that safety impacts of AVs might vary with different networks. Thus, it is important for future research to conduct further investigations with various network settings and traffic conditions. With the 100% AV penetration rate, safety benefits are slightly higher with the roundabout than with the signalised intersection (62% versus 52%). This suggests that roundabouts might be preferred in terms of safety when all vehicles are AVs.

3.1.3. Delay Results. Figure 8 shows the delay results at varying penetration rates. It is evident that as the penetration rate of AVs is increased, there is a clear decrease in delay for both the roundabout and the signalised intersection. Furthermore, the decreases in delay with increasing penetration rates are statistically significant at $p < 0.05$. This is expected given that AVs were modelled to travel with shorter headways.

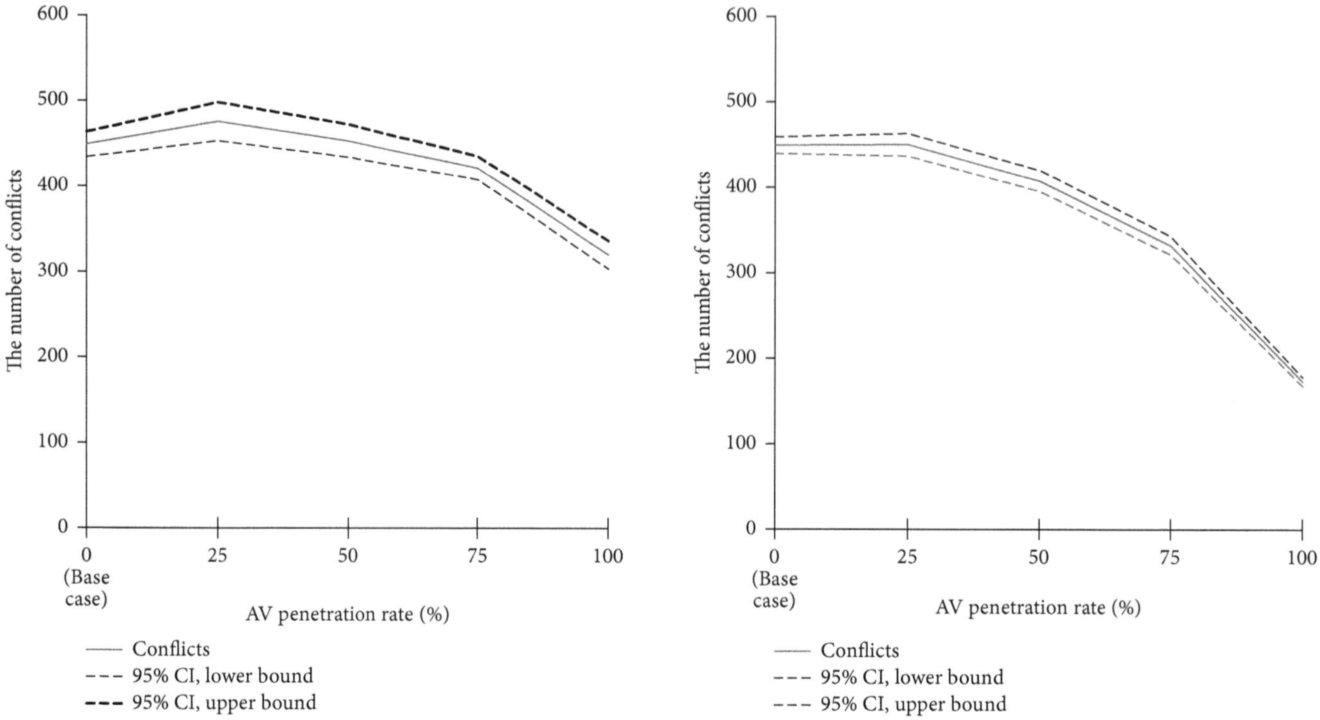

(a) 1 s TTC threshold for AV-AV conflicts

(b) 0.75 s TTC threshold for AV-AV conflicts

FIGURE 7: Total number of conflicts by AV penetration rate for the roundabout (AV-1 parameters).

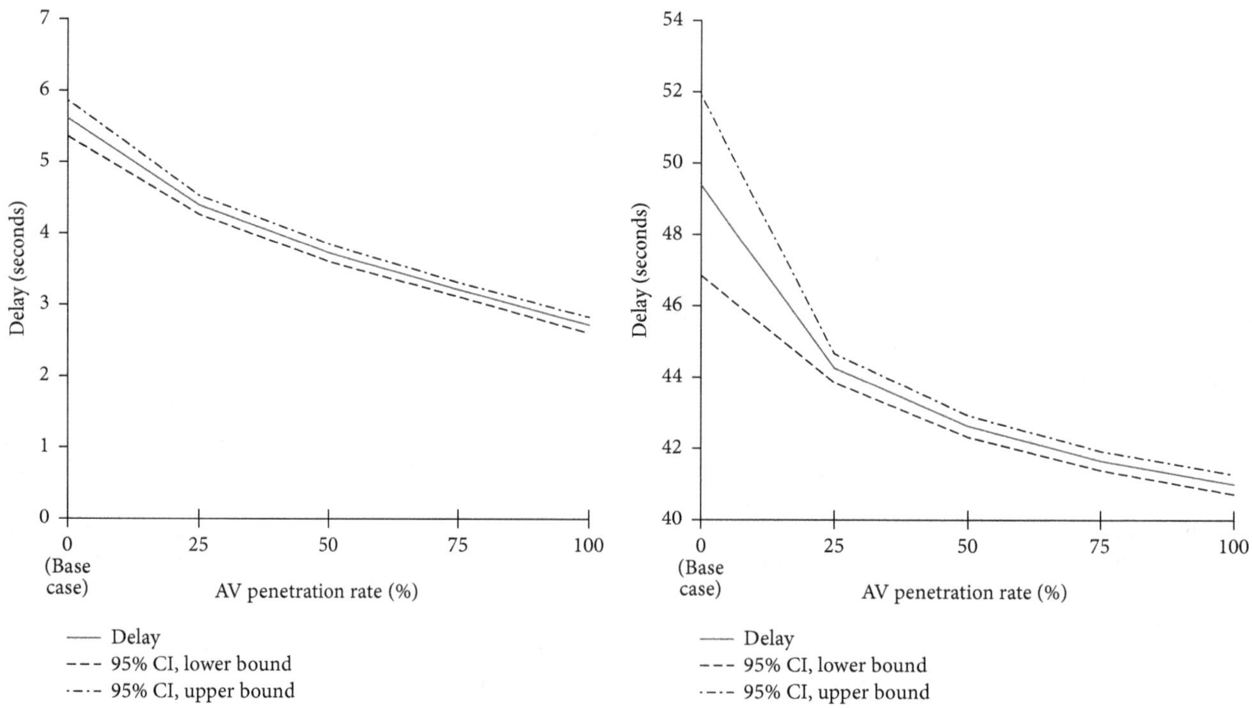

(a) Roundabout delay

(b) Signalised intersection delay

FIGURE 8: Average roundabout and intersection delay by penetration rate.

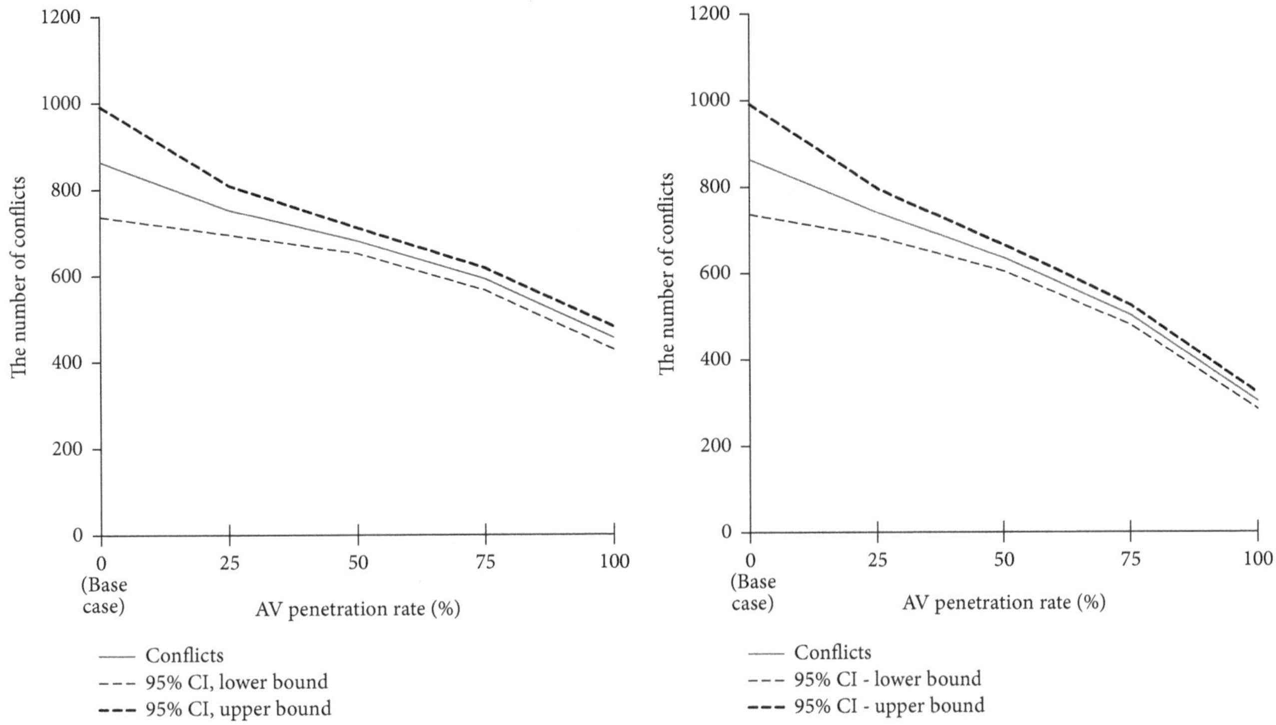

(a) 1 s TTC threshold for AV-AV conflicts

(b) 0.75 s TTC threshold for AV-AV conflicts

FIGURE 9: Total number of conflicts by AV penetration rate for the signalised intersection (AV-2 parameters).

Overall, results suggest that safety benefits of AVs can be achieved without compromising positive impacts on capacity associated with shorter headways.

3.2. AV-2 Parameters. Results with the AV-2 parameters are presented in Figures 9 and 10 for the signalised intersection and the roundabout, respectively. In general, there are similar patterns in safety benefits of AVs between the two sets of AV parameters. For the signalised intersection, the number of conflicts decreases as the AV penetration rate increases (Figure 9). Particularly with the 100% AV penetration rate, the number of conflicts continues to decrease significantly. The number of conflicts with the 100% penetration rate is significantly lower compared to all other penetration rates. Compared to the base case, the number of conflicts with the penetration rate of 50% or more is significantly lower at $p < 0.05$, with improvements ranging from 21% to 65%. For the roundabout, the number of conflicts is reduced significantly with the AV penetration rate of 75% and 100% (Figure 10). Particularly with the 100% AV penetration rate, the number of conflicts is significantly lower at $p < 0.05$ with improvements of 32% to 64% compared to the base case, depending on the AV-AV TTC threshold.

The benefits obtained from AV-2 parameters are higher compared to those from AV-1 parameters (maximum benefits of 64% versus 62% for the roundabout and 65% versus 52% for the signalised intersection). It is noted that the AV-2 parameter set has higher standstill and following variation distances whereas the AV-1 parameter set has slightly higher headway, but more aggressive acceleration. Nevertheless, both AV parameter sets suggest safety benefits of AVs with high penetration rates.

Similarly, with the AV-2 set of parameters, at an increasing penetration rate, there are delay reductions for both the roundabout and the signalised intersection cases (Figure 11). The results in both cases show that reductions in delay with increasing penetration rates of AVs are statistically significant at $p < 0.05$. This further advocates the notion that at increasing AV penetration rates the system will become more efficient as well as safer.

4. Conclusion

This paper has investigated the safety impact of autonomous vehicles (AVs) using traffic microsimulation and Surrogate Safety Assessment Model (SSAM). Safety performance of AVs (level 4 automation, fully automated) in signalised intersection and roundabout case studies was explored through the number of potential conflicts based on time to collision (TTC) and postencroachment time (PET). VISSIM was adopted as the traffic microsimulation platform to model behaviours of human-driven vehicles (HVs) and AVs. More assertive behaviours of AVs, such as shorter headways and more aggressive acceleration, were explicitly considered. Results suggest that AVs improve safety significantly with high penetration rates, even when they travel with shorter headways to improve road capacity and reduce delay. For the signalised intersection, AVs reduce the number of conflicts by 20% to 65% with the penetration rates between 50% and 100% (statistically significant at $p < 0.05$). For the roundabout,

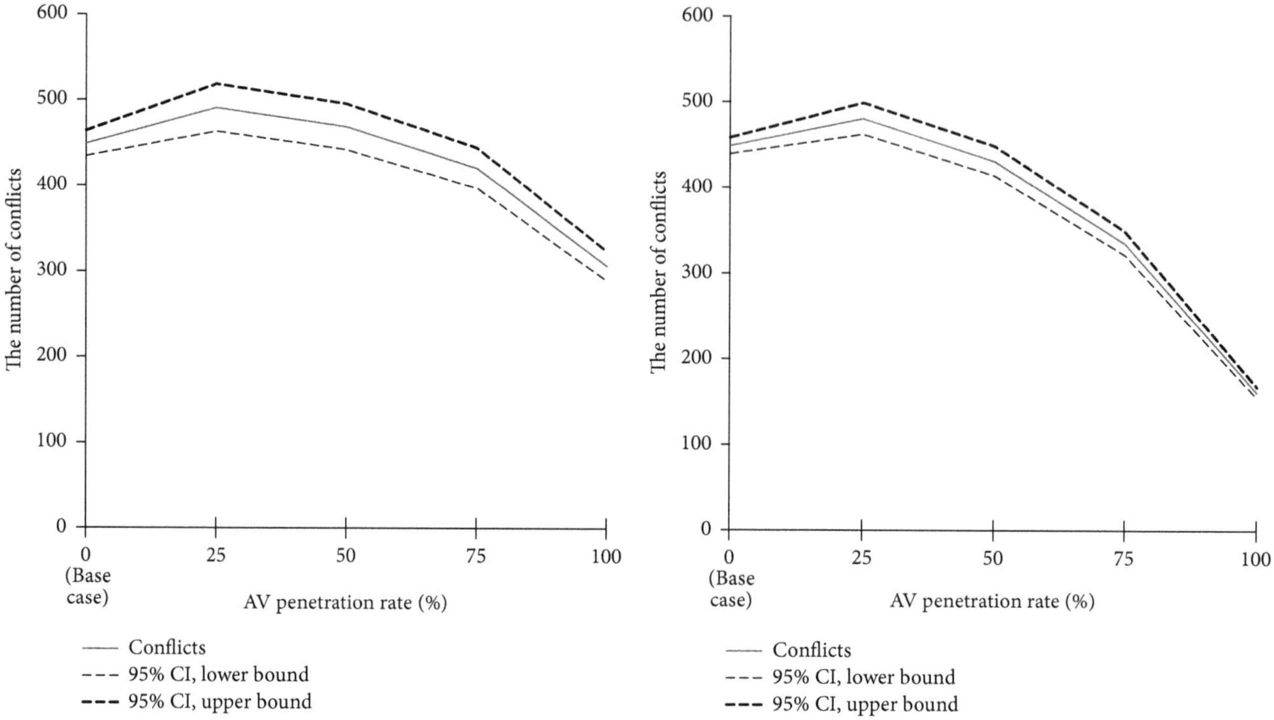

(a) 1 s TTC threshold for AV-AV conflicts

(b) 0.75 s TTC threshold for AV-AV conflicts

FIGURE 10: Total number of conflicts by AV penetration rate for the roundabout (AV-2 parameters).

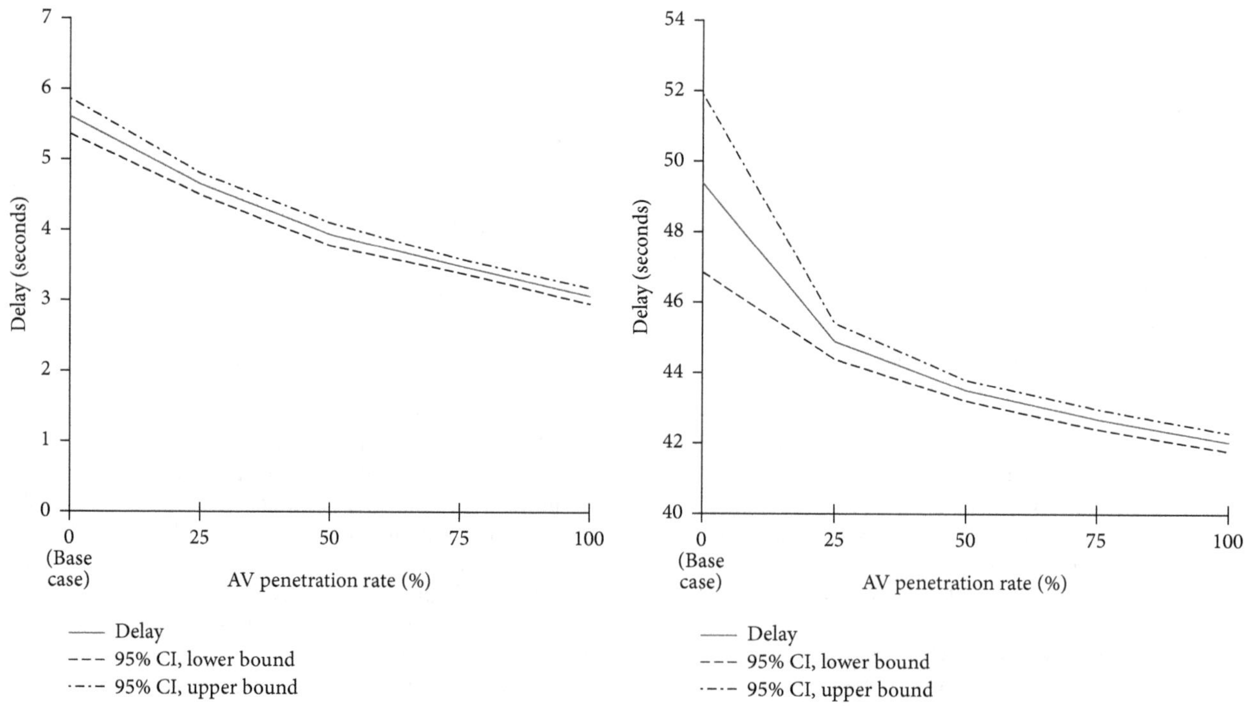

(a) Roundabout delay

(b) Signalised intersection delay

FIGURE 11: Delay results at an increasing AV penetration rate.

reductions of between 29% and 64%, in terms of conflicts, are evident with the 100% AV penetration rate (statistically significant at $p < 0.05$). An implication of these findings is that a high AV penetration rate might be required to deliver AVs' anticipated safety benefits.

The simulation-based approach presented in this paper provides an important tool to evaluate safety impacts of AVs, particularly when there has been very limited empirical data on safety performance of AVs. Nevertheless, there are limitations in the proposed approach, which should be addressed in future research. First, the ability to replicate how AVs will act within a real-world road network is limited due to the fact that AV technology is still being developed. Particularly, interactions between human drivers and AVs are largely unknown. Although this study modelled AV behaviours by modifying VISSIM's car-following model in accordance with the recent literature, there is a clear scope to develop a more realistic model for connected AVs. It is also noted that safety benefits of AVs would be higher with connected vehicle technology and control. Future research should also explore the impacts of vehicle-to-vehicle (V2V) safety technology, vehicle platooning, and V2V and vehicle trajectory control. Next, potential conflicts in this analysis were solely based on TTC and PET. The use of other emerging surrogate safety measures, such as Aggregated Crash Index (ACI) [39], should be considered to increase the validity of the approach. It might be necessary to develop new surrogate safety measures for AVs due to their different behaviours. Finally, the study only considered two case studies, a signalised intersection and a roundabout. Therefore, to draw more reliable conclusions about the overall safety impacts of AVs on a large road network, more testing with various network settings under wide-ranging traffic conditions and AV penetration rates might be required.

Disclosure

An earlier version of this paper was presented at the Australasian Transport Research Forum (ATRF 2017) in Auckland, New Zealand.

Conflicts of Interest

The authors declare that there are no conflicts of interest regarding the publication of this paper.

References

[1] HLDI, *Honda Accord collision avoidance features: an update. Bulletin Vol. 31. No. 16*, Highway Loss Data Institute, Insurance Institute for Highway Safety, Arlington, Va, USA, 2014.

[2] J. Bierstedt, A. Gooze, C. Gray, J. Peterman, L. Raykin, and J. Walters, "Effects of Next-Generation Vehicles on Travel Demand and Highway Capacity," FP Think, 2014.

[3] DPTI, *SA becomes first Australian jurisdiction to allow on-road driverless car trials*, Department of Planning, Transport and Infrastructure, 2016.

[4] T. Litman, *Autonomous Vehicle Implementation Predictions: Implications for Transport Planning*, Victoria Transport Policy Institute, 2015.

[5] P. Bansal and K. M. Kockelman, "Forecasting Americans' long-term adoption of connected and autonomous vehicle technologies," *Transportation Research Part A: Policy and Practice*, vol. 95, pp. 49–63, 2017.

[6] BITRE, *Fatal road crashes in Australia in the 1990s and 2000s: crash types and major factors*, Bureau of Infrastructure, Transport and Regional Economics, Canberra, Australia, 2011.

[7] NHTSA, *National Motor Vehicle Crash Causation Survey*, National Highway Traffic Safety Administration, U.S. Department of Transportation, 2008.

[8] S. E. Shladover, "Cooperative (rather than autonomous) vehicle-highway automation systems," *IEEE Intelligent Transportation Systems Magazine*, vol. 1, no. 1, pp. 10–19, 2009.

[9] R. Hoogendoorn, B. Van Arem, and S. Hoogendoorn, "Automated driving, traffic flow efficiency, and human factors," *Transportation Research Record*, vol. 2422, pp. 113–120, 2014.

[10] M. A. S. Kamal, J.-I. Imura, T. Hayakawa, A. Ohata, and K. Aihara, "A vehicle-intersection coordination scheme for smooth flows of traffic without using traffic lights," *IEEE Transactions on Intelligent Transportation Systems*, vol. 16, no. 3, pp. 1136–1147, 2015.

[11] L. T. Truong, C. de Gruyter, G. Currie, and A. Delbosc, "Estimating the trip generation impacts of autonomous vehicles on car travel in Victoria, Australia," *Transportation*, vol. 44, no. 6, pp. 1279–1292, 2017.

[12] D. Milakis, B. Van Arem, and B. Vanwee, "Policy and society related implications of automated driving: A review of literature and directions for future research," *Journal of Intelligent Transportation Systems: Technology, Planning, and Operations*, vol. 21, no. 4, pp. 324–348, 2017.

[13] S. Childress, B. Nichols, B. Charlton, and S. Coe, "Using an activity-based model to explore possible impacts of automated vehicles," *Transportation Research Record: Journal of the Transportation Research Board*, vol. 2493, pp. 99–106, 2015.

[14] D. J. Fagnant and K. Kockelman, "Preparing a nation for autonomous vehicles: Opportunities, barriers and policy recommendations," *Transportation Research Part A: Policy and Practice*, vol. 77, pp. 167–181, 2015.

[15] P. Koopman and M. Wagner, "Autonomous vehicle safety: an interdisciplinary challenge," *IEEE Intelligent Transportation Systems Magazine*, vol. 9, no. 1, pp. 90–96, 2017.

[16] V. V. Dixit, S. Chand, and D. J. Nair, "Autonomous vehicles: disengagements, accidents and reaction times," *PLoS ONE*, vol. 11, no. 12, Article ID 0168054, 2016.

[17] P. Rau, M. Yanagisawa, and W. G. Najm, "Target crash population of automated vehicles," in *Proceedings of the In 24th International Technical Conference on the Enhanced Safety of Vehicles (ESV)*, Gothenburg, Sweden, 2015.

[18] B. Schoettle and M. Sivak, *A Preliminary Analysis of Real-World Crashes Involving Self-Driving Vehicles*, The University of Michigan, Ann Arbor, Mich, USA, 2015.

[19] P. Bhavsar, P. Das, M. Paugh, K. Dey, and M. Chowdhury, "Risk Analysis of Autonomous Vehicles in Mixed Traffic Streams," *Transportation Research Record*, vol. 2625, pp. 51–61, 2017.

[20] F. M. Favarò, N. Nader, S. O. Eurich, M. Tripp, and N. Varadaraju, "Examining accident reports involving autonomous vehicles in California," *PLoS ONE*, vol. 12, no. 9, p. e0184952, 2017.

[21] D. Tian, J. Zhou, Y. Wang, Z. Sheng, H. Xia, and Z. Yi, "Modeling chain collisions in vehicular networks with variable penetration rates," *Transportation Research Part C: Emerging Technologies*, vol. 69, pp. 36–59, 2016.

[22] Y. Wei, C. Avcı, J. Liu et al., "Dynamic programming-based multi-vehicle longitudinal trajectory optimization with simplified car following models," *Transportation Research Part B: Methodological*, vol. 106, pp. 102–129, 2017.

[23] M. Zhou, X. Qu, and S. Jin, "On the impact of cooperative autonomous vehicles in improving freeway merging: a modified intelligent driver model-based approach," *IEEE Transactions on Intelligent Transportation Systems*, vol. 18, no. 6, pp. 1422–1428, 2017.

[24] K. Kockelman, P. Avery, P. Bansal et al., "Implications of Connected and Automated Vehicles on the Safety and Operations of Roadway Networks," Final Report, Texas Department of Transportation, Austin, Tex, USA, 2016.

[25] PTV, "VISSIM 9 - User Manual," PTV AG, Karlsruhe, Germany, 2016.

[26] D. Gettman, L. Pu, T. Sayed, and S. Shelby, "Surrogate Safety Assessment Model and Validation," Final Report, Federal Highway Administration, McLean, Va, USA, 2008.

[27] NHTSA, *U.S. Department of Transportation Releases Policy on Automated Vehicle Development*, National Highway Traffic Safety Administration, U.S. Department of Transportation, 2013.

[28] Atkins, "Research on the Impacts of Connected and Autonomous Vehicles (CAVs) on Traffic Flow," Stage 2: Traffic Modelling and Analysis Technical Report, Department for Transport, 2016.

[29] PTV, *Autonomous vehicle lateral space - example description*, PTV AG, Karlsruhe, Germany, 2017.

[30] D. Gettman and L. Head, "Surrogate Safety Measures From Traffic Simulation Models," Final Report, Federal Highway Administration, McLean, Va, USA, 2003.

[31] K. Goh, G. Currie, M. Sarvi, and D. Logan, "Experimental microsimulation modeling of road safety impacts of bus priority," *Transportation Research Record*, no. 2402, pp. 9–18, 2014.

[32] W. Young, A. Sobhani, M. G. Lenné, and M. Sarvi, "Simulation of safety: A review of the state of the art in road safety simulation modelling," *Accident Analysis & Prevention*, vol. 66, pp. 89–103, 2014.

[33] M. Essa and T. Sayed, "A comparison between PARAMICS and VISSIM in estimating automated field-measured traffic conflicts at signalized intersections," *Journal of Advanced Transportation*, vol. 50, no. 5, pp. 897–917, 2016.

[34] N. Saunier, "Surrogate Safety Analysis," *Polytechnique Montréal*, 2010.

[35] P. Chen, W. Zeng, G. Yu, and Y. Wang, "Surrogate safety analysis of pedestrian-vehicle conflict at intersections using unmanned aerial vehicle videos," *Journal of Advanced Transportation*, vol. 2017, Article ID 5202150, 12 pages, 2017.

[36] L. T. Truong, M. Sarvi, G. Currie, and T. M. Garoni, "How many simulation runs are required to achieve statistically confident results: a case study of simulation-based surrogate safety measures," in *Proceedings of the 18th IEEE International Conference on Intelligent Transportation Systems (ITSC '15)*, pp. 274–278, Las Palmas de Gran Canaria, Spain, September 2015.

[37] A. Dijkstra, P. Marchesini, F. Bijleveld, V. Kars, H. Drolenga, and M. V. Maarseveen, "Do calculated conflicts in microsimulation model predict number of crashes?" *Transportation Research Record*, no. 2147, pp. 105–112, 2010.

[38] L. T. Truong, M. Sarvi, G. Currie, and T. M. Garoni, "Required traffic micro-simulation runs for reliable multivariate performance estimates," *Journal of Advanced Transportation*, vol. 50, no. 3, pp. 296–314, 2016.

[39] Y. Kuang, X. Qu, and S. Wang, "A tree-structured crash surrogate measure for freeways," *Accident Analysis & Prevention*, vol. 77, pp. 137–148, 2015.

A Comparison Analysis of Surrogate Safety Measures with Car-Following Perspectives for Advanced Driver Assistance System

Sehyun Tak,[1] Sunghoon Kim,[2] Donghoun Lee,[2] and Hwasoo Yeo[2]

[1]*Department of The Fourth Industrial Revolution and Transport, The Korea Transportation Institute (KOTI), Republic of Korea*
[2]*Department of Civil and Environmental Engineering, Korea Advanced Institute of Science and Technology, Republic of Korea*

Correspondence should be addressed to Hwasoo Yeo; hwasoo@gmail.com

Guest Editor: Luc Chassagne

Surrogate Safety Measure (SSM) is one of the most widely used methods for identifying future threats, such as rear-end collision. Various SSMs have been proposed for the application of Advanced Driver Assistance Systems (ADAS), including Forward Collision Warning System (FCWS) and Emergency Braking System (EBS). The existing SSMs have been mainly used for assessing criticality of a certain traffic situation or detecting critical actions, such as severe braking maneuvers and jerking before an accident. The ADAS shows different warning signals or movements from drivers' driving behaviours depending on the SSM employed in the system, which may lead to low reliability and low satisfaction. In order to explore the characteristics of existing SSMs in terms of human driving behaviours, this study analyzes collision risks estimated by three different SSMs, including Time-To-Collision (TTC), Stopping Headway Distance (SHD), and Deceleration-based Surrogate Safety Measure (DSSM), based on two different car-following theories, such as action point model and asymmetric driving behaviour model. The results show that the estimated collision risks of the TTC and SHD only partially match the pattern of human driving behaviour. Furthermore, the TTC and SHD overestimate the collision risk in deceleration process, particularly when the subject vehicle is faster than its preceding vehicle. On the other hand, the DSSM shows well-matched results to the pattern of the human driving behaviour. It well represents the collision risk even when the preceding vehicle moves faster than the follower one. Moreover, unlike other SSMs, the DSSM shows a balanced performance to estimate the collision risk in both deceleration and acceleration phase. These research findings suggest that the DSSM has a great potential to enhance the driver's compliance to the ADAS, since it can reflect how the driver perceives the collision risks according to the driving behaviours in the car-following situation.

1. Introduction

Rear-end collision is one of the most frequent traffic accidents on the roads. Common contributing factors for the rear-end crashes include driver's inattention and human misjudgments on the amount of required deceleration in car-following situation. In efforts to prevent the rear-end crash and improve vehicular safety, drivers' judgments must be assisted and guided based on current or upcoming traffic situations. For such matter, various Advanced Driver Assistance Systems (ADAS) such as Forward Collision Warning System (FCWS) and Emergency Braking System (EBS) have been developed based on different data sources, including camera, radar, LIDAR, GPS, and connected vehicle network. The FCWS and EBS are designed to give warning signals or implement braking autonomously by detecting hazardous situations before a collision ahead of vehicle even occurs. One of major concerns to provide the collision warning or implement the autonomous braking in the ADAS is how the criticality of a certain traffic situation for a vehicle is assessed. To deal with the related problems, various assessment approaches have been proposed. One of the representative methods for identifying the rear-end collision risk is Surrogate Safety Measure (SSM). The SSM calculates the collision risk of a certain traffic situation with microscopic traffic parameters such as vehicle speed, acceleration, time headway, and space headway. There have been various efforts in developing SSMs for the FCWS or EBS based on the parametric method. The previous studies on the SSMs can be classified into two types. One is the perceptual approach and the other is the kinematic approach [1]. The perceptual approach-based SSM is designed to measure the collision risk based on the thresholds of perception. The representative of such approach is Time-To-Collision (TTC) [2], which estimates the expected time for two successive vehicles to collide. Some modified TTCs have been proposed, such as Inverse TTC [3], Time Exposed TTC [4], Time Integrated TTC [5], and Modified TTC [6]. On the other hand, the kinematic approach-based SSM is to estimate the rear-end collision risk based on the difference between the required stopping distances of two consecutive vehicles. There have been numerous SSMs based on this this approach, such as Potential Index for Collision with Urgent Deceleration (PICUD) [7–9], Stopping Distance Index (SDI) [10, 11], Stopping Headway Distance (SHD) [12], Crash Index (CI) [13], and Deceleration Rate to Avoid the Crash (DRAC) [14]. More recently, Deceleration-based Surrogate Measure (DSSM) was proposed by considering a human-centered design [15], and such design significantly affects the performance of the risk evaluation since the human-related parameters are strongly related to the situational awareness measure [16]. Similar to the perceptual approach-based SSMS, the kinematic approach-based SSMs are also used for assessing the safety of a vehicle and giving warning signals to the driver. The previous researches demonstrate that these SSMs show good performances in detecting critical events such as severe braking or jerk maneuvers before a collision.

On the other hand, one of the most critical factors of the SSMs employed in the FCWS and EBS is driver's compliance, which is highly correlated with the system reliability [17]. For example, a conservative SSM designed particularly for passive drivers will give frequent alerts to aggressive drivers in most cases. Then, the driver can become desensitized to the nuisance warnings [18]. On the other hand, imminent threat alerts given by an aggressive SSM may lead to missed alarms, which may not provide enough time to avoid an upcoming collision risk [19]. Therefore, drivers become more intended to ignore such system and they can nullify the effects of potential benefits from the system [20]. However, most of the previous studies on the SSMs applied to the FCWS and EBS have focused on discriminating possible collision situations in a subsequent few seconds, rather than tracking the estimated collision risk according to the drivers' driving behaviours in the entire car-following process. Therefore, there may exist inconsistency between the driver's perception of risk and the actual level of hazard.

Thus, there is a need for exploring the characteristics of existing SSMs particularly in terms of such inconsistency. For such purpose, this study aims to analyse and compare the collision risks estimated by different SSMs based on two different car-following theories, such as action point model and asymmetric driving behaviour model. DSSM is specifically is selected as the representative of human-centered design that considers the inconsistency between the driver's perception of risk and the actual level of hazard. TTC and SHD are selected as the representatives of each of the perceptual approach and kinematic approach, respectively, and they are used as the benchmarking points to be compared with DSSM. In fact, DSSM is basically based on the kinematic approaches like SHD. Compared to the SHD, which calculates the collision risk based on the variables obtained from sensor equipped in the vehicle, DSSM calculates the collision risk with more variables such as jerk rate, acceleration, and transition time for the application in Vehicle-to-Vehicle (V2V) communication environment. By comparing the SHD and DSSM, the characteristics of risk estimation with different technology bases (sensor-based and V2V communication-based) can be shown.

Considering the relationship between the driving behaviours and the collision risk estimated by the SSMs provides a foundation for monitoring the reaction of drivers to the collision risk, which can reflect the different preferences of drivers on the collision risk. The detailed explanation on the analysis method is provided in the following section. Then, Sections 3 and 4 describe the comparison results of the three SSMs according to the action point model and asymmetric driving behaviour model, respectively. Finally, brief concluding remarks are provided in the last section.

2. Analysis Approach

2.1. Car-Following Models for Analysis. To analyze the relationship between driving behaviour and collision risk in car following situation, several traffic variables are considered in this study. The traffic variable includes speed of a subject vehicle, spacing between the subject vehicle and preceding vehicle, and the relative speed between the two consecutive vehicles. There are two types of car-following processes discussed in this paper to examine the different levels of collision risk with the actual driver behaviour.

First one is analyzed in the spacing-relative speed plane, which adopts the perspectives used in action point model [21]. The action point model, which is also known as psycho-physiological car-following model, considers drivers' perception thresholds for a certain minimum value of the stimulus based on the spacing and relative speed [22]. This analysis would show how drivers react differently to the collision risk according to the changes in the spacing and relative speed. In the action point model-based analysis, the state of a vehicle is defined by using both the spacing and relative speed. Therefore, the state of a subject vehicle is not defined solely by the speed in a car-following situation of two consecutive vehicles.

Nonetheless, the speed of subject vehicle shows an important aspect of vehicle' state in a car-following process since the driver determines the acceleration depending on not only the relative speed but also the speed of the subject vehicle [23, 24]. Hence, the second method of analyzing car-following process is suggested to show a different aspect of the car-following process by focusing on a speed of subject vehicle and spacing based on the asymmetric driving behaviour [24]. The car-following process of asymmetric driving behaviour can show how a driver differently reacts to the collision risk when the vehicle is in acceleration phase and deceleration phase.

These two types of car-following processes describe how drivers perceive the collision risk in terms of the spacing, speed of subject vehicle, and relative speed. It is expected that the results of the analysis can explain why drivers show inconsistent choices of headway in car-following situations. However, these two types of car-following processes can describe only some parts of the highly complex human driving behaviours. These two types of car-following processes suit for the human driving behaviours in the stationary traffic state, particularly with the situations that both lead and following vehicles' speed are less than the free flow speed (or speed limit). Since the scope of this study is within such specific cases that have the high possibility of rear-end collision, we extract the car-following cases without any disturbances such as a vehicle cutting in or changing lane for the SSMs analysis.

2.2. Safety Surrogate Measures for Comparison. For analysis on the relationship between driving behaviour and collision risk, investigating different SSMs are necessary to be compared. This study considers three SSMs to estimate the collision risks at a given traffic situation, which includes the TTC, SHD, and DSSM. Since these three SSMs have different perspectives on the collision risk, their performances may be different from each other, even in identical traffic situations.

First introduced by Hayward [2], the TTC is one of the most representative indicators for judging the dangerous situation [25, 26]. It estimates the collision risk between two consecutive vehicles by calculating the remaining time before following vehicle crashes into a front vehicle with the assumption that the path and speed of two consecutive vehicles are maintained. The TTC is defined by

$$TTC = \frac{\left[x_{n-1}(t) - x_n(t) - s_{n-1}\right]}{\left[v_n(t) - v_{n-1}(t)\right]} \tag{1}$$

where $v_{n-1}(t)$ is the speed of leader vehicle at time t, $v_n(t)$ is the speed of following vehicle at time t, $x_{n-1}(t)$ is the location of leader vehicle at time t, $x_n(t)$ is the location of following vehicle at time t, and s_{n-1} is the length of leader vehicle. For example, a pre-determined TTC threshold value is 2.0 seconds. The current car-following situation is safe when the TTC value is greater than or equal to the threshold value, while the TTC value gets closer to zero as the rear-end collision risk increases.

For evaluating the collision risk, the concept of safe stopping distance has also been used. Methods with this concept calculate the collision risk based on the difference between stopping distances of two consecutive vehicles with full deceleration rate. The methods with such concept calculate the collision risk by assuming that the leader vehicle suddenly brakes with the maximum deceleration rate. The condition that the stopping distance of leader is smaller than sum of stopping distance of the following vehicle and space headway of following vehicle is considered as a dangerous situation. The SHD is a representative safe stopping distance-based method, which can be formulated as follows:

$$SHD = \max\left[-1.47 \times \left(v_{n-1}(t) \times h_n(t) - v_n(t) \times \tau\right) + \left[\frac{v_{n-1}(t)^2 - v_n(t)^2}{30 \times (acc/g \pm Gr)}\right], 0\right] \tag{2}$$

where $h_n(t)$ is the time headway of the following vehicle at time t, τ is the perception reaction time, *acc* is the deceleration rate, g is the gravity acceleration, and *Gr*, the grade expressed as a percentage. For example, a predetermined SHD threshold value is 20. It is risky situation when the SHD value is greater than or equal to the threshold value, while the current situation is safe when the SHD value is less than the threshold value.

The DSSM is also one of the concepts using the safe stopping distance-based method, which can well represent the individual collision risk in both acceleration and deceleration phases compared to other SSMs by adopting the transition time [15]. The DSSM is a ratio of the required deceleration and maximum deceleration performance of a subject vehicle, which is defined as follows:

$$K = \left[x_n(t) - x_{n-1}(t) + s_{n-1}\right] + \left[2 \cdot v_n(t) + a_n(t) \cdot \tau\right] \cdot \frac{\tau}{2} - M_{n-1,Tran} + M_{n,Tran} \tag{3}$$

$$b_n(t) = b_{max,n-1} \cdot \frac{\left[v_n(t) + a_n(t) \cdot \tau\right]^2}{\left[2 \cdot K \cdot b_{max,n-1} + v_{n-1}(t)^2\right]} < 0 \tag{4}$$

$$DSSM = \frac{b_n(t)}{b_{max.n}} \tag{5}$$

where $a_n(t)$ is the acceleration rate of following vehicle at time t, $a_{n-1}(t)$ is the acceleration rate of leader vehicle at time t, $b_{max.n-1}$ is the maximum braking rate of leader vehicle, which represents the vehicle's mechanical deceleration performance, $b_n(t)$ is the needed deceleration rate of following vehicle to avoid the accident at time t, $b_{max.n}$ is the maximum braking rate of following vehicle, M_{n-1} is the stopping distance of leader vehicle during transition time, and M_n is the stopping distance of following vehicle during transition time, τ is the perception reaction time. For instance, a predetermined DSSM threshold value is 1. It is an unsafe situation when the DSSM value is greater than or equal to the threshold value, while the DSSM value is less than the threshold value when the current driving situation is safe.

2.3. Data Description. For analysis on the relationship between the driving behaviours and the collision risks estimated by the different SSMs, this research uses one of the Next Generation Simulation (NGSIM) trajectory datasets, which is collected from a segment of U.S. Highway 101 in Los Angeles, California, between 7:50 a.m. and 08:35 a.m. on 15 June 2005 [27]. The length of the study site in the

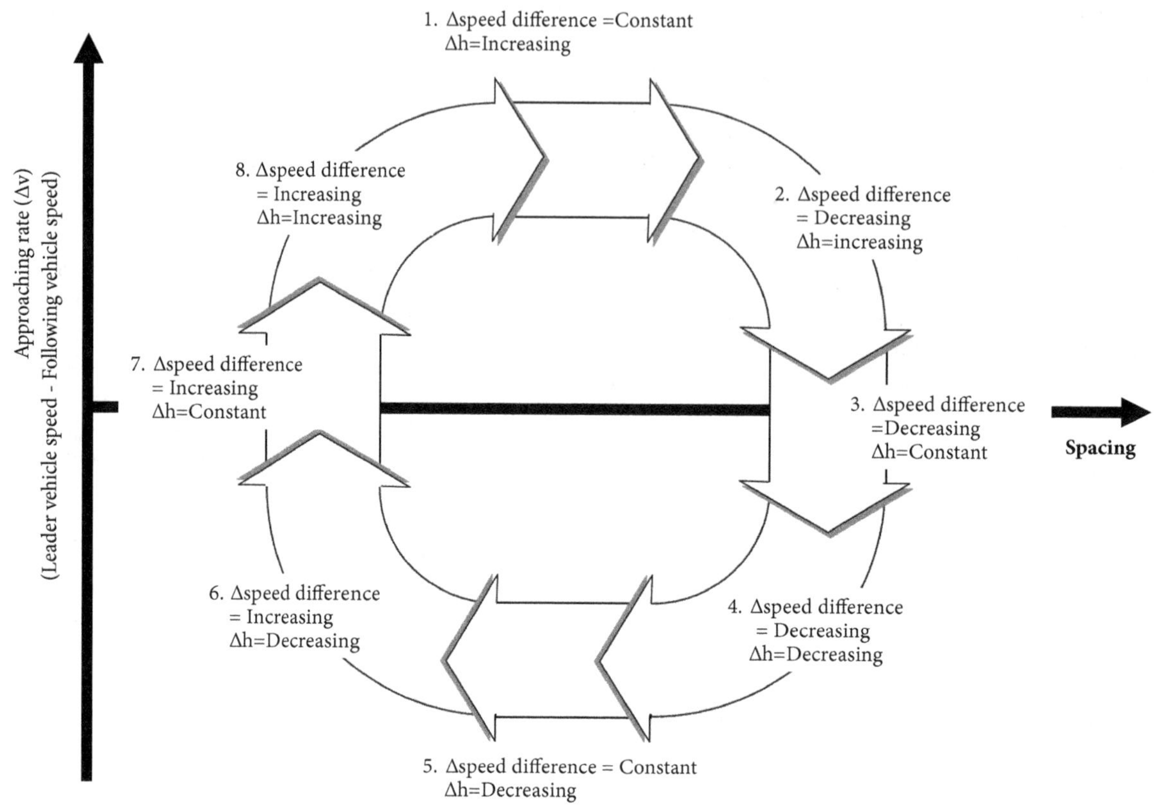

FIGURE 1: The Classification of driving state of subject vehicle.

NGSIM dataset is 640 m with five main lanes. The data contains microscopic traffic information on individual vehicular movements' trajectories, which includes location, speed, space headway, acceleration/deceleration, and vehicle type at 0.1 sec time intervals. Among the car-following cases in the dataset, 143 car-following cases that do not experience any disturbance such as cut-in and cut-out vehicle when passing through the study site are extracted, excluding motorcycles and trucks.

3. Comparison Analysis in Spacing-Relative Speed Plane

3.1. Action Point Model Perspective. As the first step of conducting comparison study, this study compares the SSMs with each other from the perspectives of the action point model. In other words, the collision risks estimated by the three SSMs will be described in the spacing-relative speed plane. As stated earlier, the car-following process of the action point model represents the driving behaviour with the psychophysical basis and shows how the driver of subject vehicle adjusts the differences in the locations and speeds between the leader and subject vehicle. In the action point model, a driver's decision is made upon certain perception threshold values. When the speed of preceding vehicle is much greater than the subject vehicle, the state of subject vehicle exceeds the perception threshold of relative speed. Then, the driver of the subject vehicle decides to accelerate. On the other hand, when the speed of preceding vehicle is much less than the subject vehicle, the state of subject vehicle exceeds the perception threshold of relative speed in negative direction. Then, the subject vehicle decreases its speed. The spacing adjustment procedure is arranged similarly in the action point model. When the spacing is much greater than the desired spacing, the state of the subject vehicle exceeds the perception threshold of spacing. Then, the subject vehicle increases the speed to reduce the spacing. In contrast, the state of the subject vehicle exceeds the perception threshold of spacing in negative direction when the spacing is much less that the desired spacing. Then, the subject vehicle reduces its current speed. Based on the two kinds of perception thresholds and driving behaviour, the driver in the subject vehicle makes a decision for either accelerating or decelerating.

Hence, in the action point car-following process, the spacing and relative speed are important variables that directly affect the decision on acceleration and deceleration action. By using these two variables, the state of the subject vehicle can be defined as shown in Figure 1.

The state of the subject vehicle can be described by eight states. The states 1, 2, and 8 represent the situations of when the preceding vehicle is faster. And in states 4, 5, and 6, the preceding vehicle is slower than the subject vehicle. The states 1 and 5 are also the points that the driver makes the decision on acceleration and deceleration. After the state 1, which is called as "catch up" action point, the collision risk is increased

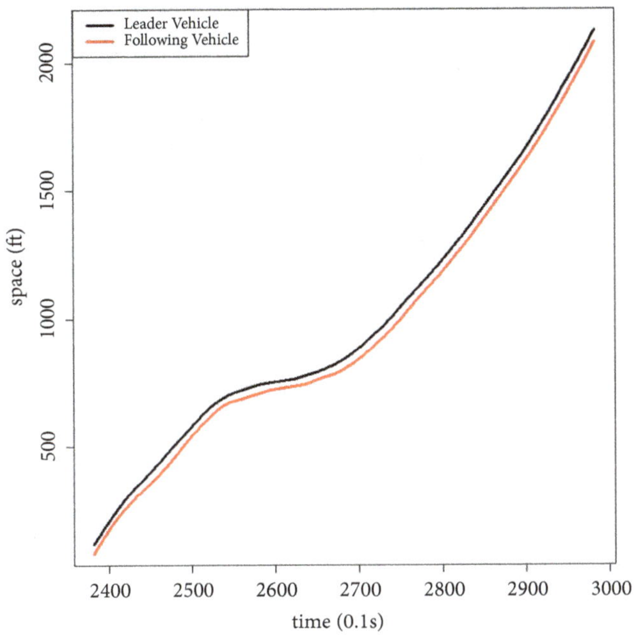

FIGURE 2: The example case for the analysis.

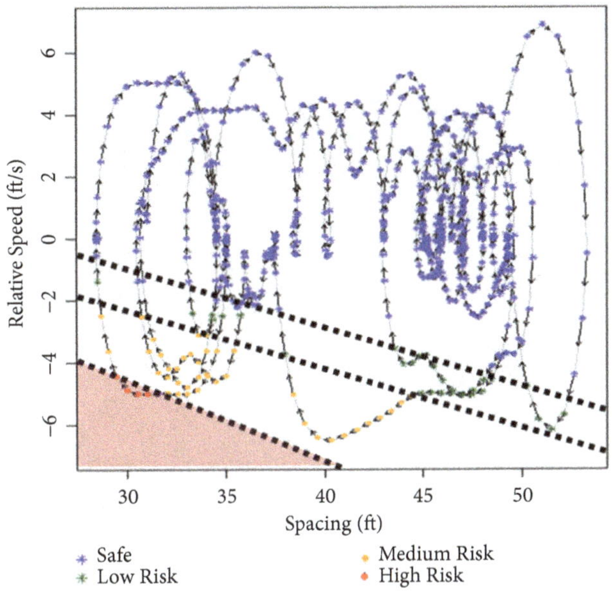

FIGURE 3: Collision risk estimated with the temporal spacing and relative speed (TTC).

due to the acceleration actions. After the state 5, which is called as "release" action point, the collision risk is decreased due to the deceleration actions. In other states, the driver maintains the decision made in the states 1 and 5.

An example case of car-following for the analysis is depicted in Figure 2. In the figure, the black and red lines are the trajectories of the leader and subject vehicle, respectively. One can easily observe that the car-following case meets a shockwave and shows the both acceleration and deceleration behaviours.

3.2. Car-Following and Collision Risk by TTC. According to the example case shown in Figure 2, the collision risks estimated by the TTC are shown in Figure 3. This figure shows the relationship between the vehicle spacing and the relative speed. The points in the plane represent the temporal measurements of the two properties and the solid lines with arrows connect the sequence of these points. As we can see in the figure, the temporal measurements draw circles recurrently on the plane during the car-following situation, and it is shown that the low relative speed values with the low spacing values tend to have the high collision risk. In this paper, the collision risks are discretionarily classified into four situations for the quantitative analysis. By the concept of TTC, the risk is considered to be high when the TTC value is low. Thus, the four classified situations considered in this paper are High Risk (TTC ≤ 2.5), Medium Risk (2.5 < TTC ≤ 5.0), low risk (5.0 < TTC ≤ 7.0), and Safe (TTC > 7.0). Note that since there have not been any previous efforts for defining the exact thresholds of the high or low risk at the current stage, the thresholds classifying the risk levels in this study are defined based on the empirical understandings with the given data. In fact, such thresholds may vary depending on the road characteristics or driving environments. Hence, some sensitivity analyses are required to investigate the issue further, but they go beyond the scope of this paper.

As shown in Figure 3, the TTC generally considers that it is safe when the preceding vehicle is faster than the subject vehicle. Hence, it only classifies the states 4, 5, and 6 into the dangerous situations. In TTC, at the similar level of spacing, the collision risk consistently increases as the relative speed decreases. At the similar level of relative speed, the collision risk also consistently increases as the spacing decreases. The level of collision risk is also indicated with different colours in the spacing and relative speed plane. Particularly, the red coloured region indicates that the driving situation is dangerous by the TTC estimation.

In terms of the action point car-following perspective, the collision risks estimated by the TTC do not match well to the action points where a driver changes acceleration and deceleration. In the near perception thresholds, which correspond to the states 1, 2, 3, and 7, the changes in the action are rarely observed. Furthermore, the TTC is not a sensitive measure that has a lack of reflecting the dynamic changes of driving situation, and it misses out some parts of the action points.

3.3. Car-Following and Collision Risk by SHD. Similar to the method of using the action point model above, the collision risks estimated by the SHD are shown in Figure 4. Unlike TTC, the risk is considered to be high when the TTC value is also high in SHD. In this paper, the four classified situations based on SHD are High Risk (SHD > 40), Medium Risk (40 ≥ SHD > 0), Low Risk (0 ≥ SHD > -50), and Safe (-50 ≥ SHD). The SHD generally considers that it is not safe when the subject vehicle is faster than the preceding vehicle. The situations with high collision risks determined by the SHD are the states 4, 5, and 6. However, unlike the TTC that

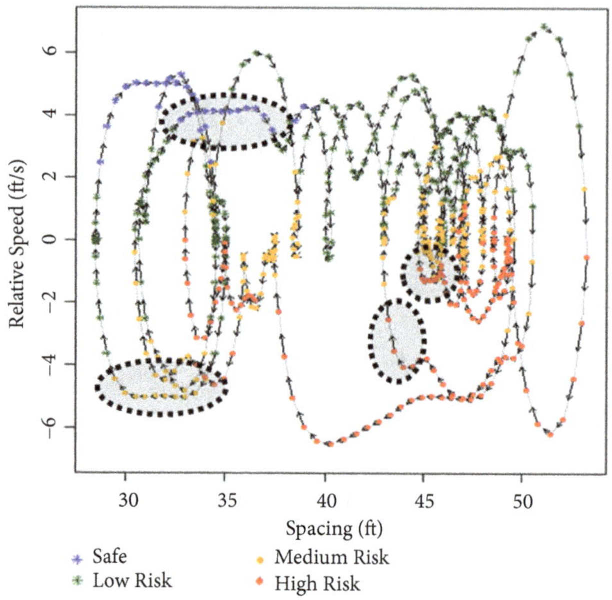

FIGURE 4: Collision risk estimated with the temporal spacing and relative speed (SHD).

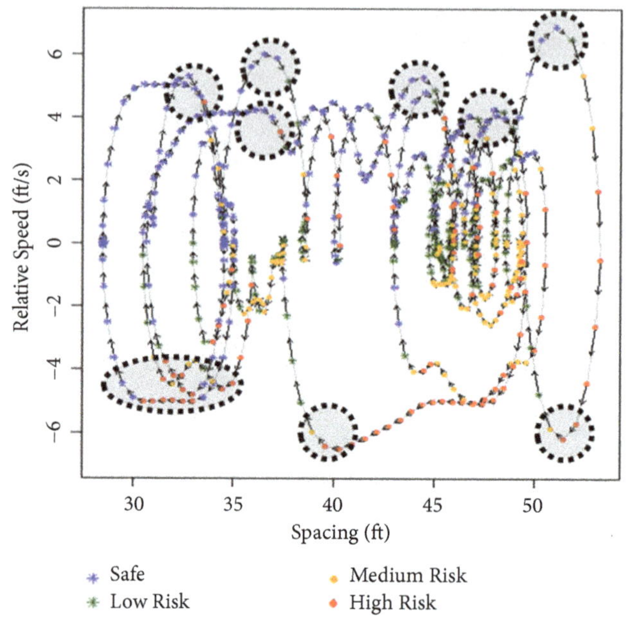

FIGURE 5: Collision risk estimated with the temporal spacing and relative speed (DSSM).

considers high risk only when the subject vehicle is faster, the high collision risk is determined by the SHD also when the preceding vehicle is faster than the subject vehicle. In terms of the spacing, there is a significant difference between the SHD and the TTC. The high collision risk in the TTC mainly occurs when the spacing is small, but it is widely distributed over the entire spacing regions in the SHD. As the result, the SHD can evaluate the collision risk in the wider range of vehicle states in the spacing and relative speed plane.

Compared to the TTC, the collision risk in the SHD is more dynamically changed during the car-following process shown in Figure 4. Some corresponding points with the action points are observed, where the driver's driving behaviour is changed due to the excess of perception threshold. The observed action points are marked with the black dotted circles in Figure 4. Before and after the observed action points, the SHD shows approximately one-step difference in terms of collision risk. However, the changing trends of estimated collision risk of the SHD do not match well to the action points of the human driver overall.

3.4. Car-Following and Collision Risk by DSSM. Similar to the TTC and SHD, the DSSM considers also that it is dangerous when the subject vehicle is faster than the preceding vehicle, as shown in Figure 5. For the DSSM cases in this paper, the four classified situations are High Risk (DSSM ≥ 1.1), Medium Risk ($1.1 >$ DSSM ≥ 0.9), Low risk ($0.9 >$ DSSM ≥ 0.75), and Safe ($0.75 >$ DSSM). The high collision risk situations are shown in the states 4, 5, and 6. Nonetheless, DSSM shows distinctive differences from other SSMs.

First, the DSSM does not show increasing or decreasing trend only with the relative speed or spacing respectively. In the TTC, the collision risk increases as the relative speed and spacing decrease overall, so that the high collision risk does not occur when there is a large value of the relative speed and spacing. In the SHD, the collision risk generally increases as the relative speed decreases, while it decreases as the spacing decreases overall. Therefore, the high collision risk hardly occurs when there is large relative speed value. On the other hand, the DSSM shows that a high collision risk occurs in all regions, even when the preceding vehicle is faster than the subject vehicle, as shown in Figure 5. Consequently, the DSSM can identify more various dangerous situations than other SSMs in all states of the action point car-following process.

Second, DSSM shows more well-matched results to the human driving behaviour in terms of the action point car-following process. As shown in Figure 5, the collision risk estimated by the DSSM is significantly changed before and after the action points, which are marked with the black dotted circles. After the action points where the preceding vehicle is slower than the subject vehicle, the estimated collision risk is sharply decreased. After the action points, which occur when the preceding vehicle is faster than the subject vehicle, the estimated collision risk of the DSSM shows increasing trend and high collision risk is also observed. This wide perceiving capability of the DSSM beyond the TTC and SHD is from the consideration of vehicles' acceleration in the evaluation of the risk.

3.5. Comparison of Average Collision Risk and High Collision Risk. Based on the classified states described as Figure 1, the trends of the estimated average collision risk and frequency of high collision risks are analyzed. The average collision risks of each SSM are calculated with the four classified situations: safe situation, low risk situation, medium risk situation, and

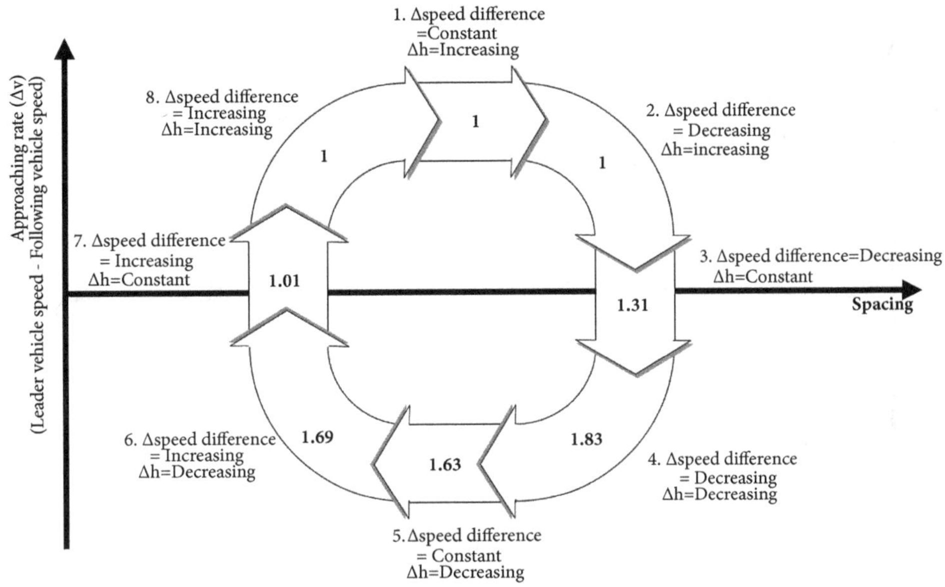

(a) Average collision risk

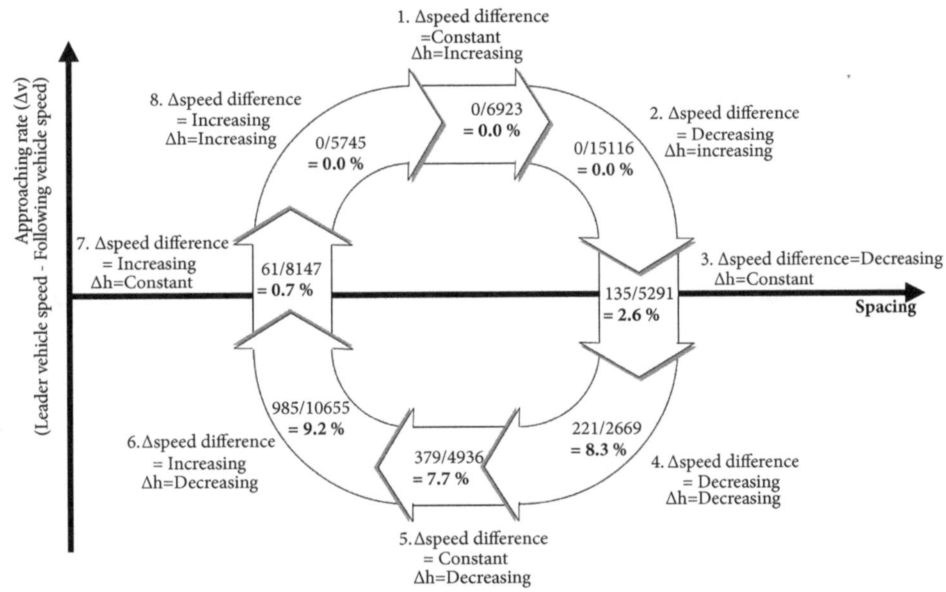

(b) Frequency of high collision risks

FIGURE 6: The TTC trends on the average collision risk and frequency of high collision risks.

high risk. The four classified situations are graded as 1, 2, 3, and 4, discretionary for the quantitative analysis. The frequency of high collision risks is computed by the same criteria used in average collision risk evaluation, but only the high-risk situations are counted. The number of high-risk situations is divided by the total number of exposure for each classified states.

Figure 6 shows the average collision risk and the frequency of high collision risks of the TTC based on the 143 car-following cases. As depicted in Figures 6(a) and 6(b), it can be seen that the TTCs do not show any common behaviours with the action points of the drivers. The average collision risk estimated by the TTC does not show any variations near the states 1 and 5, which are the situations when the drivers decide and change the actions of acceleration and deceleration. No specific relationship between the collision risk and driving actions is observed in the trend of collision risk estimated by the TTC.

Similarly, the SHD can evaluate collision risk in more various states compared to the TTC, as shown in Figure 7.

However, the high collision risks estimated by the SHD are concentrated when the subject vehicle is faster than the preceding vehicle. Moreover, the SHD tends to overestimate the collision risk in these situations, such as state 4. In terms of the action point car-following perspective, the estimated collision risk of the SHD is partially matched to the action

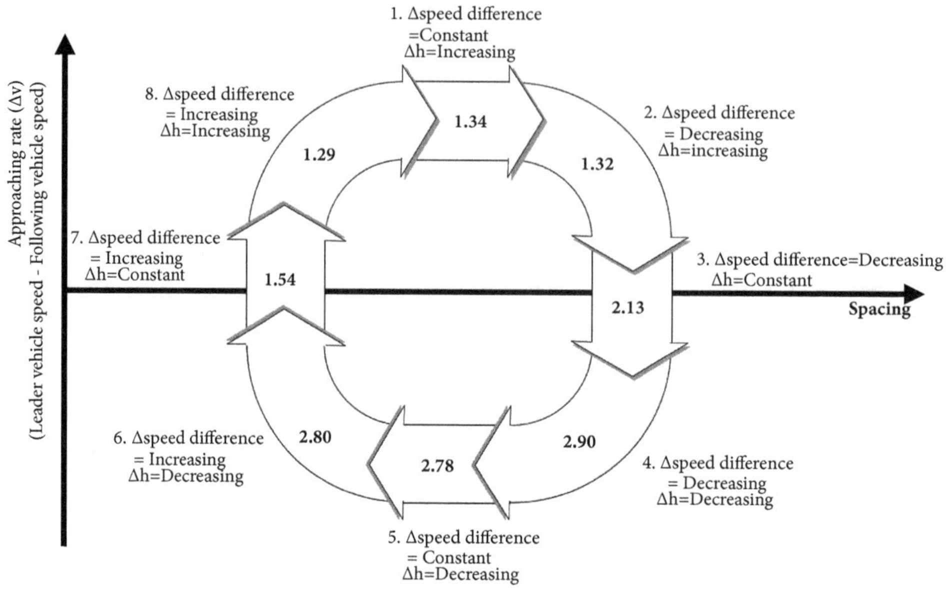

(a) Average collision risk

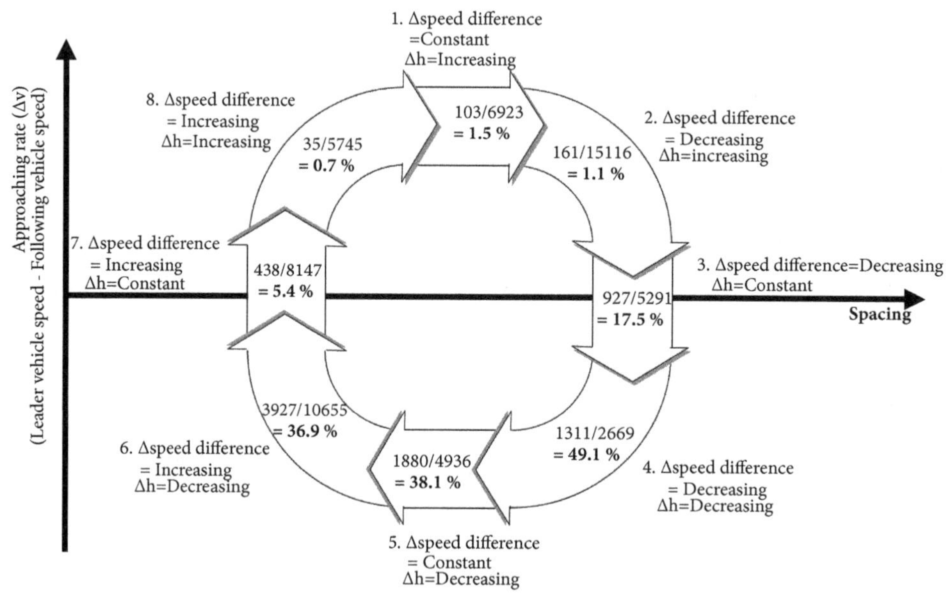

(b) Frequency of high collision risks

FIGURE 7: The SHD trends on the average collision risk and frequency of high collision risks.

point behaviours of the drivers. In the action point car-following behaviour, the driver decides and changes the actions of acceleration and deceleration near the perception thresholds, which leads to the changes in the collision risks near state1 and state 5. Near state 5, the SHD shows the similar trend to the driving action of human. When the state is changed from 4 to 5, the SHD value decreases even though the variation is not significantly large. When the state is changed from 5 to 6, the SHD value increases. Near the state 1, SHD shows the dissimilar trend to human behaviour. When the state is changed from 1 to 2, the SHD decreases even though the driver is catching up the speed.

In Figures 8(a) and 8(b), the DSSM can evaluate collision risk from more states compared to TTC and SHD. For both analyses, the maximum values are observed in state 4, while the minimum values are observed in state 8. Near these two states, the average collision risk and frequency of high collision risk show the same increasing and decreasing trends. Given this result, the DSSM is the only SSM, in which the severity grade of the collision risk is divided in balance. Therefore, it can be said that the collision risk by DSSM is well matched to the action point behaviour. Overall, the DSSM continuously increases when the state is changed from 1 to 5. And it continuously decreases when the state is changed from

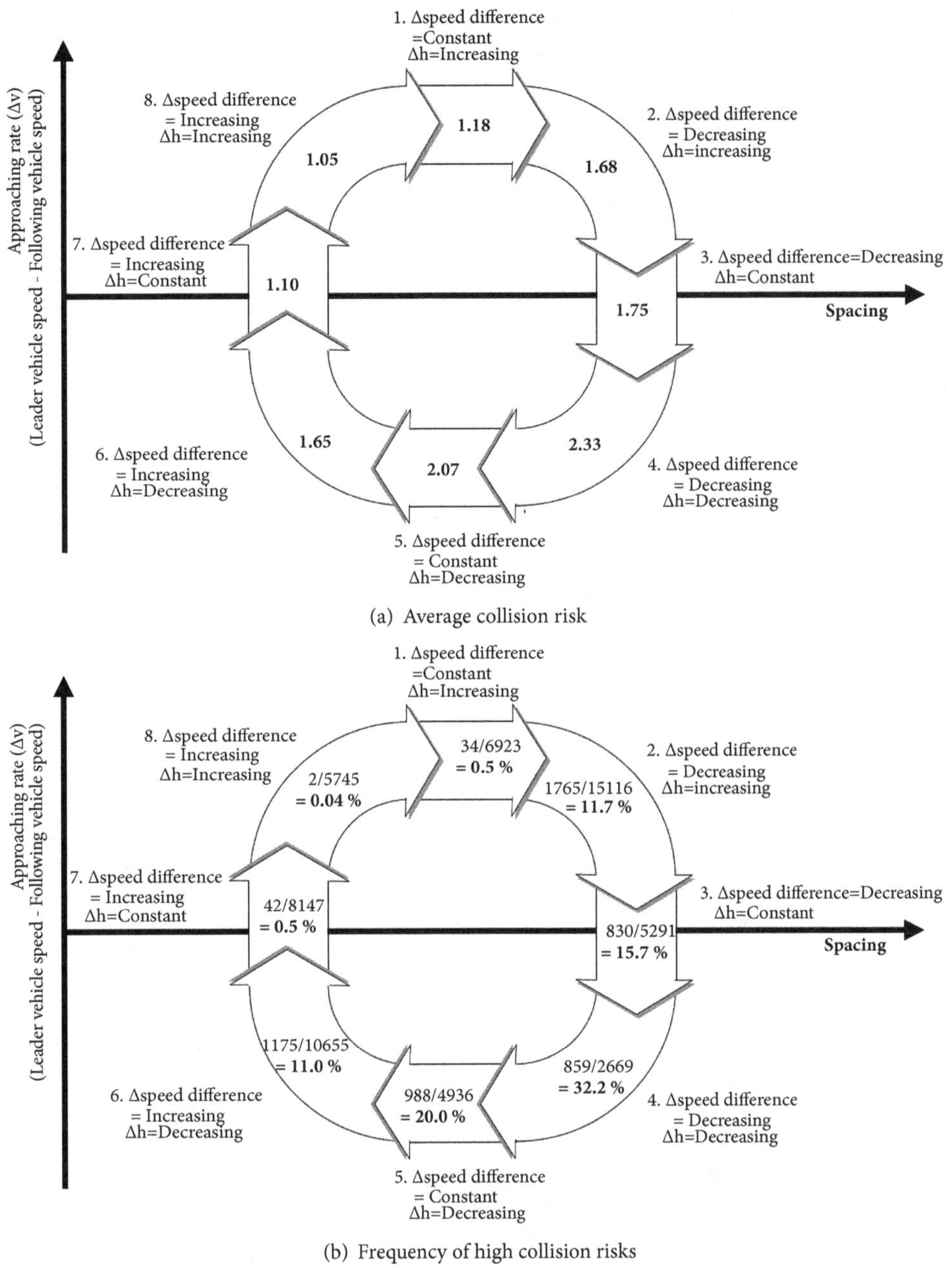

FIGURE 8: The DSSM trends on the average collision risk and frequency of high collision risks.

5 to 1. In detail, near the action points, which are states 1 and 5, the estimated collision risk of DSSM is changed with the similar process of the action point car-following behaviour. After the catch-up action points, which is state 1, the DSSM value rapidly increases. Conversely, near the release action point, which is state 5, the DSSM decreases. Especially, when the state is changed from 5 to 6, more significant decreasing trend of DSSM value is observed compared to that of SHD value due to the release action in state 5. These results coincide with the previous research findings in other car-following models, which is based on the action point driving behaviour [21, 28–32].

4. Comparison Analysis in Speed-Spacing Plane

4.1. Analysis from Asymmetric Behaviour Perspective. The action point model-based analysis shows well how a driver reacts to the external stimuli by focusing on the relationship between the leader and subject vehicle. However, in the spacing and relative speed plane, it still has a limitation to fully describe the human driving behaviour since it does not consider the absolute speed of the subject vehicle. In order to investigate the effect of such property, we analyze the relationship between the collision risk and the asymmetric driving behaviour. In the asymmetric theory, drivers show different patterns in the acceleration and deceleration phases, and such phenomenon has been observed by several researchers since the 1960's [33–38]. In this theory [24, 39, 40], drivers show two different curves of acceleration curve (A-curve) and deceleration curve (D-curve) with different spacing for the same speed. The A-curve is the boundary curve in the acceleration phase and the D-curve is the boundary curve in the deceleration phase of the car-following situations. In the A-curve, the two consecutive vehicles maintain larger spacing than in the D-curve for the same corresponding speed. The A-curve can be found by connecting the ending points of accelerating actions, and the D-curve can be found by connecting the ending points of decelerating actions. In addition, near the each of the A-curve and D-curve, an acceleration action point line and a deceleration action point line can be observed as well, respectively. The acceleration action point line can be found by connecting the starting points of accelerating actions, and the deceleration action point line can be found by connecting the starting points of decelerating actions.

These accelerating and decelerating actions in the A-curve and D-curve are highly related to the collision risk of the driver. When a driver accelerates to catch up with the leader vehicle's speed during the acceleration process, the driver increases the speed until the collision risk becomes larger than a certain threshold or the desired speed is reached. After that, if the spacing increases and the collision risk drops below certain threshold again, the driver takes acceleration again in order to keep the appropriate spacing. The accelerating actions appear repeatedly to keep the appropriate spacing and collision risk. On the other hand, during deceleration process, the driver reduces the speed until the collision risk is smaller than a certain threshold. After that, if the spacing decreases again and the collision risk is higher than a certain threshold value, the driver decelerates again. The decelerating actions also appear repeatedly in order the keep the appropriate level of collision risk. Based on such logic, a high collision risk would appear near the acceleration and deceleration action point line. Therefore, the estimated collision risk may indicate the timings of the acceleration and deceleration actions.

4.2. TTC in Speed-Spacing Plane. Based on the basic relationship between the collision risk and asymmetric driving behaviour, we analyze the three SSMs in the spacing-speed plane. Figure 9 shows the collision risks estimated by the TTC in the spacing-speed plane. The example case shown in Figure 2 is used in these analyses as well.

In the figure, the blue stars, green stars, yellow circles, and red circles represent the safe situation, low risk situation, medium risk situation, and high risk situation, respectively. The dashed lines represent the virtual deceleration action point line, D-curve, A-curve, and acceleration action point line, respectively, from left to right. The dashed lines shift to right or left depending on the driver's psychology on external stimuli and traffic state.

Figure 9 shows the collision risks estimated by the TTC in the spacing-speed plane. As shown in Figure 9, a variation

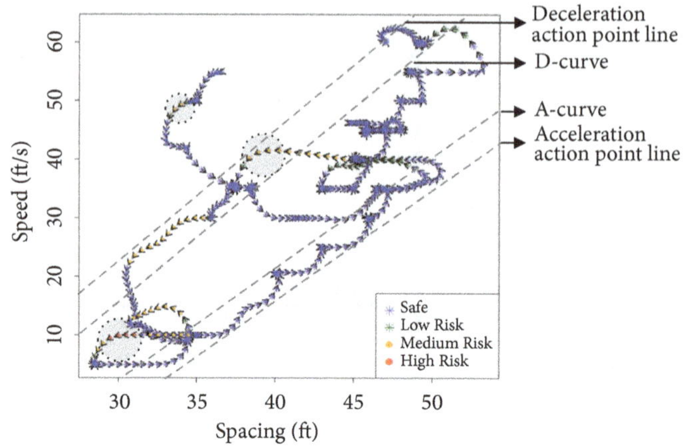

Figure 9: Collision risk estimated with the temporal spacing and speed (TTC).

of the collision risks estimated by the TTC is not large as much as the variation of driver's acceleration behaviours. This result shows that TTC is not a sensitive measure to driving actions. Some high and medium collision risk of TTC is only occurred in the deceleration process. Among these high and medium collision risk of TTC, only the three situations, which are marked with black dotted circles in Figure 9, can describe the human driving behaviour in the deceleration process. In these circled areas, it is observed that the collision risk is high until the state of driver reaches deceleration action point line. After that, the risk decreases as the driver reduces the speed overall. In the acceleration process, the high or medium level of collision risks is not observed, and most of the acceleration processes are considered as the safe situation. In the acceleration process, the leader vehicle is generally faster than the subject vehicle since the subject vehicle accelerates after the preceding vehicle accelerates with response-time delay. Therefore, in this process, the TTC does not properly evaluate the collision risk, because the TTC only evaluates the collision risk when the subject vehicle is faster than the preceding vehicle.

4.3. SHD in Speed-Spacing Plane. Figure 10 shows the collision risks estimated by the SHD. Compared to the TTC, the SHD value is more dynamically changed according to the driving action.

The SHD can identify the high collision risk during the acceleration process contrarily to what the TTC can identify only during the deceleration process. The matched results to the driving behaviour during both the acceleration and the deceleration processes are observed and they are marked by the solid and black dotted circles, respectively. In black dotted circled areas, the collision risk is decreased as the driver reduces the speed. After the state of vehicle reaches to D-curve, the driver maintains constant speeds and the SHD value is increased again. In black solid circled areas, the collision risk is increased as the driver increases the speed. After the state of vehicle reaches the A-curve, the driver maintains the constant speed and the collision risk shows the decreasing trend.

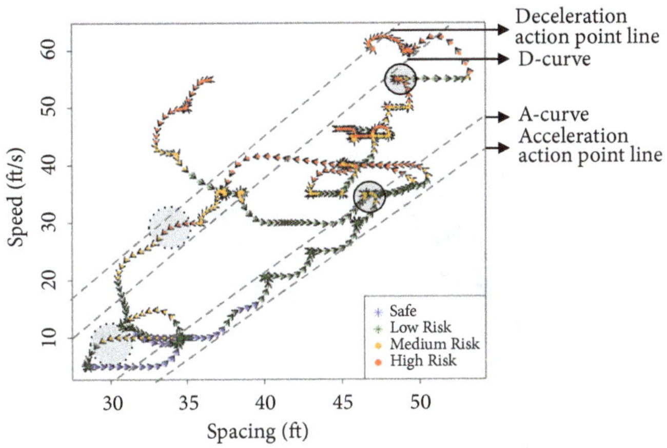

FIGURE 10: Collision risk estimated with the temporal spacing and speed (SHD).

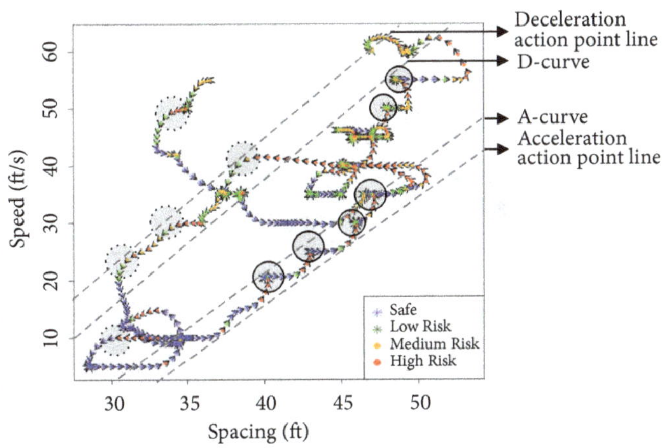

FIGURE 11: Collision risk estimated with the temporal spacing and speed (DSSM).

Even though the collision risks estimated by the SHD show more matched results to the asymmetric driving behaviour, they still show biased results due to several reasons. First, the SHD always considers the deceleration process as more dangerous situation than the acceleration process. The high collision risk is mainly distributed in the deceleration process and the sensitivity of collision risk to driving action is also high in the deceleration process. Second, the SHD is highly affected by the speed of subject vehicle itself. The collision risk estimated by the SHD generally increases as the speed of the subject vehicle increases overall. Due to these influences, the high collision risk is not observed when the speed of the subject vehicle is low, even though the state of the subject vehicle is near the acceleration action point line and deceleration action point line.

4.4. DSSM in Speed-Spacing Plane. Compared to other SSMs, the DSSM shows the most sensitive and well-matched results to the asymmetric driving behaviour, as shown in Figure 11.

The black dotted circles and black solid circles are the representative cases of the matched situations during the acceleration and the deceleration processes, respectively. In the areas near the black dotted circles in the deceleration process, the high collision risks occur right before the state reaches to the deceleration action point line. The collision risks sharply decrease after the driver starts deceleration action. In the areas near the black solid circles in the acceleration process, the high collision risks occur right before the state of the subject vehicle reaches A-curve. The collision risks decrease right after the state of the subject vehicle reaches the A-curve and as it maintains a constant speed. The collision risks gradually decrease until the state of the subject vehicle reaches to the acceleration action point line.

4.5. Comparison of Average Collision Risk and Frequency of High Collision Risk. Figure 12 compares the average collision risk and the frequency of high collision risk of each SSM. The DSSM well captures the high collision risk particularly in acceleration process compared to other SSMs. Except for the DSSM, the TTC and SHD consider the deceleration process as more dangerous situation than acceleration process. The SHD shows the highest average collision risk in both acceleration process and deceleration process with approximately three or four times higher frequency of high collision risk than the TTC and the DSSM. Considering the slightly low successful alarm ratio of the SHD compared to the DSSM in previous research [15], the SHD has a tendency of overestimating the collision risk.

5. Conclusion

In order to investigate the relationship between collision risk and human driving behaviour, this study analyzes three different SSMs based upon two different car-following theories, including the action point model and asymmetric driving behaviour model. The three SSMs, including the TTC, SHD, and DSSM, show different characteristics and trend in estimating the collision risk. In the analysis with the action-point model, the estimated collision risks of the TTC and SHD only partially match with the pattern of driver's driving behaviour. The high collision risks are concentrated where the subject vehicle is faster than preceding vehicle, so both SSMs overestimate the collision risk particularly in such situations. On the other hand, one could observe that the DSSM shows well-matched results to action point behaviour of the driver. After the catch up action point, the collision risks estimated by the DSSM rapidly increase. Conversely, near the release action point, the DSSM value shows a decreasing trend. In the analysis with asymmetric driving model, the TTC and SHD show biased collision risks in the deceleration process. Especially, the SHD is highly affected by the speed of subject vehicle itself, so the high collision risk is rarely observed when the speed of subject vehicle is low. DSSM equally evaluates the collision risk during acceleration process and deceleration process and it shows well-matched results to acceleration and deceleration actions near acceleration and deceleration action point line.

This study provides ample opportunities to design human-centered services of the FCWS and EBS in two

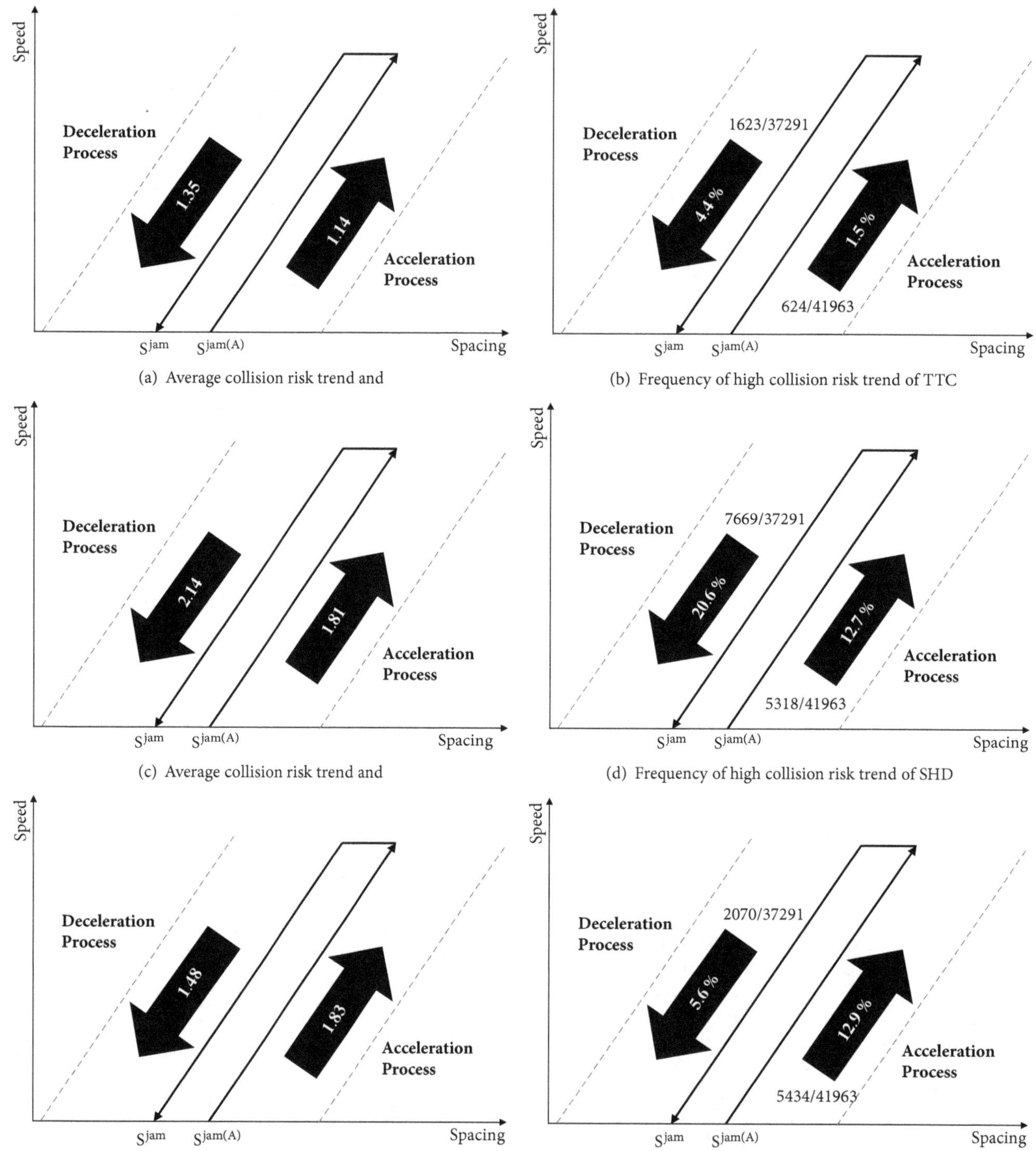

(a) Average collision risk trend and

(b) Frequency of high collision risk trend of TTC

(c) Average collision risk trend and

(d) Frequency of high collision risk trend of SHD

(e) Average collision risk trend and

(f) Frequency of high collision risk trend of DSSM

FIGURE 12: Each SSM trends on their average collision risks and frequencies of high collision risks.

aspects. First, it can provide an opportunity for significantly reducing the false-alarm ratio of a collision warning system. The SSMs can estimate a more detailed state of a subject vehicle in terms of the collision risks in various traffic situations. The effect of small changes in the collision risks varied with different driving behaviours can be tracked and analyzed continuously. Based on the research findings of this study, not only dangerous situations but also the trend of increasing (or decreasing) collision risks can be identified. By providing warnings before collision risk reaches a higher level, safety-related services can significantly improve the accuracy and give enough time for drivers to prepare for a harsh braking. Second, it can render safety-related services more acceptable to drivers. In order to make a collision warning system be acceptable to human drivers, the judgment on dangerous situations of the system needs to be similar to that of the human drivers. Microscopic analyses on SSMs and driving behaviours would provide the foundation for monitoring the reaction of human drivers to the collision risk. Furthermore, by analyzing the relationship between driving behaviour and collision risk estimated by SSMs, the safety-related services can reflect the different preferences of drivers on the collision risk.

However, this current paper still has a limitation in that the thresholds classifying the risk levels are given based on the empirical understandings with the given data. It is due to that there has not been a previous effort for defining the exact thresholds of the high or low risk so far. In fact, such thresholds for the SSMs may vary depending on the road characteristics like the nature of traffic flow (uninterrupted or interrupted flow) and the free flow speed (or speed limit). The thresholds may vary by different regions or even by different countries. Hence, the results of this current paper shall be revisited later, by considering the variations of the thresholds. Thus, the sensitivity analysis of the thresholds is an essential research topic to be addressed in further studies. Furthermore, the comparative analyses in this paper were done only with the representative SSMs of the perceptual approach (TTC) and kinematic approach (SHD and DSSM). It is suggested also to try other various SSMs while conducting further sensitivity analysis. In addition, the difference among the kinematic approaches like the SHD and DSSM is the selection of the obtainable microscopic variables to be used for the risk calculation, such as the spacing and relative speed, and these variables must be obtainable in real-time. Considering the application of the autonomous vehicles in the near future, in which more detailed information can be obtained in real-time, it is also suggested to develop the existing SSMs further into the more advanced form that can take account into various safety-related variables, such as real-time jerk rate and acceleration.

Conflicts of Interest

The authors declare that there are no conflicts of interest regarding the publication of this paper.

Acknowledgments

This research was supported by the Ministry of Land, Infrastructure, and Transport (MOLIT, KOREA) [Project ID: 18TLRP-B146748-01-000000, Project Name: Connected and Automated Public Transport Innovation (National R&D Project)] and the Ministry of Trade, Industry, and Energy (MOTIE, KOREA) [Project ID: 10063147, Project Name: Multi-Level Safety System for Intersection Collision Prevention].

References

[1] D. Lee and H. Yeo, "A study on the rear-end collision warning system by considering different perception-reaction time using multi-layer perceptron neural network," in *Proceedings of the 2015 IEEE Intelligent Vehicles Symposium (IV)*, pp. 24–30, Seoul, South Korea, June 2015.

[2] J. C. Hayward, "Near-miss determination through use of a scale of danger," *Highway Research Record*, vol. 384, pp. 24–34, 1972.

[3] R. Kiefer, D. LeBlanc, and C. Flannagan, "Developing an inverse time-to-collision crash alert timing approach based on drivers' last-second braking and steering judgments," *Accident Analysis & Prevention*, vol. 37, no. 2, pp. 295–303, 2005.

[4] H. C. Chin, S. T. Quek, and R. L. Cheu, "Traffic conflicts in expressway merging," *Journal of Transportation Engineering*, vol. 117, no. 6, pp. 633–643, 1991.

[5] H.-C. Chin and S.-T. Quek, "Measurement of traffic conflicts," *Safety Science*, vol. 26, no. 3, pp. 169–185, 1997.

[6] L. Yang, J. H. Yang, E. Feron, and V. Kulkarni, "Development of a performance-based approach for a rear-end collision warning and avoidance system for automobiles," in *Proceedings of the IEEE Intelligent Vehicles Symposium*, pp. 316–321, Columbus, OH, USA, 2003.

[7] N. Uno, Y. Iida, S. Itsubo, and S. Yasuhara, "A microscopic analysis of traffic conflict caused by lane-changing vehicle at weaving section," in *Proceedings of the 13th Mini-Euro Conference Handling Uncertainty in Transportation Analysis of Traffic and Transportation Systems*, 2002.

[8] M. Bin, N. Uno, and Y. Iida, "A study of lane-changing behavior model at weaving section considering conflicts," *Journal of the Eastern Asia Society for Transportation Studies*, no. 5, pp. 2039–2052, 2003.

[9] N. Uno and Y. Iida, "Objective Analysis of Traffic Conflict and Modeling of Vehicular Speed Adjustment at Weaving Section," *Infrastructure Planning Review*, vol. 20, no. 4, pp. 989–996, 2003.

[10] C. Oh, S. Park, and S. Ritchie, "A method for identifying rear-end collision risks using inductive loop detectors," *Accident Analysis & Prevention*, vol. 38, no. 2, pp. 295–301, 2006.

[11] C. Oh, J. Oh, and J. Min, "Real-time detection of hazardous traffic events on freeways: Methodology and prototypical implementation," *Transportation Research Record*, no. 2129, pp. 35–44, 2009.

[12] Y. Kweon, "Development of crash prediction models using real time safety surrogate measures," Tech. Rep. UVACTS-15-0-104, 2008.

[13] K. Ozbay, H. Yang, B. Bartin, and S. Mudigonda, "Derivation and Validation of New Simulation-Based Surrogate Safety Measure," *Transportation Research Record*, vol. 2083, no. 1, pp. 105–113, 2018.

[14] F. Saccomanno and F. Cunto, "Comparing safety at signalized intersections and roundabouts using simulated rear-end conflicts," *Transportation Research Record*, vol. 2078, pp. 90–95, 2008.

[15] S. Tak, S. Kim, and H. Yeo, "Development of a Deceleration-Based Surrogate Safety Measure for Rear-End Collision Risk," *IEEE Transactions on Intelligent Transportation Systems*, vol. 16, no. 5, pp. 2435–2445, 2015.

[16] F. Muehlfeld, I. Doric, R. Ertlmeier, and T. Brandmeier, "Statistical behavior modeling for driver-adaptive precrash systems," *IEEE Transactions on Intelligent Transportation Systems*, vol. 14, no. 4, pp. 1764–1772, 2013.

[17] F. Naujoks, A. Kiesel, and A. Neukum, "Cooperative warning systems: The impact of false and unnecessary alarms on drivers compliance," *Accident Analysis Prevention*, vol. 97, Article ID S0001457516303396, pp. 162–175, 2016.

[18] G. Abe and J. Richardson, "Alarm timing, trust and driver expectation for forward collision warning systems," *Applied Ergonomics*, vol. 37, no. 5, pp. 577–586, 2006.

[19] D. Lee and H. Yeo, "Real-Time Rear-End Collision-Warning System Using a Multilayer Perceptron Neural Network," *IEEE Transactions on Intelligent Transportation Systems*, vol. 17, no. 11, pp. 3087–3097, 2016.

[20] A. H. Jamson, F. C. H. Lai, and O. M. J. Carsten, "Potential benefits of an adaptive forward collision warning system," *Transportation Research Part C: Emerging Technologies*, vol. 16, no. 4, pp. 471–484, 2008.

[21] E. Todosiev, *The action point model of the driver-vehicle system*, The Ohio State University, 1963.

[22] R. M. Michaels, "Perceptual factors in car-following," in *Proceedings of the 2nd ISTTF*, pp. 44–59, London, UK, 1963.

[23] G. Newell, "A simplified car-following theory: a lower order model," *Transportation Research Part B: Methodological*, vol. 36, no. 3, pp. 195–205, 2002.

[24] H. Yeo, *Asymmetric microscopic driving behavior theory*, University of California, Berkeley, CA, USA, 2008.

[25] K. Vogel, "A comparison of headway and time to collision as safety indicators," *Accident Analysis & Prevention*, vol. 35, no. 3, pp. 427–433, 2003.

[26] Q. Liu, N. Garber, and M. Center, "Identifying the Impact of Truck-Lane Restriction Strategies on Traffic Flow and Safety Using Simulation," Tech. Rep. UVACTS-14-5-103, 2007.

[27] "NGSIM - Next Generation Simulation," 2006.

[28] M. Brackstone and M. McDonald, "Car-following: a historical review," *Transportation Research Part F: Traffic Psychology and Behaviour*, vol. 2, no. 4, pp. 181–196, 1999.

[29] D. Delorme and B. Song, "Human driver model for SmartAHS," *California Partners for Advanced Transit and Highways (PATH)*, 2001.

[30] S. Panwai and H. Dia, "Comparative evaluation of microscopic car-following behavior," *IEEE Transactions on Intelligent Transportation Systems*, vol. 6, no. 3, pp. 314–325, 2005.

[31] A. Kesting, M. Treiber, and D. Helbing, "Enhanced intelligent driver model to access the impact of driving strategies on traffic capacity," *Philosophical Transactions of the Royal Society A: Mathematical, Physical & Engineering Sciences*, vol. 368, no. 1928, pp. 4585–4605, 2010.

[32] M. Treiber and A. Kesting, *Traffic Flow Dynamics: Data, Models and Simulation*, Springer, Berlin, Germany, 2013.

[33] R. Foote, "Single lane traffic flow control," in *Proceedings of the Procedings of Second International Symposium on rthe Thory of Road Traffic Flow*, pp. 84–103, 1965.

[34] L. Edie, "Discussion of traffic stream measurements and definitions," in *Proceedings of the Procedings of Second International Symposium on rthe Thory of Road Traffic Flow*, pp. 139–154, 1965.

[35] T. Forbes, "Human factor considerations in traffic flow theory," *Highway Research Record*, vol. 15, pp. 60–66, 1963.

[36] G. F. Newell, "Theories of instability in dense highway traffic," *Journal of the Operations Research Society of Japan*, vol. 5, pp. 9–54, 1965.

[37] C. F. Daganzo, M. J. Cassidy, and R. L. Bertini, "Possible explanations of phase transitions in highway traffic," *Transportation Research Part A: Policy and Practice*, vol. 33, no. 5, pp. 365–379, 1999.

[38] H. M. Zhang, "A mathematical theory of traffic hysteresis," *Transportation Research Part B: Methodological*, vol. 33B, no. 1, pp. 1–23, 1999.

[39] H. Yeo and A. Skabardonis, "Microscopic fundamental relationships between vehicle speed and spacing in view of asymmetric traffic theory," in *Proceedings of the 14th International IEEE Conference on Intelligent Transportation Systems (ITSC)*, pp. 1410–1414, October 2011.

[40] J. Suh, H. Yeo, and A. Skabardonis, "A Study on the Wave Development and Evolution Characteristics of Stop-and-Go Traffic," in *Proceedings of the Transportation Research Board 91st Annual Meeting*, 2012.

9

Microscopic Simulation-Based High Occupancy Vehicle Lane Safety and Operation Assessment

Chao Li, Mohammad Karimi, and Ciprian Alecsandru

Department of Building, Civil and Environmental Engineering, Concordia University, Montréal, QC, Canada H3G 2W1

Correspondence should be addressed to Chao Li; chaoli0351@gmail.com

Academic Editor: Alain Lambert

This study proposes two general alternative designs to enhance the operation and safety of High Occupancy Vehicle (HOV) lanes at junctions with bus terminals or parking lots. A series of analysis tools, including microscopic simulation, video-based vehicle tracking technique, and Surrogate Safety Assessment Model (SSAM), are applied to model and test the safety and operational efficiency of an HOV road segment near a bus terminal in Québec as a case study. A metaheuristic optimization algorithm (i.e., Whale Optimization Algorithm) is employed to calibrate the microscopic model while deviation from the observed headway distribution is considered as a cost function. The results indicate that this type of HOV configurations exhibits significant safety problems (high number of crossing conflicts) and operational issues (high value of total delay) due to the terminal-bound buses that frequently need to travel across the main road. It is shown that the proposed alternative geometry design efficiently ameliorates the traffic conflicts issues. In addition, the alternative control design scheme significantly reduces the public transit delay. It is expected that this methodology can be applied to other reserved lane configurations similar to the investigated case study.

1. Introduction

The HOV lane represents a restricted usage traffic lane reserved for vehicles carrying a predetermined number of occupants. The implementation of an HOV lane system targets mobility improvement of both current and future roadway networks. Considering over forty years of deployment of HOV lanes, it has been proven that reserved lanes contribute to mitigating traffic congestion in urban areas and reduce the person-hour delay effectively [1, 2]. However, many problems related to various implementations of HOV lanes have been identified. These problems can be roughly classified into two categories, the reduction of capacity (for the non-HOV users) and potential traffic safety issues, respectively. The former category may include increased congestion on the adjacent General Purpose (GP) lanes and/or reduction of vehicle speeds due to the merging maneuvers of High Occupancy Vehicles into the GP lanes. The latter category is mainly related to the lane changes at prohibited locations, especially in the proximity of junctions with other road facilities, such as bus terminals or parking lots [3].

Currently, efforts are continually being made to explore the new ways to improve the operation and safety of HOV facilities. However, there is no universally accepted method to evaluate the effectiveness of safety of certain HOV facilities [4]. Some studies focused on the HOV safety evaluation based on the statistical analysis of accidents data during long periods [5]. Several studies examined the safety of HOV facilities with respect to different types of geometric design based on the collision and driving behavior (i.e., lane-changing) data [6, 7]. Nevertheless, obtaining reliable accident data over a long enough period is not always possible, especially for recently deployed facilities. A reliable accident-based analysis takes a long time to establish and thus is not suitable for current urban traffic system development. In addition, many characteristics of the urban traffic system may change over time (e.g., traffic demand volumes, road alignments, traffic mix, etc.), and this might require an expedited method to assess the existing traffic conditions. Accordingly, using conflict analysis as a method of safety assessment is preferable, as it makes analyzing the safety improvement before implementing any treatment in the real world possible.

However, the geometric configuration of an HOV facility has significant impacts on the safety performance [7]. For instance, conducting the before-after study of converting the continuous access to limited access of lane changes in HOV lanes has shown a significant decrease in conflict occurrence. Therefore, the HOV facilities with limited access are expected to be safer than those with continuous access. To validate this conclusion, more studies must be conducted. However, there is limited opportunity for researchers to conduct before-after studies of road facilities with respect to the geometric modification, because they are too infrequent. Therefore, utilizing simulation tools may be an effective remedial measure to overcome the limitation of data availability and to evaluate the impacts of potential geometric alignment changes of existing facilities. Several studies have introduced the evaluation of safety or capacity of HOV facilities utilizing microsimulation [3, 8, 9]. However, these studies mainly focused on the analysis results of the study areas. Therefore, it is necessary to develop a systematic assessment method for HOV lanes. In particular, the HOV deployment on arterials in the proximity of the terminals and parking lots can be conducted using real-word data to calibrate a microscopic simulation model.

In this study, a VISSIM microsimulation model is developed to test the safety and operational efficiency of an urban HOV facility near a bus terminal in Québec, Canada. This model is calibrated by employing a metaheuristic optimization algorithm–Whale Optimization Algorithm (WOA)—to minimize the deviation of simulations results from the observed data. Two general alternative network designs are proposed for comparison analysis (i.e., one modifies the existing road geometric alignment; another one proposes a change in the existing traffic control strategy). To assess the road safety impact of the proposed alternative designs, the Surrogate Safety Assessment Model (SSAM) is applied to compare the simulated vehicle conflicts between the existing network and the alternative solutions. The results indicate that the status quo of the study area exhibits a safety problem due to high interactions between buses and passenger cars. The proposed alternative geometry design efficiently eliminates the traffic conflict. In addition, the alternative control design scheme significantly reduces the public transit delay.

2. Literature Review

Traditionally, most traffic safety studies employed statistical analysis of accident records within a given study area [10–13]. Several studies pointed out the drawbacks of using authority reported crash data for safety evaluation, for example, the lack of ability to evaluate the safety of traffic facilities yet to be built or to assess the traffic remediation solution yet to be applied in the field. In addition, the seldom and random occurrence of traffic accidents lead to the slowness of establishing analysis [14] or the lack of ability to deduce the crash process [15, 16]. On account of these drawbacks, an alternative safety evaluation approach which includes the computer microsimulation modeling of vehicle interactions was developed over the past couple of decades. This approach was possible mainly due to the advancements in computing technologies that allowed the development of enhanced traffic simulation models to be able to replicate vehicle interactions through modeling complex driving behaviors [16, 17]. A significant advantage of simulation-based safety analysis is that microsimulation models can easily generate and measure various safety performance indicators [18, 19]. The typical safety performance indicator is the vehicular conflict, given that conflicts can be observed more frequently than crashes and that their frequency is expected to be correlated with the crash occurrence [14, 20–22]. Various studies have validated the statistical significance and correlation between conflicts and accidents [23–26].

A dedicated tool, namely, SSAM was developed by Federal Highway Administration (FHWA) to automatically identify, classify, and evaluate the severity of the simulated traffic conflicts [14]. Several studies showed that by combining VISSIM and SSAM a reliable tool for traffic safety evaluation can be used, provided that a consistency between the field observed and simulated conflicts is observed [27, 28]. Another study proposed a two-step calibration procedure of VISSIM (Wiedemann model) to enhance correlation between simulated and field-measured conflicts [29]. Therefore, if the simulation model is properly calibrated, it can be used to represent reliably the real-world traffic network in terms of both operation and safety parameters.

3. Methodology

3.1. Modeling of Geometry and Flow. Typically, more detailed information contained in the simulation model contributes to capturing more reliably the traffic conditions at a given study area. This is especially important for a traffic safety simulation model, which requires good accuracy of both simulated capacity and vehicle performance.

The basic input to this model is represented by the road characteristics (i.e., the number of lanes on each direction, the lane separation type, and the position of access). In this study, the links and connectors of the study area were built in VISSIM by means of an aerial photo from Google Maps®. Some details of the geometry, for example, the access position of the public transit terminal, were measured on the field and were compared with the field-recorded videos to ensure the accuracy. Similarly, the position of the reserved lane was collected on the field and included in the simulation model.

Traffic flow is another important input parameter as it relates to the road capacity, one of the potential calibration variables. Traffic flows were measured using the videos recorded at the study area—the following data was collected: the vehicle counts of each lane, vehicle routes within the study area, and the vehicle types (e.g., bus, truck, and passenger cars). In this study, in order to smooth out random variations in flows, while maintaining good precision, the vehicle flows were recorded and input into the model in five-minute increments. An additional five-minute period without vehicle demand was included at the end of each simulation scenario to avoid truncating the analysis period observed in the field. To model the observed vehicle composition, road users were identified and classified into three categories, passenger cars, buses, and trucks, respectively. The basic vehicle characteristics, for example, the acceleration rate, vehicle length, and vehicle weight of each vehicle type, can be modeled separately in VISSIM so as to reflect the traffic more realistically. To determine individual vehicle routes, vehicles were tracked in from the videos generated by three cameras that were used to cover the whole study area. The route of each vehicle in the simulation was assigned in strict accordance with the path observed on the video recordings to ensure a realistic representation of the study area.

3.2. Modeling of Traffic Signal. The peak hour traffic signal cycle length and the red, amber, and green time intervals on each direction were collected on the field and modeled in VISSIM. In this study, a fix-cycled signal program was built and set at the intersection to replicate the traffic light at the study area. Some additional signal control strategy was used in this study to improve the network performance; for example, a fix signal cycle contains a protected left-turn phase at the intersection and a pulse-triggered signal at the public transit terminal.

To improve the efficiency of public transit, a pulse-triggered signal control was implemented by adding a detector at the exit of the terminal and signal heads linked with the detector near the terminal. An add-on signals design model, namely, Vehicle Actuated Programming (VAP) was programmed to control this actuated signal. Typically, a signal phase of permanent green on the main street and permanent red on the minor road is toggled when no buses are detected. Meanwhile, when the existing buses are detected by the sensor, the signal is programmed to switch to the complementary phase (i.e., green signal on the minor road and red on the main road), thus protecting the movements of buses crossing through multiple lanes.

3.3. Modeling of Right of Way without Signal Control. In VISSIM priority rules are defined to capture the conflicting traffic flows that are not controlled by signals. In this study, the priority rules were set at the entry and the exit zones of the bus terminal, in order to realistically model the access and egress movements of buses as they were observed in the video recordings. Typically, the buses travel to and from the terminal, yielding to the vehicles traveling along the main arterial, and stop in position near the access or exit until acceptable gaps occur on both directions on the main road. Two thresholds are set for the *priority rules* to confine the crossing of the yielding vehicles, respectively, the *minimum headway* and the *minimum gap time*. A yielding vehicle will stop before the stop line until both predetermined thresholds are achieved. The values of the thresholds are determined by reviewing all the accepted gaps and headways by the crossing buses from the video.

The conflict areas are automatically generated in VISSIM where links or connectors overlap. In this study, the priority rules at the conflict areas were set to capture the vehicles approaching the conflict area from the minor road and yielding those from the main road, as typically observed in the field. The gap time needed for crossing at the conflict area was determined similarly by reviewing the video recordings.

Another important VISSIM calibration parameter is the *avoid blocking* value, which defines the ratio of vehicles that do not stop in the middle of a junction. This value is defaulted to be 100% in VISSIM; in other words, all the vehicles will follow the rule, not to block the junctions, if there is stopping traffic ahead. However, by reviewing the video recorded at the study area, no vehicle obeyed this rule. Therefore, to reflect the real conditions, this value is set to 0% for all the conflict areas in simulation models used in this study.

3.4. Modeling of Driving Behavior. Properly modeling of the field observed driving behavior is critical for road safety evaluation, since it directly influences the vehicle interactions in a micro level. Microsimulation tool VISSIM adopted Wiedemann car following model as the main portion for modeling the vehicle longitudinal movement and rule-based laws for modeling of vehicle lateral movement and lane change behavior.

In this study, the Wiedemann 74 model is selected to simulate the urban motorized traffic as suggested by the VISSIM user's manual [30]. This model contains three adjustable parameters, respectively, the *average standstill distance*, the *additive part of safety distance*, and *multiplicative part of safety distance*. *Average standstill distance* defines the average desired distance between two cars. *Additive part of safety distance* and *multiplicative part of safety distance* represent the values used for the computation of the desired safety distance. For the initial simulation, the values of these three parameters are usually defined with the default value. However, they must be calibrated later to suite the real driving behaviors of the study site.

The lane change behaviors are defined by a rule-based model in VISSIM. In this model, the critical parameter that decides whether a lane change would be executed is the *minimum headway*. A vehicle can only change lane when there is a distance gap arrival at the adjacent lane that is bigger than the predetermined minimum headway. Otherwise, it has to either travel continuously or stop and wait until the occurrence of an enough gap for it in order to merge according to its predefined route. In this study, the value of the *minimum headway* was determined by reviewing the videos.

Another noticeable parameter defined in the lane change model is the *advanced merging*; this option is selected in this study; thus more vehicles can change lanes earlier when following their routes, as encountered in the videos.

3.5. Measurement of Vehicle Speed Distribution by Feature-Based Tracking. Vehicle speed distribution is an important input parameter for safety simulation. While potentially more accurate, individual vehicle speed on multiple lanes is usually difficult to measure on the field simultaneously with radar devices. Therefore, an alternative method was applied in this study to measure the vehicle speed, which is the video-based feature tracking.

An open-sourced software project, namely, *Traffic Intelligence*, was used to automatically track and measure the speed of the vehicles caught by the video at the study

FIGURE 1: Points selected on the video frame to compute homography file.

FIGURE 2: Feature tracking process by *Traffic Intelligence*.

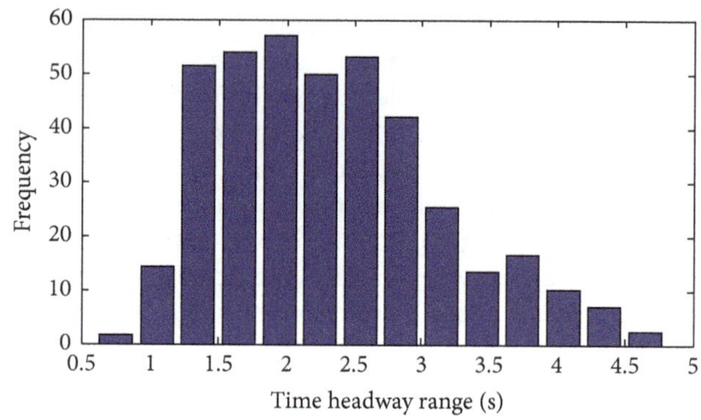

FIGURE 3: Observed vehicle gap distribution.

site [31]. *Traffic Intelligence* consists of a set of tools that work cooperatively for traffic data processing and analysis, including camera image calibration, feature tracking, and trajectory data analysis.

The feature-based tracking algorithm utilizes a homography file that projects the camera image space to the real-world ground plane. The homography file was created by utilizing a video frame and a corresponding aerial photo with known scale (pixels per meter). In this study, an aerial photo of the study site from Google Maps with known scale of 0.21 pixels per meter was adopted. In total ten noncollinear visible points on the video frame were positioned on the aerial photo; thus, the video image was projected to the aerial photo, and the vehicles tracked in the video were deemed to be tracked in the real-world plane with their speeds. Figure 1 shows the points projected to the aerial photo from the video frame.

Based on the computed homography file, the feature tracking program can be run. The predetermined number of features of each vehicle in the video was detected and tracked frame by frame until the vehicle leaves the video capture area. In order to suppress the interference of the shadows, a mask image was created and toggled with the video image; therefore only the features within the white range of the mask image can be detected, and the shadows can be filtered out. The features that move consistently were then grouped together to generate the trajectory file of each vehicle, and all the trajectories generated from the video were written into a database. The average speed of each vehicle can be easily read by processing their trajectories. Figure 2 shows the feature tracking process by *Traffic Intelligence*.

3.6. Model Calibration. In order to determine the optimum values for the calibration parameters, an objective function should be defined based on the error between observed data and simulated data. The objective function is the deviation of the simulated gap distribution from the real observed gap distribution. In order to test this goodness of fit (objective), the Chi-square test was employed. In this study, the westbound vehicle gap distribution on the GP lane near the bus terminal was taken as the criterion to calibrate the model, because the vehicle time gap directly reflects the car following behavior. The real vehicle gaps were observed manually from the video using the *MPC player* that provides milliseconds accuracy. Because the vehicles travel westbound pass through a signalized intersection before they enter the cameras field of view, to eliminate the impact of the red time at the intersection, the time gaps bigger than 5 seconds were ignored. The distribution of all the observed gaps that are smaller or equal to 5 seconds was recorded in a histogram with a sample rate of 0.3 seconds. Figure 3 shows the observed vehicle gap distribution.

In this paper, the Whale Optimization Algorithm (WOA), a metaheuristic nature-based algorithm, is applied to calibrate the model. The deviation of the simulated headway distribution from its' observed distribution is considered as the objective function to be minimized during the calibration process. WOA is inspired by hunting behavior of humpback whales. It is defined as "the simulated hunting behavior with random or the best search agent to chase the prey and the use of a spiral to simulate bubble-net attacking mechanism of humpback whales" [32]. The hunting behavior of whales is representative of the procedure of this algorithm.

The three parameters of the Wiedemann 74 model in VISSIM (i.e., average standstill distance (ASSD), additive part of safety distance (APSD), and multiplicative part of safety distance (MPSD)) which has the highest impact on the model have been selected to be calibrated. Using MATLAB, an optimization toolbox connecting to COM-interface of VISSIM by M-file programming in MATLAB, the calibration process has been accomplished. After 190 simulation runs, the optimal values of parameters were determined to be as follows: ASSD = 1.156, APSD = 0.637, and MPSD = 8.079. For diverse random seeds, the simulation results showed these optimal parameters lead to statistically matching the observed headway distributions with the simulated ones at 90% confidence level. It is worth mentioning that the simple way to optimize the cost function is exploring the whole possible region of the parameters to find the global minimum, which is extremely time-consuming. For this case study, these optimization parameters took values within the following intervals: ASSD between 0 and 2, APSD between 0 and 1, and MPSD between 0 and 10. By exploring these intervals, the optimal values were found after nearly 1000 simulation runs.

The lateral movement of buses that merge into the main traffic from HOV lane or travel across the road when an acceptable gap was identified was also calibrated by adjusting the parameters of the priority rule. The minimum gap time and distance headway were set to 6 seconds and 20 meters, respectively, similar to the values observed in the recorded videos. It is noticeable that a part of the terminal-bound buses changed lanes between the reserved HOV lane and the adjacent GP lane before the intersection; this behavior is reflected in the simulation model.

3.7. Simulation Output. VISSIM provides direct output of various kinds of simulation results. In this study, the vehicle delay and trajectory were analyzed to evaluate the operational efficiency and safety of the study area.

Vehicle delay data can be generated by setting *vehicle travel time* on the defined vehicle routes, which are defined by a *Starting Point* and an *End Point*, respectively. For the vehicles that pass through the *Starting Point* and the *End Point*, successively, the travel time delays are automatically calculated. The vehicle delays of the interested vehicle routes were then analyzed to evaluate the operational efficiency of the network.

The trajectories of all the simulated vehicles can be generated by VISSIM, and the recorded trajectory data was then analyzed using SSAM, to evaluate the vehicle conflicts within the network. For each simulation run, different simulation random seeds were applied, and the output results were averaged for analysis purposes. This simulation setup scenario accounts for the stochastic properties of the simulation model, thus reflecting real-world traffic behavior more realistically.

3.8. Analyzing Vehicle Conflicts Using SSAM. The vehicle trajectory data collected from VISSIM was used in SSAM to assess the vehicle conflicts detected in the study area. Most studies evaluate traffic safety through two surrogate measures, Time to Collision (TTC) and Postencroachment Time (PET). Values below a commonly accepted threshold of either TTC or PET value indicates a higher probability of collision. SSAM is able to automatically estimate the TTC and PET values of each vehicle interaction and thus to record all potential conflicts. In this study, the TTC and PET were set to 1.5 seconds and 5 seconds, respectively, the values frequently established by previous research studies [20, 33].

The detected conflicts were classified into three types, based on the predetermined conflict angles, namely, *crossing*, *lane changing*, and *rear ending*, respectively. The thresholds of the conflict angles were adjusted to 2 degree and 45 degree as suggested by previous studies [8]. Basically, detected conflict which has a conflict angle of 2 degrees or less is defined as *rear ending* conflict; if the conflict angle is between 2 and 45 degrees, it is detected as *lane changing*; while if the conflict angle is bigger than 45 degrees, it is recorded as the *crossing type*. However, due to the peculiarity of geometry of each study area, the link information of all the output conflicts, which was also detected by SSAM, was manually checked to properly determine their type. The three types of conflicts were recorded for subsequent comparative safety analysis.

A built-in filter of SSAM can be applied to screen out the conflicts caused by each measured movement by reading the corresponding link information. The spots where conflicts were detected can be plotted automatically on the toggled network image by positioning the VISSIM network coordinates. The conflicts of different types can be showed in different shapes or colors on the togged map to give a visual estimate of the hotspot areas (i.e., conflicts' frequency and density).

3.9. Summary. The methodology presented in this study introduces a simulation-based approach to evaluate road network safety and efficiency. To apply this methodology, the field traffic conditions are collected, and the detailed information including the field geometry, control strategy, flow, and driving behavior is reviewed. Such basic information is then integrated in a VISSIM simulation model. With an important model parameter, the vehicle speed distributions are obtained using a feature tracking program, namely, *Traffic Intelligence*. The model is properly calibrated until the output vehicle time gap distribution compared well with the field observed vehicle gap distribution by applying the Chi-square test. The model output vehicle delays are reviewed for network operational efficiency analysis, and the model output vehicle trajectory files are analyzed by SSAM to determine the conflict within the study area thus giving the safety level of the site. Figure 4 shows the flow chart of the methodology used in this study for traffic safety and operational efficiency evaluation.

4. Case Study

4.1. Study Area Description. The study area used in this study is a segment of Rte-116, a suburban highway in Lévis, Québec. Evaluations of traffic safety and operations were made at a specific location along the four-lane east-west arterial segment that includes one GP lane and one HOV lane, in both directions. The eastbound reserved lane allows buses and passenger cars with three or more passengers, while the westbound direction has a bus-only lane. The current design of this facility is such that the westbound buses arriving at or departing from the terminal have to travel across the four-lane undivided road. Figure 5 shows the current paths of the buses using the terminal.

The traffic video feeds of vehicles accessing the terminal, the commuter parking lot, and traveling along Rte-116 were collected via GoPro HD video cameras that were installed

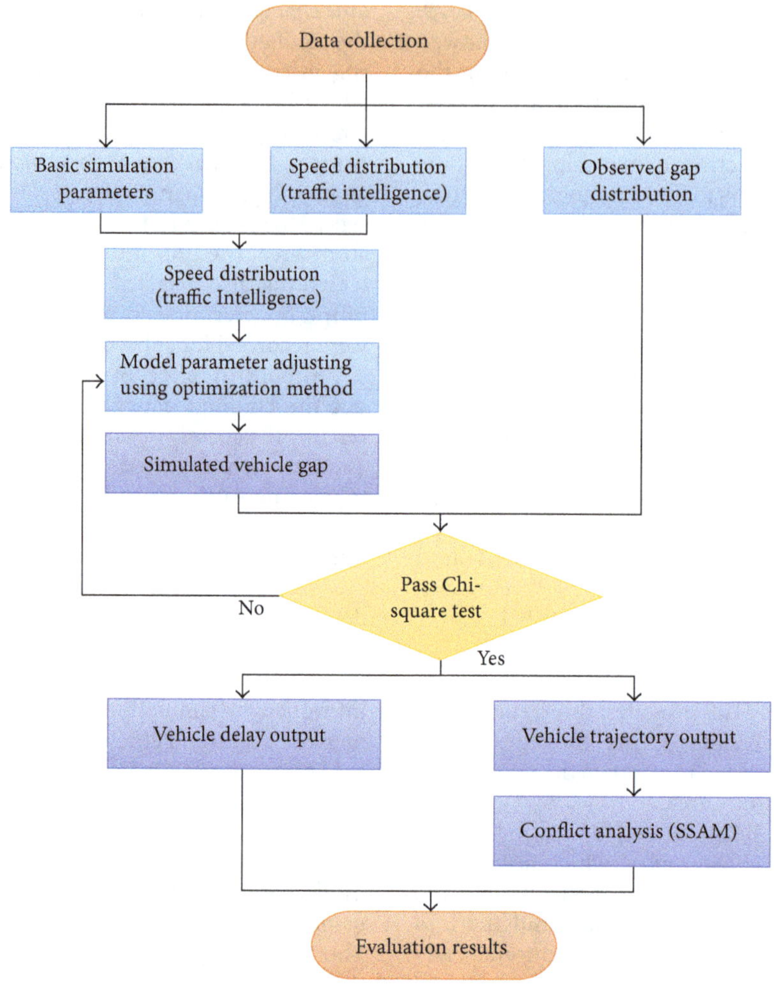

FIGURE 4: Framework of evaluation procedure.

on top of extendable masts along the roadway. Cameras 1 and 2 were both installed at the same location with views opposing each other. The orientations of these two cameras were adjusted to capture east-west traffic that interacts with both access points into and out of the bus terminal. Camera 3 was installed at the proximity of the commuter parking lot entry/exit gate, to capture interactions between main road traffic and vehicles to and from the parking. The positions of the cameras are shown in Figure 5. The video traffic data of the PM peak hour (4:30 pm~5:30 pm) was used in the final analysis of this study.

A probe vehicle was driven several times along the study segments with an arbitrarily selected constant speed. The known speed values were used to calibrate the postprocessing speed detection measuring software, *Traffic Intelligence*. A fixed 88-second cycle of the traffic signal along Rte-116 at the adjacent intersection (i.e., 40 seconds, red, 40 seconds, green, and 4 seconds, yellow) was measured in the field and used in the simulation model of the study area.

The video files from each camera were processed in 5-minute increments to manually determine the distribution of traffic flows during the analysis period. Vehicles were distinguished into four types: passenger cars (on the GP lane), buses, trucks, and reserved lane users, respectively. This definition of the traffic mix was necessary to capture more reliably the vehicle interactions in the traffic simulation model (different vehicle types exhibit different driving behaviors in terms of acceleration, minimum headway, etc.). Tables 1 and 2 show a classification of westbound and eastbound

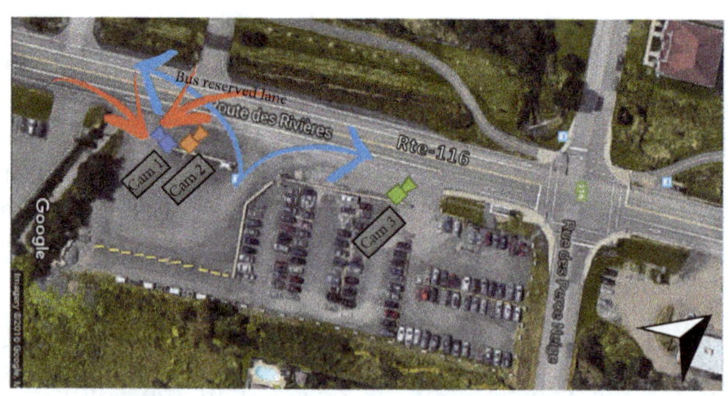

FIGURE 5: Paths of the terminal-bound buses.

TABLE 1: Observed traffic flow during the peak hour (4:30 pm~5:30 pm).

Time	Average vehicle flows (vehicles/hour)							
	Westbound				Eastbound			
	Car	Bus	Truck	HOV	Car	Bus	Truck	HOV
4:30 pm~5:30 pm	663	16	3	7	338	5	4	36

TABLE 2: Access and egress vehicles during the peak hour (4:30 pm~5:30 pm).

Time	Average vehicle flows (vehicles/hour)							
	Westbound				Eastbound			
	Bus		Parking car		Bus		Parking car	
	Access	Egress	Access	Egress	Access	Egress	Access	Egress
4:30 pm~5:30 pm	10	14	0	37	7	3	4	18

TABLE 3: Vehicles characteristics.

Vehicle type	Length (meter)	Width (meter)	Weight (ton)	Maximum acceleration (m/s^2)	Maximum deceleration (m/s^2)
Car and HOV	3.75–4.76	1.85–2.07	-	3.5	−7.5
Bus	11.54	3.17	4–12	1.24	−7.5
Truck	13.94	2.63	2.8–40	2.5	−5.5

traffic flows along the highway, as well as access/egress of the buses using the terminal during the afternoon peak period.

Traffic Intelligence was utilized to measure the vehicle speed. Calibration of the video analysis software was performed using various mask pictures to filter the shadows of the moving vehicles until the measured speeds of the probe vehicle were identical to the observed values. The vehicle speed distributions of both westbound and eastbound vehicles were recorded every five minutes and used as simulation input parameters.

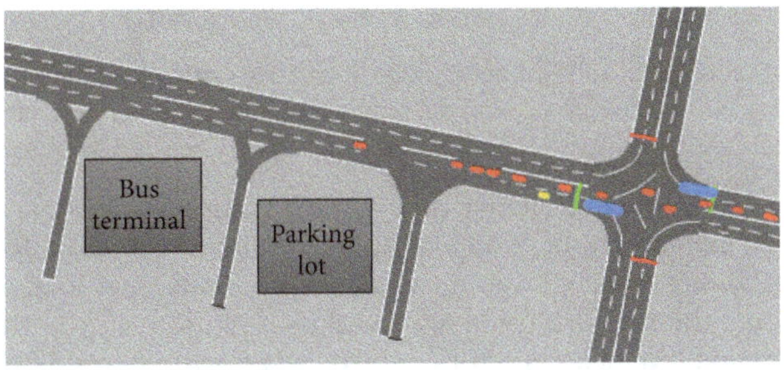

FIGURE 6: The status quo network modeled in VISSIM.

4.2. Modeling Existing Configuration and Traffic Conditions (Status Quo). The peak hour traffic was modeled in VISSIM to evaluate traffic safety and operations of the observed arterial segment. Vehicle modals used in the simulation are selected by VISSIM automatically. The vehicle characteristics of the case study are shown in Table 3. To account for the effects of stochastic variation of the model's parameters, ten different simulations with different random seeds were ran, and the average values were used in the analysis.

The average vehicle delay (excluding signal waiting time at intersection) was measured for three types of movements, using the *vehicle travel time measurements* tool. *Movement 1* identifies the westbound traffic on the GP lane. *Movement 2* is associated with westbound buses entering the terminal (i.e., buses merging from HOV lane into the GP lane and then crossing the two eastbound lanes). *Movement 3* represents westbound buses leaving the terminal (i.e., buses that cross all the four lanes to enter the highway). Vehicle trajectory files were also generated for conflict analysis.

In addition, to evaluate the impact of expected increase in traffic flow on traffic operations (i.e., average vehicle delay) and safety (i.e., conflicts frequency), the same simulation model was used to evaluate similar scenarios, assuming the traffic volume increases in the future by 10%, 20%, and 30% from the current values. Figure 6 represents a snapshot of the VISSIM simulation model using the existing geometric alignment and traffic operations conditions.

4.3. Simulations of Alternative Geometry/Control Designs. The main concern related to traffic safety at the investigated study area pertains to the placement of the reserved lanes on the outside lanes. This configuration leads to multiple lanes crossing when left turns are needed and high occurrence of vehicle interactions was observed especially during congested traffic conditions.

Two alternative designs have been tested to evaluate their potential to mitigate traffic safety and operations issues. Figure 7 shows the VISSIM network layout of the first alternative design. In this model, westbound buses were prohibited to enter or exit the terminal by crossing the highway directly. Instead, an adjacent roadway segment was inserted along the south side of bus terminal, which is directly connected to the minor road. To serve the terminal-bound buses, ten seconds of left-turning signal phase was provided at the intersection on the main road. Similarly, for each

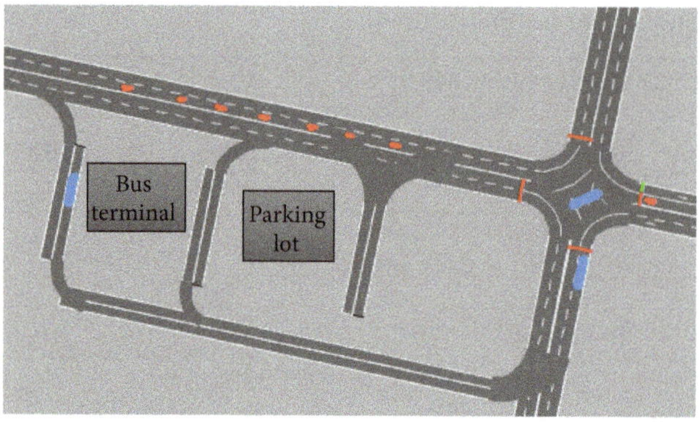

FIGURE 7: VISSIM network of alternative road geometry design.

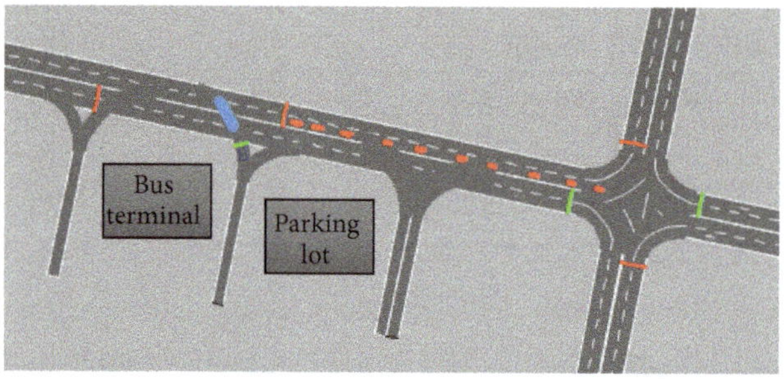

FIGURE 8: VISSIM network of alternative control design.

traffic demand alternative (i.e., current status, 10%, 20%, and 30% increments of vehicular traffic volume), the collected peak hour vehicle flows and speed distributions were used to model the network using ten simulation random seeds. The individual vehicle trajectories and delay measurements of the same movements evaluated for the status quo configuration were collected and used for comparison analysis. Figure 8 shows the VISSIM network layout of the second alternative design. In this model, a loop detector that controls a signal set was added to the existing network. This system was used to control the egress of westbound buses as they leave the terminal. The add-on signal control model VAP was created to program the signal timing. The detector was placed near the exit of the bus terminal. As long as buses are not in the proximity of the sensor, the signal indicates green for the main road to allow east-west traffic and red for the bus exit to prevent the egress buses from traveling across the road directly. When buses are detected at the terminal, exit signal turns green for them and red for traffic on the main road, which allows for protected turns. The red signal on the main road lasts for 10 seconds from the last bus detected and then turns back to green until the next detection. The same vehicle hourly flows previously processed were used in this simulation scenario, and the same ten different simulation random seeds were applied. The delay measurements of the same types of movements and trajectory data were collected for comparative analysis.

4.4. Surrogate Safety Measures of Vehicle Conflicts. SSAM was applied to assess the vehicle conflicts detected in the

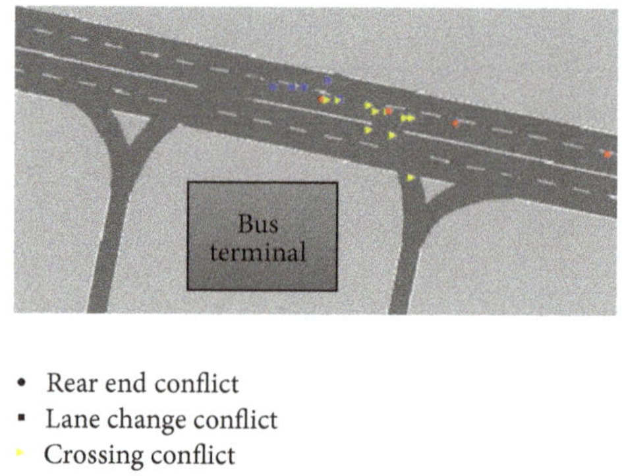

FIGURE 9: Conflicts near bus terminal plots on original network.

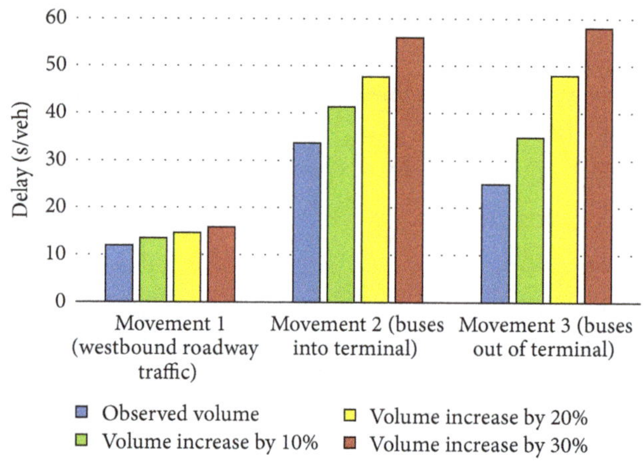

FIGURE 10: Effects of increasing traffic flow on average delay per vehicle.

study area for safety analysis. A built-in filter of SSAM was applied to screen out the conflicts caused by each measured movement by reading the corresponding link information. The spots where conflicts were detected were plotted automatically on the toggled network image by utilizing the VISSIM network coordinates. The conflicts of different types were showed in different shapes on the togged map. Figure 9 shows the spatial distribution of conflicts caused by measured movements near the bus terminal plotted on the original network.

4.5. Comparison Analysis of Safety and Operation. Figures 10 and 11 represent the impact of different traffic volumes on traffic operations (delay) and safety (conflicts).

As intuitively expected, more traffic demand leads to increased average delay. It also shows that of the three types of vehicle interactions analyzed movements labeled 2 and 3 (i.e., associated with buses entering and leaving the terminal) are affected by significantly higher delay than the vehicles moving along the east-west roadway. This is explained by the fact that buses have to make left turns from/into the arterial, and consequently they do not have the default right of way. In addition, traffic safety analysis (i.e., evaluation of vehicular interactions through the SSAM tool) shows that, for all levels of traffic demand, the majority (more than 85%) of vehicular

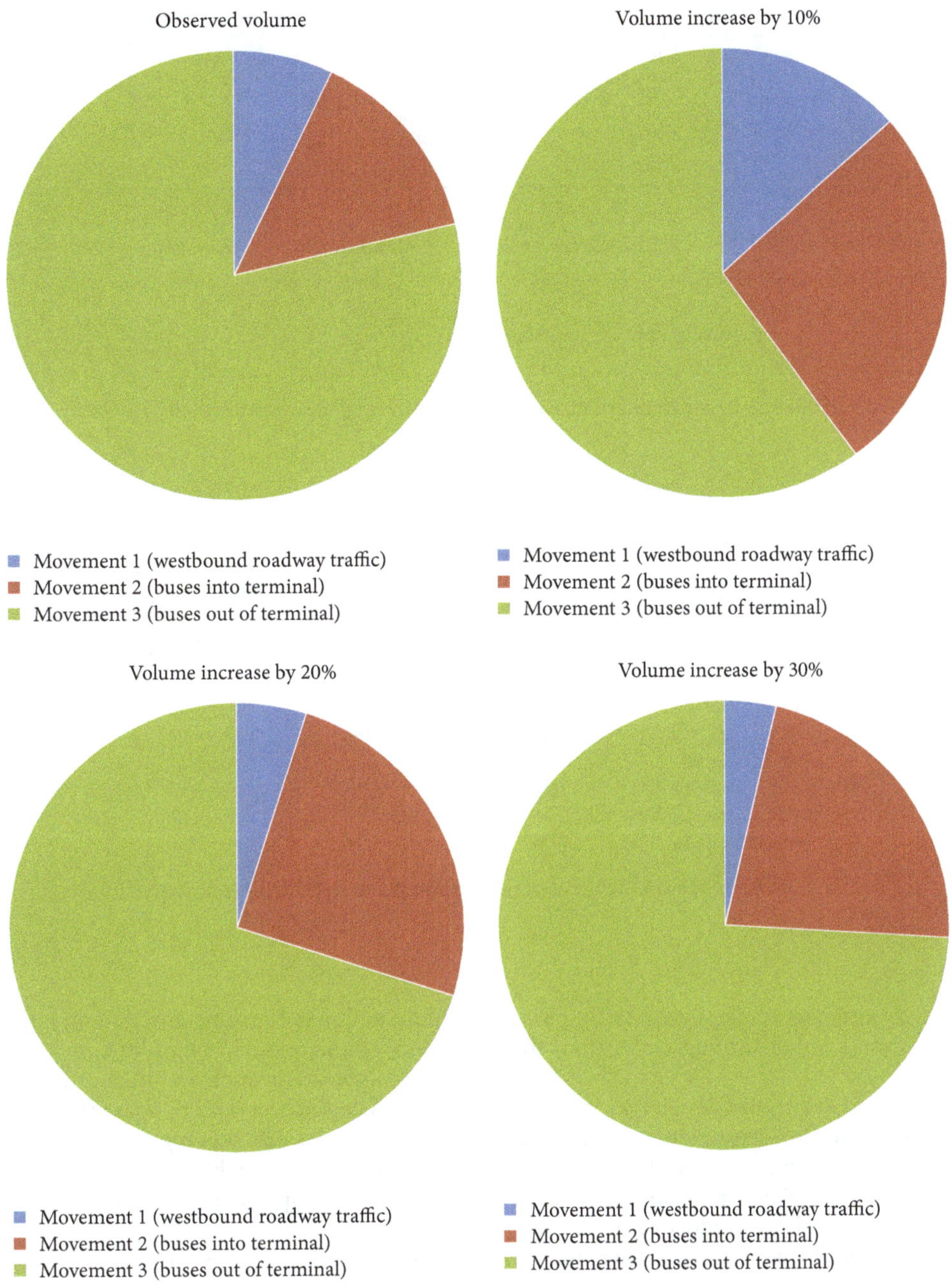

FIGURE 11: Sensitivity analysis of conflicts distribution (current configuration).

conflicts were crossing conflicts associated with the same movements of buses that enter or leave the terminal facility. Moreover, lane-changing conflicts were observed between buses moving from the reserved lane into the GP lane to engage in left-turning maneuvers towards the terminal.

Figure 12 shows the effects of different traffic volumes on traffic operations (magnitude of delay) and safety (frequency of conflicts) when the first alternative scenario was used. As expected, by including a separation median between the two directions of traffic, all vehicular conflicts associated with left-turn movements into and out of the terminal are eliminated. The sensitivity analysis demonstrates that traffic operations are not impacted by this design. It can be seen that there is a minor positive effect on the average vehicular delay for movement 1 (vehicles traveling westbound on Rte-116), but there is a significant positive effect on the average delay of buses accessing the terminal (i.e., a reduction in delay of about 85%). However, this alternative scenario brings a trade-off for the movements of buses exiting the terminal that are hindered for most traffic flow levels. The additional delay encountered by buses leaving the terminal is due to the fact that, for this design, the westbound egress buses must use

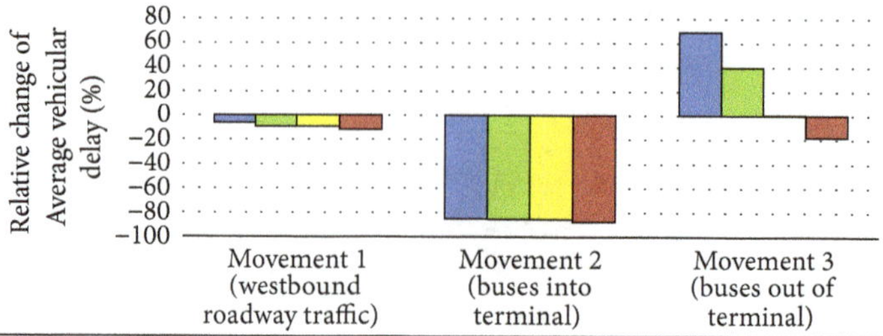

	Movement 1 (westbound roadway traffic)	Movement 2 (buses into terminal)	Movement 3 (buses out of terminal)
Observed volume	−6%	−85%	69%
Volume increase by 10%	−9%	−85%	40%
Volume increase by 20%	−9%	−85%	1%
Volume increase by 30%	−12%	−87%	−18%

FIGURE 12: Effects of first alternative design on the average delay (separation median).

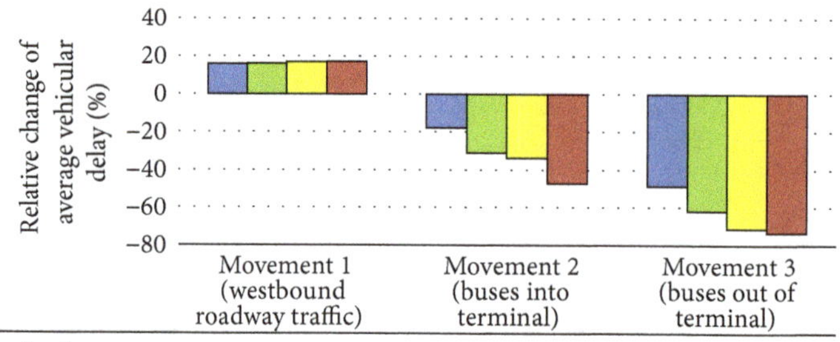

	Movement 1 (westbound roadway traffic)	Movement 2 (buses into terminal)	Movement 3 (buses out of terminal)
Observed volume	16%	−18%	−49%
Volume increase by 10%	16%	−31%	−62%
Volume increase by 20%	17%	−34%	−71%
Volume increase by 30%	17%	−47%	−73%

FIGURE 13: Effects of second alternative design on the average delay (traffic control).

the nearby intersection, and the traffic signal timing was not optimized to accommodate left-turning buses from the minor street.

The results for the second alternative design (i.e., controlling the access/egress of buses for Movements 2 and 3 via a bus-triggered traffic control signal, in order to reduce the vehicle interactions with the buses) are shown in Figure 13. It can be seen that this alternative design reduces considerably the delay of buses in and out of the terminal (Movements 2 and 3), while it increases by less than 17% the delay of vehicles traveling westbound along the arterial (Movement 1).

More importantly, the vehicular conflict analysis of these results shows the elimination of the crossing conflicts (Movements 2 and 3) related to buses accessing/leaving the terminal by turning left across the HOV and GP lanes. In addition, this design has no impact on the low conflict occurrence of Movement 1 (vehicles moving westbound on the arterial).

Several aspects of the proposed alternative designs are discussed at the end of this section. The delay of the traffic flow moving westbound on the arterial during the peak period was compared across all three simulation scenarios (i.e., current design, separation barrier, and traffic control alternative). It was found that the traffic control alternative leads to the most negative impact on the vehicular delay. In addition, conflict occurrence between the current design and the proposed traffic control design is not significantly different, due to breaking at red light; it is expected that rear ending conflicts might be more severe. On the other hand, rerouting buses through the intersection via the minor street seems to be the best option, because it eliminates completely all conflicts of left-turning vehicles while its impact on traffic operations might not be significant, since it can be mitigated with optimizing the traffic signal timing plan at the intersection.

To conclude, the existing geometric and traffic signal configurations show that there is a high occurrence of vehicular conflicts for left-turn buses approaching terminal. It can be seen from the results above that, by using the alternative designs, these types of conflicts are eliminated. In addition, the proposed alterations to existing alignment provide benefits for traffic operations because they reduce significantly the average vehicular delay. However, when traffic signals are used to control for protected left-turn buses that are rerouted through the adjacent intersection, an additional analysis of signal delay and optimization is necessary.

Similarly, the analysis of the measured movement 3 (i.e., westbound buses leaving from terminal) identifies a large number of crossing conflicts within the east-west traffic on the main arterial. Elimination of these conflicts can be achieved if this movement is protected either through the traffic signal sensitive to the buses present at the terminal exit, or by using the barrier separated geometry that reroutes the buses via the adjacent intersection. The results indicate that the network with alternative control design is the best for departing buses (i.e., the delay is the smallest).

As expected, the sensitivity analysis shows that an increased main arterial traffic volume leads to negative effects on the conflict frequency and average vehicular delay, regardless of the design used, while the alternative designs provide elimination or significant reduction in conflicts.

5. Concluding Remarks

This study proposed two general alternative geometry and control designs to improve the operation and safety of High Occupancy Vehicle (HOV) lanes near the bus terminals and parking lots. A VISSIM simulation model was created using the observed field geometry, control strategy, and vehicle flows, and then the vehicle priority rules and driving behaviors were calibrated to reflect correlated parameters observed on the field. An important model parameter, the vehicle speed distribution, was measured by feature-based tracking technique using an open-sourced program, namely, *Traffic Intelligence*. The model was calibrated using a meta-heuristic optimization algorithm (i.e., Whale Optimization Algorithm) with respect to the field-measured vehicles headway distributions. The results showed that this algorithm converged to the optimal parameters faster than searching whole parameters' intervals. The output delay data was used for operational efficiency analysis, and the output trajectory data was analyzed by SSAM to determine the number of vehicle conflicts within the study area. This procedure was applied to test the safety and operational efficiency of an HOV road segment in Lévis, Québec. The peak hour safety and operational traffic conditions of the status quo and of two alternative designs (i.e., geometry and control designs) were analyzed. The results indicate that the existing network configuration exhibits significant safety issues due to the crossing conflicts along the path of buses approaching the terminal across the four-lane arterial road. It was shown that one of the investigated alternative designs may enable the terminal-bound buses to travel on a different path to efficiently eliminate critical vehicular conflict. In addition, it was shown that the alternative control design can be used to reduce the bus delay by giving priority to public transit. It is expected that this methodology can be successfully applied to other similar reserved lanes facilities in the vicinity of the bus stations and parking lots.

Disclosure

All the viewpoints and errors expressed in this research are solely those of the authors.

Conflicts of Interest

The authors declare that there are no conflicts of interest regarding the publication of this paper.

Acknowledgments

This study was funded by Ministère des Transports du Québec (MTQ) through research Project R706.1. The authors would like to thank Matin Giahi Foomani and Gia Hung Lieu from Concordia University for the help in data collection.

References

[1] C. Fuhs and J. Obenberger, "Development of high-occupancy vehicle facilities: Review of national trends," *Transportation Research Record*, no. 1781, pp. 1–9, 2002.

[2] M. Menendez and C. F. Daganzo, "Effects of HOV lanes on freeway bottlenecks," *Transportation Research Part B: Methodological*, vol. 41, no. 8, pp. 809–822, 2007.

[3] A. Guin, M. Hunter, and R. Guensler, "Analysis of reduction in effective capacities of high-occupancy vehicle lanes related to traffic behavior," *Transportation Research Record*, no. 2065, pp. 47–53, 2008.

[4] J. I. Bauer, C. A. McKellar, J. M. Bunker, and J. Wikman, "High occupancy vehicle lanes-an overall evaluation including Brisbane case studies," in *Proceedings of the 2005 AITPM National Conference*, J. Douglas, Ed., pp. 229–244.

[5] T. F. Golob, W. W. Recker, and D. W. Levine, "Safety of high-occupancy vehicle lanes without physical separation," *Journal of Transportation Engineering*, vol. 115, no. 6, pp. 591–607, 1989.

[6] K. Jang, K. Chung, D. R. Ragland, and C.-Y. Chan, "Safety performance of high-occupancy-vehicle facilities," *Transportation Research Record*, no. 2099, pp. 132–140, 2009.

[7] X. Qi, G. Wu, K. Boriboonsomsin, and M. J. Barth, "Empirical study of lane-changing characteristics on high-occupancy-vehicle facilities with different types of access control based on aerial survey data," *Journal of Transportation Engineering*, vol. 142, no. 1, Article ID 04015034, 2016.

[8] H. Tao, M. G. Foomani, and C. Alecsandru, "A two-step microscopic traffic safety evaluation model of reserved lanes facilities: an arterial case study," in *Transportation Research Board 94th Annual Meeting No. 15-3635*, 2015.

[9] V. Thamizh Arasan and P. Vedagiri, "Microsimulation study of the effect of exclusive bus lanes on heterogeneous traffic flow," *Journal of Urban Planning and Development*, vol. 136, no. 1, Article ID 009001QUP, pp. 50–58, 2010.

[10] B. N. Persaud, R. A. Retting, P. E. Garder, and D. Lord, *Observational BeforeAfter Study of the Safety Effect of US*, Transportation Research Board, National Research Council, 2001.

[11] D. Gettman and L. Head, "Surrogate safety measures from traffic simulation models," *Transportation Research Record*, no. 1840, pp. 104–115, 2003.

[12] S. Srinivasan, P. Haas, P. Alluri, A. Gan, and J. Bonneson, "Crash prediction method for freeway segments 2 with high occupancy vehicle (HOV) lanes 3," in *Transportation Research Board 95th Annual Meeting (No. 16-6333)*, 2016.

[13] R. Elvik, "The predictive validity of empirical Bayes estimates of road safety," *Accident Analysis & Prevention*, vol. 40, no. 6, pp. 1964–1969, 2008.

[14] D. Gettman, L. Pu, T. Sayed, and S. G. Shelby, *Surrogate Safety Assessment Model and Validation: Final Report*, 2008, No. FHWA-HRT-08-051.

[15] A. Laureshyn, Å. Svensson, and C. Hydén, "Evaluation of traffic safety, based on micro-level behavioural data: theoretical

framework and first implementation," *Accident Analysis & Prevention*, vol. 42, no. 6, pp. 1637–1646, 2010.

[16] W. Young, A. Sobhani, M. G. Lenné, and M. Sarvi, "Simulation of safety: A review of the state of the art in road safety simulation modelling," *Accident Analysis & Prevention*, vol. 66, pp. 89–103, 2014.

[17] J. Archer, "Developing the potential of micro-simulation modelling for traffic safety assessment," in *Proceedings of the 13th ICTCT Workshop*, vol. 44, 2000.

[18] J. Archer, *Methods for the Assessment and Prediction of Traffic Safety at Urban Intersections and Their Application in Micro-Simulation Modelling*, Royal Institute of Technology, 2004.

[19] A. Sobhani, W. Young, and M. Sarvi, "A simulation based approach to assess the safety performance of road locations," *Transportation Research Part C: Emerging Technologies*, vol. 32, pp. 144–158, 2013.

[20] C. Hydén, *The Development of a Method for Traffic Safety Evaluation: The Swedish Traffic Conflicts Technique*, Bulletin Lund Institute of Technology, 1987.

[21] S. J. Older and B. R. Spicer, "Traffic conflicts—a development in accident research," *Human Factors: The Journal of Human Factors and Ergonomics Society*, vol. 18, no. 4, pp. 335–350, 1976.

[22] F. H. Amundsen and C. Hyden, "Proceedings of first workshop on traffic conflicts," in *Proceedings of Workshop on Traffic Conflicts*, TTI, Oslo, Norway, 1977.

[23] S. R. Perkins and J. L. Harris, "Traffic conflict characteristics-accident potential at intersections," in *Proceedings of the Traffic Safety and presented at the 47th Annual Meeting*, pp. 35–43, Highway Research Board, 1968.

[24] M. R. Parker and C. V. Zegeer, *Traffic Conflict Techniques for Safety and Operations: Engineers Guide*, 1989.

[25] E. Hauer and P. Garder, "Research into the validity of the traffic conflicts technique," *Accident Analysis & Prevention*, vol. 18, no. 6, pp. 471–481, 1986.

[26] K. El-Basyouny and T. Sayed, "Safety performance functions using traffic conflicts," *Safety Science*, vol. 51, no. 1, pp. 160–164, 2013.

[27] R. Fan, H. Yu, P. Liu, and W. Wang, "Using VISSIM simulation model and surrogate safety assessment model for estimating field measured traffic conflicts at freeway merge areas," *IET Intelligent Transport Systems*, vol. 7, no. 1, pp. 68–77, 2013.

[28] F. Huang, P. Liu, H. Yu, and W. Wang, "Identifying if VISSIM simulation model and SSAM provide reasonable estimates for field measured traffic conflicts at signalized intersections," *Accident Analysis & Prevention*, vol. 50, pp. 1014–1024, 2013.

[29] M. Essa and T. Sayed, "Simulated traffic conflicts: do they accurately represent field-measured conflicts?" *Transportation Research Record*, vol. 2514, pp. 48–57, 2015.

[30] *PTV VISSIM 6 User Manual*, Karlsrule, Germany, 2013.

[31] S. Jackson, L. Miranda-Moreno, P. St-Aubin, and N. Saunier, "Flexible, mobile video camera system and open source video analysis software for road safety and behavioral analysis," *Transportation Research Record*, no. 2365, pp. 90–98, 2013.

[32] S. Mirjalili and A. Lewis, "The whale optimization algorithm," *Advances in Engineering Software*, vol. 95, pp. 51–67, 2016.

[33] G. R. Brown, "Traffic conflicts for road user safety studies," *Canadian Journal of Civil Engineering*, vol. 21, no. 1, pp. 1–15, 1994.

10

Protected Turning Movements of Noncooperative Automated Vehicles: Geometrics, Trajectories, and Saturation Flow

Xiaobo Liu,[1] **Lijuan Lai**,[1] **You Kong**,[1] **and Scott Le Vine**[1,2,3]

[1]*School of Transportation and Logistics, Southwest Jiaotong University, Chengdu, China*
[2]*Department of Geography, State University of New York (SUNY) at New Paltz, New Paltz, NY, USA*
[3]*Department of Civil and Environmental Engineering, Imperial College London, South Kensington SW7 2AZ, UK*

Correspondence should be addressed to Scott Le Vine; levines@newpaltz.edu

Academic Editor: Emanuele Crisostomi

This study is the first to quantify throughput (saturation flow) of noncooperative automated vehicles when performing turning maneuvers, which are critical bottlenecks in arterial road networks. We first develop a constrained optimization problem based on AVs' kinematic behavior during a protected signal phase which considers both ABS-enabled and wheels-locked braking, as well as avoiding encroaching into oncoming traffic or past the edge-of-receiving-lane. We analyze noncooperative ("defensive") behavior, in keeping with the Assured Clear Distance Ahead legal standard to which human drivers are held and AVs will likely also be for the foreseeable future. We demonstrate that, under plausible behavioral parameters, AVs appear likely to have positive impacts on throughput of turning traffic streams at intersections, in the range of +0.2% (under the most conservative circumstances) to +43% for a typical turning maneuver. We demonstrate that the primary mechanism of impact of turning radius is its effect on speed, which is likely to be constrained by passenger comfort. We show heterogeneous per-lane throughput in the case of "double turn lanes." Finally, we demonstrate limited sensitivity to crash-risk criterion, with a 4% difference arising from a change from 1 in 10,000 to 1 in 100,000,000. The paper concludes with a brief discussion of policy implications and future research needs.

1. Introduction

Within the functional lifetime of much of today's road infrastructure, vehicle automation is expected to impact a range of important aspects of the transportation system, including traffic operations [1–3], safety [4], public transportation networks [5], design of arterial streets, limited-access facilities, parking facilities [6–8], activities performed while traveling [9], and possibly the dominance of the private automobile ownership business model [10, 11].

This study fits within the wider body of literature that addresses the issue of how automated vehicles (AVs) will impact traffic flow. This is a broad research question, with far-reaching consequences for future investment in the road network. This study's specific objective is to quantify the impact of AVs on the traffic flow properties of protected turning movements at signalized intersections (i.e., left- or right-turning vehicles performing turning maneuvers during a protected signal phase). We formulate a constrained optimization model of defensive driving car-following behavior during a turning maneuver that takes into account AVs' sensing/processing/actuating capabilities and then employ this model to estimate capacity (vehicles per hour per lane). The intent is to provide both a methodology and numerical results suitable for application in transportation planning models when attempting to forecast the impact of AVs on traffic flow and related outcomes (e.g., emissions). This research is timely, as the state of the art in planning studies attempting to forecast AVs impacts (cf. [12–14]) is to arbitrarily increase roadway network capacity; Guerra [15] reports that "nearly all" transportation plans in the USA do not consider AVs, which he demonstrates (on the basis of practitioner interviews) to be attributable to "uncertainties about the technology." Planners, network managers, and policymakers therefore urgently require estimates of capacity impacts based on fundamental kinematic analysis rather than guesswork.

Vehicle automation is often analyzed alongside vehicle-to-X (V2X) connectivity; the focus of this research is on noncooperative automated vehicles in which the only source of information that an individual automated vehicle uses to make safety-critical driving decisions is line-of-sight information from its own vehicle-borne sensors. While large gains in throughput can in principle be delivered through V2X connectivity, there are major barriers to delivering these benefits and there is evidence of the automotive sector preparing for an extended period of noncooperative driving (see extended discussion in Section 3.1).

Turning maneuvers account for a relatively small proportion of the time that vehicles are in motion on an arterial road network; however, they act as important bottlenecks on network-wide capacity for several distinct reasons [16–18]. A first mechanism is that a vehicle approaching a turning maneuver generally must slow as it approaches the location of the maneuver, thereby increasing the duration of the time-slot that it consumes a given physical space and hence reducing capacity. A second mechanism is that, at a typical four-way intersection, left-turning vehicles compete for signal time with through traffic and right-turning vehicles traveling in the opposing direction, in addition to traffic on the cross street. Finally, both left-turning and right-turning vehicles at arterial road-network intersections conflict with pedestrians, introducing a third mechanism that constrains capacity.

This study quantifies saturation flow rates for AVs performing turning maneuvers and is subject to a set of limitations. First, our focus is on protected turning movements; we leave the issue of "permitted" turning movements (i.e., movements that involve accepting a gap in a conflicting traffic stream) as a matter for future research, along with issues of sight line limitations and interactions between AVs and pedestrians (cf. [3]). Second, by focusing on saturation flow (i.e., maximum sustained flow), we explicitly do not address the decreased-flow regimes (termed "start-up" and "clearance" in [19]) that occur at the beginning and end, respectively, of a signal phase when vehicles are not proceeding at a steady-state rate. Third, we focus on *defensive driving* (i.e., noncooperative) behavior, as described in Section 3; this is the standard to which human drivers are held. More aggressive driving strategies such as short-headway platooning of connected AVs would tend to increase capacity; however, this behavior would be fundamentally inconsistent with the longstanding responsibility of the operator of each vehicle to not strike (i.e., "rear-end") the vehicle ahead of it.

The remainder of this paper is organized as follows: Section 2 presents further background relevant to this study. Section 3 introduces the data and methods employed in the analyses. Section 4 presents and discusses the empirical calculations of saturation flow rates, and Section 5 concludes the paper.

2. Background

Given the major policy and operational issues surrounding the issue of road-network capacity, a wide and growing body of literature has attempted to quantify the potential capacity impacts of automated vehicles. There is much interest in, for instance, whether AVs may enable increased capacity within the existing footprint of the road network, which would have the potential to reduce congestion and increase journey-time reliability without additional land consumption. A fundamental limitation of the existing literature is that, due to AVs being in their embryonic phase, the general approach to estimating capacity impacts is to employ simulation techniques rather than to perform empirical field testing on closed courses or collect observations from real-world network operations.

The body of literature on AVs' impacts on roadway network capacity (cf. [20] for a review of literature published up to year 2013) can be organized on the basis of whether the context is a freeway environment [21–25] or an arterial environment [26–30], or both [31]. The gap in the literature that this study addresses is that previous studies of AVs capacity impacts have not specified driving behavior during turning maneuvers on the basis of turning vehicles' unique kinematics.

A related body of literature that has taken shape in recent years addresses the AV motion-planning problem (i.e., programming an AV's trajectory as it performs driving tasks); compare [32] for a detailed review of this literature, with Alhajyaseen et al. [33], Gu et al. [18], and Wolfermann et al. [34] having studied the trajectories of turning vehicles at intersections. Studies in this category have focused on the task of controlling the trajectory of a single AV and do not consider the capacity impacts arising from the interactions among multiple AVs' driving behaviors (e.g., consequences on saturation flow rates).

Therefore, to the best of the authors' knowledge this study represents the first in-depth evaluation of the capacity of turning maneuvers under automated driving. Indeed, Hoogendoorn et al. [20], providing a wide-ranging year-2014 review of the literature on AVs and traffic efficiency, reported that *"no studies were found on the influence of automation of the lateral control task on traffic flow efficiency"* (p.114).

Further, in contrast to earlier studies of AVs kinematics which have specified deterministic braking system performance, we operationalize the "defensive driving" requirement with an analysis that explicitly accounts for the stochastic nature of braking system performance (i.e., achieved deceleration rates that cannot be known a priori with certainty) and which therefore allows us to perform sensitivity tests with a range of plausible crash-risk criteria.

For human drivers, the *Highway Capacity Manual* [19] is the standard traffic flow model of turning vehicles employed in the USA. Both right- and left-turning vehicles are specified to have lower saturation flow rates than vehicles traveling straight through an intersection, by 5% and 18%, respectively (see Equations 19-13 and 19-14 of [19]). In this model of intersection operations, turning radius is not explicitly taken into account in calculations of saturation flow rate. However, the lower flow rate of right-turning vehicles is an implicit recognition of the influence of turning radius, given that *ceteris paribus* a right-turning movement at a 90-degree intersection has a smaller radius than a left-turning movement (in a right-hand-drive road network). The work of [19] also contains explicit accommodation of the influence of turning radius on throughput in the special case of interchange-ramp-terminal intersections, with smaller turn radii associated with lower flow rates (see Exhibit 23-27). For instance, turning movements at ramp terminals with radii of 25, 50, and 75 feet are defined to have 18%, 10%, and 7% less capacity than through movements, *ceteris paribus*. A further implicit recognition in [19] of the effect of turn radii is the specification that when multiple turn lanes are present capacity per lane is reduced relative to when a single turn lane exists (by 3% and 12% for double-left-turn lanes and double-right turn lanes, resp.; cf. Exhibit 31-39). Empirically, Cooner et al. [35] studied the operations of triple-left-turn lanes; the results with respect to turn radii were inconclusive, with some sites showing higher capacities for larger-radius turn lanes and some for the smaller-radius lanes. Earlier studies by Sando and Mussa [36], Courage et al. [37], and Leonard [38] reported similarly inconclusive findings regarding the effects of turn radius on capacity of triple-left-turn lanes; however, Ackeret [39] reported higher saturation flow rates for larger-radius turn lanes.

3. Methods

The analysis strategy for this study incorporated a microsimulation model of the operation of two vehicles (one leading [veh_l] and one following [veh_f]) negotiating a 90-degree turning maneuver. The analysis is discrete in time (0.01-second time steps) and continuous in space. Notations are summarized in the Summary of Notations.

3.1. Calculation of Minimum Safe following Headway. Calculating minimum safe following headway for turning vehicles is a more complex problem than the same calculation for vehicles traveling only longitudinally. In the latter case, (1) can be used to determine the minimum safe following distance $x_{\min}$ that allows a following vehicle (veh_f) to avoid striking the "leading" vehicle (veh_l) that it is following in the event of veh_l unexpectedly initiating emergency braking:

$$x_{\min} = v * t_{\text{lag},f} + \frac{v^2}{2 * a_f} - \frac{v^2}{2 * a_l} + x_{\text{veh}}, \quad (1)$$

$$H_{\min} = t_{\text{lag},f} + \frac{v}{2 * a_f} + \frac{x_{\text{veh}} - v^2/(2 * a_l)}{v}, \quad (2)$$

$$C_v = \frac{1}{H_{\min}}. \quad (3)$$

Equations (1) and (2) arise from the equations governing translational motion; the derivation can be found in the Appendix of [30]. Equation (3) expresses capacity (C_v) as the reciprocal of the minimum headway value. In (1), the first term on the right-hand side (incorporating velocity v and latency time $t_{\text{lag},f}$) represents the distance that veh_f travels after veh_l begins braking and before veh_f does so. The second term accounts for the distance that veh_f will travel while it brakes. The third term accounts for the distance that veh_l will travel during its braking maneuver and is relevant only for the "weak" ACDA interpretation. This is because under the "strong" interpretation of ACDA it is assumed that veh_l will decelerate at a rate (a_l^-) that is arbitrarily large, thereby effectively removing this term from the analysis. Finally, the fourth term (x_{veh}) represents vehicle length. (*Note.* Equations (1) and (2) apply only under the conservative assumption that the rate of deceleration of the leading vehicle (a_l) is greater than or equal to the rate of deceleration of the following vehicle (a_f); that is, $a_l \geq a_f$.)

Equations (1)–(3) formalize the *Assured Clear Distance Ahead* (ACDA) legal standard, which codifies the driving behavior colloquially termed *Defensive Driving*. ACDA requires that a car's operator leave sufficient spacing behind the vehicle or object ahead of it, such that it can avoid striking the vehicle/object. In formally defining ACDA, Buchwalter et al. ([40], Automobiles and Highway Traffic, Sect. 1115) report that *"most jurisdictions follow the rule that a collision between a preceding and a following motorist gives rise to a presumption of negligence on the part of the following motorist, whether the preceding vehicle is moving, stopped, or stopping."* Though ACDA is generally applicable as a criterion, it has not been universally applied to establish negligence in a crash. A vehicle operator that strikes an object that has suddenly "darted out" (i.e., moved laterally) into their trajectory of travel is, for instance, generally not deemed to be negligent ([40], Automobiles and Highway Traffic, Sect. 498). Likewise, ACDA has not been applied uniformly by the judiciary; courts have variously interpreted it "weakly" (cf. Naffky [41], Sect.664) or "strongly," with the distinction that under the "weak" interpretation a driver is held responsible only to avoid striking the vehicle in front of itself that is traveling in the same direction, and under the "strong" interpretation a driver is responsible to avoid striking a stationary object (cf. [42]). The "strong" interpretation of ACDA results in larger following distances and hence larger headways between vehicles, to account for the fact that a vehicle's forward sight line is blocked by the vehicle in front of it. Though there is a wide body of literature that has considered cooperative automated driving (cf. [2, 25, 43]), there is also evidence of automotive manufacturers preparing for an extended period of noncooperative automated driving. In a patent application disclosed by Ford Global Technologies (a subsidiary of Ford Motor Company) in 2016, for instance, Stanek and Lockwood [44] envision AVs engaging in ACDA-compliant (noncooperative) driving behavior: *"Each of [an AV's] onboard sensors has a limitation in the range (i.e. distance) at which objects and/or conditions can be detected. . .the top speed at which the vehicle is controlled to travel under autonomous driving mode may be calculated as a factor of how far ahead the sensors can detect objects"* (p.3).

We now introduce the additional dimension of lateral (turning) motion, which is not considered in (1)–(2). Figure 1 schematically depicts the position of two vehicles engaged in left-turn maneuvers of fixed-radius r (measured from center-of-circle to vehicle center-of-gravity, consistent with AASHTO 2011 [[45]; cf. Equation (3)–(6)] in sequence (*veh_l* first, followed by *veh_f*). When traveling along a trajectory

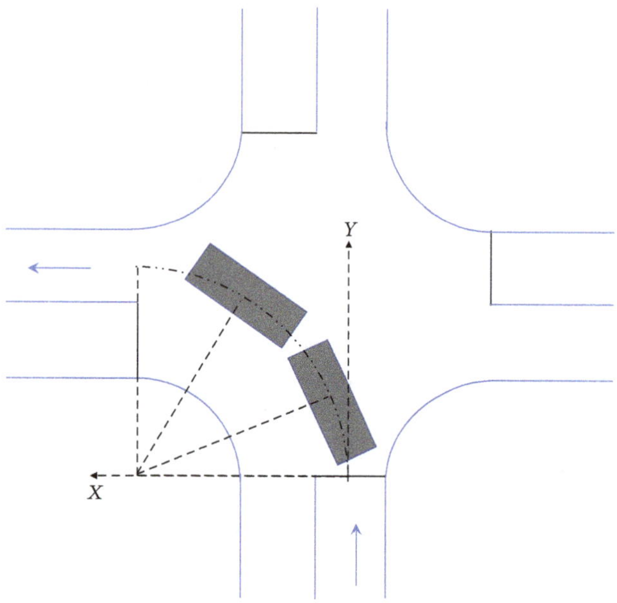

Figure 1: Schematic diagram of leading (veh_l) and following (veh_f) left-turning vehicles.

which is an arc of a circle, each vehicle's velocity has two components: translational and rotational.

The *translational* component is similar in some respects to longitudinal-only deceleration when two vehicles following one another both brake in a straight line. However, the difference in the case of turning motion is that the two vehicles are not traveling in sequence in a straight line but rather are at different points along the arc of a circle. The trajectory that the vehicles will follow after initiating emergency braking will depend on whether the braking is "wheel-locked" or is "ABS-engaged" (Automated Braking System), as is the case with new automobiles sold today in the USA (we assume that deceleration occurs at a constant linear rate, consistent with earlier models of emergency-braking maneuvers (cf. [45, 46]) and field testing results of autonomous emergency-braking systems [47]). Steering control is lost for the duration of wheel-locked braking; ABS is designed to preserve steering control (whether braking is initiated while traveling in a straight trajectory or during a turning maneuver), which is achieved by pulsing the brakes more quickly than humans' physical capabilities permit. However, ABS does not in general result in a reduction in achieved braking distance (cf. [46, 48]).

In an ABS-equipped human-driven vehicle, ABS engages automatically during emergency braking. In an AV, however, the vehicle's controller would have much better situational awareness than today's ABS systems and would therefore in principle be able to select which of "wheels-locked" or "ABS-engaged" braking is optimal in any given emergency maneuver (rather than assuming, as ABS systems do with human drivers, that engaging ABS is always optimal). In the case of two AVs (veh_l and veh_f) performing turning maneuvers, there are two separate issues: whether veh_l would engage ABS upon initiating emergency braking (which veh_f cannot know in advance or control) and whether veh_f would engage ABS (which veh_f does control). The consequence of "wheels-locked braking" is that the vehicle's translational motion will continue along the line tangent to the arc of the vehicle's prior trajectory at the point when emergency braking is initiated. By contrast, a vehicle performing ABS-engaged braking will continue its translational motion along the curved trajectory until it comes to a rest. In the case of wheels-locked braking, there is an additional constraint that the braking vehicle must not travel into the crosswalk (or past the stopbar if no crosswalk is present) or impact the curb of the exit leg of the intersection during its braking maneuver. As a practical matter, this means that, considering a northbound left-turning vehicle, the vehicle must ensure that if it must initiate emergency braking it will not encroach further north than the northern curbline of the western leg or its eastward extension across the southern boundary between the intersection "box" and the intersection's northern leg.

With regard to rotational motion, we specify constant deceleration, resulting in the rotational orientation at rest (α) of a vehicle during an ABS-enabled braking maneuver calculated via

$$\alpha = \frac{\omega * t_{\text{braking}}}{2}, \tag{4}$$

where ω is the initial rotational velocity (expressed in radians or degrees per second) and t_{braking} is the duration of the braking maneuver, which is equal to the duration required for the translational motion to come to rest.

Figures 2 and 3 show a time-lapse progression of veh_l and veh_f performing emergency braking; the parameters in this diagram are radius $r = 50'$, $v = 20$ mph, and angle β of veh_l when it initiates emergency braking of 45 degrees. Figures 2 and 3 depict the two vehicles performing ABS-engaged and wheels-locked braking maneuvers, respectively. The time-frames shown represent the following:

(i) Top-left: initiation of emergency braking by veh_l ($t = 0$)

(ii) Top-right: after a period of time $t_{\text{lag},f}$ elapses (specified to be 0.4 seconds; see Section 3.2), veh_f recognizes that veh_l has initiated emergency braking and also begins emergency braking ($t = 0.40$ seconds)

(iii) Bottom-left: veh_l has come to rest ($t = 0.97$ seconds)

(iv) Bottom-right: veh_f has also come to rest ($t = 1.52$ seconds). Note that in both Figures 2 and 3 veh_f has selected its following distance at time $t = 0$ to ensure that after the emergency-braking maneuvers it comes to rest at the precise critical location where its front bumper is infinitesimally close to the rear bumper of veh_l.

In contrast to the translation component of motion (discussed in the previous three paragraphs), the *rotational* component describes the type of motion in which a vehicle pivots around a point (in this study, we assume that the rotational motion of each vehicle is centered on the centroid of the vehicle. In practice, the rotational motion of an

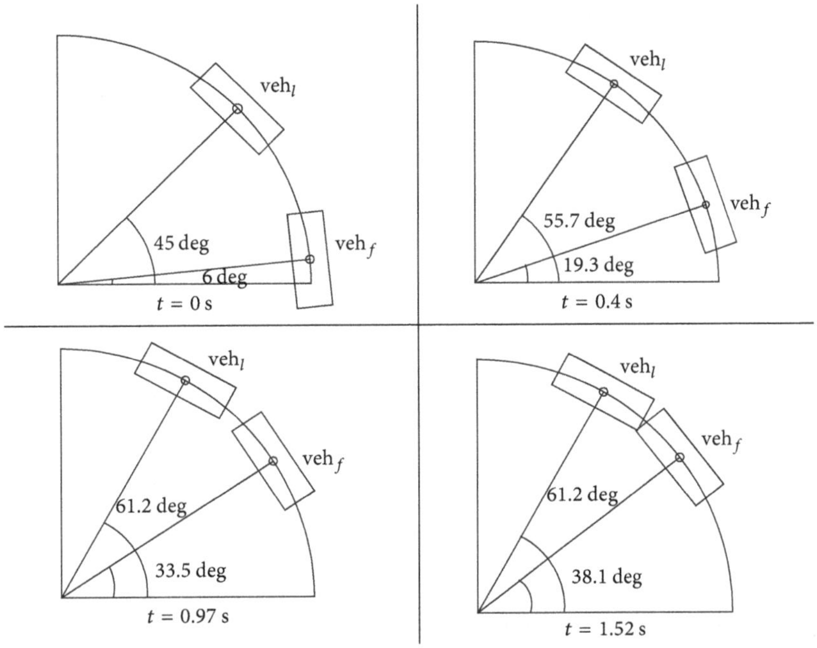

FIGURE 2: Time-lapse sequence of two left-turning vehicles, with *ABS-engaged* braking trajectories. Parameters: $r = 50'$, $\beta_l = 45°$, and $v = 20$ mph.

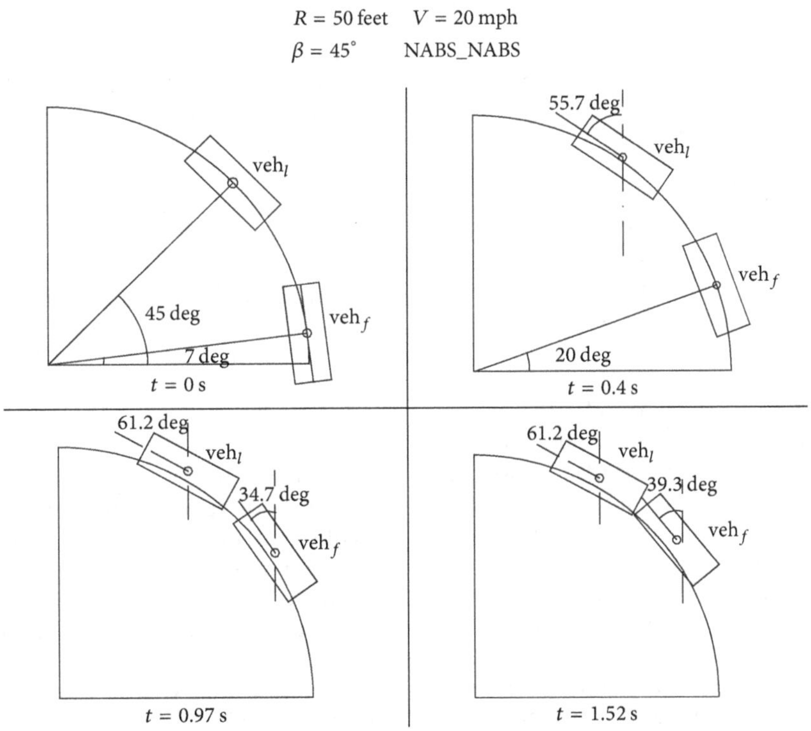

FIGURE 3: Time-lapse sequence for two left-turning vehicles, with *wheels-locked* braking trajectories. Parameters: $r = 50'$, $\beta_l = 45°$, and $v = 20$ mph.

automobile that is not heavily loaded with passengers or cargo is likely to be located closer to the front of the vehicle, due to the heavy weight of the engine and other vehicular systems located in front of the passenger compartment). When a vehicle is performing a turning maneuver, it changes orientation and therefore experiences rotational or angular velocity (degrees per unit time). Upon initiating emergency braking, this angular velocity will begin to decrease until it reaches zero. As with the translational component of motion, we specify that the rotational component also decelerates at

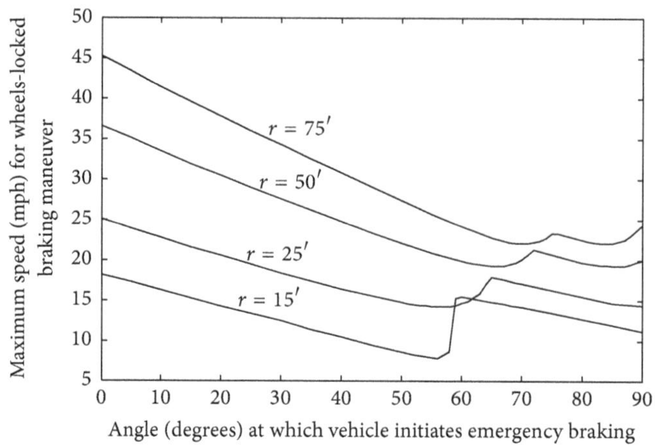

FIGURE 4: Curves depicting maximum speed (v^{WLB}) at which a *wheels-locked* braking maneuver can be performed.

a constant rate, and also that the two components of motion reach zero velocity at the same time (in this study, we assume that the rotational motion of each vehicle is centered on the centroid of the vehicle. In practice, the rotational motion of an automobile that is not heavily loaded with passengers or cargo is likely to be located closer to the front of the vehicle, due to the heavy weight of the engine and other vehicular systems located in front of the passenger compartment). The positions-at-rest of veh_l and veh_f shown in Figure 3 account for both components of motion. In general, for left-turning vehicles the wheels-locked-braking position-at-rest will be located above and to the right of the position-at-rest of ABS-engaged-braking; this is because the ABS-engaged trajectory involves the same amount of overall motion, but with the x/y-axis components of motion being distributed to a larger extent towards the left (following the arc) and therefore to a smaller extent along the y-axis towards the top of the graphic.

veh_f is the member of this vehicle-couplet that makes the choice of following distance. veh_f also has the power to decide whether it will follow the ABS-engaged or wheels-locked trajectory, but it does not know and cannot control which of these two trajectories veh_l will follow. Therefore, veh_f must assume that veh_l follows the "worst-case" of the two trajectories (i.e., the trajectory that would require that veh_f leave the largest spacing ahead of it). Because veh_f has the power to decide which trajectory it will follow, we specify that it selects the minimum of the two safe-following-distances associated with the ABS-engaged and wheels-locked trajectories. The exception to this is when multiple turn lanes exist: if the two vehicles are in any turn lane except the lane directly adjacent to through traffic in the same direction, veh_f must assume that it will follow an ABS-engaged trajectory in order to avoid encroaching into the adjacent turn lane.

Figure 4 shows the relationship between β and the maximum speed (v^{WLB}) at which a vehicle can perform a "wheels-locked" braking maneuver without encroaching into oncoming traffic or past the edge-of-lane after exiting the intersection, for selected turning radii ($r = 15'$, $25'$, $50'$, and $75'$). Two general relationships can be seen: v^{WLB} tends to decrease as β increases, and v^{WLB} also tends to decrease as r decreases. Neither of these relationships is monotonic, however, which is due to the rectangular vehicle shape. Figure 5 demonstrates this phenomenon, by showing the vehicle kinematics that result in the curve for $r = 15'$ in Figure 4 crossing the curve for $r = 25'$ in the vicinity of $\beta = 60°$. At the timeframe ($t = 0.07$ seconds) depicted in both panels of Figure 5, a vehicle traveling initially at $v = 15.2$ mph would encroach past the edge-of-lane with its front-right corner if $r = 25'$ (right panel) but would not encroach past the edge-of-lane if $r = 15'$ (left panel). In the latter case, the rear-right corner of the vehicle would be the first point of the vehicle to encroach past the edge-of-lane if speed were higher.

During a turning maneuver, a vehicle's speed is constrained by two additional factors which do not affect purely longitudinal motion: the amount of friction available to enable the vehicle to successfully navigate the desired curved trajectory, and the comfort of the vehicle's occupant(s) under lateral acceleration, which is proportional to velocity squared (cf. [29]) (for completeness, we note that a third constraint on turning-vehicle speed is the positive relationship between speed and the propensity of the vehicle to "roll-over").

We formulate car-following behavior as a constrained optimization problem. Conditional on a set of input parameters (r, v, a_l, a_f, $t_{\text{lag},f}$, and x_{veh}) that are held constant during a given instance of the optimization process, headway is minimized ($H_{\min}$):

$$H_{\min} = \min\left\{\max\left\{H_{\min}^{\text{ABS,ABS}}, H_{\min}^{\text{noABS,ABS}}\right\}, \max\left\{H_{\min}^{\text{ABS,noABS}}, H_{\min}^{\text{noABS,}noABS}\right\}\right\} \quad (5)$$

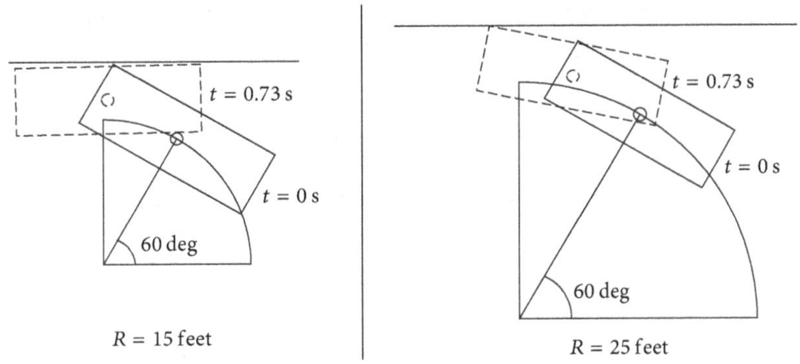

FIGURE 5: Two time-lapse sequences; each panel depicts a single vehicle's *wheels-locked* braking maneuver. Parameters: $r = 15'$ (L) and $r = 25'$ (R), $\beta = 60^{\circ}$, and $v = 15.2$ mph. Solid-outline vehicle and dashed-outline vehicle depict location at time of $t = 0$ and $t = 0.73$ seconds, respectively.

subject to the following constraint:

$$\prod_{\beta=0}^{90^{\circ}} \prod_{t=1}^{\infty} I_{\beta t} = 1. \tag{6}$$

The logic of the minimum and maximum functions in the objective function (see (5)) is explained earlier in this section. In Constraint (6), $I_{\beta t}$ is an indicator function that takes a value of 1 if no part of veh_f would strike any part of veh_l at time t if veh_l begins emergency braking at angle β and a value of 0 otherwise. The overall product function (evaluated across all values of β and t) defined by Constraint (6) therefore takes a value of 1 if and only if a candidate headway value is large enough to ensure that veh_f can be assured (conditional on the input parameters) of not striking veh_l regardless of the possibility of noncooperative driving behavior by veh_l at any point during the turning maneuver. Any candidate headway value that results in Constraint (6) evaluating to 0 is therefore rejected as "following too closely." The constrained optimization problem was solved through simulation due to the objective function not being continuously differentiable.

3.2. Values of Empirical Parameters. Several parameters in (1) require empirical specification in order to calculate the capacity of AVs performing turning maneuvers.

The latency time ($t_{\text{lag},f}$) of AVs' sensing/processing was specified to be 0.4 seconds, as in [30]. We note that assumptions of smaller values of (e.g., 0.2 seconds) would require sustained favorable visibility and weather conditions and 0% failure rates in sensing/control [4] and that larger values of latency will limit the number of instances that emergency braking is unnecessarily initiated, because a larger number of sensor readings become available.

We specified cars' length and width to be 19 feet and 7 feet, respectively, per the "passenger car" design vehicle in [45].

The deceleration values of the two vehicles (a_l and a_f for veh_l and veh_f, resp.) are drawn from empirical testing of ABS-enabled vehicles performed by the National Highway Traffic Safety Administration [49]. Empirical results of emergency-braking test runs of a Chevrolet Malibu passenger sedan on four days of dry conditions were employed (cf. [49] Tables E-4, F-4, G-4, and H-4). Of 50 braking maneuvers performed during the testing in [49], we removed 16 due to insufficient pressure applied to the brake pedal. Of the remaining 34 braking maneuvers with acceptable braking-pedal pressure, we first calculated the mean (28.3 ft/sec^2) and standard deviation of the rate of deceleration (0.67 ft/sec^2). We then tested the distribution for normality using the Shapiro-Wilk test, which supported the null assumption that the data are normally distributed, and subsequently developed estimates of various percentile points of the distribution. We specified a_l to be at the 99.9th percentile of the distribution (−30.38 ft/sec^2) and the deceleration rate of a_f to be at the 0.1th percentile (−26.21 ft/sec^2). The joint probability of these two rates being exceeded (in the case of a_l) or not achieved (in the case of a_f) is 1/1,000,000. In Section 4, we present sensitivity analyses to take into account the possibility that AVs could be programmed to take either larger or smaller risks than this criterion. We note that achievable deceleration rates during emergency braking depend on vehicular technology, which has evolved since the braking tests reported in [49] were performed in the year 2000 and is expected to continue to evolve as increasing levels of automation influence future vehicle design.

In the next section, we present results from a range of combinations of turn radius and travel speed. While alternative specifications are possible (cf. [29] regarding the relationship between lateral acceleration and occupant comfort within AVs), for the purposes of direct comparison of AV turning-movement saturation flow to human drivers we specified that AVs select their turning velocity by emulating human drivers' behavior. In the empirical analysis presented in Section 4, we characterized this behavior using the model of the relationship between radius and turning speed proposed by Wolfermann et al. [34]; for the four turning radii studied in Section 4 ($r = 15'$, $r = 25'$, $r = 50'$, and $r = 75'$), this turning speed (denoted by v^{HDM}) is 12.4, 12.9, 14.0, and 15.2 mph, respectively.

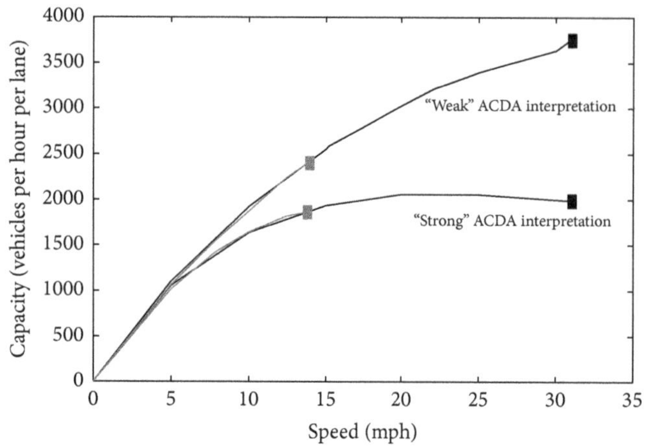

FIGURE 6: Relationship between speed, turning radius, and throughput (vehicles per hour). Gray and black curves represent $r = 15'$ and $r = 75'$, respectively; square markers indicate v^{CTF}.

4. Empirical Calculations of Saturation Flow Rates

Figure 6 shows the fundamental diagrams of traffic flow (speed plotted versus capacity) of turning vehicles based on the "weak" and "strong" ACDA interpretations (see Section 3.1). Of the four turning radii (15′, 25′, 50′, and 75′) that we analyzed, only the smallest (15′) and largest (75′) values are shown. Figure 6 shows that the effect of turning radius by itself is small, as the curves for $r = 15'$ and $r = 75'$ are nearly indistinguishable. The square markers in Figure 6 depict the maximum turning speed v^{CTF} for both turning radii that can be performed without skidding off of the turning-arc trajectory, due to the limitation of friction at the tire/roadway-interface (calculated to be a friction factor of 0.85, using the empirical value of $a = 28.3\,\text{ft/sec}^2$; cf. Section 3.2). The gray and black square markers in Figure 6 represent the v^{CTF} values for $r = 15'$ ($v^{CTF} = 13.8$ mph) and $r = 75'$ ($v^{CTF} = 30.9$ mph), respectively. Here we see that, under the "weak" ACDA interpretation, the saturation flow rate at $r = 75'$ (3750 veh/hour/lane) is much larger than the saturation flow rate for $r = 15'$ (2400 veh/hour/lane), though this difference is much smaller (both absolutely and proportionately) for the "strong" ACDA interpretation (1989 versus 1865 veh/hour/lane). Therefore, it can be concluded that the principal constraint on capacity of AVs performing turning maneuvers at intersections is the programmed turning speed (which is influenced by turning radius), rather than turning radius itself, and that the ultimate constraint on turning speed is therefore likely to be passenger comfort under lateral acceleration.

Table 1 extends from Figure 6, by presenting capacity calculations for both interpretations of ACDA, four turning radii ($r = 15', 25', 50'$, and $75'$), three turning speeds (v^{WLB}, v^{HDM}, and v^{CTF}, calculated separately for each turning radius), and two "lane contexts." Lane context refers to whether a single turn lane exists to serve a turning movement or whether multiple turn lanes are present. In the case of multiple turn lanes operating at capacity, vehicles in all turn lanes except the largest-radius turn lane (i.e., the lane directly adjacent to the "through" lanes of the same direction of travel) have the constraint that they cannot perform a "wheels-locked" braking maneuver that would intrude into the adjacent turn lane and therefore must brake along the "ABS-engaged" trajectory. It can be seen that for all combinations of radius and speed the "weak" ACDA interpretation results, as expected, in higher capacity than the "strong" interpretation. Table 1 also shows that, with turning speeds equal to the estimated average speed of human drivers negotiating the same turning maneuver (i.e., v^{WLB}), the calculated saturation flow rates consistently represent an increase in capacity relative to human drivers across all four radii. For $r = 25'$, for instance, this increase varies between a minimum of +0.2% (for a left turn, strong ACDA interpretation, and the presence of multiple turn lanes: 1809 versus 1850 veh/hour/lane) to a maximum of +43% (for a right turn, weak ACDA interpretation, single turn lane: 2308 versus 1615 veh/hour/lane). The largest calculations of capacity occur at the highest speeds conditional on available friction ($v^{CTF} = 30.9$ mph, for $r = 75'$); headways less than one second (0.96 seconds, equating to saturation flow of 3750 veh/hour/lane) can theoretically be sustained.

The analysis up to this point has specified 1 crash in 1,000,000 as the criterion for acceptable crash risk if veh_l initiates emergency braking. Table 2 shows the results of a sensitivity analysis in which this criterion is varied by two orders of magnitude both larger and smaller, for a moderate-radius turning movement ($r = 25$ feet) with a single turn lane and under the weak ACDA interpretation. It can be seen that capacity is relatively insensitive to the choice of safety standard: varying the crash risk by four orders of magnitude (between a 1 in 10,000 risk and a 1 in 100,000,000 risk) leads to only a 4% variation in turning-movement capacity (2857 versus 2748).

5. Conclusions

This study represents the first systematic evaluation of the traffic flow properties of automated vehicles when performing turning maneuvers. The driving behavior model respects the *Assured Clear Distance Ahead* criterion, which codifies "defensive driving" practice. We evaluate the kinematics of two distinct types of braking maneuvers for collision avoidance and specify that the AV trajectory-planning incorporates for the maneuver that permits the shortest following distance behind the leading vehicle while respecting the ACDA standard.

Our results demonstrate that the impact of AVs on the capacity (vehicles per hour per lane) of turning movements will be constrained primarily by the comfort of vehicle occupants (as represented in this analysis by the typical turning speed of human drivers as a function of turning radius), rather than AVs' performance limitations or the available friction at the tire/roadway-interface. We show that the rectangular shape and typical size of a passenger car lead to complex relationships between turning radius and feasible

Table 1: Calculated saturation flow rate values (veh/hour/lane) for turning movements of selected radius/speed combinations.

ACDA interpretation	Lane context		$r = 15'$			$r = 25'$			$r = 50'$			$r = 75'$	
		v^{WLB} (7.8 mph @ $r = 15'$)	v^{HDM} (12.4 mph @ $r = 15'$)	v^{CTF} (13.8 mph @ $r = 15'$)	v^{WLB} (14.3 mph @ $r = 25'$)	v^{HDM} (12.9 mph @ $r = 25'$)	v^{CTF} (17.9 mph @ $r = 25'$)	v^{WLB} (19.3 mph @ $r = 50'$)	v^{HDM} (14.0 mph @ $r = 50'$)	v^{CTF} (25.2 mph @ $r = 50'$)	v^{WLB} (22.1 mph @ $r = 75'$)	v^{HDM} (15.2 mph @ $r = 75'$)	v^{CTF} (30.9 mph @ $r = 75'$)
Weak	Single turn lane, or largest-radius turn lane if movement served by multiple turn lanes	1552	2236	2400	2449	2308	2812	3000	2449	3429	3214	2609	3750
Weak	If multiple turn lanes, all lanes except largest-radius turn lane	1552	2236	2400	2400	2250	2769	2951	2416	3396	3214	2590	3711
Strong	Single turn lane or largest-radius turn lane if movement served by multiple turn lanes	1429	1818	1865	1957	1865	2057	2045	1915	2143	2081	1989	1989
Strong	If multiple turn lanes, all lanes except largest-radius turn lane	1429	1818	1865	1946	1809	2000	2022	1885	2143	2069	1989	1989
Human driving behavior (per *Highway Capacity Manual* [19])	Single turn lane	1615 (right turn, does not vary with radius) 1805 (left turn, does not vary with radius)											
	Double turn lanes (per lane)	1425 (right turn, does not vary with radius) 1748 (left turn, does not vary with radius)											

Table 2: Sensitivity analysis of rates of deceleration of leading and following vehicle (weak ACDA interpretation).

Crash-risk criterion	a_l (ft/sec^2)	a_f (ft/sec^2)	Capacity @ $r = 25'$, $v^{\text{CTF}} = 17.9$ mph
1 in 10,000	−29.87	−26.73	2857
1 in 100,000	−30.14	−26.46	2839
1 in 1,000,000	−30.38	−26.21	2812
1 in 10,000,000	−30.60	−26.00	2801
1 in 100,000,000	−30.81	−25.79	2748

vehicle kinematics. We demonstrate that, as with human drivers, for some combinations of speed and turning radius adding additional turning lanes to serve AVs performing a turning movement would be subject to diminishing marginal returns (i.e., doubling the number of lanes from one to two would lead to less than a doubling of capacity). Finally, we demonstrate that the capacity estimates are relatively robust with regard to selection of a quantitative standard for safety; capacity is shown to vary by 4% in response to varying the safety standard by four orders of magnitude (i.e., a factor of 10,000, between a 1 in 10,000 risk and a 1 in 100,000,000 risk).

We now conclude with a brief discussion of research needs for the next phase of the research agenda.

First, we show that achievable flow rates of AVs performing turning maneuvers will have a strong dependence on AVs' turning speed, which implies that establishing guidelines for determining turning speeds may be an important policy consideration. For human drivers, turning speed is a decision made by each individual driver during each specific turning maneuver, subject only to the external constraint of a speed limit. Capacity is maximized by fast turning speeds; however individual AV occupants may desire the low rates of lateral acceleration achieved through slow turning speeds [29]. There may therefore be a role for public policy in specifying the envelope of acceptable turning speeds and rates of lateral acceleration/deceleration in a range of different circumstances. For instance, Urmson et al. [50] proposed an AV-control paradigm through which "*an autonomous vehicle may be configured with some general safe envelope driving patterns…adaptive autonomous driving [then] allows the vehicle to perform autonomous driving within the general safe envelope but still safely in the driver's preferred style of driving*" (p.2).

Second, research is needed to establish whether novel geometric design principles for turning movements can enable additional capacity gains. For instance, recent work by Kozey and colleagues [51] demonstrates that flexible use of road space enabled by vehicle automation has the potential, under certain circumstances, to allow increased capacity without large increases to the physical footprint of an intersection.

Third, given the context-specificity of whether the "strong" or "weak" Assured Clear Distance Ahead criterion is the standard to which human drivers are held, there is a need for policy clarity as to which of these criteria will govern AVs' driving behavior in various sets of circumstances. When performing turning maneuvers, an argument in favor of the applicability of the more generous "weak" ACDA interpretation is that the leading vehicle in a couplet does not physically block the sight line of the following vehicle to the same degree as when both vehicles are traveling in a straight line [52].

Fourth, further research is required to establish the operations of noncooperative AVs during "start-up" and "clearance" time at the beginning and end of a signal phase, to complement the saturation flow rates calculated in this research for the interior time points of a signal phase. Other important research questions relate to AVs' operations during permitted (as opposed to protected) turning movements and the operations of cooperative (connected) AVs when performing turning maneuvers. With regard to cooperative driving behavior, it will be necessary to establish whether connectivity between vehicles will simply reduce latency times towards zero or will allow exchange of actionable *intent* information (cf. [53]), the latter of which extends beyond the protocols of the currently planned rollout of vehicle-to-vehicle communications [54].

Finally, there is a need for integrated analysis of *uncooperative* AVs operations across a diversity of roadway geometrics and driving maneuvers. This line of enquiry would follow on from recent research by Shelton et al. [31] that studied connected-vehicle operations on a part-freeway/part-signalized-arterial network. In the coming years, it is likely that AVs operating independently of one another (as partially automated vehicles currently do) will predate the arrival of connected AVs that rely on communications links to external sensors (e.g., V2V/V2I) for making safety-critical decisions. Therefore, planners, road-network managers, and transport policymakers urgently need both standardized techniques and reliable findings regarding the impacts of "unconnected" noncooperative AVs that drive defensively, as human drivers today do.

Summary of Notations

- $\text{veh}_l, \text{veh}_f$: Leading and following vehicle, respectively
- H, H_{min}: Headway (front of veh_l to front of veh_f), units of seconds. H_{min} is minimum value
- $H_{\text{min}}^{l,f}$: H_{min}, conditional on whether veh_l and veh_f perform ABS-engaged or ABS-not-engaged braking maneuvers. For instance, $H_{\text{min}}^{\text{ABS,noABS}}$ denotes H_{min} if veh_l performs an ABS-engaged braking maneuver and veh_f performs an ABS-not-engaged maneuver
- x, x_{min}: Spacing (front of veh_l to front of veh_f), units of feet. x_{min} is minimum value
- v: Translational velocity, units of ft/sec
- ω: Rotational velocity, units of deg/sec
- a_l, a_f: Rate of deceleration of leading and following vehicle (units of ft/sec^2), respectively
- β_l, β_f: Angle (degrees) at which veh_l and veh_f, respectively, initiate emergency braking

α: Orientation (angle, units of degrees) at rest (at completion of braking maneuver)
t_{braking}: Duration (seconds) of braking maneuver
$t_{\text{lag},f}$: Time lag (seconds) between veh_l and veh_f initiating emergency braking. Fixed at 0.4 seconds throughout analysis (see discussion in Section 3.2).
x_{veh}: Length of vehicles (ft). Fixed at 19′ throughout analysis (see discussion in Section 3.2)
C_v: Capacity (vehicle per hour per lane) at velocity v
r: Turning radius (ft), defined as radius of vehicle centroid
v^{WLB}: Maximum speed at which a wheels-locked braking maneuver can be performed without encroaching into oncoming traffic or past the edge-of-lane after exiting the intersection
v^{HDM}: Average speed of human drivers during turning maneuver, assuming approach-segment speed of $v = 35$ mph (per [33])
v^{CTF}: Maximum speed to avoid skidding during turning maneuver (i.e., constrained by tire/roadway-interface friction).

Disclosure

The usual disclaimer applies: any errors in this paper are the authors' sole responsibility.

Conflicts of Interest

The authors declare that there are no conflicts of interest regarding the publication of this paper.

Acknowledgments

The authors wish to thank Catherine Herne for helpful discussions and acknowledge financial support from the National Science Foundation of China (Project no. 71671147) and the University Transportation Research Center, Region 2 (Grant no. 49198-13-28: *Simulation of Automated Vehicles' Drive Cycles*).

References

[1] S. Smith, J. Bellone, S. Bransfield et al., "Benefits Estimation Framework for Automated Vehicle Operations," U.S. Dept. of Transportation ITS Joint Program Office Report, FHWA-JPO-16-229, 2015.

[2] V. Milanes, S. E. Shladover, J. Spring, C. Nowakowski, H. Kawazoe, and M. Nakamura, "Cooperative adaptive cruise control in real traffic situations," *IEEE Transactions on Intelligent Transportation Systems*, vol. 15, no. 1, pp. 296–305, 2014.

[3] A. Millard-Ball, "Pedestrians, Autonomous Vehicles, and Cities," *Journal of Planning Education and Research*, 2016.

[4] R. Anderson, S. Doecke, J. Mackenzie, and G. Ponte, "Potential benefits of autonomous emergency braking based on in-depth crash reconstruction and simulation," in *Proceedings of the 23rd International Conference on Enhanced Safety of Vehicles*, US National Highway Traffic Safety Administration, 2013.

[5] S. Lam, J. Taghia, and J. Katupitiya, "Evaluation of a transportation system employing autonomous vehicles," *Journal of Advanced Transportation*, vol. 50, no. 8, pp. 2266–2287, 2016.

[6] NACTO, "NACTO Policy Statement on Automated Vehicles," 2016, http://nacto.org/wp-content/uploads/2016/06/NACTO-Policy-Automated-Vehicles-201606.pdf.

[7] Y. Kong, S. Le Vine, and X. Liu, "Capacity impacts and optimal geometry of automated cars parking facilities," in *Proceedings of the Automated Vehicles Symposium*, Transportation Research Board and Association for Unmanned Vehicle Systems International, San Francisco, Calif, USA, July 2016.

[8] R. Estepa, A. Estepa, J. Wideberg, M. Jonasson, and A. Stensson-Trigell, "More Effective Use of Urban Space by Autonomous Double Parking," *Journal of Advanced Transportation*, vol. 2017, pp. 1–10, 2017.

[9] A. Malokin, P. L. Mokhtarian, and G. Circella, "How Do Activities Conducted while Commuting Influence Mode Choice? Testing Public Transportation Advantage and Autonomous Vehicle Scenarios," in *Proceedings of the Annual Meeting of the Transportation Research Board*, Washington, DC, USA, January 2016.

[10] R. Krueger, T. H. Rashidi, and J. M. Rose, "Preferences for shared autonomous vehicles," *Transportation Research Part C: Emerging Technologies*, vol. 69, pp. 343–355, 2016.

[11] M. Stanley, "Shared mobility on the road of the future," 2016, http://www.morganstanley.com/ideas/car-of-future-is-autonomous-electric-shared-mobility.

[12] M. Gucwa, "Mobility and energy impacts of automated cars," in *Proceedings of the Automated Vehicles Symposium*, San Francisco, Calif, USA, 2014.

[13] S. Childress, B. Nichols, B. Charlton, and S. Coe, "Using an activity-based model to explore possible impacts of automated vehicles," in *Proceedings of the 94th Annual Meeting of the Transportation Research Board*, Washington, DC, USA, 2015.

[14] K. Kim, G. Rousseau, J. Freedman, and J. Nicholson, "The Travel Impact of Autonomous Vehicles in Metro Atlanta through Activity-Based Modeling," in *Proceedings of the 15th TRB National Transportation Planning Applications Conference*, Atlantic City , NJ, USA, 2015.

[15] E. Guerra, "Planning for Cars That Drive Themselves: Metropolitan Planning Organizations, Regional Transportation Plans, and Autonomous Vehicles," *Journal of Planning Education and Research*, vol. 36, no. 2, pp. 210–224, 2016.

[16] S. Tsao and S. Chu, "A study on adjustment factors for U-turns in left-turn lanes at signalized intersections," *Journal of Advanced Transportation*, vol. 29, no. 2, pp. 183–192, 1995.

[17] W. T. Hung, F. Tian, and H. Y. Tong, "Discharge headway at signalized intersections in Hong Kong," *Journal of Advanced Transportation*, vol. 37, no. 1, pp. 105–117, 2003.

[18] Y. Gu, Y. Hashimoto, L.-T. Hsu, M. Iryo-Asano, and S. Kamijo, "Human-like motion planning model for driving in signalized intersections," *IATSS Research*, vol. 41, no. 3, pp. 129–139, 2017.

[19] Transportation Research Board, *Highway Capacity Manual*, A Guide for Multimodal Mobility Analysis, Washington, DC, USA, 6th edition, 2016.

[20] R. Hoogendoorn, B. Van Arem, and S. Hoogendoorn, "Automated driving, traffic flow efficiency, and human factors," *Transportation Research Record*, vol. 2422, pp. 113–120, 2014.

[21] G. Arnaout and S. Bowling, "Towards reducing traffic congestion using cooperative adaptive cruise control on a freeway with a ramp," *Journal of Industrial Engineering and Management*, vol. 4, no. 4, pp. 699–717, 2011.

[22] P. A. Ioannou, *Automated Highway Systems*, Springer, New York, NY, USA, 1997.

[23] U. Karaaslan, P. Varaiya, and J. Walrand, "Two proposals to improve freeway traffic flow," in *Proceedings of the American Control Conference*, pp. 2539–2544, IEEE, June 1991.

[24] J. Vander Werf, S. E. Shladover, M. A. Miller, and N. Kourjanskaia, "Effects of adaptive cruise control systems on highway traffic flow capacity," *Transportation Research Record*, vol. 1800, no. 1, pp. 78–84, 2002.

[25] B. Van Arem, C. J. G. Van Driel, and R. Visser, "The impact of cooperative adaptive cruise control on traffic-flow characteristics," *IEEE Transactions on Intelligent Transportation Systems*, vol. 7, no. 4, pp. 429–436, 2006.

[26] K. Dresner and P. Stone, "A multiagent approach to autonomous intersection management," *Journal of Artificial Intelligence Research*, vol. 31, pp. 591–656, 2008.

[27] M. Ferreira and P. M. D'Orey, "On the impact of virtual traffic lights on carbon emissions mitigation," *IEEE Transactions on Intelligent Transportation Systems*, vol. 13, no. 1, pp. 284–295, 2012.

[28] Z. Li, L. Elefteriadou, and S. Ranka, "Signal control optimization for automated vehicles at isolated signalized intersections," *Transportation Research Part C: Emerging Technologies*, vol. 49, pp. 1–18, 2014.

[29] S. Le Vine, A. Zolfaghari, and J. Polak, "Autonomous cars: The tension between occupant experience and intersection capacity," *Transportation Research Part C: Emerging Technologies*, vol. 52, pp. 1–14, 2015.

[30] S. Le Vine, X. Liu, F. Zheng, and J. Polak, "Automated cars: Queue discharge at signalized intersections with "Assured-Clear-Distance-Ahead" driving strategies," *Transportation Research Part C: Emerging Technologies*, vol. 62, pp. 35–54, 2016.

[31] J. Shelton, S. Samant, J. Wagner, G. Goodin, E. Seymour, and T. Lomax, *Revolutionizing Our Roadways: Modeling the Traffic Impacts from Automated and Connected Vehicles in a Complex, Congested Urban Setting*, Texas A&M Transportation Institute, 2016.

[32] C. Katrakazas, M. Quddus, W. H. Chen, and L. Deka, "Real-time motion planning methods for autonomous on-road driving: State-of-the-art and future research directions," *Transportation Research*, vol. 60, pp. 416–442, 2015.

[33] W. K. M. Alhajyaseen, M. Asano, H. Nakamura, and D. M. Tan, "Stochastic approach for modeling the effects of intersection geometry on turning vehicle paths," *Transportation Research Part C: Emerging Technologies*, vol. 32, pp. 179–192, 2013.

[34] A. Wolfermann, W. K. Alhajyaseen, and H. Nakamura, "Modeling speed profiles of turning vehicles at signalized intersections," in *Proceedings of the 3rd International Conference on Road Safety and Simulation*, Indianapolis, Ind, USA, 2011.

[35] S. Cooner, S. Ranft, Y. Rathod et al., "Development of guidelines for triple left and dual right-turn lanes," Tech. Rep. 0-6112-1, Texas Transportation Institute, 2011.

[36] T. Sando and R. N. Mussa, "Site Characteristics Affecting Operation of Triple Left-Turn Lanes," *Transportation Research Record*, vol. 1852, pp. 55–62, 2003.

[37] K. Courage, B. Stephens, A. Gan, and M. Willis, Triple Left-Turn Lanes at Signalized Intersections. Florida Department of Transportation, 2017, http://www.fdot.gov/research/completed_proj/summary_te/fdot_bc131rpt.pdf.

[38] J. D. Leonard, "Operational characteristics of triple left turns," *Transportation Research Record*, vol. 1457, pp. 104–110, 1994.

[39] K. W. Ackeret, "Criteria for the geometric design of triple left-turn lanes," *ITE Journal* , vol. 64, no. 12, pp. 27–33, 1994.

[40] J. D. Buchwalter, W. H. Danne, J. J. Dvorske et al., *American Jurisprudence: Automobiles and Highway Traffic*, 2nd edition, 2016.

[41] J. E. Naffky, *Corpus Juris Secundum: A Contemporary Statement of American Law as Derived from Reported Cases and Legislation*, vol. 60A, Motor Vehicles, 2012.

[42] "Louisiana Farm Bureau Mutual Insurance v. Dunn. 484 So.2d 853," 1986.

[43] S. Shladover, D. Su, and X. Lu, "Impacts of Cooperative Adaptive Cruise Control on Freeway Traffic Flow," *Transportation Research Record*, vol. 2324, pp. 63–70, 2012.

[44] J. F. Stanek and J. A. Lockwood, "Automotive Drone Deployment System. United States Patent Application #2016/0347452," *Assigned to Ford Global Technologies LLC*, 2016.

[45] American Association of State Highway and Transportation Officials, *A Policy on Geometric Design of Highways and Streets*, 6th edition, 2011.

[46] D. B. Fambro, K. Fitzpatrick, and R. J. Koppa, "Determination of Stopping Sight Distances," National Cooperative Highway Research Program (NCHRP) Report, 400, National Academy Press, Washington DC, USA, 1997.

[47] W. Hulshof, I. Knight, A. Edwards, M. Avery, and C. Grover, "Autonomous Emergency Braking Results," in *Proceedings of the 23rd International Conference on Enhanced Safety of Vehicles*, US National Highway Traffic Safety Administration, Washington, DC, USA, 2013.

[48] D. Burton, A. Delaney, S. Newstead, D. Logan, and B. Fildes, *Effectiveness of ABS and Vehicle Stability Control Systems. ISBN*, Royal Automobile Club of Victoria, Victoria, Australia, 2004.

[49] National Highway Traffic Safety Administration, *Final Report for the Methodology study of the Consumer Braking Information Initiative - Phase II*, 2000.

[50] C. P. Urmson, D. A. Dolgov, and P. Nemec, "Driving Pattern Recognition and Safety Control," United States Patent 8634980, 2014.

[51] P. Kozey, Y. Xuan, and M. J. Cassidy, "A low-cost alternative for higher capacities at four-way signalized intersections," *Transportation Research Part C: Emerging Technologies*, vol. 72, pp. 157–167, 2016.

[52] C. de Santos-Berbel, M. Essa, T. Sayed, and M. Castro, "Reliability-Based Analysis of Sight Distance Modelling for Traffic Safety," *Journal of Advanced Transportation*, vol. 2017, Article ID 5612849, 12 pages, 2017.

[53] A. C. Mersky and C. Samaras, "Fuel economy testing of autonomous vehicles," *Transportation Research Part C: Emerging Technologies*, vol. 65, pp. 31–48, 2016.

[54] J. Harding, G. Powell, R. Yoon et al., *Vehicle-to-Vehicle Communications: Readiness of V2V Technology for Application*, Report No. DOT HS 812 014, National Highway Traffic Safety Administration, Washington, DC, USA, 2014.

11

Characterization of Road Condition with Data Mining based on Measured Kinematic Vehicle Parameters

Johannes Masino,[1] Jakob Thumm,[1] Guillaume Levasseur,[2] Michael Frey,[1] Frank Gauterin,[1] Ralf Mikut,[3] and Markus Reischl[3]

[1] *Institute of Vehicle System Technology, Karlsruhe Institute of Technology, Germany*
[2] *Artificial Intelligence Research Laboratory, Université Libre de Bruxelles, Belgium*
[3] *Institute for Automation and Applied Informatics, Karlsruhe Institute of Technology, Germany*

Correspondence should be addressed to Johannes Masino; johannes.masino@kit.edu

Guest Editor: Petru Andrei

This work aims at classifying the road condition with data mining methods using simple acceleration sensors and gyroscopes installed in vehicles. Two classifiers are developed with a support vector machine (SVM) to distinguish between different types of road surfaces, such as asphalt and concrete, and obstacles, such as potholes or railway crossings. From the sensor signals, frequency-based features are extracted, evaluated automatically with MANOVA. The selected features and their meaning to predict the classes are discussed. The best features are used for designing the classifiers. Finally, the methods, which are developed and applied in this work, are implemented in a MATLAB toolbox with a graphical user interface. The toolbox visualizes the classification results on maps, thus enabling manual verification of the results. The accuracy of the cross-validation of classifying obstacles yields 81.0% on average and of classifying road material 96.1% on average. The results are discussed on a comprehensive exemplary data set.

1. Motivation

In 2006, bad conditions of road infrastructure were one of the causes of 50% of fatal accidents in France [1]. In 2016, four accidents in Germany were caused exclusively by road surface damage [2]. Road traffic authorities are aimed to improve and automate monitoring the road state to detect and repair road damages to enhance the safety of road traffic. Based on the detection results, specific and cost-optimized maintenance of roads can be ensured. Furthermore, suppliers of navigation systems can profit from the available information of the road state, because roads in bad condition may be neglected in route planning [3]. Automotive manufacturers can use the collected data to control adaptive vehicle suspensions and to display warnings in real time [4].

Contrary to physical modeling, data-based estimation of the road state does not require any comprehensive system characterization, such as vehicle, road, sensor, and environment. Moreover, modeling of a full vehicle requires five acceleration sensors or gyroscopes to measure vertical accelerations of unsprung masses and accelerations and rotations of the vehicle body [5]. To monitor road sections, a vehicle can be used as a mobile sensor platform that records both vehicle dynamics and the environment, such as the road state. The road state can be estimated using cameras or inertial measurement units to record rotation speeds and accelerations of the vehicle. Such sensors are already integrated into modern vehicles having an active or adaptive body control or new lighting systems. Inertial sensors are even part of the standard equipment of new vehicles and data can be fused with GPS data for more accurate positioning, an example being the new Audi A7. Previous studies revealed that measurements made by these sensors allow for the derivation of road features, such as potholes or mends of asphalt roads [6–8].

The inertial sensor is inexpensive and part of the standard equipment of many vehicles. Its data include information of minor unevenness of road surfaces that causes the vehicle to vibrate. Inertial sensors, however, only provide data on the road section just crossed. Cameras, by contrast, record the complete road section in front of the vehicle, including the neighboring lane. However, they are integrated into high-class vehicles only. Cameras currently used in vehicles are of limited accuracy and can detect potholes with a minimum depth of about 3 cm only.

Presently, the state of motorways is measured automatically using expensive and complex measurement vehicles, while that of roads in urban and rural areas is determined manually [9]. These methods are associated with a high expenditure. Due to manual evaluation, it takes a long time until the road network quality is updated. Safety-relevant damage may be detected too late. This may have severe consequences, such as traffic accidents or cost-intensive and complete renewal of the road.

For road maintenance, some countries determine the stochastic road profile depth or international roughness index (IRI), as outlined in [10]. However, the latter is often calculated for 100 m intervals only. As a result, certain obstacles, such as potholes, are not detected. In countries pursuing a systematic road maintenance scheme, not only the IRI but also individual obstacles are measured. This also is the objective of the present study.

Approaches to automatic road state monitoring using inertial sensors exist, e.g., [6, 11–13]. They only concentrate on single road features (such as potholes), do not have any representative dataset, or are based on data measured under restricted conditions, e.g., in speed limit areas or on certain sections only. Moreover, the validation phase only covers checks as to whether the road damage detected actually is damage or not (true or false positives), but not whether road damage was overseen (false negatives).

Road construction offices also need to know the material (road surface), as repairs on different surfaces produce different results and may cause different types of damage [14]. It is also important to distinguish between safety-relevant damage that has to be repaired within 24 hours and damage that is not relevant to safety and the repair of which can be planned and postponed.

The main contribution of this paper is to evaluate the principle feasibility of automatic road surface and road damage measurement with an inertial sensor in the vehicle body. Therefore, this work is aimed at

(i) designing a processing chain to evaluate road data based on measurements of inertial sensors,

(ii) automatically recording an adequate dataset,

(iii) developing and evaluating a method to estimate road surfaces and damage, and at

(iv) integrating the algorithms developed into a graphic user interface for evaluation of datasets with alternative parameterizations by nonexperts as well.

The methodology will be presented in Section 2. Section 3 will outline the implementation derived, while Section 4 will explain the results based on a first dataset. The result, its applicability, and open problems will be discussed in Section 5.

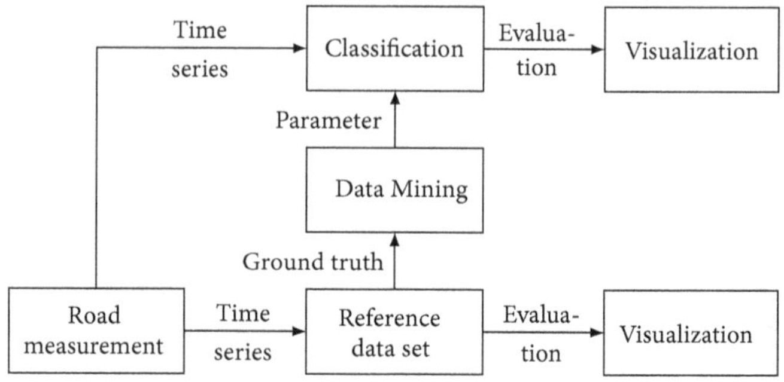

FIGURE 1: Overview of method.

2. Methods

2.1. Design. Figure 1 presents an overview of the method to evaluate the road state [15]. In a first step, the road state is to be measured by suitable sensors. For measurement, acoustic sensors, such as the sensors described in [16], acceleration sensors and gyroscopes, cameras, and similar devices, can be used. As a result, several synchronized time series will be obtained. To obtain a representative reference data set, sensor data have to cover a maximum of framework conditions, e.g., variations of external temperature, driver, and speed. Every point of time/road section has to be assigned a label, e.g. type of road surface, simultaneously or afterwards. In this way, a data set with correct allocations of sensor data to labels is obtained (ground truth). By means of data mining, models can be designed (offline) for retrospective evaluation (offline) or classification during driving operation (online). The results of the classification models then have to be visualized and evaluated on the basis of map material. To estimate the information on the road surface and event or damage that is of relevance to road construction offices, two separate classification routines have to be developed.

2.2. Data Acquisition. The data measured by the sensors installed in the vehicle, e.g., GPS and inertial sensors, are encoded on the CAN bus and cannot be read without the communication matrix that is available to the control system developer and automotive manufacturer only. Hence, an inexpensive measurement system similar to the inertial sensor incorporated in the vehicle is proposed for the easy measurement and readout of data. Measurements cover the position and dynamics of the vehicle, in particular vertical dynamics caused by unevenness [17]. In addition, the data may be assigned labels during measurement already. The measurement system (Figure 2) mainly consists of a GPS receiver (Adafruit ultimate GPS Hat) and a MEMS inertial sensor (LSM9DS1) measuring accelerations and rotation rates of the vehicle along all three axes. The sensor data are acquired using a Raspberry Pi and stored as a csv-table in fused form. As soon as the engine of the vehicle is turned off, the UPS is activated and data can be transmitted via WiFi to a central data base, if the Raspberry Pi is connected to a known WiFi network.

The GPS receiver has a sample rate of 10 Hz, a position resolution of 3 m, and a speed resolution of 0.1 m/s. As the

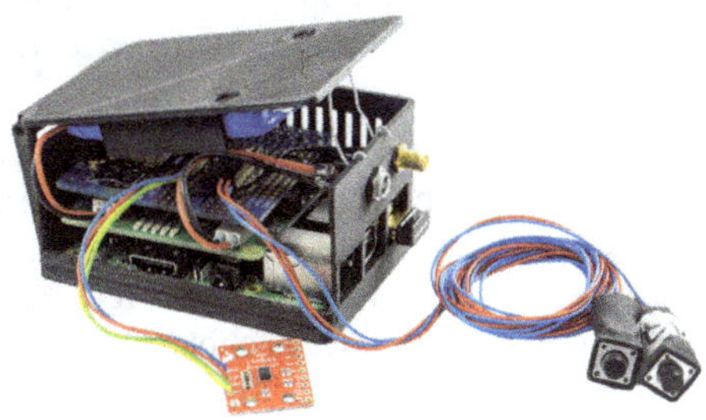

FIGURE 2: Measurement unit, consisting of Raspberry Pi, inertial sensor, Adafruit ultimate GPS Hat, UPS Hat, and two buttons for labeling the data. In service, the inertial sensor is fixed under the Raspberry Pi on the bottom of the case. The case is cut open for visualization purposes.

TABLE 1: Code and corresponding name of the classes for material and event.

m-Material	*e*-Event
0- Unknown	0- Unknown
1- Smooth surface	1- Good
2- Damaged asphalt	2- Light damages
3- Damaged concrete	3- Pothole
4- Cobblestone	4- Manhole cover
	5- Railway crossing
	6- Speed bump

inertial sensor is a low-cost MEMS sensor, the sample rate is not uniform. This has to be compensated by a filter in data processing. The sample rate is about 220 Hz. The accuracy of the acceleration sensor is 0.05 m/s^2 and of the gyroscope 0.003 degree/s. Without data transfer, a 32 GB memory card can record data for up to 1000 h.

For allocating labels to data, different approaches are presented in literature. For example, a microphone records the (road) damage report of the passenger [13]. This method, however, is subject to several deficiencies. Among others, the labels are recorded much later than the actual road damage and the soundtrack is not synchronized with the sensor data. Reference [6] uses "loosely labeled" training data. Here, only the number of classes but not the exact position is recorded for large road segments.

The measurement system developed for this study is based on two buttons. A pressed button annotates the damage class (event) or the change of road surface (material). For every measurement drive, data with a certain material (e.g., if asphalt 1, otherwise 0) and an event (e.g., if pothole 1, otherwise 0) are recorded. After the measurement drives, binary coding of the data of the respective files is transformed into the coding given in Table 1.

The unix time t, ID for the sensor, speed v, position and time stamp of the GPS lat, lon, and t_{GPS}, accelerations a and rotation rates ω along all three axes, and the two labels for the event e and material m are recorded and stored in a csv-table on the measurement system (Figure 2). The measurement system is installed near the static center of mass of a BMW 116d in the console between the driver's seat and passenger's seat. Orientation of the sensor axes corresponds to the vehicle axes according to ISO 8855:2011. For the method to be generally applicable, measurement data are recorded on randomly selected roads in the region of Karlsruhe, Germany. The speed, road condition, and environmental conditions (e.g., measurement drives in good and rainy weather) are varied strongly. In total, reference data are recorded for a period of three months on a distance of more than 200 km. The data are recorded on eight days (three times a whole day) by three different drivers. Acquisition of reference data is a time-consuming process, as the materials and events have to be crossed under variable environmental conditions and at variable speeds. In particular, individual events, such as potholes, of various types have to be found in the road network and crossed several times with variable approach angles and vehicle tracks.

Lacking GPS data due to variable scanning rates are reconstructed by linear interpolation. As the measurement series are not recorded at a constant sample rate, resampling is required. By resampling, the data are converted from the time domain (s^{-1}) to the space domain (m^{-1}). In [18, 19], it was shown that the response of the vehicle to the excitation of the road depends on speed and that presentation in the space domain reduces this effect. All-time series are resampled with a (spatial) frequency of 100 m^{-1}. The section driven is calculated from the time stamp and speed with the help of the implicit Euler method. Calculation via GPS would also be possible but lacks precision.

The classes of materials m and events e are encoded by natural numbers (Table 1). Light damages are general types of unevenness, which are not safety-relevant and include minor faults and repairs. Manhole cover, railway crossing, and speed bump are construction obstacles. A speed bump is defined as an elevated construction transverse to the driving direction. The pothole represents a fault of at least 2 cm in depth. The latter event is safety-relevant and should be repaired within a maximum of 24 h. For every sample point k of the reference data set, two labels are annotated for the material $m[k]$ and event $e[k]$.

2.3. Signal Processing

2.3.1. Overview. To derive information on the road surface or material and event/damage from the sensor data recorded, the data streams first have to be transferred to a feature space. Feature extraction calculates representative and useful individual features from complete or partial measurement series. Without knowing the physical model of effects of asphalt changes or road damage on the sensor, it is recommended to calculate a large set of features and to check their suitability for the classification problem based on data with the corresponding labels (ground truth). Efficient feature calculation is needed for calculation on mobile devices (e.g., microcontrollers).

2.3.2. Generation of New Time Series. To describe the road state, acceleration in vertical direction (a_z) and rotation speeds in longitudinal and transverse direction (ω_x and ω_y) are very important [6]. Furthermore, the roll and pitch acceleration ($\dot{\omega}_x$ and $\dot{\omega}_y$) as well as the jerk ($\dot{a}_z$) of the vehicle are done using the derivation of the vertical acceleration in time domain. The space series data of the vertical roll and pitch acceleration is transformed into frequency domain with the short-time Fourier transform, which contains the short-term distance-localized frequency content of the signal. Hereby, features based on specific frequency bands can be investigated. Hence, the three distance series data are extended by the following data streams, which lead us to 7 data streams in total:

(i) vertical acceleration,
(ii) roll acceleration,
(iii) pitch acceleration,
(iv) deviation of vertical acceleration,
(v) short-time Fourier transformed vertical acceleration,
(vi) short-time Fourier transformed pitch acceleration and
(vii) short-time Fourier transformed roll acceleration.

2.3.3. Feature Extraction. The features are calculated for windows with a specific length in distance domain and a specific overlap. A window U_i denotes all indexes $[n_i - L/2, n_i + L/2 - 1]$ with the running index n, window index i, and window length L. The window overlap r corresponds to those values from the window U_i that are contained in the previous window U_{i-1}, i.e., $n_{i+1} = n_i + L(1 - r)$. If a longer distance is chosen, short amplitudes, for example, due to potholes, have a weaker impact on the value of features, which incorporate the overall signal, such as the standard deviation. These short amplitudes can be captured by shortening the window size or using features, which calculate extrema.

For the feature extraction for material and events we use window sized of 50 m or 5 m, respectively, and an overlap of 20%.

From the distance series data, we calculate the standard deviation as well as peak-to-peak. The root mean square value or effective value for specific frequencies and the spectral centroid is extracted from the short-time Fourier transformed data streams for the following spatial frequency bands ($1\,\mathrm{m}^{-1}$):

$$[0.1,\ 0.5]\,[0.5,\ 15]\,[15,\ 20]\,[0.1,\ 25]\,[0.1,\ 50] \quad (1)$$

The vehicle velocity has a strong sensitivity on the vehicle vibration. Previous research suggests performing a linear regression with each feature as the dependent variable and the velocity as the independent variable [12]. The velocity dependency is then removed by subtracting the estimated linear equation from the corresponding feature. However, the vehicle vibration and the extracted features are not linear dependent on the velocity. The dependent parameters are incorporated and the mean velocity is calculated for each window as additional feature. To allow nonlinear relationships a kernel function of higher order can be applied for the classification.

Of the GPS latitude *lat* and longitude *lon* time series, the medians in every window are used for later visualization.

2.3.4. Classification. Based on the extracted individual features and the corresponding labels, two classifiers are designed for material and event. For the design and application of classification, a combination of feature selection, feature aggregation, and classifier is chosen.

For the surface classification, the five best individual features each are determined using the multivariate analysis of variances (MANOVA) method, for event classification the ten best features are selected. For visualization purposes, the selected individual features are then aggregated to two features using linear discriminant analysis (DA), which can also minimize the calculation expenditure. A support vector machine (SVM) classifier with polynomial kernel function with order 2 is used. Validation is carried out with the help of cross-validation with 5-folds.

2.3.5. Performance Measures. From the correct and false predicted instances, we can calculate a confusion matrix $M = (m_{ij}) \in \mathbb{N}^{k\times k}$ for classes K_i, $i = 1,\ldots,k$. In the confusion matrix, m_{ii} presents the *true positives* for class i. The other elements in column j are called *false negatives*, in row i *false positives* and in the diagonal *true negatives*.

From the confusion matrix, one can calculate multiple performance measures to evaluate the model, such as recall with $m_{ii}/\sum_{j=1}^{n} m_{ji}$ for class K_i, the overall accuracy of the classifier with $\sum_{i=1}^{n} m_{ii}/\sum_{i=1}^{n}\sum_{j=1}^{n} m_{ij}$, or the precision $m_{ij}/\sum_{j=1}^{n} m_{ij} = \pi_{ij}$. The precision presents the fraction of retrieved instances that are relevant and can be seen as the probability π_{ij} of the classifier to predict class i as class j for $i, j = 1,\ldots,l$. An overview for performance measures for different calculation problems can be found in [20].

3. Implementation

To facilitate operation by non-experts, the methods are implemented in a graphical user interface called *Vehicle Learner Toolbox*, which is available in [21]. It is based on MATLAB and implements several machine learning operations of the freely available toolbox SciXMiner [22] (formerly, Gait-CAD [23]). The *Vehicle Learner Toolbox* provides the possibility to

(i) import vehicle sensor data in different file formats,
(ii) compress the imported data and automatically extract various features,
(iii) train a classifier model with a wide-ranging set of options,
(iv) test the trained classifier with a test set,
(v) visualize the results with the help of plots and maps.

A project folder can be selected and sensor data can be imported in the corresponding frame *Data* (Figure 3). There is the option to assign the sensor data to specific vehicles, since they vary in suspensions, damping, and other

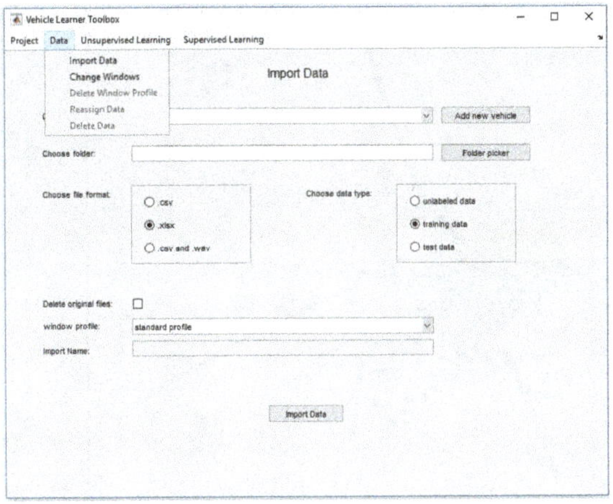

FIGURE 3: The import data frame of the *Vehicle Learner Toolbox*. Different vehicles can be chosen and the data type can be set.

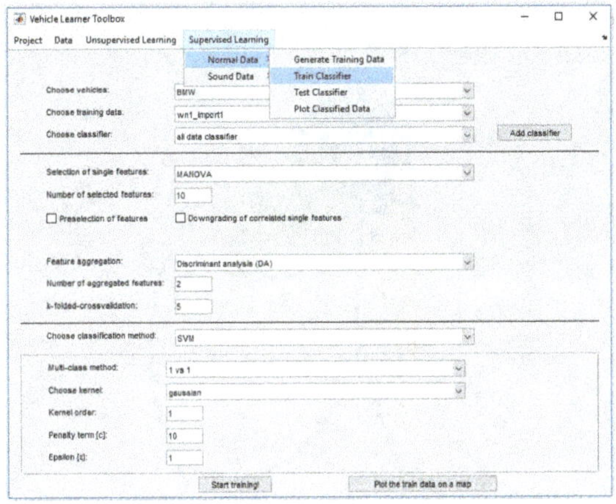

FIGURE 4: The train classifier frame of the *Vehicle Learner Toolbox*. Multiple classifiers can be trained with a wide range of options.

parameters, which have an impact on the vibration behaviour. Therefore, in the following data processing, feature selection and classification can be performed for data from specific vehicles. The import allows .csv and .xlsx file format with the following column headers:

(i) timestamp (unix timestamp)

(ii) x-, y-, z-accel (the acceleration values in each direction)

(iii) x-, y-, z-gyro (the gyroscope values in each direction)

(iv) gps-timestamp (format: YYYY-MM-DDThh:mm:ss, 000Z)

(v) lat, lon (position in latitude and longitude)

(vi) speed (in m/s)

(vii) m, e (material m and event e labeling, if the data is not labeled, these columns should only contain zeros).

Since the GPS data is acquired with a lower sample rate compared to the inertial sensor, these data are automatic interpolated. Furthermore, the sensor signals are subject to noise [24] and are automatic smoothed during the import process with the following filters. Despite the noise of a MEMS gyroscope visible as spikes in the signal, it is well known for its good accuracy in short term [25]. A suitable filter for this purpose is the median filter, which is robust against outliers and removes noise while preserving high frequency content. Since the data from the MEMS accelerometer do not show such spikes but contains more noise in the short term [25], a Savitzky-Golay FIR smoothing filter is applied. It fits a polynomial of a specified degree to frames of noisy data and minimizes the least-squares error [26]. Therefore, the filter outperforms standard averaging FIR filters, which might remove high frequency content with the noise.

There is also the possibility to import tire cavity sound data along the inertial sensor data for road roughness estimation, as presented in [16, 27], but is not substance in this paper. Moreover, the imported data set can be categorized as training, testing or unlabeled data.

Furthermore, the parameters for the window profile, such as length of road segments and overlapping factor of these windows, can be determined, as well as the resampling frequency. The standard window profiles are *material* with a window length of 50 m and *event* with a window length of 5 m.

After the import and preprocess of the data, new time series data are calculated and features are automatically extracted, as proposed in Sections 2.3.2 and 2.3.3. The code to calculate new data series or features can be easily added in the corresponding MATLAB function.

The proposed data mining methods (Section 2.3.4) can be applied in the toolbox under the menu *Supervised Learning* (Figure 4). In the first step, a training data set must be generated. There are two different ways to accomplish this. Either an external data set containing features can be imported or the imported data within the toolbox can be used and modified by choosing the time interval of the data acquisitions or the area. Furthermore, data annotated with specific labels can be excluded from the classification. For our example, all data with labels *0-unknown* were deleted (Table 1). Another option is to thin out classes with significant more data points than other classes to allow an approximately uniform distribution of data points among the classes to prevent over-fitting of specific classes. Furthermore, systematic errors during labeling the data can be removed; e.g., if the trigger to annotate the data was activated too early or too late the annotation can be moved or data points with the wrong annotation can be excluded. After the generation of the data set to be processed, the settings for the classifier can be determined under the tab *Train Classifier* (Figure 4).

In the first step the vehicle and the training data set must be set. Afterwards a new classifier model can be created or an existing model can be selected. The next section contains the settings of feature selection (e.g,. MANOVA) and aggregation (e.g., discriminant analysis), as proposed in Section 2.3.4. Reducing the amount of features highly influences the classification result by reducing the chances for overfitting. It is possible to cross-validate the training process

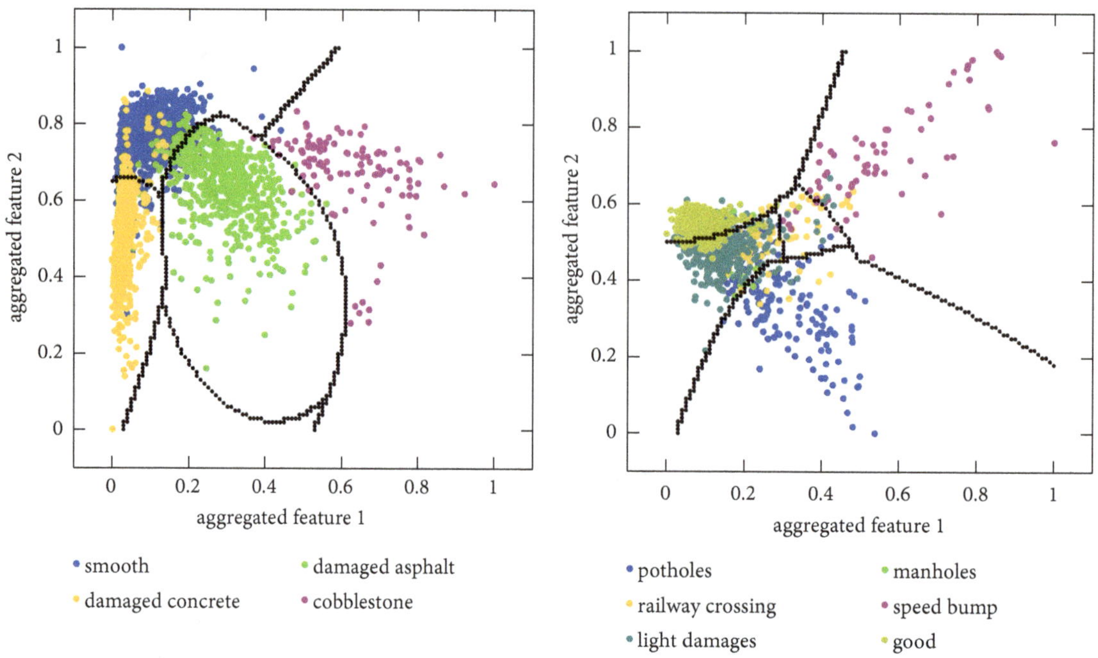

(a) Classification of material m

(b) Classification of events e

FIGURE 5: Classification results with two aggregated features and borders of the classifier in black.

by setting the k-folded-cross-validation value to higher than 1. The last section offers a variety of settings for the classifier, e.g. for a SVM, including the kernel function and penalty term. Afterwards, the classifier can be trained and data can be plotted on open street maps. Furthermore, the confusion matrix and the total loss is shown in the MATLAB console.

For testing new data, a data set with modifications in time range and area to be analyzed can be generated as described for training, and a trained classifier must be selected. If the test data set is labeled, the output of the prediction is again a confusion matrix and the classification error. Moreover, the results can be visualized and plotted on open street maps, as it will be presented in Section 4. The trajectories will be cut into segments of different color referring to the corresponding classes, which are predicted.

4. Results

4.1. Event Classification. The accuracy of the cross-validation of classifying events yields 81% on average without feature aggregation. The aggregated feature space and the lines of the function to classify the events is shown in Figure 5(b).

The illustration of the classification shows that road segments in *good condition*, with *light damages*, *speed bumps*, and *potholes*, can be separated well. This indication is proofed by the quantitative results, listed in Table 2.

The precision and recall for the mentioned classes is above 70%, whereas the performance measures for *manhole cover* and *railway crossing* is below 62% on average.

The most important features, determined with MANOVA, are

(i) peak-to-peak of pitch acceleration

(ii) peak-to-peak of roll acceleration

(iii) maximum of jerk in vertical direction

(iv) root mean square (RMS) of the vertical acceleration

(v) speed

By comparing each class with each other, it emerges that the peak-to-peak value of pitch and roll acceleration are mainly responsible to separate events, which occur on

(i) both vehicle lanes (railway crossing, speed bump),

(ii) on only one side of the vehicle (manhole cover, pothole),

(iii) or have only little impact on the vehicle vibration (light damages, road segments in good condition).

In addition, the average RMS of the vertical acceleration is important to separate light damages and road segments in good condition. Furthermore, *potholes* and *manhole covers* are dividable through the maximum RMS of the roll acceleration for the frequency range 15 to 25 m^{-1}. However, latter events are often misclassified as segments in *good condition* or *light damages*. *Speed bumps* and *railways crossings* are separable by the value of the peek-to-peek of the pitch rate, whereas *railways crossings* are also often misclassified as light damages.

To test the classifier, a data set of more than 200 km of street data is classified and plotted on open street maps. The results are promising and represent the actual street condition in many occasions. A few examples of classified areas are shown below.

Table 2: Results of single features classification.

Material	Precision (% ± SD)	Recall (% ± SD)	Events	Precision (% ± SD)	Recall (% ± SD)
smooth surf.	95.8 ± 0.7	97.6 ± 0.7	good	83.9 ± 1.5	92.4 ± 2.7
damaged asph.	97.3 ± 2.0	94.9 ± 1.4	light dam.	78.5 ± 3.8	70.8 ± 3.3
damaged con.	95.6 ± 0.8	92.9 ± 1.1	pothole	85.3 ± 6.6	77.4 ± 7.2
cobblestones	99.1 ± 1.8	100.0 ± 0.0	manhole	40.9 ± 13.6	29.1 ± 15.7
			railway cro.	61.6 ± 4.9	57.0 ± 7.9
			speed bump	94.9 ± 7.1	91.0 ± 12.2
Average	96.4 ± 0.8	96.4 ± 0.8		74.2 ± 6.3	69.6 ± 8.1
Accuracy	96.1 ± 0.4			81.0 ± 1.7	

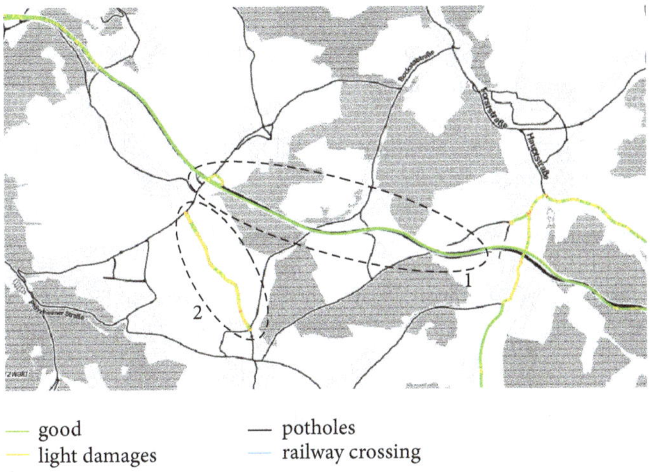

Figure 6: Event classification results of two high speed roads in the south west of Karlsruhe, Germany.

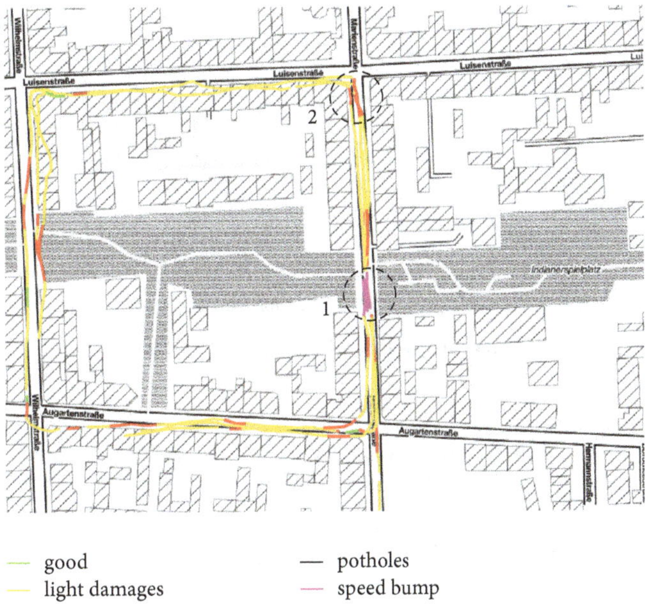

Figure 7: Event classification results for road segments in the city of Karlsruhe, Germany.

The first example shows the event classification results on two different high speed roads (Figure 6). The upper one with Label 1 is a freshly renovated asphalt highway with close to no damages and the lower one with Label 2 is a poorly patched asphalt road with a lot of medium and severe damages. The classification successfully predicted the upper roadway as *good street*. Most parts of the lower street were predicted as *light damage* and some points even as *potholes*. The results represent the road condition very accurate. The only noticeable misclassification is *railway crossing* that was predicted once (Label 3).

The second example presents data acquired in an urban area in Karlsruhe, the predictions are shown in Figure 7. The roads in this area are poorly preserved and there is a *speed bump* at a pedestrian crossing (Label 1). The classification model correctly predicts the *speed bump* (Label 1) for all overdrives and a *pothole* (Label 2) on both driving directions.

The third interesting sector is shown in Figure 8. Potholes (Labels 2 and 3), which were at the edge of the driving line, were overdriven multiple times and the classifier predicts the severe damage accordingly. Sometimes the output at the road segments is not *pothole* but *light damages* or even *good road condition*. The reason might be that the pothole was avoided by the driver.

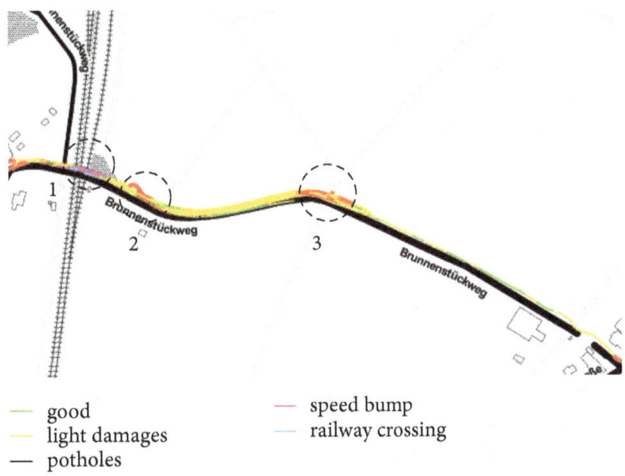

Figure 8: Event classification results of road segments outside of Karlsruhe, Germany, for the events *potholes* and *railways crossing*.

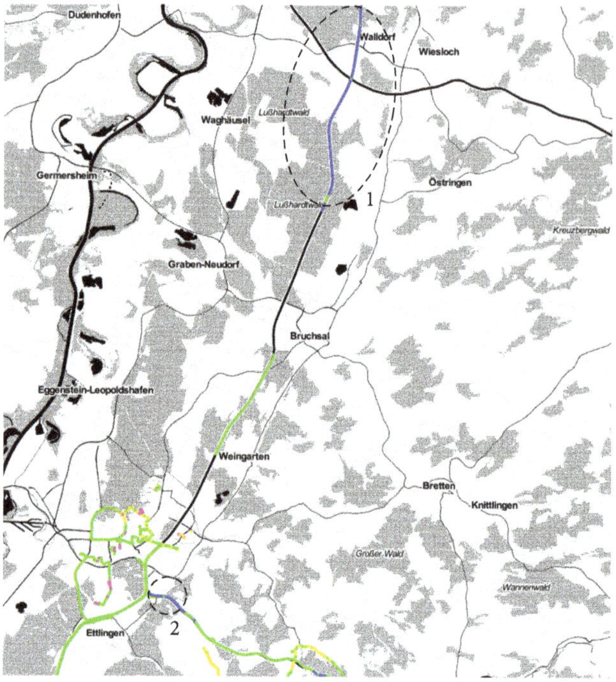

— smooth asphalt
— damaged asphalt
— damaged concrete
— cobblestones

FIGURE 9: Material classification of an aged concrete highway.

The *railway crossing* (Label 1) is more elevated than other crossings and miss-classified as *speed bump* in few cases.

4.2. Road Surface Classification. When classifying road surfaces, cross-validation yields 96.1% accuracy on average without aggregation of features. The aggregated feature space is shown in Figure 5(a). The figure indicates, that the misclassifications are asphalt classified as *damaged asphalt* or *damaged concrete* and vice versa. The illustrated results are underlined by Table 2, where the precision and recall for *cobblestone* is above 99.0%, whereas the performance measures for *asphalt*, *damaged asphalt*, and *damaged concrete* are between 92.0 and 97.6% percent.

The three best individual features for the classification of road surface according to MANOVA are

(i) RMS of the roll acceleration for frequency range from 5 to 15 m^{-1} on average

(ii) standard deviation of the pitch rate

(iii) stand deviation of the RMS of the vertical acceleration for frequency range 15 to 25 m^{-1}

The values of RMS of the roll acceleration and vertical acceleration separate the classes *smooth surfaces*, *damaged asphalt* and *cobbled stone*. The values are greatest for *cobbled stone* and low for *smooth surface*.

Standard deviation of pitch rate separates the classes *damaged concrete* from all other classes. The reason are probably poor and aged concrete joints.

— smooth asphalt
— damaged asphalt
— damaged concrete
— cobblestones

FIGURE 10: Material classification of cobblestone and road segments in urban area.

The material classifier was applied to the same data set described for event classification. The classifier was able to reflect the road surface very precisely. The following figures display the performance on different surfaces. Analog to the event classification shown in Figure 6, the material classifier could distinguish between both roads and correctly classified them as *smooth surface* and *damaged asphalt*, respectively.

In contrast, Figure 9 shows a long highway segment with aged concrete and distinctive concrete joints, which have to be maintained shortly. Except for one short segment, which was classified as *smooth surface*, the road state was correctly predicted.

The classification results of data acquired in the urban area of Karlsruhe (Figure 10) show two correctly predicted areas of cobblestone (Label 1 and 2). The remaining road segments are correctly classified as segments with *light damages* or in *good condition*. Especially latter class was correctly predicted for a road segment, which was recently renewed (Label 3). One miss-classification of *cobblestone* can be found close to Label 1. However, this road segment is highly damaged with multiple potholes, which have a high impact on the vehicle vibration similar to cobblestones.

5. Conclusions and Outlook

The results show that the system presented in this paper can classify both road materials and events. The features selected by MANOVA are in agreement with the theory of vehicle excitation. Material classification performs well according to the results of the cross-validation and the test data.

There are multiple miss-classifications of the prediction of events, especially for structural obstacles, such as *manhole covers* and *railway crossing*. However, these events might be marked on a map and excluded from classification and investigation, as the main objective is to detect road damage.

One reason for misclassifications of the events *good condition*, *light damages*, and *pothole* might be false manual annotating, since there is sometimes only a fine line between the degree of damages, or the events were not fully overdriven, especially for *potholes*.

As the system is of modular design, the number and type of sensors and sensor modality can be varied. When adapting feature extraction, also camera recordings might be useful. Transferability to other vehicles with different chassis and dimensions has not been examined so far. Presumably, the algorithms of parameters are adapted to the vehicle with which the learning data set was recorded. Here, fusion of learning data sets from several vehicles and an accordingly adapted classification routine might help. It can be assumed that the results will be slightly worse.

Generally, the inertial sensor represents a very good option to collect information on the tire/road contact at low costs and over wide areas. Use of information of several vehicles can compensate the drawback of some drivers passing by safety-relevant damage that, hence, is not measured by the sensor. Moreover, obstacles at the roadside are not crossed and, hence, cannot be detected.

Fusion of camera and inertial sensor data probably would be the optimum solution for a mobile determination of the state of road traffic infrastructure. For road construction offices, use of a low-cost and computationally efficient system, consisting of an inertial sensor, Raspberry Pi, and simple signal processing, is sufficient and can be recommended.

Conflicts of Interest

The authors declare that they have no conflicts of interest.

Acknowledgments

This work was funded by the Federal Ministry for the Environment, Nature Conservation, Building and Nuclear Safety, Germany, within the Environmental Research Plan 2014 (Project no. 3714541000).

References

[1] P. De Gonneville and G. Martin, "Le mécanisme d' accident," in *Sétra: Service d' études techniques des routes et autoroutes*, Bagneux, 2006.

[2] "Destatis. Verkehrsunfälle. Destatis, Mar. 2017. Artikelnummer: 2080700161124".

[3] P. Mohan, V. N. Padmanabhan, and R. Ramachandran, "Nericell: Rich monitoring of road and traffic conditions using mobile smartphones," in *Proceedings of the The ACM Conference on Embedded Networked Sensor Systems*, 2008.

[4] M. Schade, "Der intelligente Löcher-Sucher," *Auto Bild*, 2016.

[5] E. Esmailzadeh and F. Fahimi, "Optimal adaptive active suspensions for a full car model," *Vehicle System Dynamics*, vol. 27, no. 2, pp. 89–107, 1997.

[6] J. Eriksson, L. Girod, B. Hull, R. Newton, S. Madden, and H. Balakrishnan, "The pothole patrol: using a mobile sensor network for road surface monitoring," in *Proceedings of the 6th International Conference on Mobile Systems, Applications, and Services (MobiSys '08)*, pp. 29–39, Breckenridge, CO, USA, June 2008.

[7] C. Koch and I. Brilakis, "Pothole detection in asphalt pavement images," *Advanced Engineering Informatics*, vol. 25, no. 3, pp. 507–515, 2011.

[8] S. C. Radopoulou and I. Brilakis, "Patch detection for pavement assessment," *Automation in Construction*, vol. 53, pp. 95–104, 2015.

[9] S. C. Radopoulou and I. Brilakis, "Improving Road Asset Condition Monitoring," pp. 3004–3012.

[10] ISO, *ISO 8608: Mechanical Vibration - Road Surface Profiles - Reporting of Measured Data*, International Organization for Standardization, 2016.

[11] K. Chen, G. Tan, M. Lu, and J. Wu, "CRSM: a practical crowdsourcing-based road surface monitoring system," *Wireless Networks*, vol. 22, no. 3, pp. 765–779, 2016.

[12] M. Perttunen, O. Mazhelis, F. Cong et al., "Distributed road surface condition monitoring using mobile phones," in *Proceedings of the International Conference on Ubiquitous Intelligence and Computing*, pp. 64–78, Springer, 2011.

[13] F. Seraj, A. Dilo, T. Luarasi et al., "RoADS: A Road Pavement Monitoring System for Anomaly Detection Using Smart Phones," in *Big data analytics in the social and ubiquitous context*, pp. 128–146, 2016.

[14] I. Tekin, *Economic investigations of the current state of the road system in order to determine the advantages and areas of application of a road monitoring system [Bachelors Thesis]*, Karlsruhe Institute of Technology, Institute of Vehicle Systems Technology.

[15] J. Masino, G. Levasseur, M. Frey, F. Gauterin, R. Mikut, and M. Reischl, "Charakterisierung der Fahrbahnbeschaffenheit durch Data Mining von gemessenen kinematischen Fahrzeuggrößen," *Automatisierungstechnik*, vol. 65, no. 12, 2017.

[16] J. Masino, J. Pinay, M. Reischl, and F. Gauterin, "Road surface prediction from acoustical measurements in the tire cavity using support vector machine," *Applied Acoustics*, vol. 125, pp. 41–48, 2017.

[17] J. Masino, M. Frey, F. Gauterin, and R. Sharma, "Development of a highly accurate and low cost measurement device for Field Operational Tests," in *Proceedings of the 3rd IEEE International Symposium on Inertial Sensors and Systems, ISS 2016*, pp. 74–77, USA, February 2016.

[18] M. Sayers, T. Gillespie, and C. Queiroz, "The international road roughness experiment: Establishing correlation and a calibration standard for measurements," *World Bank Technical Paper*, 1986.

[19] C. C. Ward and K. Iagnemma, "Speed-independent vibration-based terrain classification for passenger vehicles," *Taylor & Francis, Vehicle System Dynamics*, vol. 47, no. 9, 2009.

[20] M. Sokolova and G. Lapalme, "A systematic analysis of performance measures for classification tasks," *Information Processing & Management*, vol. 45, no. 4, pp. 427–437, 2009.

[21] J. Thumm and J. Masino, *Vehicle Learner Toolbox for Road Condition Estimation, Matlab files*, 2018.

[22] R. Mikut, A. Bartschat, and W. Doneit, "The MATLAB toolbox SciXMiner: User's manual and programmer's guide," 2017, https://arxiv.org/abs/1704.03298.

[23] R. Mikut, O. Burmeister, S. Braun, and M. Reischl, "The open source MATLAB toolbox Gait-CAD and its application to bioelectric signal processing," in *Proceedings of the DGBMT-Workshop Biosignalverarbeitung*, pp. 109–111, Potsdam, Germany, 2008.

[24] C. Halfmann and H. Holzmann, *Adaptive Modelle für die Kraftfahrzeugdynamik*, Springer, Berlin, Germany, 2003.

[25] O. J. Woodman, "An introduction to inertial navigation," Tech. Rep., University of Cambridge, Computer Laboratory, 2007.

[26] S. J. Orfanidis, *Introduction to signal processing*, Prentice-Hall, Inc, 1995.

[27] J. Masino, M.-J. Foitzik, M. Frey, and F. Gauterin, "Pavement type and wear condition classification from tire cavity acoustic measurements with artificial neural networks," *The Journal of the Acoustical Society of America*, vol. 141, no. 6, pp. 4220–4229, 2017.

[28] J. Masino, "Characterization of Road Condition with Data Mining Based on Measured Kinematic Vehicle Parameters, Data," 2018.

12

Acceptance of Driverless Vehicles: Results from a Large Cross-National Study

Sina Nordhoff,[1,2] **Joost de Winter**,[3] **Miltos Kyriakidis**,[4] **Bart van Arem**,[1] **and Riender Happee**[1,5]

[1]*Department of Transport & Planning, Delft University of Technology, Delft, Netherlands*
[2]*Innovation Centre for Mobility and Societal Change, Berlin, Germany*
[3]*Department of Biomechanical Engineering, Delft University of Technology, Delft, Netherlands*
[4]*ETH Zurich, Future Resilient Systems, Singapore ETH Centre, Singapore*
[5]*Department of Cognitive Robotics, Delft University of Technology, Delft, Netherlands*

Correspondence should be addressed to Sina Nordhoff; s.nordhoff@tudelft.nl

Academic Editor: Emanuele Crisostomi

Shuttles that operate without an onboard driver are currently being developed and tested in various projects worldwide. However, there is a paucity of knowledge on the determinants of acceptance of driverless shuttles in large cross-national samples. In the present study, we surveyed 10,000 respondents on the acceptance of driverless vehicles and sociodemographic characteristics, using a 94-item online questionnaire. After data filtering, data of 7,755 respondents from 116 countries were retained. Respondents reported that they would enjoy taking a ride in a driverless vehicle (mean = 4.90 on a scale from 1 = disagree strongly to 6 = agree strongly). We further found that the scores on the questionnaire items were most appropriately explained through a general acceptance component, which had loadings of about 0.7 for items pertaining to the usefulness of driverless vehicles and loadings between 0.5 and 0.6 for items concerning the intention to use, ease of use, pleasure, and trust in driverless vehicles, as well as knowledge of mobility-related developments. Additional components were identified as thrill seeking, wanting to be in control manually, supporting a car-free environment, and being comfortable with technology. Correlations between sociodemographic characteristics and general acceptance scores were small (<0.20), yet interpretable (e.g., people who reported difficulty with finding a parking space were more accepting towards driverless vehicles). Finally, we found that the GDP per capita of the respondents' country was predictive of countries' mean general acceptance score ($\rho = -0.48$ across 43 countries with 25 or more respondents). In conclusion, self-reported acceptance of driverless vehicles is more strongly determined by domain-specific attitudes than by sociodemographic characteristics. We recommend further research, using objective measures, into the hypothesis that national characteristics are a predictor of the acceptance of driverless vehicles.

1. Introduction

1.1. The Rise of Driverless Vehicles. Driverless vehicles are currently being developed in a number of commercial and research projects worldwide [1, 2]. Shuttles that function as shared transport systems are a promising business case [3]. Such shuttles may contribute to environmentally friendly mobility and tackle the inefficiencies of today's transport systems [4, 5]. However, driverless vehicles will only become a success if they are accepted by their users [6]. User acceptance of these vehicles needs to be studied at an early stage, preferably before the technology is publicly available.

1.2. Individual Predictors of Acceptance. Previous questionnaire studies on the acceptance of automated vehicles have identified several predictors of acceptance. It has been shown that men are more favorable towards automated vehicles than women. For example, it has been found that men were more likely to have a positive attitude towards automated vehicles [7], were more willing to pay more and were less concerned about automated vehicles [8], were more comfortable to allow a fully automated car to perform all functions [9], reported more pleasure and less anxiety with automated vehicles [10], and were more likely to express a positive intention to use and own automated vehicles than women [11].

The effect of age on acceptance is mixed. Becker and Axhausen [12] performed a review on questionnaire studies about automated vehicles and found that 6 out of 10 studies (i.e., [13–18]) that examined age effects reported that younger people were more accepting of automated vehicles than older people. In contrast, one online survey study in their review [19] found that people aged 36 to 65 had a more positive attitude and a stronger intention to use automated vehicles than people aged 18 to 35. In Nordhoff et al. [20], older people were more likely to intend to use driverless vehicles and were more positive towards the vehicle characteristics, but gave lower ratings to the effectiveness of the vehicle compared to their existing travel mode.

It has also been found that people with a higher income are willing to pay more for vehicles equipped with automated driving features [8, 13]. Furthermore, individuals with a higher driving mileage were more positive towards automated vehicles [8, 21], and had a higher willingness to pay for high automation levels [13].

Several researchers [22–24] have studied the role of personality-related attitudes as predictors of acceptance. For example, Bansal et al. [13] found that technology-savvy individuals were more positive towards automated vehicles, which is in agreement with Zmud and Sener [24] who found that individuals with a higher intent to use automated vehicles were the ones using smartphones, text messaging, Facebook, and transportation apps, and with Lavieri et al. [25] who found that tech-savvy individuals are likely to be early adopters of automated vehicles.

Some of the above-mentioned predictor variables are also used in technology acceptance models, which have been developed to explain and predict user behavior across a variety of domains. The Unified Theory of Acceptance and Use of Technology (UTAUT [26]) represents a synthesis of eight influential technology acceptance models, including the Theory of Reasoned Action (TRA) [27], the Technology Acceptance Model [28, 29], and the Theory of Planned Behavior (TPB) [30]. Using the UTAUT model as baseline, a conceptual model of the acceptance of driverless vehicles by Nordhoff et al. [6] postulates that acceptance is the result of the domain-specific attitudes performance expectancy and effort expectancy, as well as the symbolic-affective construct social influence and the pleasure-arousal-dominance framework [31]. The suitability of technology acceptance models to predict the acceptance of automated vehicles was shown by Madigan et al. [22, 32] and Zmud and Sener [24].

1.3. National Predictors of Acceptance of Driverless Vehicles. In addition to individual differences in the acceptance of automated vehicles, national differences have been studied as well. Participants from higher-income countries were found to be less comfortable with the transmission of their data [8], and less likely to express a positive comment about automated driving in an open-ended question [33]. Similarly, a cross-national survey by the World Economic Forum (WEF) and the Boston Consulting Group (BCG) [34] showed that the question "How likely would you be to let your children ride alone in a fully self-driving car?" was answered more positively by respondents from lower-income countries. In the same vein, the willingness to share a self-driving taxi was higher in low-income than in high-income countries [34], and respondents from China and India were more positive about automated vehicles than Japan, the US, UK, and Australia. However, both low- and high-income countries were equally concerned about safety issues related to fully automated cars [35].

1.4. Objectives of This Research. Although a large number of survey studies on automated driving exist (see [12], for a review), most of these studies have used relatively small and specific populations from Western countries (e.g., France, Switzerland, Germany, US, UK, and Australia). These studies examined the influence of respondents' demographics (e.g., age and gender) or attitudes (e.g., technology acceptance) on the acceptance of automated vehicles.

Based on findings from previous research (e.g., [8, 36]), it may be expected that correlations between sociodemographics and technology acceptance will be small at best (i.e., around $r = 0.10$). For example, a review of thirty years of acceptance research on wind energy suggests that demographic variables only explain a small amount of variance in attitudes towards wind energy [37]. Large sample sizes are thus needed to be able to detect significant correlations. For example, for a sample size of 400, the 99% confidence interval of a correlation of 0 is −0.13 to 0.13. Substantially larger sample sizes are needed to achieve statistical power for differentiating whether one correlation is stronger than another.

In summary, the above studies have contributed to knowledge of the acceptance of automated vehicles, but there is a paucity of knowledge regarding their acceptance in large cross-national samples. As the mobility sector faces a shift from motorized to sustainable and collaborative forms of mobility [38], this paper examines the role of individuals' wish for a car-free future, their knowledge of mobility-related developments, and their attitudes towards driverless vehicles (e.g., transport-related, symbolic-affective, usefulness, and ease of use) among 7,755 respondents from 116 countries. To examine cross-national effects, correlations between countries' GDP per capita and respondents' mean ratings were examined. The present study did not test hypotheses, but used an exploratory analysis to detect patterns in a large-sample dataset. More specifically, respondents completed a large number of diverse items, and we subjected the items to descriptive analyses to examine for which items the respondents showed low or high agreement. We subsequently performed a data reduction (principal component analysis) to examine the sources of variation in the data.

2. Methods

2.1. Instrument and Procedure. A 94-question survey was created on CrowdFlower (https://www.crowdflower.com). The survey instructions informed the respondents that their

FIGURE 1: Photos included at the top of the questionnaire [39, 40].

answer would be anonymous and that the completion of the survey would take around 20 minutes.

The instructions informed the respondents about a typical usage scenario with driverless vehicles as follows:

> *Automated vehicles are now being extensively tested on public roads by auto builders, suppliers and software companies. These vehicles still have steering wheel and pedals, and require a qualified driver to monitor the automation and to take back control when needed. This questionnaire is about the next level of automation being driverless vehicles. Driverless vehicles operate without a driver and do not have a steering wheel, gas or brake pedals. In the beginning, they will not operate on all roads and not in all traffic situations. With this survey, I would like to find out what do you think about these driverless vehicles and whether you would be ready and willing to accept and use them.*
>
> *Imagine that a driverless vehicle is waiting for you outside the train station or some other public transport stop (e.g. bus, tram, metro) to drive you to your final destination, providing last-mile transport. It can also drive you back to your original destination. The driverless vehicle is 100% electric. You may also book it on-demand via your smartphone-app. When you enter the vehicle, you may need to give in your destination via an interface – for example a keyboard – so that the vehicle knows where you would like to go. As the vehicle can accommodate 6–8 passengers, it may be the case that you share the vehicle with a few unknown travelers having the same destination like you. The pictures below show the interior and exterior of such a driverless vehicle.*

Next, two pictures of a driverless vehicle were shown from the French company Easymile [39, 40] (Figure 1) to ensure that respondents had an idea of the type of vehicle being the topic of our research.

No requirements regarding respondents' country were set. Each respondent was paid €0.20 for the participation in the survey.

2.2. Questionnaire Content

2.2.1. Survey Instructions and Sociodemographic Characteristics. The first two questions asked respondents whether they had read and understood the questionnaire instructions (Q1) and whether the instructions were clear (Q2). Questions Q3–Q27 and Q71 asked respondents about their sociodemographic characteristics and their travel behavior, including questions such as gender (Q3), year of birth (Q4), type of residential situation (Q7), and monthly net household income (Q12).

2.2.2. Domain-Specific Attitudes: Usefulness and Ease of Use. Questions Q49 and Q53 asked respondents to provide their agreement with whether they would use a driverless vehicle for their daily travel because it would be better and more convenient, or more useful, than their existing form of travel, respectively. Respondents were also asked to rate their agreement with whether using a driverless vehicle would be easier than existing travel (Q50), and whether learning to operate a driverless vehicle would be easy for them (Q52).

2.2.3. Transport-Related Attitudes: Satisfaction with Daily Travel and Enjoyment of Manual Driving. Respondents were asked whether they were satisfied with their daily travel in terms of the ability to organize their day flexibly with public transport (Q28) and whether they are satisfied with the possibilities available to cover their daily travel needs (Q31). In questions Q29, Q30, Q66, and Q69, respondents were asked whether they need a vehicle to be flexible (Q29), whether they consider driving especially fun (Q30), whether they often feel like a racing driver when driving manually (Q66), and whether they would like to learn to drive vehicles exceeding a speed of 300 km/h (Q69).

2.2.4. Symbolic-Affective Attitudes: Pleasure and Social Influence. Respondents were asked whether they would enjoy taking a ride in a driverless vehicle (Q37), and whether they would find it important that driverless vehicles are aesthetic in terms of styling and design (Q39). They were further asked questions relating to whether they would like to have their friends or family or other important people to them adopt the driverless vehicle before they do (Q60), and whether people who are important to them would like it when they would use a driverless vehicle (Q63).

2.2.5. Personality-Related Attitudes: Trust, Liking of Being in Control, Enjoyment of Technology, Knowledge of Mobility, Future Orientation, Wish for Car-Free Future, and Skepticism. Further questions concerned the trust in driverless vehicles (Q44, Q56, Q58, Q64) and the preferred level of control and supervision in driverless vehicles (Q36, Q46, Q48, Q68, Q70). Respondents were also asked questions regarding their enjoyment of technology in general (Q72–Q74) and their knowledge of mobility-related developments (Q75–Q78, Q87). In Q83–Q86 and Q88–Q94, respondents were asked to indicate their level of agreement with items pertaining to their future orientation and their wish for a car-free future in cities. Questions Q42, Q57, Q59, and Q65 asked respondents to rate their level of skepticism with driverless vehicles. Specifically, respondents indicated whether they agreed with the statement that driverless vehicles would take away the driving pleasure or enjoyment (Q42), whether they would feel uncomfortable entrusting the safety of a family member to a driverless vehicle (Q57), whether they would refrain from using driverless vehicles because technology can sometimes fail (Q59), and whether they think that there will always be accidents, even with driverless vehicles on the road (Q65).

2.2.6. Intention to Use Driverless Vehicles. Respondents were presented with questions that addressed whether they would use a driverless vehicle that is 100% electric (Q38), whether they would share a driverless vehicle with around 6–8 fellow travelers having the same route like them (Q40), how often they intend to use a driverless vehicle when it is available on the market (Q41), and whether they would use a 100% electric driverless vehicle from the train station or some other public transport stop to their final destination or vice versa (Q43). Q54 asked respondents at which usage rate they would use a driverless vehicle. Respondents were further asked to indicate whether they would buy a driverless vehicle (Q80), and use it with other passengers as part of public transport (Q81), or in a carsharing scheme (Q82).

2.2.7. Miscellaneous Questions. As automated shuttles can accommodate 8 to 10 passengers in a small space at the same time, respondents were asked to indicate whether they prefer to keep a physical distance between themselves and strangers (Q61). They were also asked whether environmental protection is crucial for their choice of transportation in Q62. Q67 asked respondents to rate their agreement with whether driving without accidents is mainly a matter of luck.

2.2.8. Questions Not Included in the Analysis. Respondents were asked why they would use a driverless vehicle (e.g., to pick up kids from school or bring them to soccer practice) (Q32), which activities they would perform in a driverless vehicle (e.g., reading a book) (Q33), for which travel purposes they would use driverless vehicles (e.g., commuting to work) (Q34), which features they find attractive in driverless vehicles (e.g., button to stop the vehicle) (Q35), and how driverless vehicles should be operated (e.g., on a fixed schedule) (Q45). Q47 asked respondents about their preferred level of human supervision (e.g., remote supervision from external control room). In Q55, respondents were asked to indicate under which financial condition they would use a driverless vehicle. In Q79, respondents were asked to indicate the potential factors or concerns that would discourage them from using a driverless vehicle (e.g., loss of driving enjoyment). These questions were not included in the analysis because they had multiple options in a checkbox format and were not measured on an ordinal scale.

2.2.9. Measurement of Questionnaire Items. The above survey questions were based on literature about automated vehicles and technology acceptance [8, 17, 22, 23, 26, 41–44]. More specifically, the relevance of each response category was demonstrated by prior work on technology acceptance models, most notably the UTAUT model (Section 1.3). The role of transport-related attitudes (Section 2.2.3) was considered pivotal, as the success of driverless vehicles hinges on individuals' willingness to change their travel mode. Personality-related attitudes (Section 2.2.5) have been investigated before (e.g., [8]), but may need further investigation as their potential relevance for predicting preferences to use automated vehicles has been indicated before (e.g., [23, 24]). With developments such as the electrification of transport, the redesign of cities to promote sustainable modes of transport, and the restriction of private car use, people's attitudes towards a car-free future (Section 2.2.5) are important to be able to implement the necessary changes.

Questions Q28–Q31, Q36–Q40, Q42–Q44, Q48–Q53, Q56–Q70, Q72–Q78, Q80–Q83, Q87, Q92, and Q93 were measured on a six-point scale from 1 *(Disagree strongly)* to 6 *(Agree strongly)*. Questions Q84–Q86 were measured on a four-point scale from 1 *(Unlikely)* to 4 *(Probable)*. Questions Q88–Q91 were measured on a four-point scale from 1 *(I would not appreciate this development at all)* to 4 *(I would really appreciate this development)*. Q41 was on a scale from 1 *(Never or almost never)* to 5 *(Daily or almost daily)*. Q54 was on a scale from 1 *(Never)* to 6 *(When it is used by 95% to 100% of the people)*. Finally, Q94 was on a scale from 1 *(All parts of the city must be accessible by car. Therefore, more car parking space should be provided)*, to 2 *(The number of parking spaces should stay as it is)*, to 3 *(The number of parking spaces should be reduced to make more space available for other uses (e.g., pedestrians, bike parking, playgrounds or parks))*. All items had a response option "I prefer not to respond." The survey was administered in English.

2.3. Analyses of Responses. Descriptive statistics (i.e., means, standard deviations, and distribution of responses) were calculated per questionnaire item, and the mean ratings of items that were measured on a scale from *Disagree strongly* to *Agree strongly* were compared. Next, a principal component analysis (PCA) was conducted on all questions that were measured on an ordinal scale (except for the questions on the instructions Q1-Q2 and the sociodemographic questions Q3–Q27, Q71) to investigate the major sources of variation in the items. For the PCA, missing data due to respondents selecting "I prefer not to respond" were replaced with the value from the single "nearest neighbor" variable (1NN), using Euclidean distance. According to Beretta and Santaniello [45], 1NN imputation preserves the original variability of the data, which is why we selected this method instead of a multiple neighbors approach (kNN).

Spearman correlation coefficients were calculated between respondents' sociodemographics (Q3–Q27, Q71) and the component scores. These correlations were assessed in three ways: (1) across the whole sample of respondents, (2) across the respondents within each respondent's country, and then sample size weighting the correlations across the countries, and (3) across the whole sample while partialling out (i.e., controlling for) the time to complete the survey and whether participants found the instructions clear (Q2). Note that the second (i.e., "within-country") correlation coefficient is similar to the estimated fixed-effect coefficient when fitting a mixed-effects model to the data, with the respondents' country as grouping variable. The correlations between sociodemographics and component scores were deemed robust only if all three correlations were similar (i.e., all three correlations being 0.05 or higher, or −0.05 or lower). This threefold approach to assessing correlations with sociodemographics was intended to protect against the ecological fallacy (i.e., national differences may contribute to first correlation, but not to the second) and response style (i.e., the third correlation statistically controls for how quickly people answer the survey and whether they found the instructions clear).

Finally, it was investigated whether the acceptance of driverless vehicles is associated with the countries' developmental status. Spearman rank-order correlation coefficients (ρ) were calculated between countries' GDP per capita and the countries' mean component scores. Only countries with at least 25 respondents were selected for this cross-national analysis [8]. All analyses were conducted in MATLAB 2016a.

3. Results

3.1. Number of Respondents and Respondents' Satisfaction with the Questionnaire. In total, 10,000 questionnaires were completed. Responses were gathered between April 13 and April 19, 2015. CrowdFlower enables participants to rate their satisfaction with the questionnaire. Respondents were overall satisfied with the survey, with a score of 4.0 on a scale from 1 *(very dissatisfied)* to 5 *(very satisfied)*.

3.2. Data Filtering. We applied a strict data screening to enhance data quality. Participants were excluded if they indicated that they had not read the instructions (n = 107), if they reported a birth year yielding an age younger than 18 (n = 70) or older than 110 (n = 111), or if they did not indicate their age or gender (n = 156). This upper limit of age was selected as a reasonable maximum human lifespan [46]. Only strings that contained a four-digit birth year were retained for calculating the participants' age.

Respondents were also excluded if their country of origin was not identified by CrowdFlower (n = 1), and if they were affiliated with the same IP address (n = 172; the first response was kept, but subsequent items from the same IP address were removed), were faster than the fastest 5% (n = 497; see [47] for rationale), had missing data due to database/recording errors (n = 24), or responded "I prefer not to respond" or "I don't know" to 9 or more questions (i.e., 10% of the questions) (n = 731). Furthermore, we excluded respondents (n = 848) who selected the same answer *(Disagree strongly, Disagree moderately, Agree moderately, or Agree strongly)* to the questions "I would feel comfortable in a vehicle without a steering wheel, gas or brake pedals" (Q44) and "I would not use a driverless vehicle because technology can sometimes fail" (Q59) as these questions have opposite meaning. This exclusion was performed to filter out respondents with an acquiescent response style. In total, 2,245 respondents were excluded, leaving 7,755 respondents from 116 countries in the analysis.

3.3. Descriptive Statistics of the Questionnaire Items. The mean age of respondents was 32.49 years (SD = 10.53, Q4). 31.20% of the respondents were female, and 68.80% were male (Q3). 7,032 of 7,755 respondents answered the question about their net household income. Of those 7,032 respondents, 35% had a net monthly household income below $ 1,000, 23% between $ 1,000 and $ 1,599, 20% between $ 1,600 and $ 2,899, 11% between $ 2,900 and 3,999, 6% between $ 4,000 and $ 5,000, and 5% more than $ 5,000 (Q12).

Respondents lived on average 9.43 miles away from their workplace, training post, or school (SD = 15.56, median = 5, Q9), and on average had 1.34 vehicles in their household (SD = 0.86, Q15). Their most frequent mode of transport was "walking more than 500 meters per trip" (M = 3.96, Q22), and the "conventional vehicle, as a driver or passenger" (M = 3.90, Q25), followed by "public transport for distances below and over 100 km per one way" (M = 2.91, 2.15; Q26-Q27), "cycling" (M = 2.14, Q23), and a "moped or motorcycle as a driver" (M = 1.76, Q24).

3.3.1. Domain-Specific Attitudes: Usefulness and Ease of Use. Table 1 shows the means (M), standard deviations (SD), and distribution of the questionnaire items. For usefulness (Q49, Q53), the higher rating was obtained for using a driverless vehicle for daily travel because it would be better and more convenient than existing travel (M = 4.48, Q49), and the lower rating for thinking that driverless vehicles would be more useful than existing travel (M = 4.35, Q53). For ease of use (Q50, Q52), the higher rating was obtained for thinking that learning to operate a driverless vehicle would be easy (M = 4.76, Q52), while the lower rating was obtained for thinking that driverless vehicles would be easier to use than existing travel (M = 4.46, Q50).

3.3.2. Transport-Related Attitudes: Satisfaction with Daily Travel and Enjoyment of Manual Driving. For satisfaction with daily travel (Q28, Q31), the higher rating was obtained for being satisfied with the possibilities available to cover daily travel needs (M = 4.36, Q31), and the lower rating for being able to organize the day flexibly with public transport (M = 3.73, Q28). For enjoyment of manual driving (Q29, Q30, Q66, Q69), the highest rating was obtained for needing a vehicle to be flexible (M = 4.77, Q29), and the lowest rating for liking to learn to drive vehicles that can exceed the speed of 300 km/h (M = 3.27, Q69).

3.3.3. Symbolic-Affective Attitudes: Pleasure and Social Influence. For pleasure (Q37, Q39), respondents gave the higher rating for thinking that they would enjoy taking a ride in a driverless vehicle (M = 4.90, Q37), and the lower rating for finding it important that driverless vehicles are aesthetic in terms of styling and design (M = 4.62, Q39). For social influence (Q60, Q63), respondents gave the higher rating for believing that people important to them would like it when they use driverless vehicles (M = 4.33, Q63), and the lower rating for liking to have their friends or family or other important people to them adopting the driverless vehicle before they do (M = 3.77, Q60).

3.3.4. Personality-Related Attitudes: Trust, Liking of Being in Control, Enjoyment of Technology, Knowledge of Mobility, Future Orientation, Wish for Car-Free Future, and Skepticism. Similar mean ratings were also obtained for trusting driverless vehicles (3.80–4.36; Q44, Q56, Q58, Q64), liking of being in control (3.92–5.18; Q36, Q48, Q68, Q70), enjoyment

Table 1: Means (M), standard deviations (SD), and distribution of questionnaire items on a scale from 1 (disagree strongly) to 6 (agree strongly), unless indicated otherwise*.

Semantic content	Item	M	SD	1	2	3	4	5	6
Liking of being in control	Q36. Would like to have button inside DV.	5.18	1.15	146	144	297	1299	1532	4296
Intention to use	Q38. Would use 100% electric DV.	5.09	1.20	200	168	301	1316	1804	3924
Enjoyment of technology	Q74. Fun to use electronic device.	4.99	1.10	79	179	402	1656	2143	3261
Pleasure	Q37. Think I would enjoy taking a ride in DV.	4.90	1.22	226	213	350	1610	2213	3114
Enjoy manual	Q29. Need vehicle to be flexible.	4.77	1.38	294	355	601	1539	1700	3230
Ease of use	Q52. Learning to operate DV would be easy for me.	4.76	1.18	162	218	536	2015	2228	2535
Intention to use	Q43. Would use 100% electric DV from train station to final destination.	4.72	1.25	256	227	506	1965	2217	2546
Enjoyment of technology	Q73. Rapidly and intuitively learn to handle unfamiliar electronic devices.	4.67	1.17	136	255	605	2201	2304	2210
Miscellaneous	Q62. Environmental protection crucial for transportation.	4.64	1.21	143	291	697	2267	1999	2312
Knowledge of mobility	Q75. Often think about how mobility in city could be improved.	4.63	1.22	163	320	681	2179	2110	2268
Pleasure	Q39. Would find it important that DVs are aesthetic in terms of styling and design.	4.62	1.21	201	263	601	2252	2286	2101
Intention to use	Q40. Would share DV with 6–8 fellow travelers.	4.48	1.29	296	343	752	2235	2192	1905
Usefulness	Q49. Would use DV for daily travel because would be better and more convenient.	4.48	1.28	289	329	754	2293	2168	1894
Ease of use	Q50. Think DV would be easier to use than existing travel.	4.46	1.28	273	328	849	2280	2118	1879
Enjoy manual	Q30. Driving is especially fun for me.	4.44	1.45	458	445	741	1899	1812	2263
Skeptical	Q65. There will always be accidents, even with DVs on the road.	4.38	1.27	227	434	908	2463	1976	1698
Trust	Q56. Trust that DV can drive without assistance from me.	4.36	1.25	280	370	872	2456	2280	1479

Table 1: Continued.

Semantic content	Item	M	SD	1	2	3	4	5	6
Satisfaction daily travel	Q31. Satisfied with possibilities available to cover daily travel needs.	4.36	1.28	272	442	886	2348	2189	1582
Usefulness	Q53. Think that DV would be more useful than existing travel.	4.35	1.30	308	391	948	2376	2033	1663
Liking of being in control	Q48. Would like to take over control from DV.	4.34	1.40	438	425	832	2314	1755	1940
Liking of being in control	Q70. Careful driver can prevent any accident on road.	4.33	1.40	322	600	1059	1865	1996	1886
Knowledge of mobility	Q87. Would like to use mobility flat-rate for mobility services in city.	4.33	1.12	191	296	731	3194	2036	1173
Social influence	Q63. People who are important to me would like it when I use DV.	4.33	1.18	221	328	896	2725	2193	1253
Intention to use	Q81. Would use SDVs together with passengers in public transport.	4.32	1.29	366	384	793	2505	2214	1453
Intention to use	Q80. Would like to buy SDV.	4.22	1.37	446	504	900	2452	1914	1504
Wish for car-free future	Q93. Roads be redesigned with bicycle lane that replaces car lane.	4.20	1.35	407	461	1077	2491	1741	1501
Trust	Q58. Trust DV to be safe and reliable in severe weather conditions.	4.15	1.34	384	532	1212	2346	1891	1354
Future orientation	Q83. SDVs will be legally accepted as independent drivers.	4.14	1.24	345	414	1122	2816	1935	1045
Intention to use	Q82. Would like to use SDVs in carsharing scheme.	4.09	1.27	402	492	1013	2807	1950	975
Miscellaneous	Q61. Prefer to keep physical distance between myself and strangers.	4.09	1.32	369	601	1193	2538	1797	1201
Trust	Q44. Would feel comfortable in vehicle without steering wheel, gas or brake pedals.	4.01	1.43	563	587	1388	2232	1581	1368
Knowledge of mobility	Q76. Friends and acquaintances often consult me on mobility options.	4.00	1.30	401	656	1208	2633	1883	929

Table 1: Continued.

Semantic content	Item	M	SD	1	2	3	4	5	6
Enjoyment of technology	Q72. Friends and acquaintances often ask for advice, when they have technical problem.	4.00	1.31	450	655	1066	2719	1886	921
Knowledge of mobility	Q78. Often provide others with information regarding mobility options.	4.00	1.30	396	662	1207	2670	1868	916
Knowledge of mobility	Q77. Often first to make people aware of new mobility.	4.00	1.33	428	656	1267	2542	1798	1014
Liking of being in control	Q68. When driver is involved in accident, (s)he did not drive properly.	3.92	1.41	471	877	1415	2100	1748	1099
Wish for car-free future	Q92. Would like to live in car-free neighbourhood.	3.88	1.48	703	730	1330	2236	1471	1238
Trust	Q64. Would trust driving skills of DV more than own driving skills.	3.80	1.47	732	803	1446	2060	1637	1019
Skeptical	Q42. DVs would take away the pleasure or enjoyment.	3.78	1.42	693	780	1340	2497	1477	920
Skeptical	Q57. Would feel uncomfortable entrusting safety of family to DV.	3.78	1.33	492	830	1621	2503	1461	806
Social influence	Q60. Would like to have friends or family adopt DV before I do.	3.77	1.30	520	740	1592	2676	1450	692
Intention to use	Q54. Indicate when you would use DV. Only most relevant option.*	3.74	1.15	221	768	2228	2555	1408	507
Satisfaction daily travel	Q28. Can organize my day flexibly with public transport.	3.73	1.49	884	821	1149	2308	1638	883
Intention to use	Q51. Even if more expensive than existing travel, would prefer DVs.	3.72	1.45	756	850	1537	2170	1504	904
Intention to use	Q41. Indicate when you would use DV. Only most relevant option.*	3.67	1.27	633	894	1399	2163	2573	
Skeptical	Q59. Would not use DV because technology can fail.	3.48	1.32	636	1128	1995	2429	946	594
Future orientation	Q86. SDVs will be normal part of everyday mobility.*	3.43	0.79	231	665	2102	4242		
Enjoy manual	Q66. Often feel like racing driver when driving manually.	3.40	1.55	1314	932	1388	2036	1250	687

TABLE 1: Continued.

Semantic content	Item	M	SD	1	2	3	4	5	6
Wish for car-free future	Q91. In cities, EVs will completely replace combustion engines within next 20–30 years.*	3.37	0.82	307	598	2284	3845		
Future orientation	Q84. SDVs will mostly be shared vehicles.*	3.36	0.85	338	671	2050	3796		
Enjoy manual	Q69. Like to learn to drive vehicles that exceed speed of 300 km/h.	3.27	1.72	1836	1011	1164	1604	1077	1014
Future orientation	Q85. SDVs will mostly be privately owned.*	3.22	0.93	426	1133	1870	3516		
Wish for car-free future	Q90. Roads in cities will be redesigned to privilege non-motorized travel.*	3.21	0.88	411	913	2590	3253		
Miscellaneous	Q67. Driving without accidents is mainly matter of luck.	3.16	1.53	1439	1445	1417	1849	989	575
Wish for car-free future	Q88. Many car-free neighborhoods with exceptional car use and no parking space.*	2.94	0.96	672	1258	2563	2226		
Wish for car-free future	Q89. City centers closed to car-traffic.*	2.89	0.99	789	1399	2439	2242		
Wish for car-free future	Q94. Think of your own neighbourhood. Which of following statements do you agree to most? Select only one option.*	1.83	0.82	3375	2302	2078			
Liking of being in control	Q46. Mind being transported by DV supervised by ECR?*	1.58	0.49	3136	4390				

Note. (E) DV = (electric) driverless vehicle, SDV = self-driving vehicle, and ECR = external control room.

of technology (4.00–4.99; Q73–Q75), knowledge of mobility (4.00–4.63; Q76–Q78, Q87), future orientation (3.22–4.14; Q83–Q86), wish for a car-free future (1.83–4.22; Q88–Q94), and skepticism (3.48–4.38; Q42, Q57, Q59, Q65).

For trusting driverless vehicles, respondents gave the highest rating for trusting that a driverless vehicle can drive without their assistance (M = 4.36, Q56), and the lowest rating was obtained for trusting the driving skills of a driverless vehicle more than one's own driving skills (M = 3.80, Q64).

The highest rating for liking of being in control and the highest overall mean rating were obtained for liking to have a stop button inside the driverless vehicle (M = 5.18, Q36), and the lowest rating for believing that when a driver is involved in an accident, (s)he did not drive properly (M = 3.92, Q68).

For skepticism, the highest rating was obtained for believing that there will always be accidents, even with driverless vehicles on the road (M = 4.38, Q65), and the lowest rating for not using a driverless vehicle because technology can sometimes fail (M = 3.48, Q59).

3.3.5. Intention to Use Driverless Vehicles. For intention to use (Q38, Q40, Q43, Q51, Q80–Q82), the highest rating was obtained for using a driverless vehicle that is 100% electric (M = 5.09, Q38), and the lowest rating for preferring driverless vehicles even if they are more expensive than existing travel (M = 3.72, Q51).

3.3.6. Miscellaneous Questions. For the miscellaneous items (Q62, Q67), the higher mean rating was obtained for environmental protection being crucial in the choice of transportation (M = 4.64, Q62), and the lower rating for believing that driving without accidents is mainly a matter of luck (M = 3.16, Q67).

In summary, the results indicate that driverless vehicles are regarded as fun, useful, and easy to use. Lower mean ratings, yet on the "agree" end of the scale (i.e., >3.5), were obtained for items relating to trusting those vehicles more than one's own driving skills.

3.4. Principal Component Analysis of the Questionnaire Items: General Acceptance Component. A PCA was performed of the responses on the 58 items that were measured on an ordinal scale (i.e., Q28–Q31, Q36–Q44, Q46, Q48–Q54, Q56–Q70, Q72–Q78, Q80–Q94). The Kaiser-Meyer-Olkin (KMO) index of sampling adequacy was 0.941, indicating that the data are suitable for factor-analytic purposes [48].

Based on the scree plot (Figure 2), and based on the interpretability of the loadings, we decided to retain one general component. This component was interpreted as "general acceptance." The Cronbach alpha for the 58 variables was 0.899 and 0.928 if selecting the 32 from 58 variables that loaded higher than 0.4, a common cut-off value [49]. The participants' scores on the component were standardized, so that the mean was equal to 0 and the standard deviation was equal to 1 (min = −4.71, max = 2.36, and N = 7,755).

Table 2 provides an overview of the 58 items and their corresponding loadings on the general acceptance component.

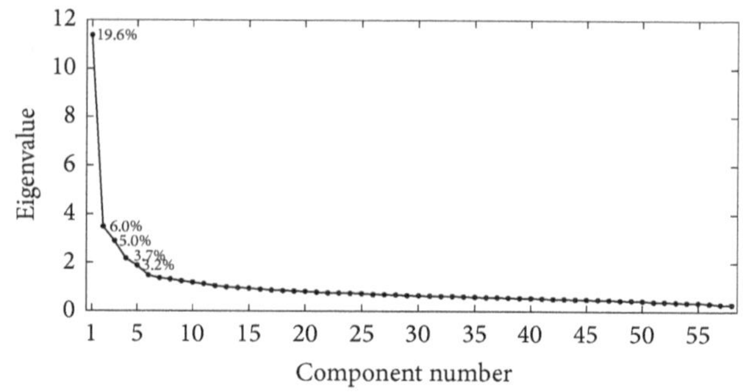

FIGURE 2: Eigenvalues of the correlation matrix sorted in descending order ("scree plot"). Also shown are the percentages of variance explained (being proportional to the eigenvalue) for the first five components. It can be seen that one dominant component emerged, explaining 19.6% of the variance.

Loadings are referred to as negative if smaller than 0, small if between 0 and 0.40, medium if between 0.40 and 0.60, and high if larger than 0.60.

(i) Negative and small mean loadings were obtained for skepticism towards driverless vehicles (−0.35 to −0.08; Q42, Q57, Q59, Q65; mean loading = −0.16).

(ii) Small mean loadings were obtained for items capturing the satisfaction with daily travel (0.11–0.22; Q28, Q31; mean loading = 0.16), enjoyment of manual driving (0.16–0.19; Q29–Q30, Q66, Q69; mean loading = 0.17), and the liking of being in control (0.08–0.31; Q36, Q46, Q48, Q68, Q70; mean loading = 0.20).

(iii) Small to medium mean loadings were obtained for the wish for a car-free future (0.14–0.47; Q88–Q94; mean loading = 0.37), future orientation (0.24–0.56; Q83–Q86; mean loading = 0.39), and enjoyment of technology (0.33–0.46; Q72–Q74; mean loading = 0.39).

(iv) Medium mean loadings were obtained for items measuring social influence (0.29–0.66; Q60, Q63; mean loading = 0.47), intention to use (0.11–0.65; Q38, Q40-Q41, Q43, Q51, Q54, Q80-Q81; mean loading = 0.52), knowledge of mobility-related developments (0.49–0.58; Q75–78, Q87; mean loading = 0.53), pleasure of driverless vehicles (0.46–0.67; Q39, Q37; mean loading = 0.57), trust in driverless vehicles (0.49–0.65; Q44, Q56, Q58, Q64; mean loading = 0.57), and ease of use of driverless vehicles compared to respondents' existing form of travel (0.49–0.71; Q50, Q52; mean loading = 0.60).

(v) High mean loadings were obtained for items pertaining to the perceived usefulness of driverless vehicles (0.72–0.73; Q49, Q53; mean loading = 0.73).

We have also done a supplementary analysis in which we retained five components instead of one component, because the scree plot suggests that although a single dominant component exists, a five-component solution may also be appropriate. After oblique rotation of the loadings (Promax,

TABLE 2: Component loadings.

Semantic content	Item	GAC	Component 1	Component 2	Component 3	Component 4	Component 5
Usefulness	Q49. Would use DV for daily travel because would be better and more convenient.	**0.73**	**0.80**	0.08	0.03	−0.03	−0.05
Usefulness	Q53. Think that DV would be more useful than existing travel.	**0.72**	**0.75**	0.15	−0.03	−0.01	−0.01
Ease of use	Q50. Think DV would be easier to use than existing travel.	**0.71**	**0.77**	0.14	−0.01	−0.02	−0.04
Ease of use	Q52. Think that learning to operate DV would be easy for me.	**0.49**	**0.43**	−0.12	0.30	−0.08	0.11
Pleasure	Q37. Think I would enjoy taking a ride in DV.	**0.67**	**0.79**	−0.11	0.18	−0.06	−0.11
Pleasure	Q39. Would find it important that DVs are aesthetic.	**0.46**	**0.43**	0.12	0.36	−0.05	−0.03
Social influence	Q63. People who are important to me would like it when I use DV.	**0.66**	**0.57**	0.20	0.02	0.03	0.09
Social influence	Q60. Would like to have friends or family adopt DV before I do.	0.29	0.28	**0.49**	0.21	−0.02	−0.13
Trust	Q56. Trust that DV can drive without assistance from me.	**0.65**	**0.75**	−0.03	−0.03	−0.07	−0.02
Trust	Q58. Trust DV to be safe and reliable in severe weather conditions.	**0.58**	**0.64**	0.18	−0.11	−0.07	0.02
Trust	Q44. Would feel comfortable in vehicle without steering wheel, gas or brake pedals.	**0.56**	**0.65**	0.10	−0.18	−0.02	−0.03
Trust	Q64. Would trust driving skills of DV more than own driving skills.	**0.49**	**0.52**	0.34	−0.22	0.08	−0.08
Intention to use	Q80. Would like to buy SDV.	**0.65**	**0.62**	0.10	−0.03	−0.02	0.09
Intention to use	Q38. Would use 100% electric DV.	**0.61**	**0.69**	−0.19	0.30	0.01	−0.15
Intention to use	Q43. Would use 100% EDV from train station to final destination.	**0.59**	**0.67**	−0.16	0.20	−0.02	−0.10
Intention to use	Q81. Would use SDVs together with passengers in public transport.	**0.59**	**0.53**	0.05	−0.06	0.07	0.06

Table 2: Continued.

Semantic content	Item	GAC	Component 1	Component 2	Component 3	Component 4	Component 5
Intention to use	Q82. Would like to use SDVs in carsharing scheme.	**0.57**	**0.47**	0.18	−0.08	0.06	0.13
Intention to use	Q51. Even if more expensive than existing travel, would prefer DVs.	**0.55**	**0.51**	**0.42**	−0.12	−0.03	0.08
Intention to use	Q41. Indicate when you would use DV. Only most relevant option.	**0.53**	**0.60**	−0.17	−0.09	−0.04	0.04
Intention to use	Q40. Would share DV with 6–8 fellow travelers with same route like.	**0.52**	**0.51**	0.00	0.11	0.03	−0.03
Intention to use	Q54. Indicate when you would use DV. Only most relevant option.	0.11	0.17	0.26	0.14	0.00	−0.20
Knowledge of mobility	Q87. Would like to use mobility flat-rate for mobility services in city.	**0.58**	0.38	−0.05	0.12	0.17	0.13
Knowledge of mobility	Q75. Often think about how mobility in city could be improved.	0.56	0.19	−0.07	0.12	0.13	**0.42**
Knowledge of mobility	Q78. Often provide others with information regarding mobility options.	**0.52**	−0.02	0.27	−0.10	0.03	**0.78**
Knowledge of mobility	Q77. Often first to make people aware of new mobility options.	**0.51**	−0.05	0.28	−0.12	0.02	**0.80**
Knowledge of mobility	Q76. Friends and acquaintances often consult me on mobility options.	**0.49**	−0.06	0.26	−0.08	0.02	**0.78**
Future orientation	Q83. SDVs will be legally accepted as independent drivers.	**0.56**	**0.52**	0.14	−0.10	0.03	0.08
Future orientation	Q86. SDVs will be normal part of everyday mobility.	**0.45**	0.31	0.04	0.02	0.23	0.00
Future orientation	Q84. SDVs will mostly be shared vehicles.	0.31	0.16	0.04	0.10	0.23	−0.03
Future orientation	Q85. SDVs will mostly be privately owned.	0.24	0.12	0.27	0.08	0.12	−0.02
Wish for car-free future	Q93. Roads be redesigned with bicycle lane that replaces car lane.	**0.47**	0.05	−0.01	0.04	**0.58**	0.08

Table 2: Continued.

Semantic content	Item	GAC	Component 1	Component 2	Component 3	Component 4	Component 5
Wish for car-free future	Q91. In cities, EVs will completely replace combustion engines within next 20–30 years.	0.43	0.08	−0.19	0.04	**0.60**	−0.02
Wish for car-free future	Q90. Roads and streets in cities will be redesigned to privilege non-motorized travel.	0.41	−0.06	−0.07	−0.04	**0.76**	0.01
Wish for car-free future	Q92. Would like to live in car-free neighbourhood.	**0.41**	0.01	−0.04	−0.04	**0.67**	0.00
Wish for car-free future	Q88. Many car-free neighborhoods with exceptional car use and no parking space.	0.39	−0.07	0.10	−0.10	**0.72**	0.04
Wish for car-free future	Q89. City centers closed to car-traffic.	0.32	−0.15	0.02	−0.10	**0.78**	−0.01
Wish for car-free future	Q94. Think of your own neighbourhood. Which of following statements do you agree to most? Select only one option.	0.14	−0.08	−0.27	−0.13	**0.46**	0.00
Enjoyment of technology	Q74. Fun to use electronic device.	**0.46**	0.19	−0.34	0.31	0.00	0.37
Enjoyment of technology	Q73. Rapidly and intuitively learn to handle unfamiliar electronic devices.	0.39	0.04	−0.27	0.22	−0.02	**0.53**
Enjoyment of technology	Q72. Friends and acquaintances often ask for advice, when they have technical problem.	0.33	−0.12	0.00	0.03	−0.03	**0.69**
Liking of being in control	Q70. Careful driver can prevent any accident on road.	0.31	0.08	0.19	0.20	0.09	0.15
Liking of being in control	Q36. Would like to have button inside DV.	0.27	0.28	−0.26	**0.58**	−0.01	−0.16
Liking of being in control	Q68. When driver is involved in accident, (s)he did not drive properly.	0.22	0.01	0.32	0.09	0.10	0.12
Liking of being in control	Q46. Mind being transported by DV supervised by ECR?	0.12	0.01	0.20	−0.04	0.06	0.08
Liking of being in control	Q48. Would like to take over control from DV when want this.	0.08	−0.02	0.17	**0.53**	−0.06	−0.03
Enjoy manual	Q30. Driving is especially fun for me.	0.19	0.01	0.19	**0.48**	−0.16	0.19
Enjoy manual	Q66. Often feel like racing driver when driving manually.	0.18	−0.05	**0.60**	0.10	−0.07	0.24

TABLE 2: Continued.

Semantic content	Item	GAC	Component 1	Component 2	Component 3	Component 4	Component 5
Enjoy manual	Q29. Need vehicle to be flexible.	0.17	0.19	−0.01	**0.43**	−0.22	0.04
Enjoy manual	Q69. Like to learn to drive vehicles that exceed speed of 300 km/h.	0.16	−0.05	**0.45**	0.05	−0.13	0.32
Satisfaction daily travel	Q28. Can organize my day flexibly with public transport.	0.22	0.12	0.11	0.10	0.22	−0.11
Satisfaction daily travel	Q31. Satisfied with possibilities available to cover daily travel.	0.11	0.05	0.13	**0.47**	0.01	−0.13
Skeptical	Q42. DVs would take away the pleasure or enjoyment.	−0.08	−0.26	0.30	**0.42**	−0.03	0.06
Skeptical	Q65. There will always be accidents, even with DVs on the road.	−0.09	−0.17	−0.10	**0.41**	0.07	−0.10
Skeptical	Q57. Would feel uncomfortable entrusting safety of family to DV.	−0.13	−0.34	0.29	0.39	0.10	−0.01
Skeptical	Q59. Would not use DV because technology can fail.	−0.35	**−0.58**	0.30	0.34	0.13	−0.02
Miscellaneous	Q61. Prefer to keep physical distance between myself and strangers.	−0.03	−0.09	0.19	0.37	0.05	−0.15
Miscellaneous	Q62. Environmental protection crucial for transportation.	**0.50**	0.26	0.06	0.18	0.21	0.10
Miscellaneous	Q67. Driving without accidents is mainly matter of luck.	0.05	−0.03	**0.54**	0.00	0.00	0.00

Note. DV = driverless vehicle; SDV = self-driving vehicle. The general acceptance component is the first principal component. Components 1–5 are the first five components, after Promax rotation. Component loadings of magnitude > 0.40 are listed in boldface.

with kappa = 4), the five components were interpreted as (1) general acceptance, (2) thrill seeking, (3) wanting to be in control manually, (4) supporting a car-free environment, and (5) being comfortable with technology. The loadings of the five-component solution are shown in Table 2. Cronbach's alpha for these five components (after selecting the items that loaded higher than 0.4) were 0.916, 0.587, 0.580, 0.763, and 0.800, indicating that the first component is most consistent.

3.5. Correlations between Sociodemographic Characteristics and the General Acceptance Score. Table 3 shows the correlations between sociodemographic characteristics and the general acceptance score. It can be seen that the survey time and whether the respondents found the survey instructions clear (Q2) correlated relatively strongly with the general acceptance score ($\rho = 0.20$ and 0.22, resp.). Hence, these two variables were partialled out. These partial correlations were similar to the zero-order correlations (Table 3).

Furthermore, it can be seen that whole-sample and within-country correlations were similar, but not for all items. For example, across the whole sample a near-zero correlation was found between respondents' income (Q12) and general acceptance ($\rho = -0.01$), while within countries this correlation was slightly positive ($\rho = 0.06$). Here, it is possible that the large national differences in income masked the positive correlation between income and general acceptance within nations.

Overall, the correlations between sociodemographic characteristics and general acceptance component scores were small (<0.20). The strongest correlation with the general acceptance score was found with the difficulty of finding a parking place ($\rho = 0.17$, Q16), and the frequency of use of public transport (<100 km) ($\rho = 0.14$, Q26). Having a monthly pass or annual travel card for public transport ($\rho = 0.11$, Q19), living in a city environment ($\rho = 0.12$, Q7), frequency of walking more than 500 m per trip ($\rho = 0.08$, Q22), frequency of cycling ($\rho = 0.06$, Q23), and distance of living from workplace, training post, or school ($\rho = 0.07$, Q9) also correlated with general acceptance. No robust correlations (i.e., stronger than 0.05 for all three correlation types) were found between age (Q4) and gender (Q3) and the general acceptance score. Males had a higher general acceptance score than females across the whole sample, but this effect was not statistically significant within countries.

3.6. National Differences in the General Acceptance Score. This section presents results from 7,188 respondents from 43 countries with 25 or more respondents. The three countries with the highest GDP per capita were the USA ($ 52,980), Canada ($ 52,305), and the Netherlands ($ 50,793), whereas the three countries with the lowest GDP per capita were India ($ 1,455), Pakistan ($ 1,282), and Bangladesh ($ 954).

We found that the developmental status of the respondents' country (GDP per capita) was predictive of the countries' mean general acceptance score ($\rho = -0.48$, $n = 43$; Figure 3(a)). For example, respondents from higher-income countries gave more negative ratings to intention to use (e.g., Q51 "Even if it were more expensive than my existing form of travel, I would prefer driverless vehicles to my existing form of travel", $\rho = -0.64$), the perceived effectiveness (e.g., Q50, "I think driverless vehicles would be easier to use than my existing form of travel", $\rho = -0.57$), and pleasure of driverless vehicles (e.g., Q39, "I would find it important that driverless vehicles are aesthetic in terms of styling and design", $\rho = -0.57$).

As a validity check of the self-reports, we observed that the mean self-reported income (Q12) correlated strongly with GDP per capita ($\rho = 0.71$, $n = 43$; Figure 4(a)), whereas the median self-reported annual mileage (Q17) correlated strongly with GDP per capita on a national level ($\rho = 0.71$, $n = 43$; Figure 4(b)).

Table 4 shows cross-national correlations for a number of additional variables (selected from [50]). It can be seen that the general acceptance score correlates substantially not only with GDP per capita but also with other developmental indexes, including national performance in educational tests ([51]; $\rho = -0.60$), average life expectancy in 2013 ([52]; $\rho = -0.43$), motor vehicle density ([52]; $\rho = -0.59$), median age in 2014 ([53]; $\rho = -0.52$), and road traffic death rate per 100,000 population ([54]; $\rho = 0.56$). The second component (i.e., thrill seeking) correlates negatively ($\rho = -0.72$; see Figure 3(b)) with GDP per capita, meaning that people in lower-income countries are more thrill seeking. Also, respondents in higher-income countries reported to be less supportive of a car-free environment ($\rho = -0.44$), and to be less comfortable with technology ($\rho = -0.44$).

4. Discussion

4.1. Main Results at the Individual Level. A variety of studies have previously examined the acceptance of automated vehicles. However, there is limited knowledge on the acceptance of automated vehicles without steering wheel and brake and gas pedals across countries. The present study surveyed 7,755 respondents from 116 countries on their acceptance of driverless vehicles, attitudes towards technology, knowledge of mobility-related developments, and sociodemographic characteristics, using a 94-item online questionnaire.

We found that respondents considered driverless vehicles easy to use and convenient (Q49, Q50, Q52). Furthermore, respondents could imagine using 100% electric driverless vehicles in connection with public transport (Q43). The perceived enjoyment of taking a ride in driverless vehicles was also rated positively by respondents (M = 4.90/6, Q37). This corresponds with Nordhoff et al. [20], who found that respondents strongly agreed with the statement that the driverless vehicle was fun and enjoyable after having taken a ride in the vehicle (M = 5.40/6). Our respondents gave the highest rating (M = 5.18/6, Q36) to being able to take over control from a driverless vehicle by a button inside the vehicle to stop it, which indicates that participants want to be able to retain some degree of control over the driverless vehicle. This finding is in line with Schoettle and Sivak [18] who found that 96.2% of respondents preferred the availability of vehicle controls.

By means of a PCA, we reduced the scores on all questionnaire items into one general component of acceptance. A general acceptance component for driverless vehicles is a

Table 3: Means (M) and Spearman rank-order correlations (ρ) between sociodemographic characteristics and the general acceptance score (GAC) for the whole sample, for respondents within countries, and for the whole sample after partialling out the time to complete survey and whether respondents found instructions clear ($N = 7{,}755$).

Item	M	ρ with GAC (whole sample)	ρ with GAC (within country)	ρ with GAC (survey time & instructions clear partialled out)
Survey time ranking 0-1	0.54	0.20	0.14	—
Survey start time	—	0.05	0.02	0.05
Survey end time	—	0.05	0.02	0.05
Q2. The definitions given in the instructions are clear to me	5.36	0.22	0.26	—
Q3. What is your gender? 1 (female), 2 (male)	1.69	0.07	0.01	0.09
Q4. Year of birth (converted to age) (open question)	32.49	0.02	0.08	−0.02
Q7. Which of the following possibilities describe your current residential situation the most? 1 (Outside city in house in countryside) to 4 (In apartment in immediate city centre)	2.95	0.12	0.08	0.10
Q8. Number of people in household between 14 and 17 years (open question)	0.27	0.05	0.03	0.06
Q8. Number of people in household between 6 and 13 years (open question)	0.36	0.04	0.02	0.04
Q8. Number of people in household younger than 6 years (open question)	0.35	0.04	0.02	0.04
Q8. Number of people older than 18 years old (open question)	2.67	0.09	0.01	0.08
Q9. Distance between home and workplace, training post or school (in miles) (open question)	9.43	0.07	0.06	0.08
Q12. Net monthly household income 1 (<$ 1,000) to 6 (>$ 5,000)	2.45	−0.01	0.06	−0.01
Q13. Having valid driver license 1 (No), 2 (Yes)	1.80	0.03	0.05	0.02
Q15. Number of vehicles in household 1 (0), 2 (1), 3 (2), 4 (3), 5 (>3)	1.34	0.00	0.01	−0.01
Q16. Difficulty of finding parking space 1 (Not at all difficult) to 4 (Very difficult)	2.52	0.17	0.12	0.16

TABLE 3: Continued.

Item	M	ρ with GAC (whole sample)	ρ with GAC (within country)	ρ with GAC (survey time & instructions clear partialled out)
Q17. Annual driving mileage as driver or passenger (in miles) (open question)	8,190	−0.00	0.07	−0.04
Q18. Involvement in accidents in last three years 1 (0), 2 (1), 3 (2), 4 (3), 5 (4), 6 (5), 7 (>5)	0.43	0.02	−0.01	0.04
Q19. Having monthly pass or annual travel card for public transport 1 (No), 2 (Yes)	1.34	0.11	0.07	0.13
Q20. Distance between home and nearest public transport stop (in min.) 1 (<5 min.), 2 (5–10 min.), 3 (10–20 min.), 4 (20–30 min.), 5 (>30 min.)	1.93	0.03	0.01	0.05
Q21. Number of carsharing memberships 1 (0), 2 (1), 3 (2), 4 (>2)	0.22	0.03	0.01	0.08
Q22. Frequency of walking more than 500 meters per trip 1 (Never or almost never) to 5 (Daily or almost daily)	3.96	0.08	0.08	0.07
Q23. Frequency of cycling 1 (Never or almost never) to 5 (Daily or almost daily)	2.14	0.06	0.06	0.09
Q24. Frequency of using moped or motorcycle as driver 1 (Never or almost never) to 5 (Daily or almost daily)	1.76	0.08	0.01	0.11
Q25. Frequency of using conventional vehicle as driver or passenger 1 (Never or almost never) to 5 (Daily or almost daily)	3.90	0.08	0.07	0.05
Q26. Frequency of using light transit (<100 km per one way) 1 (Never or almost never) to 5 (Daily or almost daily)	2.91	0.14	0.09	0.15
Q27. Frequency of using public transport (>100 km per one way) 1 (Never or almost never) to 5 (Daily or almost daily)	2.15	0.10	0.03	0.14
Q71. Severity of motion sickness 1 (I do not experience motion sickness), 2 (Moderate), 3 (Severe)	1.70	−0.09	−0.08	−0.05

Note. For a sample size of 7,755, correlations of 0.03 and higher, or −0.03 and lower, are statistically significantly different from zero, $p < 0.01$.

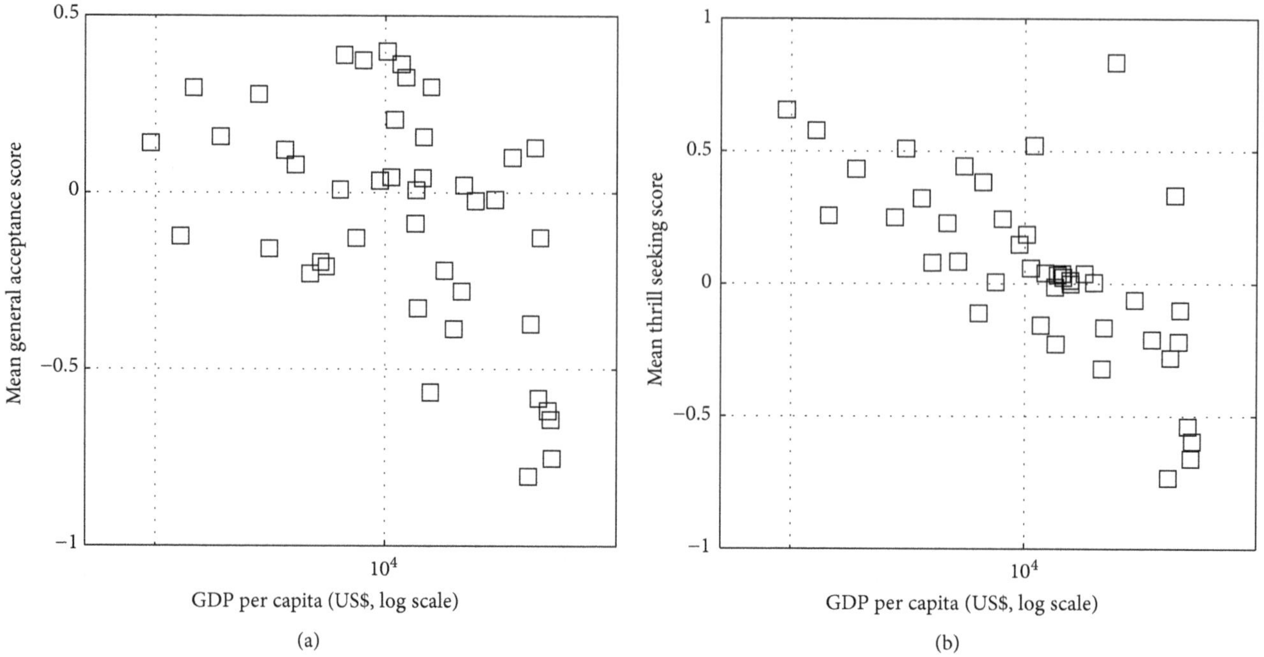

FIGURE 3: Correlation between countries' gross domestic product (GDP) per capita and participants' mean general acceptance score (a) and participants' mean thrill seeking score (b). Each marker represents a country with 25 or more respondents (n = 43).

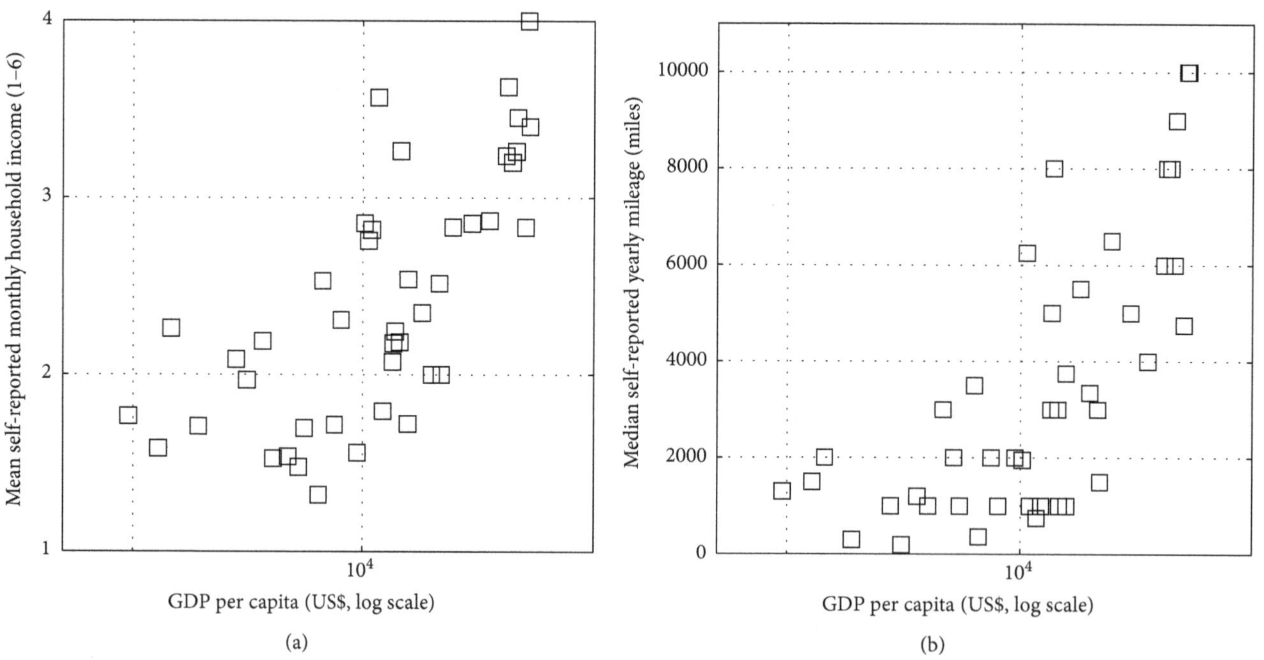

FIGURE 4: Correlation between countries' gross domestic product (GDP) per capita and participants' mean self-reported net household income (a) and participants' median self-reported yearly mileage (b). Each marker represents a country with 25 or more respondents (n = 43).

novel idea, which could move research on the acceptance of driverless vehicles in a new direction. The decision to retain one component may come at a price of oversimplification. However, we argue that the general acceptance component is a parsimonious reflection of the diverse items that concern acceptance of driverless vehicles. Also, the single component is in accordance with the percentage of variance explained by the components (scree plot) and was clearly interpretable. The notion of a single factor or component has also been proposed in various psychological domains with the general intelligence factor and the general personality factor being notable examples [55, 56]. Future studies should explore the hierarchical structure of the acceptance component. Thus, while we argue that a single acceptance component exists

TABLE 4: Spearman correlation matrix at the national level (n = 43).

	1	2	3	4	5	6	7	8	9	10	11
(1) General acceptance score											
(2) Component 1 score (general acceptance)	0.98										
(3) Component 2 score (thrill seeking)	0.50	0.41									
(4) Component 3 score (wanting to be in control manually)	0.04	0.05	−0.32								
(5) Component 4 score (supporting a car-free environment)	0.82	0.78	0.38	0.10							
(6) Component 5 score (being comfortable with technology)	0.92	0.88	0.55	−0.04	0.72						
(7) Road traffic death rate per 100,000 population	0.56	0.52	0.67	−0.22	0.26	0.58					
(8) Gross domestic product (GDP) per capita	−0.48	−0.42	−0.72	0.21	−0.44	−0.44	−0.58				
(9) National performance in educational tests	−0.60	−0.56	−0.78	0.35	−0.39	−0.67	−0.71	0.66			
(10) Average life expectancy	−0.43	−0.37	−0.70	0.16	−0.25	−0.45	−0.67	0.82	0.67		
(11) Motor vehicle density	−0.59	−0.53	−0.86	0.27	−0.45	−0.60	−0.71	0.88	0.81	0.81	
(12) Median age	−0.52	−0.45	−0.79	0.11	−0.25	−0.55	−0.73	0.60	0.73	0.66	0.80

Note. The general acceptance score is the first principal component score. The component 1–5 scores are the scores for the first five components, after Promax rotation.

at the top of the hierarchy, the existence of lower-level constructs is also likely. While the current general acceptance component is mainly an assembly of attitudinal constructs (e.g., knowledge of mobility, wish for car-free future) and behavioral beliefs (e.g., pleasure, social influence), the role of further entities such as values or norms (e.g., ecological or proenvironmental attitudes [57]) could be investigated.

High mean loadings (>0.60) on the general acceptance component were obtained for perceived usefulness. This aligns with the UTAUT model [26], which found that usefulness ("performance expectancy") is a main determinant of the acceptance of driverless vehicles. Medium mean loadings (~0.5) occurred for social influence, trust, and intention to use. These findings support the role of social influence, intention to use [22], and trust [58] as determinants of the acceptance of driverless vehicles.

The correlations between sociodemographic characteristics and general acceptance scores were small (<0.20), which is in agreement with studies examining the influence of sociodemographics on proenvironmental behavior [36], or with acceptance research on wind energy [37]. However, the small correlations were interpretable: The strongest correlation was found for the difficulty of finding a parking place, which is consistent with the literature on acceptance of transport-related measures (e.g., [59, 60]). The difficulty of finding a parking place may be indicative of the severity of transport-related problems and could therefore be a factor that influences people's willingness to accept driverless vehicles as a solution to this problem. The present study further found that living in the city and frequency of public transport use were positively correlated with the general acceptance score. These positive correlations may be explained because the driverless shuttle as depicted in the survey (Figure 1) is a form of public transport, and some people are more accustomed to using public transport and sharing space with strangers than others. In summary, our findings indicate that sociodemographic characteristics are less influential than domain-specific attitudes (e.g., performance expectancy) in predicting self-reported acceptance of driverless vehicles.

4.2. Main Results at the National Level. The countries' mean general acceptance score was negatively correlated with national GDP per capita and other developmental indexes (e.g., average life expectancy, motor vehicle density). In an additional analysis, we retained five instead of one component (Table 2), and we observed substantial negative correlations between countries' GDP per capita and the mean thrill seeking score, and between GDP per capita and the mean score for supporting a car-free environment (Table 4). These national differences may reflect national differences in thrill seeking personality, or effects of differences in road infrastructure (see also [50]). For example, low-income countries suffer more from transport-related problems [54, 61], which may make technological solutions and a car-free infrastructure in cities appealing for people living in these countries. It is recommended for future research to examine the mechanisms that explain the national differences in the acceptance of driverless vehicles. The relationships between the general acceptance component and the four additional components, as well as with other variables identified in this study, could be more closely examined in confirmatory studies using multiple well-defined scales (e.g., sensation seeking scale by Zuckerman et al. [62], in addition to our thrill-related items) and multivariate analyses (e.g., structural equation modeling).

Our study revealed differences between within-country and across-countries correlations. For example, the correlation between gender and the general acceptance scores was stronger for males across countries than within countries. This can be explained by a confounding effect, in the sense that the lower-income countries contained a relatively high proportion of males ($\rho = -0.49$, $n = 43$, Q2, see supplementary material), and respondents in lower-income countries were more accepting of driverless cars. We recommend that future research makes a distinction between within-country and between-country effects.

4.3. Study Limitations. Although it is important to study the acceptance before the technology is commercially available, the respondents did not physically get to see driverless vehicles, which may bias results. For example, it is possible that the respondents had overly positive attitudes that may have been nurtured by the portrayal of automated vehicles in the media.

Second, the survey instructions did not include a reference to the capabilities of the driverless vehicle, nor its speed. Current prototypes of driverless vehicles drive at a speed of 8 to 20 km/h [20]. It can be assumed that these speeds are too slow for integrating these vehicles into traffic without jeopardizing traffic flow efficiency or the acceptance by potential users.

Third, the crowdsourcing participants are not necessarily representative of the general population as they are younger and more highly educated compared to the general population [63]. Future research should more closely examine the effects in representative cross-national populations, using gender- and age-stratified samples.

Fourth, the survey was conducted in April 2015, meaning that the possibility exists that our data are not representative of today's public opinion, for example, due to changes in media coverage about automated vehicles. Abraham et al. [64] found that respondents were less comfortable with self-driving vehicles than a sample from one year ago. On the other hand, self-driving vehicles have not been commercialized yet, but are only used as part of various experiments and demonstration projects worldwide. Therefore, there has been little opportunity for respondents to actually experience self-driving vehicles and to adjust their opinion accordingly.

Finally, the use of self-reported measures in technology acceptance studies has been criticized, as the data that are collected are not independent of the method that is used to collect the data [65]. Our study indeed found evidence for common method effects, as the general acceptance score correlated relatively strongly with the time to complete the survey, and with whether participants found the survey instructions clear ($\rho \sim 0.20$ at the individual level, $n = 7{,}755$). At the national level ($n = 43$), it is possible that the negative correlation between the countries' mean general acceptance scores and GDP per capita ($\rho = -0.48$) reflects a common method bias due to social desirability or response style. For example, it is possible that respondents in higher-income countries gave lower acceptance ratings because of their better command of the English language. Better English language skills may be a reason why respondents in higher-income countries took less time to complete the survey ($\rho = -0.65$), and found the instructions clearer ($\rho = 0.43$, Q2; see supplementary material). Indeed, it cannot be ruled out that language barriers jeopardized the validity of our results, in the sense that it is possible that respondents in non-English speaking countries had difficulty with understanding the meaning of certain questions, as well as the notion of driverless shuttles. Then again, we did find strong correlations between GDP per capita and mean self-reported income (Q12), and median mileage (Q17) ($\rho = 0.71$ and 0.71, resp.), suggesting that self-reports are largely valid. Furthermore, GDP per capita correlated *positively* with some items (e.g., $\rho = 0.39$ for "Q36. I would like to have a button inside the driverless vehicle which I can press to stop it"), but *negatively* with others (e.g., $\rho = -0.64$ for "Q51. Even if it were more expensive than my existing form of travel, I would prefer driverless vehicles to my existing form of travel"), which suggests that respondents in different countries meaningfully responded to item content and showed no strong acquiescence bias by "agreeing" with all survey items. More fundamentally, our research points to interesting issues when it comes to measuring "acceptance", as acceptance is inherently a subjective construct that is substantively related to bias and preconceptions. Similar discussions have been held regarding the interpretation of the general factor of personality (GFP). Musek ([66, page 120]) pointed out that "considering the fact that some variance of the social desirability itself represents a substantive trait, the obvious conclusion is that GFP can only partly be explained by social desirability as a mere response style." Future research should use self-reports together with objective usage data of driverless vehicles.

4.4. Conclusions. In conclusion, our survey showed that respondents believe that driverless shuttles are easy to use and convenient. Our study also revealed cross-national differences and found that lower-income countries were more accepting of driverless vehicles than higher-income countries. Finally, we extracted a general acceptance component, which is an innovative measure that comprises pivotal items concerning the acceptance, and hence future success, of driverless vehicles.

Conflicts of Interest

The authors declare that there are no conflicts of interest regarding the publication of this paper.

Acknowledgments

The authors would like to thank Andreas Graff, Dr. Christian Hoffmann, and Dr. Lisa Ruhrort from the Innovation Centre for Mobility & Societal Change (InnoZ) in Berlin for their valuable contributions to design the questionnaire.

References

[1] G. Eden, B. Nanchen, R. Ramseyer, and F. Evéquoz, "On the road with an autonomous passenger shuttle: integration in public spaces," in *Proceedings of the 2016 CHI Conference Extended Abstracts on Human Factors in Computing Systems*, pp. 1569–1576, ACM, Denver, Colo, USA, 2017.

[2] S. Shladover, "The Truth about Self-Driving Cars: They are coming, but not the way you may have been led to think," *Scientific American*, vol. 314, no. 6, pp. 52–57, 2016.

[3] Attias D., "The autonomous car, a disruptive business model?" in *The Automobile Revolution*, Attias D., Ed., Springer International Publishing, Switzerland, 2017.

[4] A. Scheltes and G. Correira, "Exploring the use of automated vehicles as last mile connection of chain trips through an agent-based simulation model: An application to Delft," in *Proceedings*

of the International Journal of Transportation Science and Technology, vol. 6, pp. 28–41, Netherlands, 2017.

[5] B. Van Arem, N. Van Oort, M. Yap, B. Wiegmans, and G. H. Correira, *Opportunities and Challenges for Automated Vehicles in the Zuidvleugel*, 2015, http://nielsvanoort.weblog.tudelft.nl/files/2015/03/TUD2103-essay-final.pdf.

[6] S. Nordhoff, B. Van Arem, and R. Happee, "Conceptual model to explain, predict, & improve user acceptance of driverless podlike vehicles," *Transportation Research Record*, vol. 2602, pp. 60–67, 2016.

[7] L. M. Hulse, H. Xie, and E. R. Galea, "Perceptions of autonomous vehicles: Relationships with road users, risk, gender and age," *Safety Science*, vol. 102, pp. 1–13, 2018.

[8] M. Kyriakidis, R. Happee, and J. C. F. De Winter, "Public opinion on automated driving: Results of an international questionnaire among 5000 respondents," *Transportation Research Part F: Traffic Psychology and Behaviour*, vol. 32, article no. 813, pp. 127–140, 2015.

[9] M. Regan, M. Cunningham, V. Dixit et al., *Preliminary Findings from The First Australian National Survey of Public Opinion about Automated And Driverless Vehicles*, 2017, http://advi.org.au/wp-content/uploads/2017/07/ADVI-Public-Opinion-Survey-7June2016.pdf.

[10] C. Hohenberger, M. Spörrle, and I. M. Welpe, "How and why do men and women differ in their willingness to use automated cars? The influence of emotions across different age groups," *Transportation Research Part A: Policy and Practice*, vol. 94, pp. 374–385, 2016.

[11] J. Piao, M. McDonald, N. Hounsell, M. Graindorge, T. Graindorge, and N. Malhene, "Public views towards implementation of automated vehicles in urban areas," *Transportation Research Procedia*, vol. 14, pp. 2168–2177.

[12] F. Becker and K. W. Axhausen, "Literature review on surveys investigating the acceptance of automated vehicles," *Transportation*, vol. 44, no. 6, pp. 1293–1306, 2017.

[13] P. Bansal, K. M. Kockelman, and A. Singh, "Assessing public opinions of and interest in new vehicle technologies: An Austin perspective," *Transportation Research Part C: Emerging Technologies*, vol. 67, pp. 1–14, 2016.

[14] J. Missel, *Ipsos MORI Loyalty Automotive Survey*, 2014, http://www.ipsos-mori.com/researchpublications/researcharchive/3427/Only-18-per-cent-of-Britons-believe-driverless-cars-to-be-an-important-development-for-the-car-industry-to-focus-on.aspx?utm_campaign=cmp_325684&utm_source=Getanewsletter.

[15] Seapine Software, "Study Finds 88 Percent of Adults Would Be Worried about Riding in a Driverless Car," http://www.seapine.com/about-us/press-release-full?press=217, 2014.

[16] Power JD, *Automotive Emerging Technologies Study Results*, 2012, http://autos.jdpower.com/content/press-release/gGOwCnW/2012-u-s-15automotive-emerging-technologies-study.htm.

[17] B. Schoettle and M. Sivak, *A Survey of Public Opinion about Autonomous and Self-Driving Vehicles in the U.S., the U.K., and Australia*, 2014, https://deepblue.lib.umich.edu/bitstream/handle/2027.42//103024.pdf.

[18] B. Schoettle and M. Sivak, *Motorists' Preferences for Different Levels of Vehicle Automation*, 2015, https://deepblue.lib.umich.edu/bitstream/handle/2027.42//103217.pdf.

[19] C. Rödel, S. Stadler, A. Meschtscherjakov, and M. Tscheligi, "Towards autonomous cars: The effect of autonomy levels on Acceptance and User Experience," in *Proceedings of the 6th International Conference on Automotive User Interfaces and Interactive Vehicular Applications, AutomotiveUI 2014*, USA, September 2014.

[20] S. Nordhoff, J. De Winter, R. Madigan, N. Merat, B. Van Arem, and R. Happee, "User acceptance of automated shuttles in Berlin-Schöneberg: A questionnaire study," *Manuscript Submitted for Publication*, 2017.

[21] R. Shabanpour, S. N. D. Mousavi, N. Golshani, J. Auld, and A. Mohammadian, "Consumer preferences of electric and automated vehicles," in *Proceedings of the 5th IEEE International Conference on Models and Technologies for Intelligent Transportation Systems, MT-ITS 2017*, pp. 716–720, IEEE, Naples, Italy, June 2017.

[22] R. Madigan, T. Louw, M. Dziennus et al., "Acceptance of Automated Road Transport Systems (ARTS): An adaptation of the UTAUT model," in *Proceedings of the 6th Transport Research Arena*, pp. 2217–2226, Elsevier, Warsaw, Poland, 2016.

[23] M. D. Yap, G. Correia, and B. Van Arem, "Valuation of travel attributes for using automated vehicles as egress transport of multimodal train trips," *Transportation Research Procedia*, vol. 10, pp. 462–471, 2015.

[24] J. P. Zmud and I. N. Sener, "Towards an understanding of the travel behavior impact of autonomous vehicles," *Transportation Research Procedia*, vol. 25, pp. 2500–2519, 2017.

[25] P. S. Lavieri, V. M. Garikapati, C. R. Bhat, R. M. Pendyala, S. Astroza, and F. F. Dias, "Modeling individual preferences for ownership and sharing of autonomous vehicle technologies," *Transportation Research Record: Journal of the Transportation Research Board*, pp. 1–10, 2017.

[26] V. Venkatesh, M. G. Morris, G. B. Davis, and F. D. Davis, "User acceptance of information technology: toward a unified view," *MIS Quarterly: Management Information Systems*, vol. 27, no. 3, pp. 425–478, 2003.

[27] M. Fishbein and I. Ajzen, *Belief, Attitude, Intention, and Behavior: An Introduction to Theory and Research*, Addison-Wesley, Boston, Mass, USA, 1975.

[28] F. D. Davis, "Perceived usefulness, perceived ease of use, and user acceptance of information technology," *MIS Quarterly*, vol. 13, no. 3, pp. 319–339, 1989.

[29] F. D. Davis, R. P. Bagozzi, and P. R. Warshaw, "User acceptance of computer technology: a comparison of two theoretical models," *Management Science*, vol. 35, no. 8, pp. 982–1003, 1989.

[30] I. Ajzen, "From intentions to actions: A theory of planned behavior. In," in *Action-Control: from Cognition to Behavior*, J. Kuhl and Beckmann, Eds., pp. 11–39, Springer, Heidelberg, Germany, 1985.

[31] A. Mehrabian, "Pleasure-arousal-dominance: a general framework for describing and measuring individual differences in temperament," *Current Psychology*, vol. 14, no. 4, pp. 261–292, 1996.

[32] R. Madigan, T. Louw, M. Wilbrink, A. Schieben, and N. Merat, "What influences the decision to use automated public transport? Using UTAUT to understand public acceptance of automated road transport systems," *Transportation Research Part F: Traffic Psychology and Behaviour*, vol. 50, pp. 55–64, 2017.

[33] P. Bazilinskyy, M. Kyriakidis, and J. de Winter, "An international crowdsourcing study into people's statements on fully automated driving," *Procedia Manufacturing*, vol. 3, pp. 2534–2542, 2015.

[34] World Economic Forum (WEF) and Boston Consulting Group (BCG), "Self-driving vehicles in an urban context," http://www3.weforum.org/docs/WEF_Press%20release.pdf, 2015.

[35] B. Schoettle and M. Sivak, *Public Opinion about self-driving vehicles in China, India, Japan, the U.S., the U.K., and Australia*, 2014, http://deepblue.lib.umich.edu/bitstream/handle/2027.42/109433/103139.pdf?sequence=1&isAllowed=y.

[36] P. C. Stern, "New environmental theories: Toward a coherent theory of environmentally significant behavior," *Journal of Social Issues*, vol. 56, no. 3, pp. 407–424, 2000.

[37] J. Rand and B. Hoen, "Thirty years of North American wind energy acceptance research: What have we learned?" *Energy Research and Social Science*, vol. 29, pp. 135–148, 2017.

[38] S. Barr and G. Shaw, *Knowledge Co-Production And Behavioural Change: Collaborative Approaches for Promoting Sustainable Mobility*, D. Hopkins and J. Higham, Eds., Goodfellow Publishers, Oxford, UK, 2016.

[39] Easymile, *Video of EZ10 at EPFL: Demonstration of the EZ10 driverless vehicle at the École Polytech-nique Fédérale de Lausanne (EPFL)*, Springer, Lausanne, Switzerland, 2015, http://easymile.com/easymile-on-youtube/.

[40] H. Schuhmacher, "WEPODS: Autonome Minibusse," http://androidmag.de/report/wepods-autonome-minibusse/, 2015.

[41] A. Parasuraman, "Technology readiness index (TRI): a multiple-item scale to measure readiness to embrace new technologies," *Journal of Service Research*, vol. 2, no. 4, pp. 307–320, 2000.

[42] S. Vlassenroot, K. Brookhuis, J. De Mol, and F. Witlox, "Acceptability of ISA: results from a large-scale survey in Belgium and the Netherlands," in *Proceedings of the 8th ITS European Congress and Exhibition on Intelligent Transport Systems and Services*, Lyon, France, 2011.

[43] D. Yagil, "Reasoned action and irrational motives: A prediction of drivers' intention to violate traffic laws," *Journal of Applied Social Psychology*, vol. 31, no. 4, pp. 720–740, 2001.

[44] R. Krueger, T. H. Rashidi, and J. M. Rose, "Preferences for shared autonomous vehicles," *Transportation Research Part C: Emerging Technologies*, vol. 69, pp. 343–355, 2016.

[45] L. Beretta and A. Santaniello, "Nearest neighbor imputation algorithms: A critical evaluation," *BMC Medical Informatics and Decision Making*, vol. 16, article no. 74, pp. 197–208, 2016.

[46] X. Dong, B. Milholland, and J. Vijg, "Evidence for a limit to human lifespan," *Nature*, vol. 538, no. 7624, pp. 257–259, 2016.

[47] J. C. F. De Winter and P. A. Hancock, "Reflections on the 1951 Fitts list: Do humans believe now that machines surpass them?" in *Proceedings of the 6th International Conference on Applied Human Factors and Ergonomics (AHFE)*, pp. 5334–5341, Las Vegas, Nev, USA, 2015.

[48] H. F. Kaiser, "A second generation little jiffy," *Psychometrika*, vol. 35, no. 4, pp. 401–415, 1970.

[49] R. A. Peterson, "A meta-analysis of variance accounted for and factor loadings in exploratory factor analysis," *Marketing Letters*, vol. 11, no. 3, pp. 261–275, 2000.

[50] J. C. F. de Winter and D. Dodou, "National correlates of self-reported traffic violations across 41 countries," *Personality and Individual Differences*, vol. 98, pp. 145–152, 2016.

[51] H. Rindermann, "The g-factor of international cognitive ability comparisons: The homogeneity of results in PISA, TIMSS, PIRLS and IQ-tests across nations," *European Journal of Personality*, vol. 21, no. 5, pp. 667–706, 2007.

[52] World Bank, http://data.worldbank.org/indicator, 2015.

[53] Central Intelligence Agency, "The World Factbook," Retrieved from https://www.cia.gov/library/publications/the-world-factbook/fields/2177.html, 2015.

[54] World Health Organization, *Global Status Report on Road Safety 2015*, World Health Organization, Geneva, Switzerland, http://www.who.int/violence_injury_prevention/road_safety_status/2015/GSRRS2015_Summary_EN_final2.pdf?ua=1, 2015.

[55] A. R. Jensen, *The g Factor: The Science of Mental Ability*, Praeger, Westport, Connecticut, USA, 1998.

[56] J. Musek, "A general factor of personality: evidence for the Big One in the five-factor model," *Journal of Research in Personality*, vol. 41, no. 6, pp. 1213–1233, 2007.

[57] L. Steg and C. Vlek, "Encouraging pro-environmental behaviour: An integrative review and research agenda," *Journal of Environmental Psychology*, vol. 29, no. 3, pp. 309–317, 2009.

[58] J. K. Choi and Y. G. Ji, "Investigating the Importance of Trust on Adopting an Autonomous Vehicle," *International Journal of Human-Computer Interaction*, vol. 31, no. 10, pp. 692–702, 2015.

[59] L. Steg and C. Vlek, "The role of problem awareness in willingness-to-change car-use and in evaluating relevant policy measures," in *Traffic and Transport Psychology*, J. A. Rothengatter and E. Carbonell Vaya, Eds., pp. 465–475, Pergamon, Oxford, UK, 1997.

[60] S. A. Rienstra, P. Rietveld, and E. T. Verhoef, "The social support for policy measures in passenger transport. A statistical analysis for the Netherlands," *Transportation Research Part D: Transport and Environment*, vol. 4, no. 3, pp. 181–200, 1999.

[61] S. N. Forjuoh, "Traffic-related injury prevention interventions for low-income countries.," *The International Journal of Injury Control and Safety Promotion*, vol. 10, no. 1-2, pp. 109–118, 2003.

[62] M. Zuckerman, S. Eysenck, and H. J. Eysenck, "Sensation seeking in England and America: cross-cultural, age, and sex comparisons," *Journal of Consulting and Clinical Psychology*, vol. 46, no. 1, pp. 139–149, 1978.

[63] J. C. F. De Winter, M. Kyriakidis, D. Dodou, and R. Happee, "Using crowd flower to study the relationship between self-reported violations and traffic accidents," in *Proceedings of the 6th International Conference on Applied Human Factors and Ergonomics (AHFE)*, pp. 2518–2525, Las Vegas, Nevada, USA, 2015.

[64] H. Abraham, B. Reimer, B. Seppelt, C. Fitzgerald, B. Mehler, and J. F. Coughlin, "Consumer interest in automation: Change over one year," in *Proceedings of the Transportation Research Board 97th Annual Meeting*, Washington Convention Center, Washington DC, USA, 2018.

[65] D. W. Straub Jr. and A. Burton-Jones, "Veni, vidi, vici: Breaking the TAM logjam," *Journal of the Association for Information Systems*, vol. 8, no. 4, pp. 223–229, 2007.

[66] J. Musek, *The General Factor of Personality*, Academic Press, Cambridge, Mass, USA, 2017.

13

A Low-Cost Consistent Vehicle Localization based on Interval Constraint Propagation

Zhan Wang and Alain Lambert

Laboratoire de Recherche en Informatique (LRI), CNRS, Univ Paris-Sud, Université Paris-Saclay, 91403 Orsay, France

Correspondence should be addressed to Alain Lambert; alain.lambert@u-psud.fr

Academic Editor: Krzysztof Okarma

Probabilistic techniques (such as Extended Kalman Filter and Particle Filter) have long been used to solve robotic localization and mapping problem. Despite their good performance in practical applications, they could suffer inconsistency problems. This paper proposes an interval analysis based method to estimate the vehicle pose (position and orientation) in a consistent way, by fusing low-cost sensors and map data. We cast the localization problem into an Interval Constraint Satisfaction Problem (ICSP), solved via Interval Constraint Propagation (ICP) techniques. An interval map is built when a vehicle embedding expensive sensors navigates around the environment. Then vehicles with low-cost sensors (dead reckoning and monocular camera) can use this map for ego-localization. Experimental results show the soundness of the proposed method in achieving consistent localization.

1. Introduction

Using sensory information to locate the robot in its environment is one of the most fundamental problem providing a mobile robot with autonomous capabilities. A vast number of works [1, 2] dedicated to such problems use both proprioceptive and exteroceptive sensors. Proprioceptive sensors such as odometers and gyrometers are used to calculate the elementary movements, which are used to estimate the robot pose (position and orientation). However, this generates cumulative errors as the robot moves [3, 4]. To cope with this problem, exteroceptive sensors (lasers, sonars, GPS, cameras, etc.) are used to gather information either from the surrounding environment or from a global reference, aiming at eliminating the accumulated errors and improving the estimation of the robot pose [5, 6].

Probabilistic techniques have long been used in solving robotic localization and mapping problem. The most commonly used methods are Kalman filtering and Particle filtering [7, 8]. These methods associate a probability to a position, but nothing ensures that the robot is indeed in the position with the highest probability. It has long been noticed by the research community that probabilistic methods suffer the inconsistency problem [9]. It is due to the fact that they are based on the assumption that the sensor noises are described by probability density function. However, in real world, many sensors do not possess Gaussian noise model or their distributions are simply not available [10].

Interval analysis [11] is an alternative and less known method which allows us to solve nonlinear problems in a guaranteed way. Instead of making hypothesis on the probability distribution of sensor noises, interval methods take the assumption that all the noises are bounded within known limits. This seems to be a realistic representation as most sensor manufacturers provide the maximum and minimum measurement errors under suitable working conditions. These extreme values of error can then be regarded as the error bounds. Interval based algorithms can be used to recursively propagate such bounded errors by using consistency techniques and systematic search methods. Contrary to probability methods, interval methods provide guaranteed sets which enclose the real state, without losing any feasible value.

Interval methods have achieved promising results in parameter and state estimation tasks [12–14], and they found their way to the mobile robotic localization and mapping area. Instead of considering Gaussian noise, all model and measurement noises are assumed to be bounded. The aim is to guarantee the configurations of the robot pose that are consistent with the given measurements and noise bounds. Kieffer [15] and Jaulin [16] provided simulation results for a robot equipped with a belt of on-board ultrasonic sensors. The feasibility of interval methods on localization applications had been shown. Followed-up researchers focused on real-time implementation of such localization technique. Kieffer [17] and Seignez [18] presented results for a mobile robot navigating in an indoor environment, with odometers mounted on each rear wheel to measure the motion and sonars located in different directions to perceive the environment. The ability of interval method to cope with erroneous data and to obtain accurate estimations of the robot pose was demonstrated. Lambert [19] extended such work by providing results for an outdoor vehicle equipped with odometers, gyro, and GPS receiver. Comparison was made with the Particle Filter localization, showing the better performance of interval method in terms of consistency. Gning [2] and Kueviakoe [20] dealt with the outdoor localization problem in the framework of Interval Constraint Satisfaction Problem (ICSP). Those works used constraint propagation techniques to fuse the redundant data of sensors. Drevelle [21] uses relaxed constraint propagation approach to deal with erroneous GPS measurements. Bonnifait [22] combined constraint propagation and set inversion techniques and presented a cooperative localization method with significant enhancement in terms of accuracy and confidence domains. Experimental results illustrate that constraint propagation techniques are well adapted to real-time implementation and provide consistent result for localization. Some other researchers proposed specific contractors [23] or separators [24] to solve the localization problem.

Indeed, when considering consistency issue, interval methods are advantageous in vehicle localization. However, they are prone to large output imprecision, which can make them unsuitable for high level task. We improve on it and present an extension of others' works dealing with the localization problem by using ICP. Instead of using ultrasonic sensors [16], sonar [18, 25], or GPS [19, 26] as exteroceptive sensors, we propose to use a monocular camera. The aim is to achieve both consistent and low imprecision localization result. We propose to drive a vehicle embedding precise but expensive sensors around the environment in order to build up a map represented in an interval way. Then the proposed method enables a set or a single vehicle to navigate consistently in such environment using this map and low-cost sensors. By constructing the problem as an ICSP, the proposed method is able to handle both mapping and localization in a unified form using Interval Constraint Propagation techniques.

The paper is organized as follows. Section 2 introduces the basics of interval analysis. Section 3 illustrates how the mapping and localization problem can be formulated as an Interval Constraint Satisfaction Problem. Sections 4 and 5 present the simulation and experimental results in terms of mapping and localization. Section 6 draws a conclusion of the paper.

2. Overview of Interval Analysis and Constraint Propagation

Interval analysis and constraint propagation are the theoretical tools used in interval based methods for robotic localization and mapping. In this section, we give a brief overview of these tools.

2.1. Interval Analysis. Interval analysis [11, 27] is a numerical method which allows us to solve nonlinear problems in a guaranteed way. One of the pioneers of interval analysis, Ramon E. Moore, proposed to represent a solution of a problem by an interval in which the real solution is guaranteed. An interval $[x]$ is a connected subset of $\mathbb{R}$, defined by its lower bound $\underline{x}$ and upper bound $\overline{x}$:

$$[x] = [\underline{x}, \overline{x}] = \{x \in \mathbb{R} \mid \underline{x} \le x \le \overline{x}\} \tag{1}$$

The width of a nonempty interval $[x]$ is calculated by $w([x]) = \overline{x} - \underline{x}$ and the midpoint (or center) is $mid([x]) = (\overline{x} - \underline{x})/2$. If $[x]$ is an interval containing the x position of a vehicle, $mid([x])$ can be taken as an estimation of x and $w([x])/2$ can be regarded as the estimation error [28].

Let $\mathbb{IR}$ be the set of the intervals of $\mathbb{R}$. A *box* (also called *interval vector*) $[\mathbf{x}]$ ($[\mathbf{x}] \in \mathbb{IR}^n$) is a generalization of the interval concept. It is a vector whose components are intervals:

$$[\mathbf{x}] = [x_1] \times [x_2] \times \cdots \times [x_n] \tag{2}$$

For instance, the configuration of a vehicle's pose usually contains 3 parameters: position x, position y, and heading angle θ. Consequently, for vehicle localization, the solution is a three-dimensional box: $[x] \times [y] \times [\theta]$.

The width of a nonempty box $[\mathbf{x}]$ can be computed by $w([\mathbf{x}]) = \max_{i=1\ldots n} w([x_i])$. We define the volume of $[\mathbf{x}]$ as

$$vol([\mathbf{x}]) = \prod_{1 \le i \le n} w([x_i]) \tag{3}$$

The volume of the box is usually used to evaluate the estimation uncertainty of a state vector [28].

2.2. Operations of Interval Arithmetic. Rules have been defined to apply the basic arithmetical operations on intervals [27]. Let us consider two intervals $[x]$ and $[y]$ and a binary operator $\diamond \in \{+, -, \times, \div\}$; the smallest interval which contains all feasible values for $[x] \diamond [y]$ is defined as follows:

$$[x] \diamond [y] = \{x \diamond y \mid x \in [x], y \in [y]\} \tag{4}$$

Example 1. Addition: $[-4, 3] + [2, 5] = [-2, 8]$; subtraction: $[3, 5] - [1, 2] = [1, 4]$; multiplication: $[-1, 8] \times [-3, 1] = [-24, 8]$; division: $[-2, 8] \div [2, 4] = [-1, 4]$.

The set-theoretic operations can be applied to intervals. The *intersection* of two intervals is defined by

$$[x] \cap [y] = \{z \in \mathbb{R} \mid z \in [x] \text{ and } z \in [y]\} \tag{5}$$

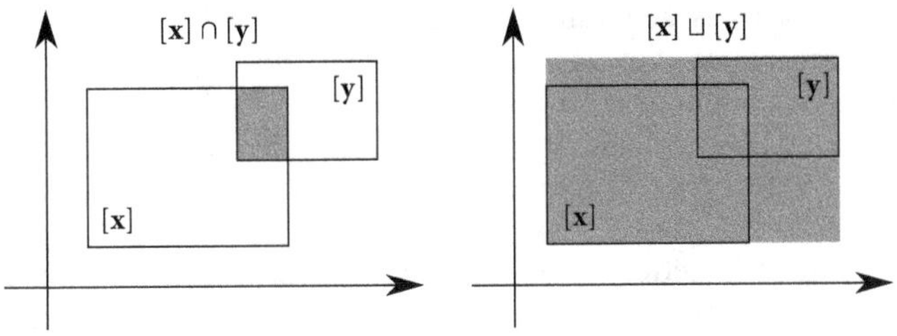

FIGURE 1: Intersection and interval hull of interval boxes.

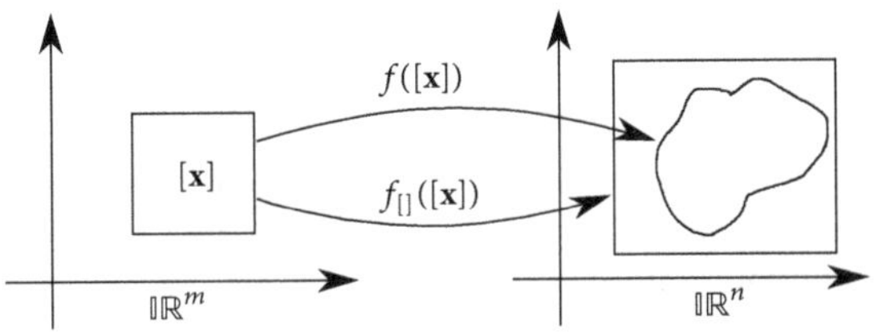

FIGURE 2: Images of an interval box by function f and its inclusion function $f_{[]}$.

The result of *intersection* is always an interval, but it is not the case for their *union*

$$[x] \cup [y] = \{z \in \mathbb{R} \mid z \in [x] \text{ or } z \in [y]\} \qquad (6)$$

Consequently, an *interval hull* is defined as the smallest interval that contains all the subsets of $[x] \cup [y]$, denoted by $[x] \sqcup [y]$:

$$[x] \sqcup [y] = [[x] \cup [y]] \qquad (7)$$

For instance, the interval hull of $[2,4] \cup [5,8]$ is the interval $[2,8]$. Figure 1 gives an example of *intersection* and *union* for 2 interval boxes; it is realized by applying the intersection and interval hull operator to each component of the box.

2.3. Inclusion Function. The image of a vector function $f : \mathbb{IR}^m \rightarrow \mathbb{IR}^n$ (defined by arithmetical operators and elementary functions) over an interval box $[\mathbf{x}]$ can be evaluated by its inclusion function $f_{[]}$, whose output contains all possible values taken by $f(\cdot)$ over $[\mathbf{x}]$ (see Figure 2):

$$\forall [\mathbf{x}] \in \mathbb{IR}^m, \quad f([\mathbf{x}]) \subset f_{[]}([\mathbf{x}]) \qquad (8)$$

The simplest way to create an inclusion function is to replace all the variables and operators by their interval equivalents. The resulting interval function is called "natural inclusion function". For instance,

$$\forall x \in \mathbb{R}, \text{âNĹ} f(x) = 4x^4 - 7\ln(x) - 3 \qquad (9)$$

has the natural inclusion function:

$$\forall [x] \in \mathbb{IR}, \quad f_{[]}([x]) = 4[x]^4 - 7\ln_{[]}([x]) - 3 \qquad (10)$$

Inclusion function can also be used to represent equations defined by interval variables. Such equations, also called constraints, are the core of Interval Constraint Satisfaction Problems.

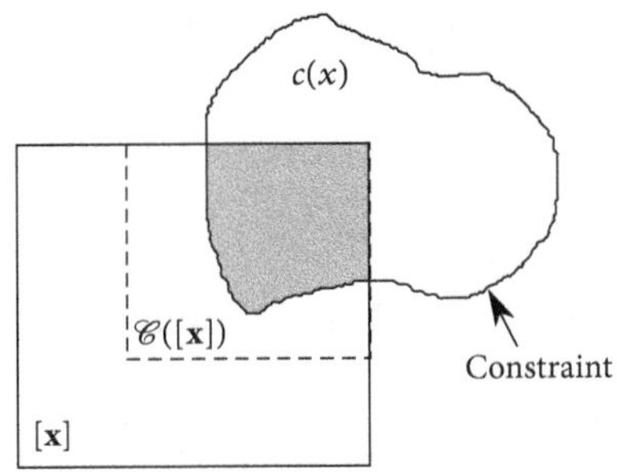

FIGURE 3: Contraction of a box by a contractor $\mathscr{C}$.

2.4. Contractor. The concept of contractor is directly inspired from the ubiquitous concept of filtering algorithms in constraint programming. Given a constraint c relating a set of variables x, an algorithm $\mathscr{C}$ is called a contractor (or a filtering algorithm) if $\mathscr{C}$ returns a subdomain of the input domain $[\mathbf{x}]$ and the resulting subdomain $\mathscr{C}([\mathbf{x}])$ contains all the feasible points (the gray domain in Figure 3) with respect to the constraint c:

$$\mathscr{C}([\mathbf{x}]) \subseteq [\mathbf{x}] \quad \text{and } \forall x \in [\mathbf{x}],\ c(x) \Longrightarrow x \in \mathscr{C}([\mathbf{x}]) \qquad (11)$$

2.5. Interval Constraint Satisfaction Problem (ICSP). The concept of Interval Constraint Satisfaction Problem (ICSP) was introduced by Hyvönen [29] in 1992. It is typically defined as

(i) a set V of variables $\{v_1, v_2, .., v_n\}$,

(ii) a set D of domains $\{d_1, d_2, .., d_n\}$, such that for each variable v_i, a domain d_i with the possible values for that variable is given; d_i is an interval or union of intervals,

(iii) a set C of p constraints $\{c_1, c_2, .., c_p\}$; each constraint c_i defines the relationships of a number of variables from V; e.g., $c_1(v_1, v_2, v_3) = 0$ restricts the possible domains of v_1, v_2, and v_3.

An ICSP is a mathematical problem whose solution is the minimal domain of the variables satisfying all the constraints.

2.6. Interval Constraint Propagation (ICP). To solve an ICSP, an Interval Constraint Propagation (ICP) algorithm iterates domain reductions until no domain can be contracted. To satisfy the set of p constraints, it removes, from the domain of the variable, every value that is not compatible with the constraints and the other variables. ICP reduces the size of the domains in a consistent way by repeating this removal operation. This solving process is called *contraction* and the corresponding algorithm (e.g., Forward/Backward) used for contraction is regarded as a *contractor*.

Let us consider a simple ICSP instance which is defined by 3 variables and 2 constraints:

$$\begin{aligned} z &= x + \ln(y) \\ y &= z^2 \end{aligned} \qquad (12)$$

The domains of x, y, and z are, respectively, $[x]$, $[y]$, and $[z]$. An inclusion function of (12) is

$$\begin{aligned}[z] &= [x] + \ln_{[]}([y]) \\ [y] &= [z]^2\end{aligned} \tag{13}$$

The Forward/Backward Propagation (FBP), an ICP algorithm, can be applied to solve the ICSP. FBP executes the contraction process in two phases: firstly, the Forward propagation reduces the left terms of (13) via

$$[z] = [z] \cap ([x] + \ln_{[]}([y])) \tag{14a}$$

$$[y] = [y] \cap [z]^2 \tag{14b}$$

Secondly, Backward propagation reduces the right terms of (13) by

$$[x] = [x] \cap ([z] - \ln_{[]}([y])) \tag{15a}$$

$$[y] = [y] \cap \exp_{[]}([z] - [x]) \tag{15b}$$

$$[z] = [z] \cap \sqrt{}_{[]}[y] \tag{15c}$$

Forward and Backward equations, obtained from the constraints, are computed one after the other (here we compute from (14a) to (15c)). When all the equations have been computed, the algorithm restarts from the beginning equation until the domains of $[x]$, $[y]$, and $[z]$ are no more contracted (or contracted less than a specified parameter).

3. Localization and Mapping

3.1. Problem Statement. Let us consider a vehicle navigating in a 2D environment. At time step k, the vehicle pose is represented by a three-dimensional vector $\mathscr{X}_k = (x_k, y_k, \theta_k)^T$ with x_k, y_k being the vehicle's position and θ_k the orientation. The n stationary landmarks are represented by $\mathscr{M} = (\mathscr{L}_1, \mathscr{L}_2, \cdots, \mathscr{L}_n)^T$. $\mathscr{L}_i$ is a state vector defining the landmark position (see Section 3.3). For the sake of simplicity, we assume that the pose evolution follows *Markov chain*, where the current pose $\mathscr{X}_k$ depends only on the very last pose $\mathscr{X}_{k-1}$ and the current control input vector $\mathscr{U}_k$. The input vector $\mathscr{U}_k$ consists of linear and angular elementary movements, which were collected by proprioceptive sensors (e.g., odometer and gyro). If these odometric measurements are available, the state vector $\mathscr{X}_k$ can be evaluated through the dead reckoning function (also called motion model) expressed as follows:

$$\mathscr{X}_k = f(\mathscr{X}_{k-1}, \mathscr{U}_k) + w_k \tag{16}$$

where w_k is the measurement noise affecting the odometric measurements.

The vehicle is equipped with exteroceptive sensors, recording information about the environment. At each time step k, the vehicle observes a number of landmarks which can be regarded as a subset of the entire map; the measurement $\mathscr{Z}_k$ depends only on the current vehicle pose $\mathscr{X}_k$ and the map state $\mathscr{M}$:

$$\mathscr{Z}_k = \mathscr{H}(\mathscr{X}_k, \mathscr{M}) + r_k \tag{17}$$

where $\mathscr{H}(\cdot)$ is generally a nonlinear function; it differs for different types of exteroceptive sensors. The output of $\mathscr{H}(\cdot)$ is a geometric parameterization of the landmark (e.g., distance, angle, or pixel coordinates), and r_k is the measurement noise of the exteroceptive sensors which introduce uncertainty to the landmark state.

Interval methods are based on the assumption that the support of sensor noises is unknown but bounded by real intervals. Mathematically, $\forall k > 0$, $w_k \in [\mathbf{w}]$ and $r_k \in [\mathbf{r}]$, where $[\mathbf{w}]$ and $[\mathbf{r}]$ are real intervals defined by lower and upper bounds. Then the localization problem can be regarded as an ICSP: the set of variables is defined by the vehicle pose and the landmark position; the motion and observation functions ((16) and (17)) define the constraints. ICP techniques allow us to propagate the interval uncertainties in a consistent way without any probability hypothesis or linearization process. As a result, all possible solutions are found and guaranteed.

3.2. Motion Model. The motion estimation process uses the measurements of proprioceptive sensors to predetermine the localization vector $[\mathscr{X}_k]$. EKF use a probabilistic motion model to fuse sensor data [30]. We make an extension of this model to obtain an interval based model defined by Seignez et al. [18]:

$$[\mathscr{X}_k] = f_{[]}([\mathscr{X}_{k-1}], [\mathscr{U}_k]) \tag{18}$$

$$\begin{pmatrix}[x_k]\\ [y_k]\\ [\theta_k]\end{pmatrix} = \begin{pmatrix}[x_{k-1}] + [\delta s_k]\cos_{[]}\left([\theta_{k-1}] + \left[\dfrac{\delta\theta_k}{2}\right]\right)\\ [y_{k-1}] + [\delta s_k]\sin_{[]}\left([\theta_{k-1}] + \left[\dfrac{\delta\theta_k}{2}\right]\right)\\ [\theta_{k-1}] + [\delta\theta_k]\end{pmatrix} \tag{19}$$

where $[\mathscr{U}_k] = ([\delta s_k], [\delta\theta_k])^T$ is the input vector. δs_k and $\delta\theta_k$ represent, respectively, the elementary displacement and rotation.

The elementary movements can be deduced from raw odometric data (odometer and gyro) using an interval based static fusion method:

$$\begin{aligned}[\delta s_k] &= \frac{[\pi]([\omega_l][\delta p_l] + [\omega_r][\delta p_r])}{[P]} \\ [\delta\theta_k] &= \frac{[\pi]([\omega_l][\delta p_l] - [\omega_r][\delta p_r])}{[e]\cdot[P]} \cap [\delta\theta_{gyro}]\end{aligned} \tag{20}$$

where δp_l and δp_r are the odometer measurements of the left and right wheels. ω_l and ω_r represent the radius of the wheels. P is the odometer resolution and e is the length of the rear axle. $\delta\theta_{gyro}$ is the gyro measurement. For further information about interval based odometric sensor integration, the reader might refer to [31]. Once we have constituted intervals from sensor data, the vehicle pose $[\mathscr{X}_k]$ can be estimated by applying FBP algorithm (Section 2.6) over constraints deduced from the motion model.

3.3. Landmark Parameterization. A landmark (or feature point) is a 3D point in the global world. Probabilistic methods generally represent landmark with an inverse depth parameterization [32]. They model the uncertainty of each parameter by Gaussian distributions. However, it turns out to be not so efficient to represent the linear depth uncertainty of monocular vision. Interval analysis provides an easy and efficient way to parameterize landmarks. Each landmark $\mathscr{L}_i$ is defined as a six-dimensional state vector: $([x_o],[y_o],[z_o],[\alpha_i],[\varphi_i],[d_i])^T$, which models the estimated landmark position at

$$[\mathscr{L}_i] = ([x_o],[y_o],[z_o])^T + [d_i]\cdot[\mathbf{m}]([\alpha_i],[\varphi_i]) \quad (21)$$

where coordinates $[x_o]$, $[y_o]$, and $[z_o]$ represent the optical center of the camera when the landmark was seen for the first time, $[\alpha_i]$ and $[\varphi_i]$ represent azimuth and elevation angle for the ray which traces the landmark, and $[d_i]$ is the depth of the landmark. $[\mathbf{m}]([\alpha_i],[\varphi_i])$ is a unitary vector pointing from the camera to the landmark $[\mathscr{L}_i]$:

$$[\mathbf{m}]([\alpha_i],[\varphi_i]) = \begin{pmatrix} \cos_{[]}([\alpha_i])\cos_{[]}([\varphi_i]) \\ -\sin_{[]}([\alpha_i])\cos_{[]}([\varphi_i]) \\ \sin_{[]}([\varphi_i]) \end{pmatrix} \quad (22)$$

Since all the parameters are represented by intervals, and $[d_i]$ can be initialized as $[0,+\infty]$, the landmark's uncertainty is modeled as an infinite cone which combines the vehicle pose uncertainty and the observation uncertainty. It is a realistic representation for the monocular vision uncertainty. The major advantage is that the initialization of $[d_i]$ is undelayed, guaranteed, and efficient for landmarks over a wide range as it always includes all possible value without any prior information.

3.4. Observation Model. The observation of a landmark is the coordinates of the pixel where the landmark is projected on the image. In order to detect and extract landmark information from an image, we used Speed Up Robust Features (SURF) [33]. The work [34] recommended it after a comparison between several algorithms, showing that SURF provide results with a very low error rate. The projection of the ith landmark to the image frame can be formulated as follows:

From (21), the landmark's position in the world coordinate $\mathbf{O}_w$ can be expressed:

$$[\mathscr{L}_i^w] = ([x_o],[y_o],[z_o])^T + [d_i]\cdot[\mathbf{m}]([\alpha_i],[\varphi_i]) \quad (23)$$

In the camera coordinate $\mathbf{O}_c$, the landmark $[\mathscr{L}_i^w]$ will be seen in the position $[\mathscr{L}_i^c]$:

$$[\mathscr{L}_i^c] = [R^{wc}]([\mathscr{L}_i^w] - [R^{rw}][\mathscr{X}_k]) \quad (24)$$

where $[\mathscr{X}_k]$ is the current vehicle pose. R^{wc} and R^{rw} are, respectively, the transformational matrix between the world and the camera coordinate and the robot and world coordinate, respectively.

The camera does not observe $[\mathscr{L}_i^c]$ directly but its projection in the image frame $(u,v)^T$. According to the pin-hole camera model [35], the observation $\mathscr{Z}_k^i$ of $[\mathscr{L}_i^c]$ on the image frame can be predicted by

$$[\mathscr{Z}_k^i] = \mathscr{H}_{[]}([\mathscr{X}_k],[\mathscr{L}_i^c]) \quad (25)$$

$$\begin{pmatrix} [u_{pre}^i] \\ [v_{pre}^i] \end{pmatrix} = \begin{pmatrix} [c_u] + [f]\,[k_u]\dfrac{[\mathscr{L}_i^c(x)]}{[\mathscr{L}_i^c(z)]} \\ [c_v] + [f]\,[k_v]\dfrac{[\mathscr{L}_i^c(y)]}{[\mathscr{L}_i^c(z)]} \end{pmatrix} \quad (26)$$

where f, k_u, k_v, c_u, c_v are the camera intrinsic parameters. Equation (26) is also used as the projection function.

3.5. Data Association. Data association focuses on finding out the correspondence between the map and observations. Different from probabilistic method, which deals with uncertainty ellipsoid in projection procedure, interval method proceeds with bounding box. The matching process is carried out as follows:

(i) Project the landmarks to the current image plane; discard those which lay out of the image bounds. Each landmark defines a rectangle searching area.

(ii) In the searching area, search for the candidate feature that is matched to the landmark. We adopt Dan's method [36] which performs a two-stage matching process (considering both the Euclidean distance and dominant orientation information of feature descriptors) to provide a robust and accuracy matching result. For each pair of matched features, a Zero Normalized Cross Correlation (ZNCC) [37] score is computed in the mapping stage to help to evaluate the robustness of the landmarks.

(iii) Filtering step is necessary to keep a desirable amount of matched points. The landmark uncertainty and ZNCC score can be used for the selection process. We make the matched results distributed uniformly over the image, providing enough parallax for the localization.

3.6. ICSP Formulation. The data association process generates a set of 3D to 2D correspondences, which we can use to build the ICSP constraints. Let us consider the ith landmark $[\mathscr{L}_i]$ that has been matched. The observation of $[\mathscr{L}_i]$ is $([u_{obs}^i],[v_{obs}^i])^T$. The predicted position on the image plane, denoted by $([u_{pre}^i],[v_{pre}^i])^T$, is computed via (23), (24), and (26). The observation can be used to correct the system state through the predicted observation:

$$\begin{aligned} [u_{pre}^i] &= [u_{obs}^i] \\ [v_{pre}^i] &= [v_{obs}^i] \end{aligned} \quad (27)$$

Equation (27) creates a link between the prediction and observation data of the ith landmark $[\mathscr{L}_i]$. This link (thanks to (23), (24), (26), and (27)) imposes constraints on $[\mathscr{X}_k]$ and $[\mathscr{L}_i]$, which can be used to update both the vehicle pose and landmark position.

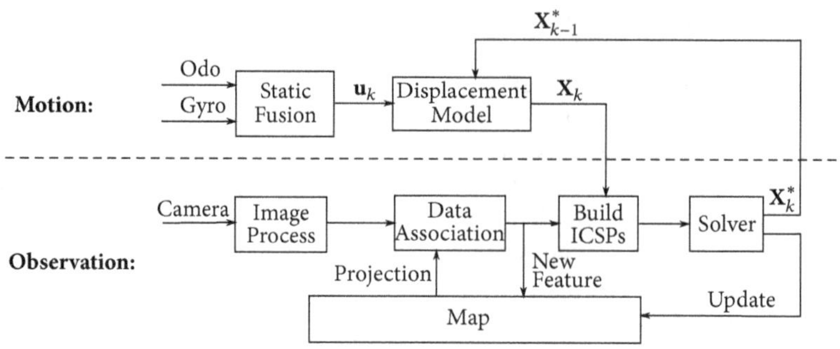

FIGURE 4: Mapping framework.

Based on the above derivation, an ICSP composed of 2 top level constraints and a set of 9 variables can be set up. For each matched landmark i, the $\text{ICSP}_k(V_k, D_k, C_k)$ is generated and defined as follows:

(i) A set of 9 variables:

$$V_k = \{x_k, y_k, \theta_k, x_o, y_o, z_o, \alpha_i, \varphi_i, d_i\} \tag{28}$$

(ii) A set of 9 domains:

$$D_k = \{[x_k], [y_k], [\theta_k], [x_o], [y_o], [z_o], [\alpha_i], [\varphi_i], [d_i]\} \tag{29}$$

(iii) A set of 2 top level constraints:

$$C_k = \left(C_k^1, C_k^2\right)^T \tag{30}$$

$$\begin{aligned} C_k^1 &\Longleftrightarrow \left[u_{pre}^i\right] = \left[u_{obs}^i\right] \\ C_k^2 &\Longleftrightarrow \left[v_{pre}^i\right] = \left[v_{obs}^i\right] \end{aligned} \tag{31}$$

If the data association process found N matched landmarks, then we get a multiple ICSP to solve:

$$\nu\text{ICSP}_k = \left(\text{ICSP}_k^1, \text{ICSP}_k^2, \cdots, \text{ICSP}_k^N\right) \tag{32}$$

Each ICSP contains constraints between the vehicle pose and a unique landmark position. The good point is that they are all linked together via the same variables x_k, y_k, and θ_k (the vehicle pose at current time k). This correlation makes ICP algorithms very powerful to jointly contract the domains of all variables.

3.7. System Overview. Figure 4 depicts the framework of the mapping stage. It is an interval SLAM process which is cast into two parts: motion and observation. In the motion part, the preliminary pose of the vehicle is predicted by fusing the measurements of the odometers and gyro with bounded error model; in the observation part, features are firstly extracted from the image, data association task aims at finding out the correspondences between the features and the map, new features will be initialized as new landmarks and added to the map, and old ones will be processed to generate new ICSPs; then these ICSPs are jointly solved by an ICP algorithm, which correct the vehicle pose and update the map.

(1) **while** Localization is required **do**
(2) Input: $[\mathcal{X}_{k-1}], [\mathcal{U}_k], [\mathcal{M}], I_k$
(3) Output: $[\mathcal{X}_k]$
(4) $[\mathcal{X}_k] \leftarrow f_{[]}([\mathcal{X}_{k-1}], [\mathcal{U}_k])$
(5) $S_k \leftarrow$ ExtractFeature(I_k)
(6) $[(u,v)_{obs}, \mathcal{L}]^N \leftarrow$ DataAssociate$([\mathcal{X}_k], [\mathcal{M}], S_k)$
(7) **for** $j \leftarrow 1$ to N **do**
(8) $\text{ICSP}_j \leftarrow$ Build$([\mathcal{X}_k], [(u,v)_{obs}, \mathcal{L}]^j)$
(9) **end for**
(10) $\nu\text{ICSP}_k = \bigcup_{j=1 \text{ to } N} \text{ICSP}_j$
(11) $[\mathcal{X}_k^*] \leftarrow$ Solver(νICSP_k)
(12) **end while**

ALGORITHM 1: Localization.

The localization stage is quite similar to the mapping stage. There are two main differences: high cost sensors are no longer used and no new feature is added to the map. Algorithm 1 gives the pseudocode of our localization method:

(i) Line (4): motion estimation. The current vehicle pose is calculated based on the motion function $f_{[]}$ and input vector $[\mathcal{U}_k]$.

(ii) Line (5): image processing. A set of SURF S_k are extracted from the new image I_k.

(iii) Line (6): data association. The correspondence between the features and the map is found.

(iv) Line (8): ICSP generating. New ICSP are generated based on the matched landmarks.

(v) Line (10)-(11): Solving. νICSP_k is composed by the ICSPs and solved by an ICP algorithm.

4. Simulation Results

A simulation experiment is set up to evaluate the performance of our proposed ICSP based mapping and localization method. The vehicle starts from the origin (see Figure 5) and moves straight forward in the positive direction of x-axis with constant linear and angular velocity ($v = 0.1m \cdot s^{-1}$, $\omega = 0$). 5 landmarks are positioned in the surrounding environment without any prior knowledge of their positions. At each time

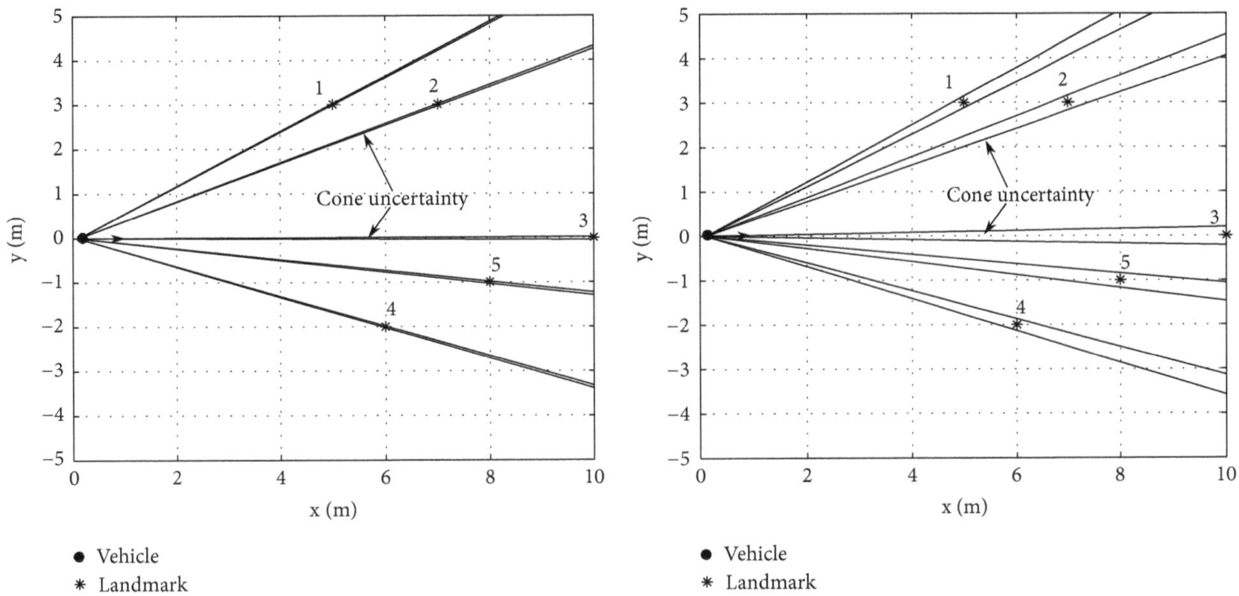

(a) Without vehicle pose uncertainty

(b) With vehicle orientation uncertainty

FIGURE 5: Interval based landmark initialization.

step, the vehicle detects the 5 landmarks with a simulated camera model $[f, k_u, k_v, c_u, c_v] = [40, 8, 8, 320, 240]$) and focuses on estimating their position. The experiment is first carried out without odometric sensor noise (the vehicle pose is perfectly known) to verify the feasibility of our method. This condition (no odometric noise) is then released in the experiment presented in Section 4.3.

4.1. Landmark Initialization. At time step $k = 1$, the 5 landmarks are first detected and the initialization procedure is executed. Figure 5(a) shows the top view of the initialization result of landmarks (see Section 3.3). Each landmark's uncertainty is initialized as an infinite cone because of the lacking information about the landmark depth. As there is no uncertainty on vehicle pose, the landmarks are well initialized and the opening angles which indicate the observation uncertainty are small (due to the bounded observation noise which is 1 pixel). If the vehicle has an uncertainty on its orientation (the heading angle is not perfectly defined: $[\theta_k] = [-\pi/180, \pi/180]$), then the cones of uncertainty are much larger than the previous ones (see Figure 5(b)). This is reasonable since the cone uncertainty is the fusion of observation uncertainty and vehicle pose uncertainty. Because our bounded error hypothesis and the use of interval analysis, the initialization result is guaranteed to contain the true landmark position.

4.2. Landmark Uncertainty Evolution. As the vehicle moves, the landmarks are repeatedly observed with different parallaxes. By building and solving the associated ICSP (see Section 3.6), each of the landmark's parameters can be estimated. Figure 7 shows, since initialization, the evolution of the poses uncertainties of landmarks 1, 2, and 5 in top view. Each row plots the estimation result at time steps 1, 5, 10, and

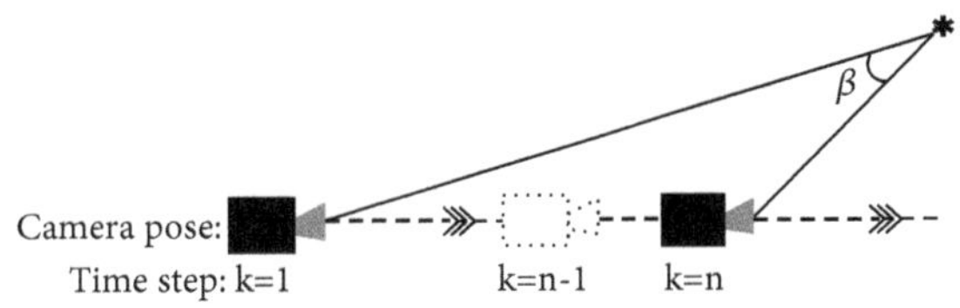

FIGURE 6: Parallax observations.

20. The parallax β (defined as the angle difference between the initial observation and current observation, see Figure 6) is displayed as the title of each subfigure. With the estimation proceeds, the landmark uncertainty estimate improves. As we can see, the bigger the parallax is, the better the contraction of the landmark uncertainty will be. When enough parallax is eventually available, the infinite interval uncertainty is significantly reduced. The uncertainty of landmark 3 keeps infinite because it locates on axis x and no parallax has been detected during the horizontal translation.

To visualize the evolution of a landmark state, it is necessary to convert the 6-dimensional representation to a global 3D point using (21): $(x_o, y_o, z_o, \alpha_i, \varphi_i, d_i)^T \rightarrow (x_i, y_i, z_i)^T$. The interval uncertainty (IU) can be evaluated by computing the box volume [31]: $U(i) = w([x_i]) \times w([y_i]) \times w([z_i])$. As the depth of each landmark is initialized as $[0, +\infty]$, the x, y, and z coordinates have infinite value, and the uncertainty turns out to be infinite at the first hand. After multiple observations (at different times) from different parallaxes, the landmark's state is updated. Figure 8 depicts the evolution of the uncertainty associated with the landmarks 1, 2, 4, and 5. The 4 landmarks encounter different observation parallaxes, so the contraction of their uncertainties is not simultaneous. However, they exhibit the same performance:

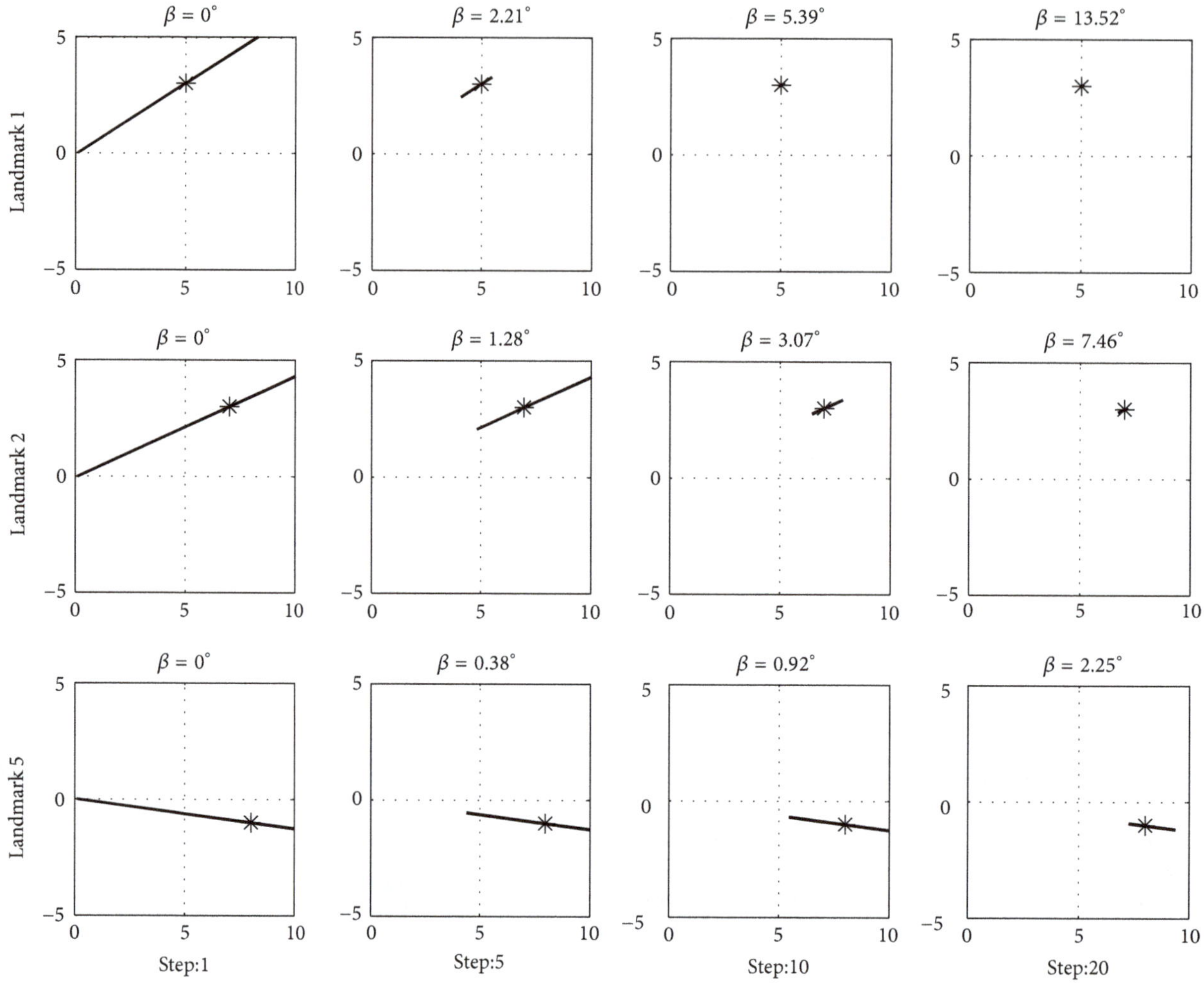

FIGURE 7: Interval uncertainty evolution.

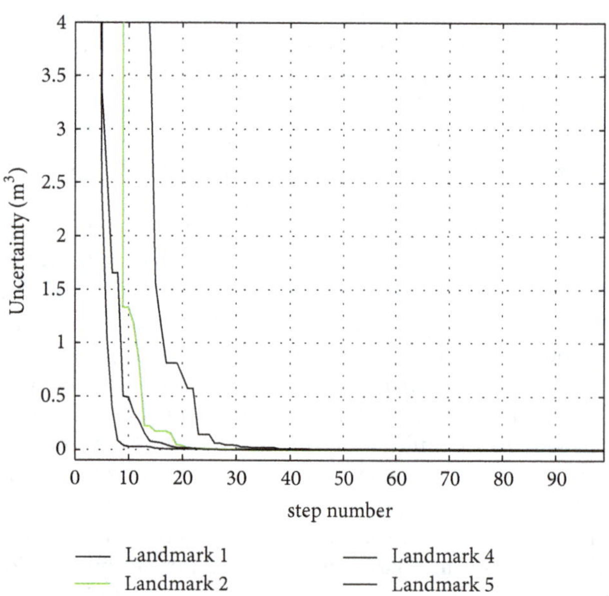

FIGURE 8: Landmark position uncertainty evolution.

the interval uncertainty of each landmark is well propagated and decreases exponentially with the vehicle navigating in the environment.

4.3. Estimation Accuracy. In real world, the vehicle pose is always accompanied with uncertainties due to the cumulative odometric noise. The vehicle pose uncertainty will affect the estimation accuracy of the landmarks (a smaller vehicle pose uncertainty means higher landmarks estimation accuracy). In our simulation, the error bound of the odometric measurement is supposed to be proportional to elementary displacement, such that $w([\mathscr{U}_k]) \leq \gamma \cdot \mathscr{U}_k + \Gamma$, $\Gamma = [0.0001m, 0.01°]^T$, with $\mathscr{U}_k$ being the real odometric measurement, γ denoting the ratio value (bigger γ value indicates a relative bigger uncertainty of odometric measurements), and Γ accounting for the steady system error. The average position uncertainty of the 4 landmarks has been computed at each time step. Figure 9 displays the results with different γ value ($\gamma = 0.01, 0.1$ and 0.2). As it is shown, the estimation accuracy decreases when γ value becomes bigger. This result supports the fact that the landmark estimation accuracy is strongly correlated with the vehicle pose uncertainty. Bigger

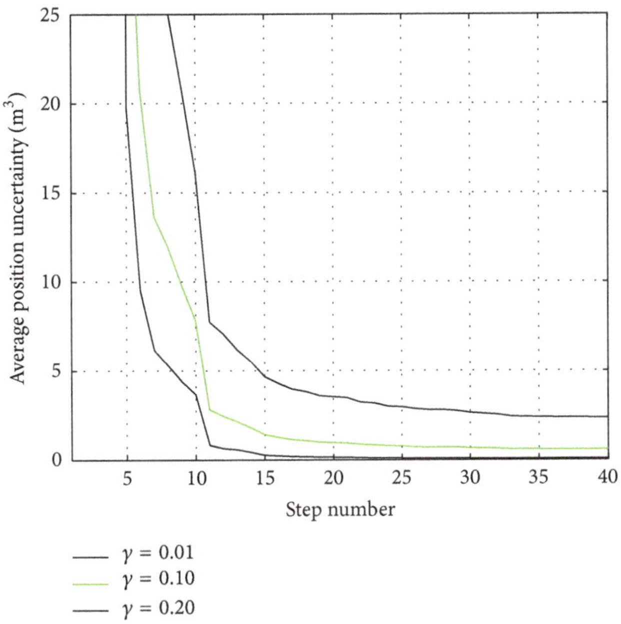

FIGURE 9: Average landmark uncertainty under different γ values.

odometric uncertainty will result in a worse estimation of vehicle pose; as a result, the landmark estimation accuracy decreases.

5. Real Data-Set Experiment

5.1. Experiment Set-Up. To evaluate the performance of our method in realistic circumstance, we used the experimental data set from Institute Pascal [38]. It contains multisensory timestamped data which can be used in a large variety of applications. It includes data from low-cost sensors, ground truth, and multiple sequences as well as data for calibration. Data were collected on the VIPALAB platform (see Figure 10(a)), driving on the PAVIN track (see Figure 11(a)), an experimental site located on the French campus of Blaise Pascal University. The environment is composed of scaled street, road markings, functional traffic lights, painted walls, buildings vegetation, and so on (see Figure 11(b)). The platform was equipped with multiple sensors (see Figure 10(b)). In our experiment, we need to use 4 sets of timestamped sensor data:

(i) Odometers: each rear wheel is equipped with an odometer in the form of a ring gear providing 64 "top" per wheel revolution.

(ii) Gyro (Melexis MLX90609-N2): it measures the rotational speed in an inertial reference system. Vibrating silicon structures, that use the Coriolis force, output the yaw rate from which we can deduce the elementary rotation.

(iii) Camera (FOculus FO432B camera + Pentax C418DX lens): it provides information about the perceived environment. We use the camera which located in the left-front of the vehicle and looking ahead.

TABLE 1: Average bounding boxes size.

Parameter	x	y	z
Mean width (m)	0.207	0.221	0.013

(iv) RTK-GPS: the data acquired with an embedded ProFlex500 RTK-GPS receiver from NavtechGPS coupled with a Sagitta Magellan GPS base station are considered as the ground truth. This system provides an accurate absolute localization measurement to within 2 cm (±1 cm).

5.2. Mapping Stage. A mapping stage is first conducted by our method in order to build a map of the environment. The vehicle follows 6 simple consecutive loops around the PAVIN environment. Figure 12(a) shows the reference trajectory that the vehicle passed by and Figure 12(b) details the 6 loops in order. As we have discussed in Section 4.3, the estimation of landmark is strongly correlated with the vehicle pose uncertainty. So it is necessary to maintain the pose uncertainty in a reasonable size during the mapping stage in order to pursue a precise map.

Figure 13 shows the vehicle position uncertainty (calculated by $(w([x])/2) \cdot (w([y])/2)$) with and without RTK-GPS correction during the mapping process. As we can see, without correction, the position uncertainty is cumulative and increases very quickly until loop closure is realized. This is because when navigating, the odometric noise is cumulative since there is no absolute measurement. Interval analysis is a pessimistic method maintaining all possible solutions, the cumulative errors will result in interval expansion, and the pose uncertainty thus becomes larger and larger. This is a hinder to pursue a good map. To overcome this problem, our method makes use of the RTK-GPS measurement, and the GPS error bounds are characterized as the maximum imprecision ($\leq$ $2cm$) of the GPS receiver. The cumulative errors can be eliminated and a precise estimation of the vehicle position is obtained (see Figure 13(b)). The result is assumed to be consistent since only ICP technique was used.

At the end of the loop, 3450 images are processed and a map of 1140 landmarks is created. Figure 14 shows part of the map projected on $x - y$ plane. The black bounding boxes denote the estimated landmark's position. The average size of these bounding boxes is shown in Table 1, which can be regarded as a factor to evaluate the quality of the map. The z coordinate gets much higher precision than x and y; this is because the estimation of z is only correlated with the observation elevation angle which is invariant w.r.t the vehicle's heading angle, while x and y are affected by the heading uncertainty. A good way to improve the map is to get a better estimation of the vehicle's orientation. Note that the map building process could be done offline such that computation intensive ICP algorithms and manually guidance could be used to obtain a better map.

5.3. Localization Result. The localization stage benefits from the map data which offer position constraints. With the built map, the vehicle seeks consistent localization using the

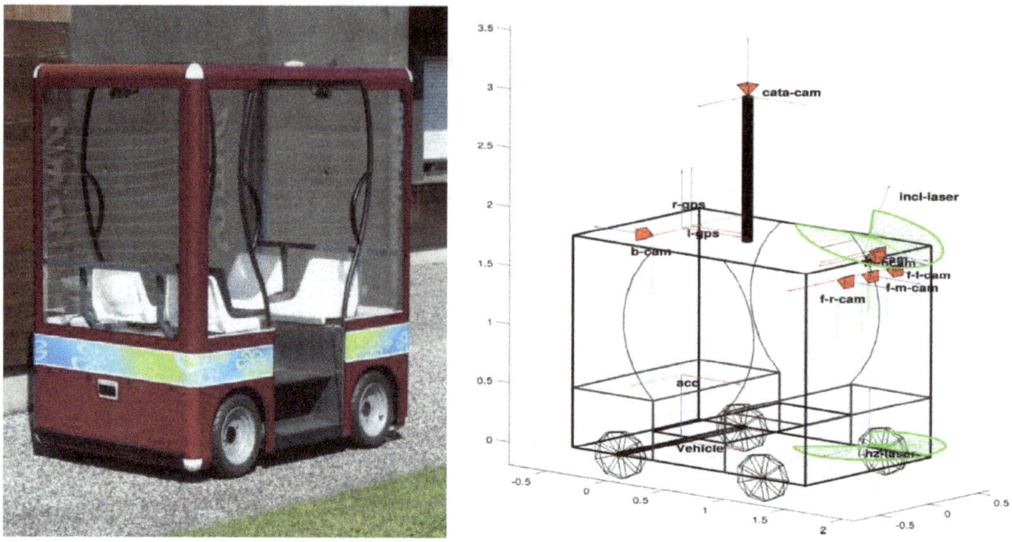

(a) VIPALAB platform (b) Vehicle structure

FIGURE 10: VIPALAB Experimental platform.

(a) The PAVIN track (b) Few images of the environment

FIGURE 11: Institute Pascal Experimental Environment.

current image captured by the monocular camera. Three sequences of data from the whole data set are used to test our localization method. The reference path (RTK-GPS) of the vehicle when these data are collected is illustrated in Figures 15, 18(a), and 18(b).

Our proposed localization method is firstly tested on Sequence 1, and the vehicle follows a nearly 200m trajectory; see Figure 15. The blue path is the reference trajectory from RTK-GPS, the red path is the dead reckoning result, and the black squares are the localization boxes output by our proposed interval method. The dead reckoning result is obtained via a geometric evolution model using the odometer and steering angle data. It can be seen from the figure that the localization boxes followed the reference trajectory in a consistent way (intersect with the blue line). Table 2 (DR: dead reckoning; IM: interval method) presents the localization result of both methods at $t = 200$ and $t = 400$ and at the end of the track. By fusing the map, our interval method gives a consistent estimation of the vehicle pose: the localization boxes well enclose the ground truth.

To compare the vehicle position estimated by our proposed method and the dead reckoning, the root of sum square error (RSSE) is computed at each time step. It is expressed by the following formula:

$$\mathrm{RSSE}_k = \sqrt{E_k^2(x) + E_k^2(y)} \tag{33}$$

where $E_k(x)$ and $E_k(y)$ are the estimation error of mid($[x]$) position and mid($[y]$) position at time step k. RTK-GPS data is regarded as the ground truth.

Figure 16 shows the RSSE results of both methods. As it could be expected, the cumulative odometric error of dead reckoning is significant and caused an increasing discrepancy (at the end, the discrepancy reaches about 3 meters). Our method takes advantage of the map data and corrects the odometric error, providing a very small discrepancy.

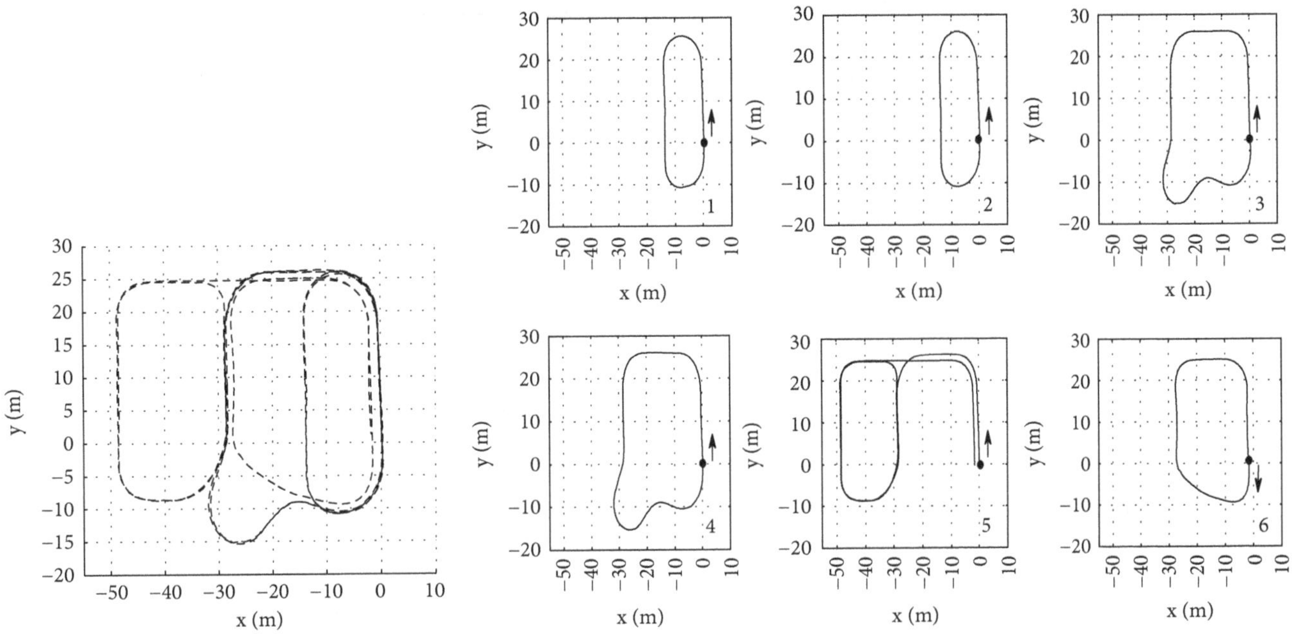

(a) Full mapping path

(b) Consecutive single loop

FIGURE 12: The reference path (RTK-GPS) of mapping sequence.

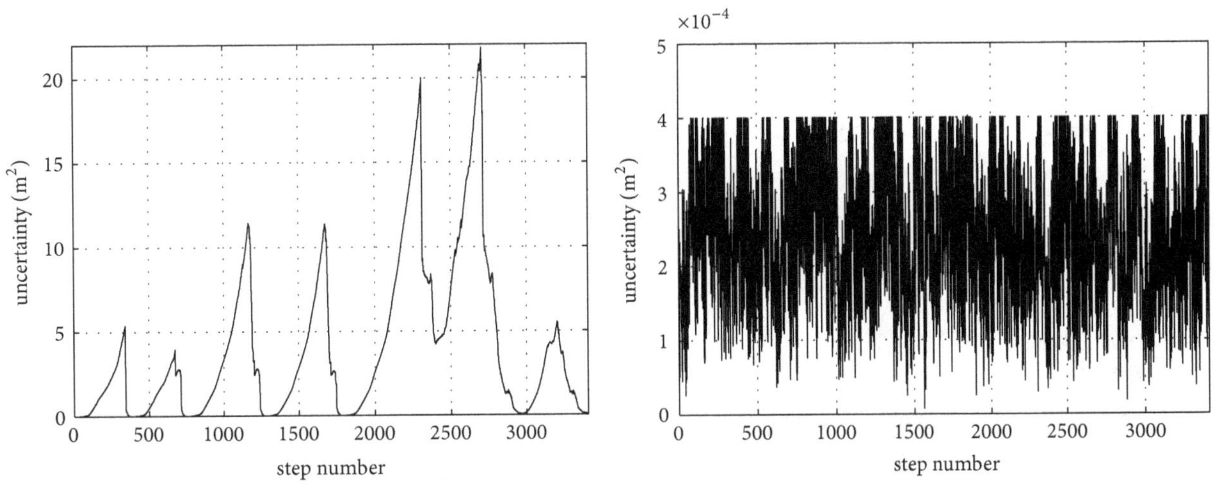

(a) Without RTK-GPS correction

(b) With RTK-GPS correction

FIGURE 13: Vehicle position uncertainty.

In order to verify the consistency of the localization results, interval error is defined by the upper and lower bounds of the estimated state minus the corresponding reference (real value), mathematically,

$$[E_k(\alpha)] = \left[\underline{\alpha_k} - \alpha_k^*, \overline{\alpha_k} - \alpha_k^*\right] \tag{34}$$

where $[\underline{\alpha_k}, \overline{\alpha_k}]$ is the estimated state of variable α at time step k and α_k^* is the corresponding reference. The estimated state is said to be consistent if and only if $0 \in [E_k(\alpha)]$ is a true statement. Moreover, an estimation method is consistent and precise if $\forall k \in \{1, 2, \cdots, \infty\}, 0 \in [E_k(\alpha)]$ is always satisfied and the interval error width $w([E_k(\alpha)])$ is thin.

We compute the real-time interval error of our localization result. At each time instant k when the new captured image had been processed, $E_k(x)$ and $E_k(y)$ are calculated, respectively. Figure 17 depicts the interval corridors consisting of the upper and lower bounds of E_k. The zero line is well included by those corridors, proving that the localization results are consistent all along the track. This is a significant result in vehicle localization field where safety is a crucial issue.

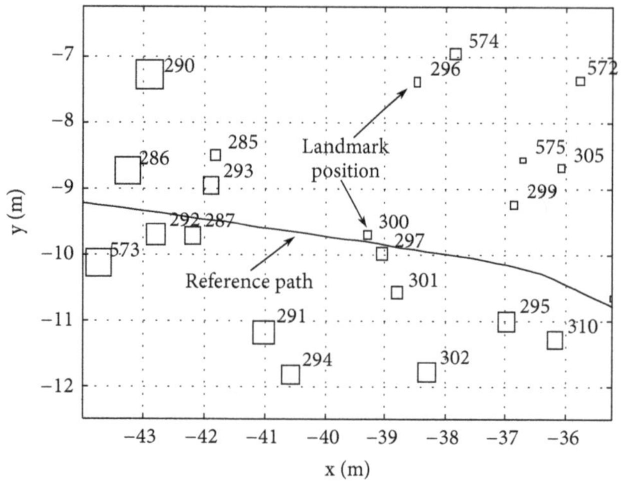

FIGURE 14: Map building result.

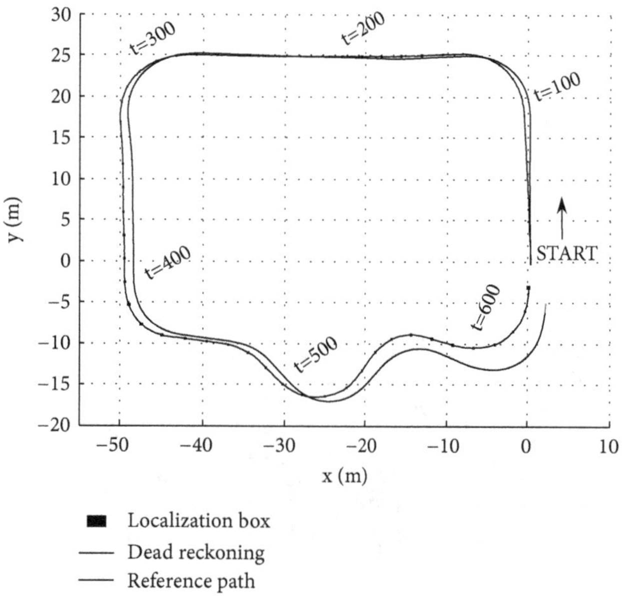

FIGURE 15: Localization trajectory of Sequence 1.

TABLE 2: Localization result.

Time	Position	RTK	DR	IM
t=200	x	-22.32	-22.72	[-22.35, -22.26]
	y	25.17	25.23	[25.14, 25.22]
t=400	x	-49.23	-48.13	[-49.29, -49.18]
	y	-2.51	-2.31	[-2.63, -2.45]
End	x	0.14	1.89	[0.10, 0.18]
	y	-3.26	-5.23	[-3.32, -3.21]

Similar results are obtained when performing the localization process on Sequences 2 and 3 with the same map; the output localization boxes by our method and the corresponding reference trajectory are displayed in Figure 18. It shows the robustness of our method providing consistent

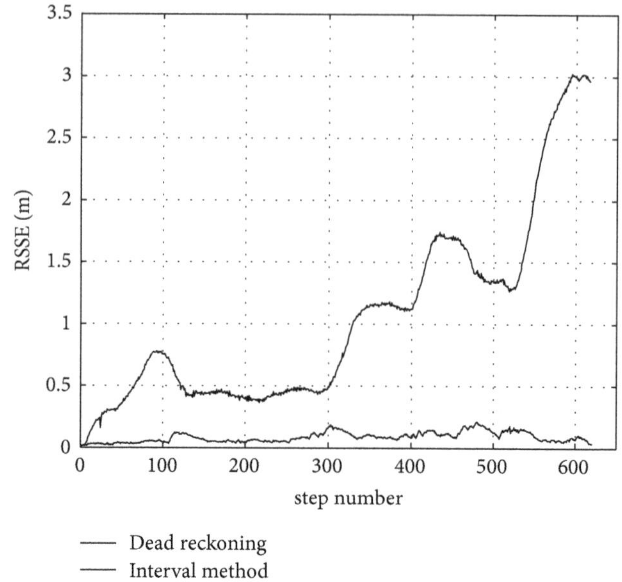

FIGURE 16: The error of localization results by DR and IA.

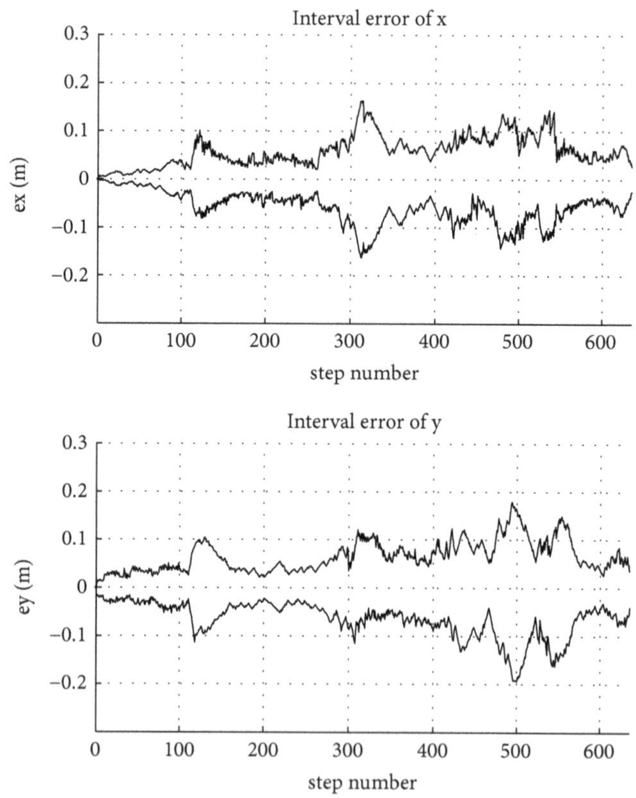

FIGURE 17: Interval error of real-time localization result.

localization result. The width of the interval error corridors (see Figure 17) with respect to the vehicle's x position and y position is displayed in Figure 19, reporting a maximum value of 0.6 m. The average interval error width and mean RSSE are computed to evaluate the localization accuracy. Table 3 details the results of the 3 experiments. Computing an average value for the 3 experiments gives 12.7 cm accuracy for x and

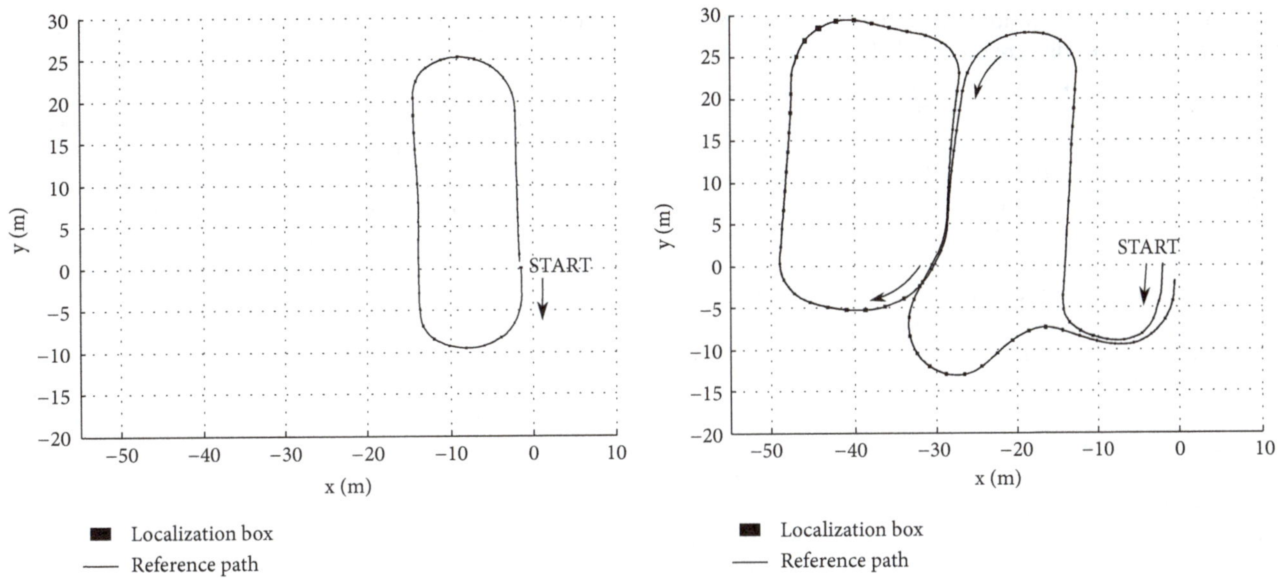

(a) Sequence 2 (b) Sequence 3

FIGURE 18: Localization boxes compare with reference path.

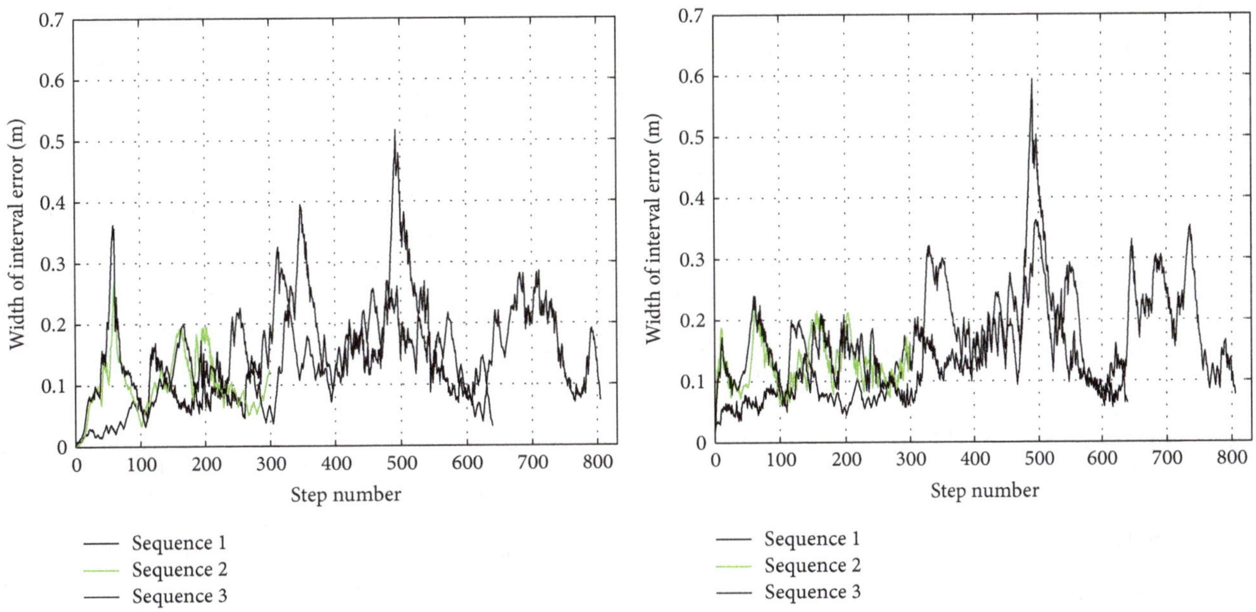

(a) Width of $E_k(x)$ (b) Width of $E_k(y)$

FIGURE 19: Interval error width of localization result.

14.4 cm for y. The average RSSE value is about 9.66 cm (less than 10 cm). Note that these localization results are consistent, and the localization boxes are guaranteed to enclose the real positions of the vehicle.

TABLE 3: Mean localization interval error and RSSE.

Experiment	x (cm)	y (cm)	RSSE (cm)
Sequence 1	11.7	13.4	8.90
Sequence 2	10.3	12.7	8.29
Sequence 3	16.1	17.2	11.8

5.4. Discussion. In this paper, we present a complete methodology for localizing a vehicle within a prebuilt map. We bring forward a two-stage framework (mapping and localization) to solve the problem. Closer approaches have already been proposed by different researchers [39–41]. They all involve a visual learning step to reconstruct a map of the environment, and they use this map for localization and navigation task.

Eric Royer et al. [39] uses bundle adjustment in mapping step and the localization results are obtained via Newton iteration method. Hyon Lim et al. [41] uses a structure from motion (SFM) algorithm to reconstruct the 3D environment; then in the localization step, two discrete Kalman filters are employed to estimate the camera trajectory. Courbon's method [40] does not involve the 3D reconstruction procedure during the map building step; instead, it records some key views and the performed path as references and uses them for future navigation missions. Our method uses totally different theory and techniques from these approaches, we cast the two-stage problem into an ICSP, and deterministic techniques (ICP) are used in order to find the solution.

The localization accuracy of these works is close to ours. The paper [39] has an average localization error of 15 cm over 6 localization experiments on 3 outdoor sequences. The reported mean indoor localization errors of [41] are ranged between 5.6 cm and 10.8 cm among 6 experiments. The outdoor mean position error is less than 15 cm. The paper [40] presents an average lateral error of 25 cm on an urban vehicle navigating along a 750 m trajectory. Our method has been tested on 3 outdoor sequences; the position error (RSSE) ranges between 8.29 cm and 11.8cm. The strength of our method is that the localization boxes are guaranteed to contain the real position of the vehicle. Such a guarantee is due to the modeling of the noises as intervals and the computing of constraints using interval functions.

The formulas of our algorithm work whatever the camera manufacturer is. As for different cameras, the observation model always holds; only the camera intrinsic parameters $(f, k_u k_v, c_u, c_v)$ change. Furthermore, the feature points extracted from images of various cameras would be the same since the SURF algorithm is scale invariant, rotation invariant, and affine transformation invariant [33]. This property makes it possible to find correct correspondence between two images from different cameras. This feasibility has been shown by [39] in two sequences of images taken by a pair of cameras: one sequence was used in the map learning step, and the other one was used for vehicle localization; the result turns out to be robust.

6. Conclusion

In order to overcome the inconsistency problem of probabilistic methods, this paper proposed a consistent vehicle localization method based on interval analysis. We propose a 2-stage process: mapping and localization. The map is built by a leader vehicle equipped with a sensor or a set of sensors providing an accurate positioning. The built map can be beneficial for a fleet of vehicles intending to achieve a consistent localization in the environment. By fusing low-cost dead reckoning sensors and map data, the proposed localization method is able to locate the vehicle consistently in the environment using the current image taken by the camera. The cumulative odometric error is eliminated and the localization consistency is maintained. Our method is validated with an outdoor car like vehicle equipped with odometers, gyro, and monocular camera. The experiments highly illustrate the consistency of the localization results provided by our method. Future work will deal with a formal proof of the obtained consistency.

Our methodology can be used to solve various problems, even problems with higher DOF. Indeed, the used ICP algorithms express a polynomial time with respect to the number of variables in the ICSP [42], so our ICP based method is capable of scaling up to higher DOF cases without high CPU load. In order to adapt to other applications, we have only to define a new motion model.

Conflicts of Interest

The authors declare that they have no conflicts of interest.

References

[1] M. Adams, S. Zhang, and L. Xie, "Particle filter based outdoor robot localization using natural features extracted from laser scanners," in *Proceedings of the Proceedings- 2004 IEEE International Conference on Robotics and Automation*, pp. 1493–1498, May 2004.

[2] A. Gning and P. Bonnifait, "Guaranteed dynamic localization using constraints propagation techniques on real intervals," in *Proceedings of the Proceedings- 2004 IEEE International Conference on Robotics and Automation*, pp. 1499–1504, May 2004.

[3] D. Gruyer, R. Belaroussi, and M. Revilloud, "Map-aided localization with lateral perception," in *Proceedings of the 25th IEEE Intelligent Vehicles Symposium, IV 2014*, pp. 674–680, June 2014.

[4] A. Howard, M. Matark, and G. Sukhatme, "Localization for mobile robot teams using maximum likelihood estimation," in *Proceedings of the IROS 2002: IEEE/RSJ International Conference on Intelligent Robots and Systems*, vol. 1, pp. 434–439, Lausanne, Switzerland, 2002.

[5] D. Gruyer, A. Lambert, M. Perrollaz, and D. Gingras, "Experimental comparison of bayesian positioning methods based on multi-sensor data fusion," *International Journal of Vehicle Autonomous Systems*, vol. 12, no. 1, pp. 24–43, 2014.

[6] S. Thrun, W. Burgard, and D. Fox, "A Probabilistic Approach to Concurrent Mapping and Localization for Mobile Robots," *Autonomous Robots*, vol. 5, no. 3-4, pp. 253–271, 1998.

[7] W. Burgard, D. Fox, D. Hennig, and T. Schmidt, "Estimating the absolute position of a mobile robot using position probability grids," in *Proceedings of the 1996 13th National Conference on Artificial Intelligence*, pp. 896–901, August 1996.

[8] F. Dellaert, D. Fox, W. Burgard, and S. Thrun, "Robust Monte Carlo localization for mobile robots," *Artificial Intelligence*, vol. 128, no. 1-2, pp. 99–141, 2001.

[9] S. Huang and G. Dissanayake, "A critique of current developments in simultaneous localization and mapping," *International Journal of Advanced Robotic Systems*, vol. 13, no. 5, pp. 1–13, 2016.

[10] M. Di Marco, A. Garulli, A. Giannitrapani, and A. Vicino, "A set theoretic approach to dynamic robot localization and mapping," *Autonomous Robots*, vol. 16, no. 1, pp. 23–47, 2004.

[11] R. E. Moore, *Interval Analysis*, Prentice-Hall, Englewood Cliffs, NJ, USA, 1966.

[12] A. Gning and P. Bonnifait, "Constraints propagation techniques on real intervals for guaranteed fusion of redundant data. application to the localization of car-like vehicles," *Automatica*, vol. 42, no. 7, pp. 1167–1175, 2006.

[13] L. Jaulin, *Applied Interval Analysis: with Examples in Parameter And State Estimation, Robust Control And Robotics*, Springer, 2001.

[14] T. Rassi, N. Ramdani, and Y. Candau, "Set membership state and parameter estimation for systems described by nonlinear differential equations," *Automatica*, vol. 40, no. 10, pp. 1771–1777 (2005), 2004.

[15] M. Kieffer, L. Jaulin, É. Walter, and D. Meizel, "Robust autonomous robot localization using interval analysis," *Reliable Computing. An International Journal Devoted to Reliable Mathematical Computations Based on Finite Representations and Guaranteed Accuracy*, vol. 6, no. 3, pp. 337–362, 2000.

[16] L. Jaulin, M. Kieffer, E. Walter, and D. Meizel, "Guaranteed robust nonlinear estimation with application to robot localization," *IEEE Transactions on Systems, Man, and Cybernetics, Part C (Applications and Reviews)*, vol. 32, no. 4, pp. 374–381, 2002.

[17] M. Kieffer, E. Seignez, A. Lambert, E. Walter, and T. Maurin, "Guaranteed robust nonlinear state estimator with application to global vehicle tracking," in *Proceedings of the 44th IEEE Conference on Decision and Control, and the European Control Conference, CDC-ECC '05*, pp. 6424–6429, esp, December 2005.

[18] E. Seignez, M. Kieffer, A. Lambert, E. Walter, and T. Maurin, "Real-time bounded-error state estimation for vehicle tracking," *International Journal of Robotics Research*, vol. 28, no. 1, pp. 34–48, 2009.

[19] A. Lambert, D. Gruyer, B. Vincke, and E. Seignez, "Consistent outdoor vehicle localization by bounded-error state estimation," in *Proceedings of the 2009 IEEE/RSJ International Conference on Intelligent Robots and Systems, IROS 2009*, pp. 1211–1216, October 2009.

[20] I. K. Kueviakoe, A. Lambert, and P. Tarroux, "A real-time interval constraint propagation method for vehicle localization," in *Proceedings of the 2013 16th International IEEE Conference on Intelligent Transportation Systems: Intelligent Transportation Systems for All Modes, ITSC 2013*, pp. 1707–1712, nld, October 2013.

[21] V. Drevelle and P. Bonnifait, "Robust positioning using relaxed constraint-propagation," in *Proceedings of the 23rd IEEE/RSJ 2010 International Conference on Intelligent Robots and Systems, IROS 2010*, pp. 4843–4848, twn, October 2010.

[22] K. Lassoued, P. Bonnifait, and I. Fantoni, "Cooperative Localization with Reliable Confidence Domains Between Vehicles Sharing GNSS Pseudoranges Errors with No Base Station," *IEEE Intelligent Transportation Systems Magazine*, vol. 9, no. 1, pp. 22–34, 2017.

[23] B. Desrochers and L. Jaulin, "A minimal contractor for the polar equation: Application to robot localization," *Engineering Applications of Artificial Intelligence*, vol. 55, pp. 83–92, 2016.

[24] L. Jaulin, A. Stancu, and B. Desrochers, "Inner and outer approximations of probabilistic sets," in *Proceedings of the American Society of Civil Engineers (ASCE) Second International Conference on Vulnerability and Risk Analysis and Management and Sixth International Symposium on Uncertainty Modelling and Analysis*, pp. 13–16.

[25] L. Jaulin, "Range-only SLAM with occupancy maps: A set-membership approach," *IEEE Transactions on Robotics*, vol. 27, no. 5, pp. 1004–1010, 2011.

[26] P. T. I. K. Kueviakoe and A. Lambert, "Vehicle localization based on interval constraint propagation algorithms," in *Proceedings of the IEEE International Conference on Cyber Technology in Automation, Control and Intelligent Systems*, pp. 179–184, May 2013.

[27] R. E. Moore, R. B. Kearfott, and M. J. Cloud, *Introduction to Interval Analysis*, Society for Industrial Mathematics, 2009.

[28] F. Le Bars, E. Antonio, J. Cervantes, C. De La Cruz, and L. Jaulin, "Estimating the trajectory of low-cost autonomous robots using interval analysis: Application to the eurathlon competition," in *Marine Robotics and Applications*, pp. 51–68, Springer, 2018.

[29] E. Hyvönen, "Constraint reasoning based on interval arithmetic: the tolerance propagation approach," *Artificial Intelligence*, vol. 58, no. 1-3, pp. 71–112, 1992.

[30] A. Lambert and N. Le Fort-Piat, "Safe task planning integrating uncertainties and local maps federations," *International Journal of Robotics Research*, vol. 19, no. 6, pp. 597–611, 2000.

[31] B. Vincke, A. Lambert, D. Gruyer, A. Elouardi, and E. Seignez, "Static and dynamic fusion for outdoor vehicle localization," in *Proceedings of the 11th International Conference on Control, Automation, Robotics and Vision, ICARCV 2010*, pp. 437–442, December 2010.

[32] J. M. M. Montiel, J. Civera, and A. J. Davison, "Unified inverse depth parametrization for monocular SLAM," in *Proceedings of the 2nd International Conference on Robotics Science and Systems (RSS '06)*, pp. 81–88, August 2006.

[33] H. Bay, A. Ess, T. Tuytelaars, and L. Van Gool, "Speeded-up robust features (surf)," *Computer Vision and Image Understanding*, vol. 110, no. 3, pp. 346–359, 2008.

[34] A. Schmidt, M. Kraft, and A. Kasiński, "An Evaluation of Image Feature Detectors and Descriptors for Robot Navigation," in *Computer Vision and Graphics*, pp. 251–259, 2010.

[35] R. Hartley and A. Zisserman, *Multiple View Geometry in Computer Vision*, Cambridge University Press, 2nd edition, 2003.

[36] Z. Dan and H. Dong, "A robust object tracking algorithm based on surf," in *Proceedings of the IEEE International Conference on Wireless Communications & Signal Processing (WCSP)*, pp. 1–5, 2013.

[37] J. P. Lewis, "Fast normalized cross-correlation," *Vision Interface*, vol. 10, pp. 120–123, 1995.

[38] H. Korrapati, J. Courbon, S. Alizon, and F. Marmoiton, "The institut pascal data sets: un jeu de données en extérieur, multicapteurs et datées avec réalité terrain, données d'étalonnage et outils logiciels," in *Orasis, Congrès des jeunes chercheurs en vision par ordinateur*, 2013.

[39] E. Royer, M. Lhuillier, M. Dhome, and T. Chateau, "Localization in urban environments: Monocular vision compared to a differential GPS sensor," in *Proceedings of the 2005 IEEE Computer Society Conference on Computer Vision and Pattern Recognition, CVPR 2005*, vol. 2, pp. 114–121, June 2005.

[40] J. Courbon, Y. Mezouar, and P. Martinet, "Autonomous navigation of vehicles from a visual memory using a generic camera model," *IEEE Transactions on Intelligent Transportation Systems*, vol. 10, no. 3, pp. 392–402, 2009.

[41] H. Lim, S. N. Sinha, M. F. Cohen, M. Uyttendaele, and H. J. Kim, "Real-time monocular image-based 6-DoF localization," *International Journal of Robotics Research*, vol. 34, no. 4-5, pp. 476–492, 2015.

[42] G. Chabert and L. Jaulin, "Contractor programming," *Artificial Intelligence*, vol. 173, no. 11, pp. 1079–1100, 2009.

14

Speed Behavior of Passenger Car on Helical Ramps and Helical Bridges in Mountain Riverside City

Jin Xu[1], **Jinghou Fu**[2], **Xiaoming Liu**[2], **and Yiming Shao**[3]

[1]*Professor, Chongqing Key Laboratory of "Human – Vehicle -Road" Cooperation & Safety for Mountain Complex Environment, Chongqing Jiaotong University, No. 66, Xuefu Ave., Chongqing 400074, China*
[2]*Postgraduate Student, College of Traffic and Transportation, Chongqing Jiaotong University, No. 66, Xuefu Ave., Chongqing 400074, China*
[3]*Professor, College of Traffic and Transportation, Chongqing Jiaotong University, No. 66, Xuefu Ave., Chongqing 400074, China*

Correspondence should be addressed to Jin Xu; yhnl272727@163.com

Academic Editor: Md. Mazharul Haque

Helix alignment can allow for a rapid change in road elevation in size-constrained spaces, and it is becoming increasingly popular in interchange design throughout the world. However, driving patterns and vehicle operating characteristics have not been clearly defined on helical ramps. This work conducted field driving tests on four helical ramps located in Chongqing, China. The trajectory, speed, and acceleration of vehicles under normal driving conditions were collected. Thus speed characteristics and speed patterns on helical ramps, as well as their affecting factors, were obtained and analyzed. The findings in this study can provide basic data referencing for designers and engineers, to help them understand how a multilayered helical ramp works, thus to essentially improve the safety level of helix ramps. Moreover, our findings allow bridge designers to understand whether the actual vehicle operating conditions acted in accordance with their expected design requirements, i.e., whether the expected design requirements are achieved.

1. Introduction

Helix curves are effective for overcoming elevation differences in size-constrained spaces and for allowing the route to ascend rapidly [1]. A helical ramp is the most effective design for achieving traffic transition in areas with highly limited space and large differences in elevation. Therefore, in recent years, helix curves are being used more extensively in highway and municipal engineering projects, such as expressways, tunnels, interchanges, and urban roads. Moreover, as the number of cities and towns in mountainous areas increases and the land use in cities intensifies, the use of helical ramps and helical bridges will become more extensive. In addition, the increasing navigational clearance requirements for river-spanning bridges, along with increasing environmental protection awareness, also promote the use of helical ramps. In China Shanghai Nanpu Bridge is a famous site where helical ramps are used in the interchange at cross-river bridge, as shown in Figure 1(a). Chongqing is a large and flourishing mountain city and is famous for its numerous complex bridges and roadways, the helical bridge in Figure 1(b) is a ramp of Chongqing Sujiaba Interchange, and the helical bridge in Figure 1(c) is a segment of Hongyun Road, which is under construction and is to be opened to traffic in Apr. 2019. The helical ramp in Figure 1(d) is an emergency lane linked to Tianxingzhou Yangtze River Bridge, located in Wuhan, China. Similarly, in other countries out of China helical bridges also were adopted in mountain roads and interchanges; for instance, Nanadaru Bridge Loop in Figure 1(e) is a helical bridge constructed in Kawazu, Japan, and Tokyo Rainbow Bridge Loop in Figure 1(f) is another helical bridge in Japan.

Compared to conventional ramps with general configurations, helical ramps and bridges have the following distinctive features:

(i) Lane keeping and steering control may have particular maneuvering demands for a driver when driving on a long, curvature-fixed helix curve.

FIGURE 1: Helical bridges and helical ramps in and out of China. (a) Shanghai Nanpu Bridge, China; (b) Chongqing Sujiaba Interchange, China; (c) helical bridge of Hongyun Road in Chongqing, China; (d) helical emergency ramp of Tianxingzhou Yangtze River Bridge, China; (e) Nanadaru Bridge Loop, Japan; (f) Tokyo Rainbow Bridge Loop, Japan.

(ii) A one-directional steep slope (single gradient), meaning that an ascending ramp has a continuously increasing elevation and a descending ramp has a continuously decreasing elevation.

(iii) Large turning-angle loops with a curve angle exceeding 720° (currently, all helical ramps have two or more layers) and a continually changing vehicle direction while driving. In contrast, hairpin curves on a curvy mountainous road have a deflection angle of approximately 180°, and the loop ramps of a cloverleaf interchange, which are generally considered difficult to navigate, have angles of only approximately 270°.

Therefore, because of the distinctive alignment characteristics and driving conditions of helical ramps, there is an urgent need to investigate several related issues, such as whether the driving behavior and vehicle operating modes have any particularities, whether the driving behavior matchs the design expectations and driving assumptions, and, if any deviations exist, whether the extent of the deviation is serious. It is obvious that these issues are also the basis for improving the intrinsic safety of helical ramps at the design stage.

2. Literature Review

At present, relevant studies closely related to helical ramps mainly focus on the structural design of the ramp itself, such as the pier and beam layout [2], consolidation methods [3], structural force calculations [4, 5], and the treatment of foundations [6] and rarely focus on the driving behavior and vehicle operating performance within the helical ramps.

Existing studies are mainly devoted to investigating the vehicle operation/driving characteristics on conventional interchange ramps with general configurations. Among them, based on data obtained from a vehicle mounted GPS, Zhang et al. [7] established an operating speed model for interchange ramps as a function of curvature change rate and superelevation. Zhang et al. [8] used a radar speed gun to measure the speed of passenger cars in a deceleration lane and established an operating speed model at the theoretical gore and physical gore. Using a similar approach, Wang and Yang [9] measured the velocities of passenger cars at the theoretical gore, the mid-point of the deceleration lane, and the physical gore of an expressway interchange and obtained a model for calculating the lengths of the deceleration lanes. Guo et al. [10] used a driving simulator to simulate the driving process on the entrance and exit ramps of expressways and analyzed speed behavior in the deceleration and acceleration lanes based on the use of the accelerator/brake pedals. Wang et al. [11] used vehicle-mounted equipment to continuously measure driver eye movement, vehicle speed, and lane usage in the expressway interchange area and analyzed the driving behavior in a merging area, ramp area, and diverging area.

Polus et al. [12] used aerial cameras to record the continuous driving behavior of a vehicle in a freeway interchange merging area and obtained the merging point of vehicles exiting from the acceleration lane (entering the mainline traffic flow), as well as their continuous speed and acceleration. Using similar methods, they also investigated the driving behavior in the diverging area and determined the exact location and spatial distribution of the vehicles exiting the mainline traffic flow (entering the deceleration lane) [13], as well as their deceleration rates. These two works provided valuable data support for distinguishing the design parameters for parallel and tapered deceleration/acceleration lanes. Ahammed et al. [14] used a radar gun to track the speed of vehicles traveling in the merging area of an interchange and developed the model of operating speed, acceleration,

FIGURE 2: Helical ramps and helical bridges selected in this study.

and distance of the acceleration lane. Fitzpatrick et al. [15] analyzed the rationality of the basic assumptions for vehicle deceleration in the merging areas and the applicability of the deceleration rate in current design standards in U.S. Taking roundabouts as the research object, Zirkel et al. [16] discussed the effect of 85th percentile speed distribution and sight distance of the drivers on crash rates for low-volume single-lane roundabouts. Perco et al. [17] established an operating speed model for the merging area and diverging area of two types of interchanges by using the data obtained from radar guns. Yang [18] investigated the acceleration characteristics for metered on-ramps with various geometric features and also figured out speed profiles to guide the design of acceleration-lane length, using automatic video processors. Li and Sun [19] collected merging maneuvers of 370 drivers from the NGSIM (Next Generation Simulation Program) dataset and developed a novel data mining tool, called two-step cluster analysis, and the merging maneuver was divided into four clusters by this tool. Lvu et al. [20] analyzed the effect of driver's gender, occupation, and experience on behavior while driving on a freeway deceleration lane based on field operational test data. However, all the above studies concentrate on merging/diverging traffic and deceleration/acceleration areas and do not concentrate on the ramps.

In summary, at present, there are no reports available on driving behavior on helical ramps or helical bridges. Thus, researchers, designers, and managers are not able to understand the operating characteristics of vehicles traveling on helical ramps (bridges). Among driving behavior parameters, speed is a basic response of vehicle operation and driver behavior; at the same time, it is the core parameter that controls the interchange geometry and configuration in design. For this reason, a natural driving test involving passenger cars traveling on helical ramps was carried out in this study. Operating parameters were collected continuously using on-board equipment, and the speed amplitude and change patterns of a passenger car on a helical ramp (bridge) were determined. The findings of this study answered the following questions:

(i) How does vehicle speed change when a car is driving on a helical ramp? Will the continually decreasing or increasing elevation of the helical ramp cause a change in driving speed?

(ii) What kind of speed changes occur at the two ends of the helical ramp?

(iii) Do there exist diversified patterns in speed change along a helical ramp for individuals?

(iv) Are the speed amplitude consistent with the design expectations of the helical ramp, and do any serious deviations exist?

3. Experiment Design

In view of the structural form and geometric feature of helical ramps, vehicles with on-board apparatuses were used to conduct field driving test in this study. The operational parameters and the steering input of the driver were collected continuously while the car was traveling on the helical ramp and helical bridge. The following section provides a detailed description of the experimental design.

3.1. Experimental Objects. Four sites within the jurisdiction of Chongqing, China, were selected: the Rongqiao Helix Bridge, which is located in an urban area of Chongqing City, and the Wujiang Bridge helix ramp, Yangtze River Bridge helix ramp, and Jinkai ring bridge, which are located in the Fuling District, a city approximately 130 km from Chongqing. The Jinkai ring bridge loops around the Jinkai Ornamental Shopping Mall in two loops; that is, the building is encircled by the helix bridge. Table 1 lists the main technical parameters of the four experimental objects, and Figure 2 shows a panoramic view of the two helical ramps and two helical bridges, and their surrounding road network.

Table 1: Main technical parameters of the test objects.

Test object name	Location	V_d (km/h)	Traffic lane number	Travel direction	Structural form	G (%)	Horizontal Radius (m)	Number of layers	Test vehicles	Participant identification number
Yangtze River Bridge (YZ)	Fuling District, Chongqing	20	1	one-way	Single column pier	4.5%	39.525 m for ascending ramp, 50 m for descending ramp	2	Buick GL8 Firstland, Buick Buddha	D6–D15
Wujiang Bridge (WJ)	Fuling District, Chongqing	30	2	one-way	Single pier continuous cantilever beam	4.5%	51 m for ascending ramp, 38.25 m for descending ramp	2	Buick GL8 Firstland, Buick Buddha	D6–D13
Rongqiao Helix Bridge (RQ)	Urban area of Chongqing	40	2	one-way	Door-type multi-span continuous beam	6%	Median strip radius 55 m	2	Benz Vito, Toyota Hiace, Buick Buddha	D1–D5; D10–D11; D13–D15
Jinkai Ring Bridge (JK)	Fuling District, Chongqing	20	2	two-way	Door-type multi-span continuous beam	5–6%	30 m (centerline)	2	Buick Buddha	D8–D15

V_d: design speed; G: gradient of the ramp.

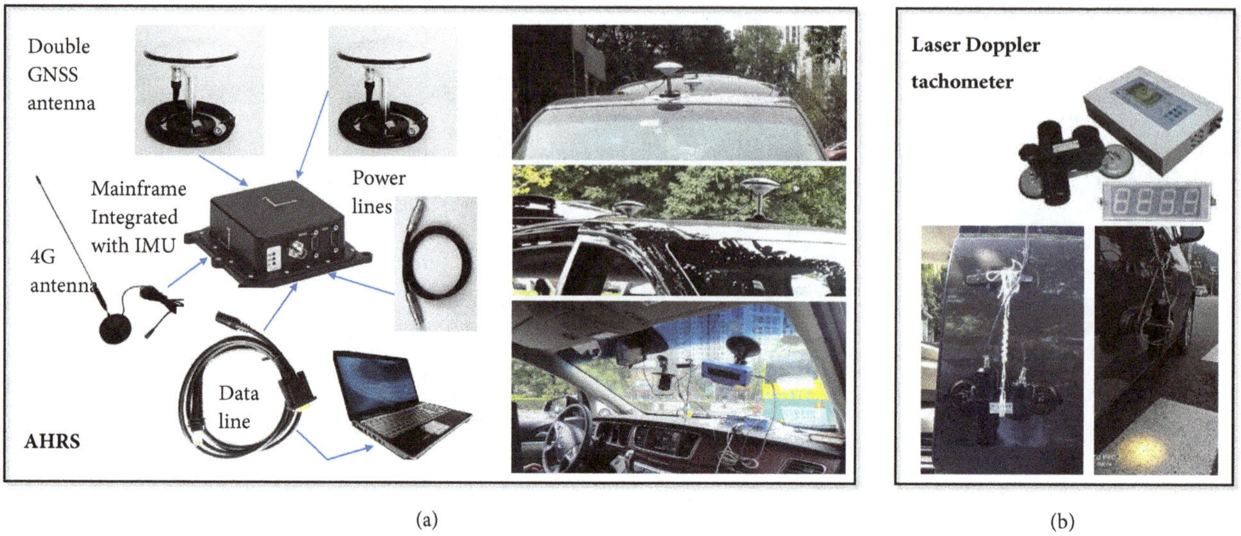

FIGURE 3: On-board equipment used in naturalistic data collection. (a) The composition of AHRS, fixed on the test car; (b) Laser Doppler Tachometer.

The pavement of the experimental objects was clean, smooth, and labelled with clear and complete markings. Field driving tests were carried out from 10:00 am to 17:30 pm to avoid peak traffic times in the morning and evening, and the drivers steered their cars in free flow traffic in all test rounds (occasionally, the subject car was obstructed by slow-moving forward vehicles, such as trucks). The driving tests were conducted from March 2016 to October 2017.

3.2. Testing Equipment and Vehicles. Radar gun and GPS are the two most commonly used methods for obtaining vehicle speed. The former measures the vehicle speed at a particular location. This method has limitations because the angle between direction of travel and the longitudinal axis of the gun cannot exceed 10–30°, the observation distance cannot exceed 400 m, and there are also minimum speed requirements. Therefore, radar guns are only suitable for measuring speed in the merging and diverging areas of interchanges and cannot be used for continuous speed acquisition on large-angle ramps. GPS is mainly used for recording the continuous speed of a single vehicle. However, as the vehicle travels on a helical ramp, the upper layer decks significantly obstruct the GPS signal, causing a disruption in the real-time positioning data and a gap in the recorded speed data.

In this study, we recorded the vehicle trajectory and speed using two types of apparatuses. One is an Attitude and Heading Reference System (AHRS), which consisted of an inertial measurement unit (IMU, an integration of triple-direction accelerometer and gyroscope) combined with a dual GPS antenna, as illustrated in Figure 3(a). When the GPS signal was lost, the algorithm integrated the acceleration signal measured by the IMU to obtain the vehicle position and performed an offline data compensation. The other is a Laser Doppler Tachometer, as shown in Figure 3(b). The speed results from the ARHS measured on the tangent section were used for calibrating the Laser Doppler Tachometer, and the speed data output from the latter was used in the analysis. In addition, cameras were fixed to the front and rear windows to record the environment in front of and behind the vehicle, and videos were used in postprocessing to analyze the reasons for fluctuations in the speed. We rented seven-seat commercial vehicles as the test vehicles, which included a Buick Buddha, Buick GL8 Firstland, Benz Vito, and Toyota Hiace, of which the first three are MPVs and the last is minivan. The distribution of these vehicles on the four sites is listed in Table 1.

3.3. Participants (Drivers). A total of 15 healthy drivers participated in our field driving test. The age range of the participants was 23–57 years, with an average age of 38. Seven drivers were 20–26 years old, three drivers were 35–39 years old, three drivers 40-49 years old, and two drivers were 55–59 years old. The length of their driving experience ranged from 1 to 31 years, averaging 10 years overall. Eight participants had 1–4 years of experience, two had 5–9 years, two had 15–19 years, two had 20–24 years, and one had 30–34 years. The subjects were recruited from an urban area of Chongqing and were familiar with the vehicle model used in this study. Before the test, the participants were instructed to drive the vehicles according to their daily driving style and habits; no requirements were imposed and no suggestions were given regarding the driving process. If a participant inquired about any requirements or speed control limits before or during the driving test, the given answer was "no requirements." This condition was intended to preserve the natural driving habits of the participants to the greatest extent possible.

3.4. Procedure of Driving Test. The route for each trial site was arranged before the field test was conducted, including the location where the personnel started to record and save the data from onboard instruments and the location where the driver turned his/her car after driving off the helical ramp. For example, the helical ramp on the Fuling Yangtze River Bridge is presented in Figure 4 to illustrate the path

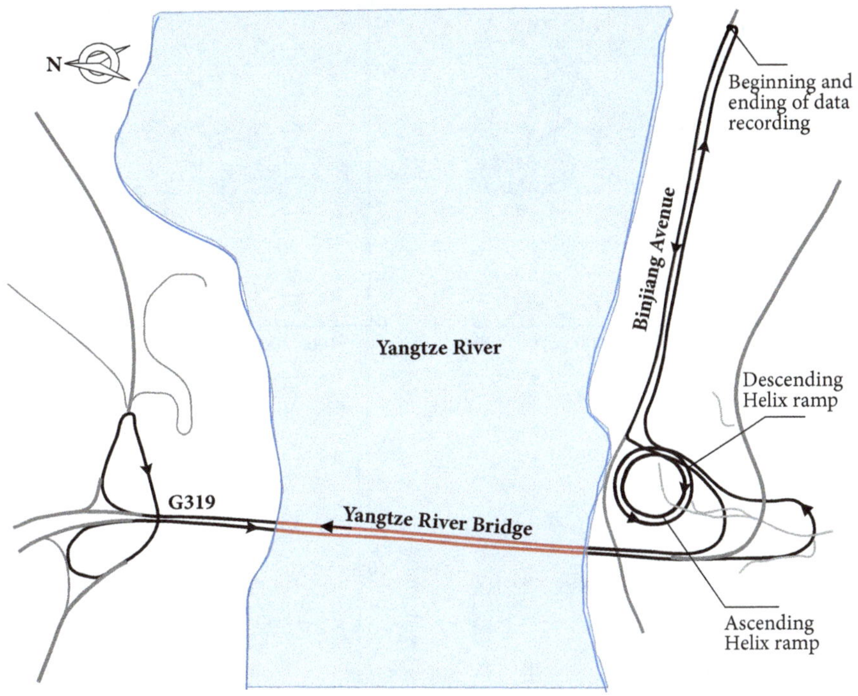

FIGURE 4: Route of field driving test for helical ramp of Fuling Yangtze River Bridge.

of the vehicle in one loop of the driving test. At each site, before we began recording data, the system time of each instrument was calibrated with the host computer. After each subject completed his/her required driving round, the on-board apparatus stopped recording.

For each test object, we designated one course on the ascending ramp and one course on the descending ramp as a complete round. We requested each driver to complete 4–6 rounds. The purpose was to ascertain whether the first lap experience was significantly different from the subsequent laps. At the same time, it also ensured an adequate number of data samples and prevented a situation in which valid data could not be collected due to unforeseen circumstances occurring during a single round.

For each participant, we first segmented the continuous speed profile measured from a driver based on videos recorded by the front camera, one speed profile for one round, and secondly cut the speed data within the area of helical ramp and separate it from the original data, to ensure each ramp travel (ascending or downward) corresponding to one speed curve. Finally, the fluctuations and changing trends of the speed along the ramp were analyzed.

4. Overall Speed Characteristics on Helical Ramps (Bridges)

Design speed is the key parameter that controls the appearance, floor space, and operating performance of an interchange. In this section, we examine the coordination between the design of the helical ramp and the actual driving requirements based on the measured speed and analyze the speed behavior at two ends of the ramps.

4.1. Ascending Helical Ramps

4.1.1. General Characteristics. Figure 5 shows the continuous speed profiles within the range of the helix for each of the four test objects. Figure 5(a) presents the speed profiles of ten participants from the ascending ramp of the Fuling Yangtze River Bridge. When entering the ascending ramp from Binjiang Avenue, the driver had to wait at a traffic signal before making a 90° left turn onto the ramp, as shown in Figure 4. Therefore, the vehicles started from a very low speed and accelerated onto the helical ramp; i.e., there was an "initial acceleration" stage, i.e., Stage I, for approximately 60–90 m. Stage II, the main part of speed profiles that cover the range of helix ramp, seems wholly to show a slight increase along the distance travelled.

Figure 5(b) shows the speed profiles of eight participants on the ascending helical ramp of the Fuling Wujiang Bridge. Before entering the helical ramp, participants had to drive on a 60-m tangent after waiting for the left turn signal light on the mainline, as shown in Figure 2(b). The speed change trend in stage II merits our attention. We can see that, after a relatively intense acceleration period, the participants basically maintained a stable speed.

Figures 5(c) and 5(d) are the speed profiles for the Chongqing Rongqiao Helix Bridge and the Jinkai Ring Bridge, respectively. These two ramps are part of the road sections and not connecting ramps between two different roads. Therefore, no route change is necessary. We can see from the two figures that even though the speed fluctuates within the helical bridge, the overall amplitude is stable for the most part. In Figure 5(c), the reduction at end of speed profiles (Stage III) is caused by the fact that the vehicle needed to stop and wait at a traffic signal before it could enter the

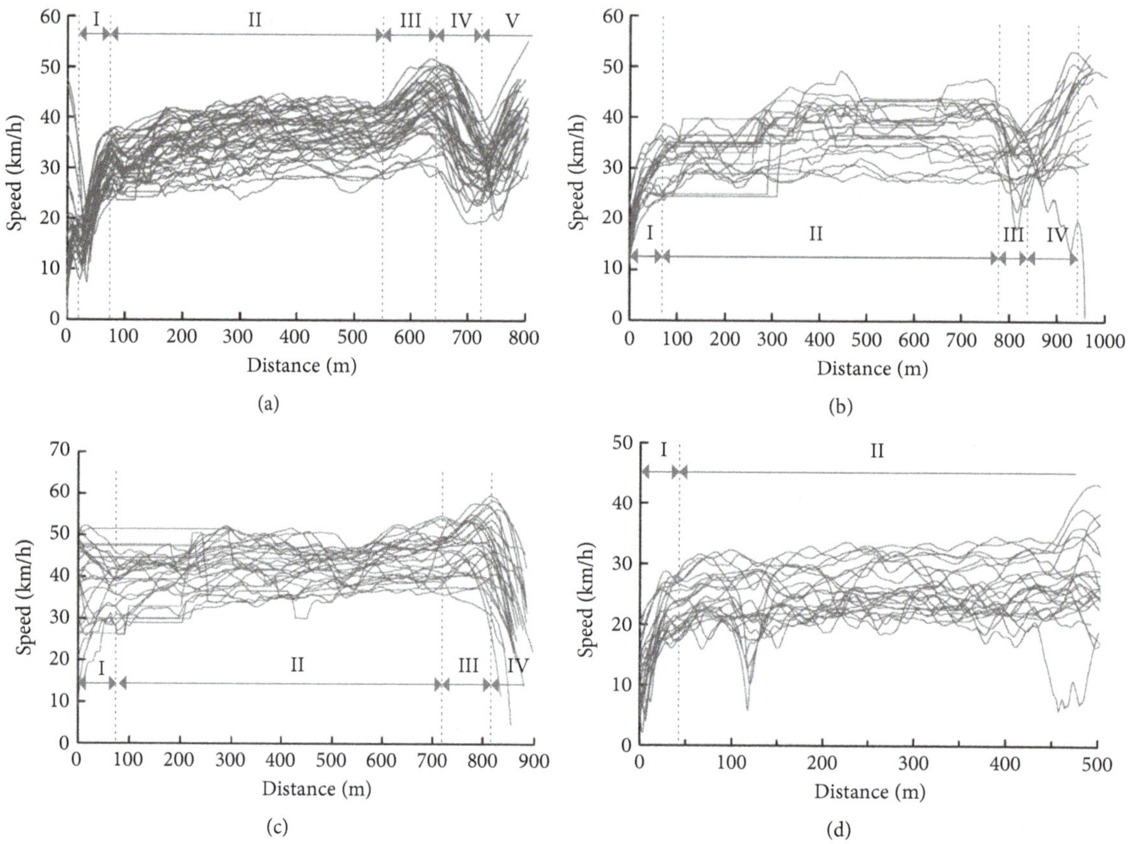

Figure 5: Driving speed on ascending ramp. (a) Ramp of Fuling Yangtze River Bridge, design speed (V_d) 20 km/h. (b) Ramp of Fuling Wujiang Bridge, V_d = 30 km/h. (c) Ramp of Chongqing Rongqiao Helix Bridge, V_d = 40 km/h. (d) Jinkai Ring Bridge, V_d = 20 km/h.

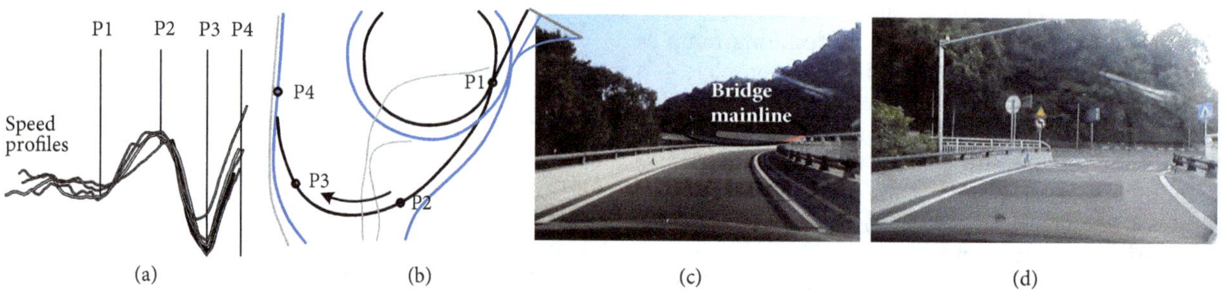

Figure 6: Feature points of speed change. (a) Driving speed; (b) feature points of speed change at merging area; (c)-(d) driver's line-of-sight environment corresponding to points P2 and P3.

intersected road, Jiaming Road. And in Figure 5(d) the speed profiles exhibit a slightly rising tendency along the distance travelled, which is similar to the speed change within Stage II in Figure 5(a).

After a comprehensive comparison of the measured speed of all participants, we can draw the following conclusions: First, the continuously increasing elevation did not cause any speed reduction for passenger cars; second, even though the design speed of the helical ramp of the Yangtze River Bridge was 20 km/h, the measured speed of most of participants ranged from 30 to 45 km/h. This indicates that the speed perception provided by the ramp environment to the driver is seriously inconsistent with the design expectations.

4.1.2. Feature Points. In Figure 5(a), while approaching the end of the ascending ramp, the speed curves for all participants show an initial increase (Stage III), followed by a rapid decrease (Stage IV), and finally another sharp increase (Stage V). Figures 6(a) and 6(b) show the feature points where the change in speed occurs, as well as their corresponding positions on the ramp. It is evident that the driver started to accelerate immediately after exiting the end point of the helix, P1. When arriving at P2, the mainline (bridge) and its traffic flow enter the sight line of the driver, as shown in Figure 6(c), causing the driver to decelerate subconsciously. When the speed experiences the lowest value at P3, shown in Figure 6(d), the driver begins to accelerate again. Location P4

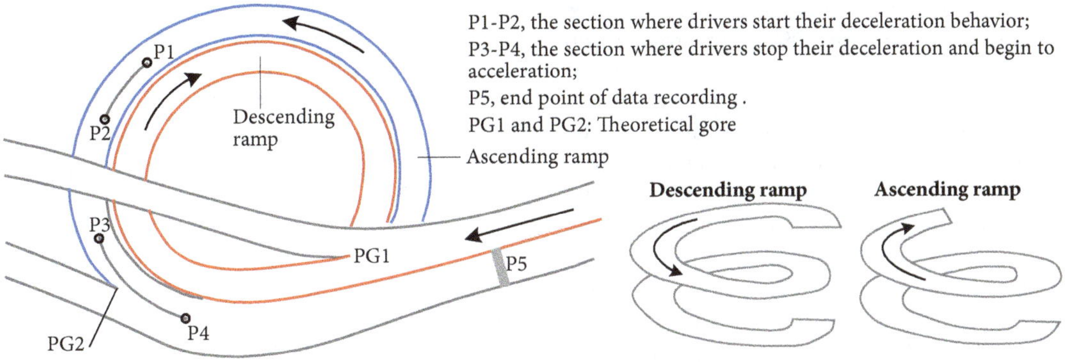

FIGURE 7: Feature points of speed change on the ascending ramp of Fuling Wujiang Bridge.

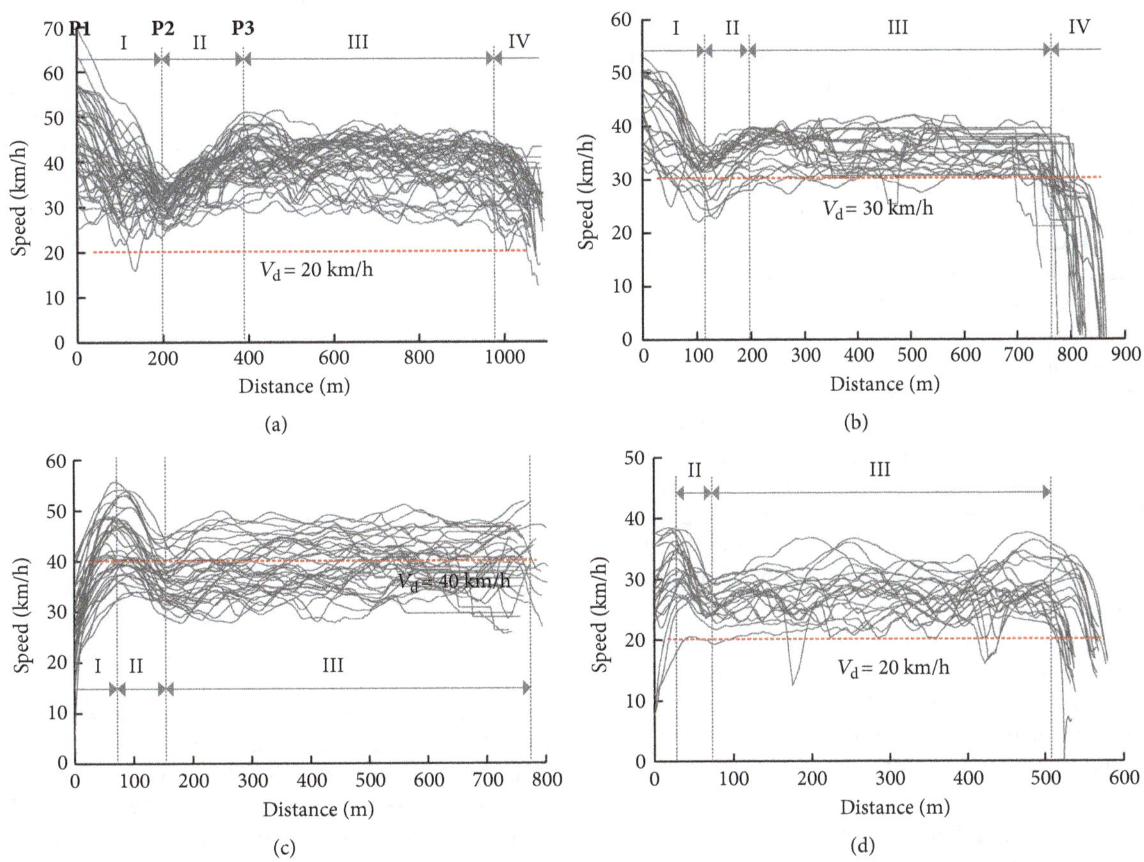

FIGURE 8: Driving speed on descending ramps. (a) Ramp of Fuling Yangtze River Bridge, design speed (V_d) 20 km/h. (b) Ramp of Fuling Wujiang Bridge, V_d = 30 km/h. (c) Ramp of Chongqing Rongqiao Helix Bridge, V_d = 40 km/h. (d) Jinkai Ring Bridge, V_d = 20 km/h.

is the end point of the taper, i.e., the theoretical gore, and also the end point for data recording.

Moreover, the speed profiles in Figure 5(b) show different levels of decreases in speed at the end of the ramp of Wujiang Bridge, i.e., Stage III. Based on the videos from the front camera and the diction of the driver, we suspected that the deceleration was caused by the driver observing the traffic conditions on the main bridge. The drivers began to accelerate again at the merge point, namely, Stage IV. The feature points of speed change along this ramp are present in Figure 7. Based on the above observations and the conditions observed on the end section of the helical ramp of Fuling Yangtze River Bridge, we conclude that the deceleration operation is a common driving behavior just before leaving the ramp.

4.2. Descending Helical Ramps

4.2.1. General Characteristics. Figure 8 shows the speed profiles for the four descending ramps. Figure 8(a) presents the speed on the ramp of Fuling Yangtze River Bridge. This ramp consists of two parts: part one is the curved ramp section before the starting point of the helix and consists of three horizontal curves that account for 35% of the total ramp length; part two is the helical ramp section. It is evident from the figure that vehicle speed on the first part decreases significantly, starting from the theoretical gore. Therefore, the deceleration distance can be regarded as reflecting the natural habits of drivers. Moreover, the change in speed within the helical ramp range (Stage III) exhibited an overall descending trend in general.

Figure 8(b) shows the speed for the descending helical ramp of the Fuling Wujiang Bridge. The starting point of the speed profiles is located at the end of bridge mainline, and the end point is the position where the ramp meets Binjiang Avenue after reaching the ground level. Similarly, the speed curve trend shows a decrease first (Stage I), followed by an increase (Stage II), while remaining constant in the next section (Stage III). The decline of the speed curve in its end is caused by the fact that there is a traffic signal at the traffic merge point, so the vehicle should wait for the green signal before entering the mainline traffic flow.

Figure 8(c) shows the speed on the descending ramp of the Rongqiao Helix Bridge. Although the speed fluctuates slightly within the helical ramp (Stage III), it is obvious that the overall trend is constant. The intersection of Rongqiao Road and Mingjia Road is a T-junction; moreover, there is a pedestrian crossing at the beginning of the helical ramp (Figure 2(c)), so the vehicles need to decelerate or stop before accelerating onto Rongqiao Road. Therefore, the speeds in Figure 8(c) all begin with an acceleration until a peak value is reached, and then a deceleration is exhibited when entering the helical ramp. Finally, a constant speed is shown at the end. Figure 8(d) shows the speed on the descending ramp of the Jinkai Ring Bridge. The profiles show a similar trend to that of the Rongqiao Helix Bridge, except that the speed amplitude is slightly lower.

Based on Figure 8, we did not observe an increasing trend in speed when traveling on descending ramps, whether it was a helical ramp or a helical bridge, indicating that the continuously decreasing elevation along the distance did not result in an increase in the traveling speed of passenger cars.

4.2.2. Deceleration and Acceleration Length. From Figure 8(a) we can find that the deceleration distance (the distance between point P1 and point P2) of most participants is about 200 m. The feature points P1 to P3 on the ramp are marked in Figure 9, and we can see that the ending point of deceleration, P2, is just located on the middle of the hairpin curve, while the point P3, the ending point of followed acceleration, coincides with the start point of the helical ramp. Therefore, we may recognize the fact that drivers' deceleration and acceleration behavior in this case is highly depending on the geometric features of the ramp.

It is evident from Figure 8(b) that drivers decelerate when entering the ramp from the mainline (a 650 m long tangent) of Wujiang second bridge. We extracted the deceleration distance from the speed-distance curves for all participants at this location, and the results are shown in Figure 10(a). The distribution interval of the discrete points was 95–165 m,

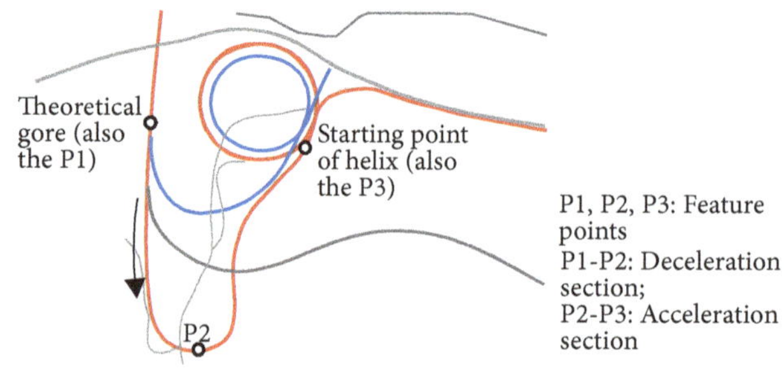

FIGURE 9: Feature points of speed change on the ascending ramp of Ramp of Fuling Yangtze River Bridge.

and the range of an average line was 110–129 m. Figure 10(b) shows the cumulative frequency of the deceleration distance, which obeys the Gamma distribution, shown in (1) and (2). The estimated values of parameters α and β are 65.778 (95CI: 39.01-111.92) and 1.841 (95CI: 1.09-3.11), respectively, and the 85th percentile deceleration distance is 138 m. The distribution of the end points of the deceleration distance was marked on this ramp, as shown in Figure 10(c), where L_{dmin} and L_{dmax} are the minimum and maximum deceleration distance, respectively. It is evident from the figure that the deceleration behavior lasts over the range of helical ramp, at approximately 1/3 of the circumference of the ramp.

$$f(x) = \frac{\alpha^{\beta}}{\Gamma(\alpha)} x^{\alpha-1} e^{-\beta x}, \quad x > 0 \tag{1}$$

$$\Gamma(\alpha) = \int_0^{+\infty} t^{x-1} e^{-t} dt \tag{2}$$

Figure 11(a) shows the deceleration distances (the vehicle travelled in Stage II of Rongqiao Helical Bridge) of ten participants in an increasing order. The differences between different drivers are significant, with the average line ranging from 60 to 140 m. Figure 11(b) shows the cumulative frequency of the acceleration and deceleration distance, where parameters α and β of the Gamma distribution function for deceleration distance are estimated at 6.619 (95CI: 3.81-11.49) and 13.068 (95CI: 7.36-23.19), respectively. Figure 11(c) shows the feature points of the acceleration and deceleration behavior, where L1 and L2 are the minimum and maximum acceleration distance, respectively, while L4 marks the area within which the end points of driver deceleration behavior are distributed. Therefore, we can conclude that the deceleration will continue into the helical ramp range for a distance approximately 40% of the ramp circumference.

5. Speed Patterns on Helical Ramps (Bridges)

From the speed profiles set presented in Figures 5 and 8, we can get general understanding on speed change within the range of helical ramp, i.e., rising, falling, or stabilizing. However, for a specific participant, his/her speed preference might be the opposite of the overall trend in speed tendency; i.e., individual speed change might be masked and hide in the overall trend, while, for highway design and traffic safety, the

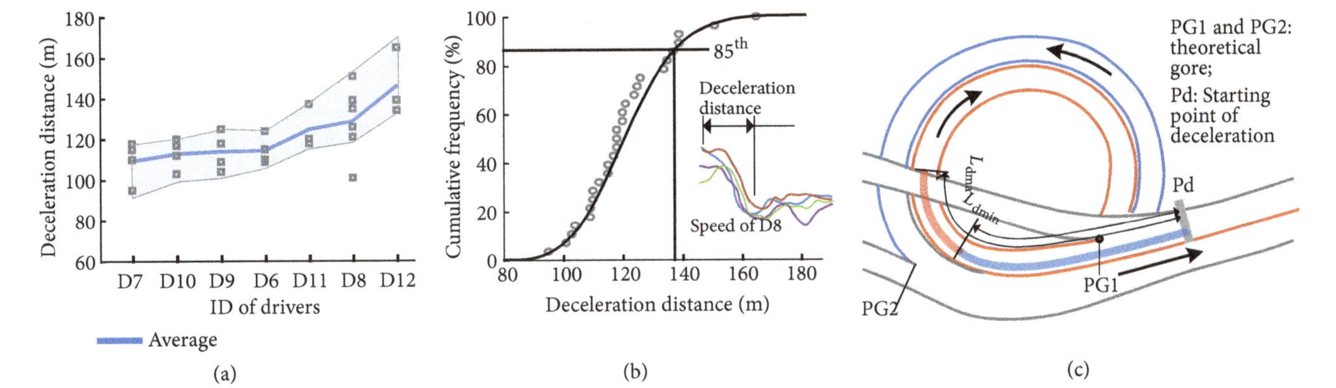

FIGURE 10: Deceleration distances on the descending ramp of Fuling Wujiang Bridge. (a) Deceleration distances of different test participants; (b) deceleration distance cumulative frequency curve; (c) deceleration distance distribution area.

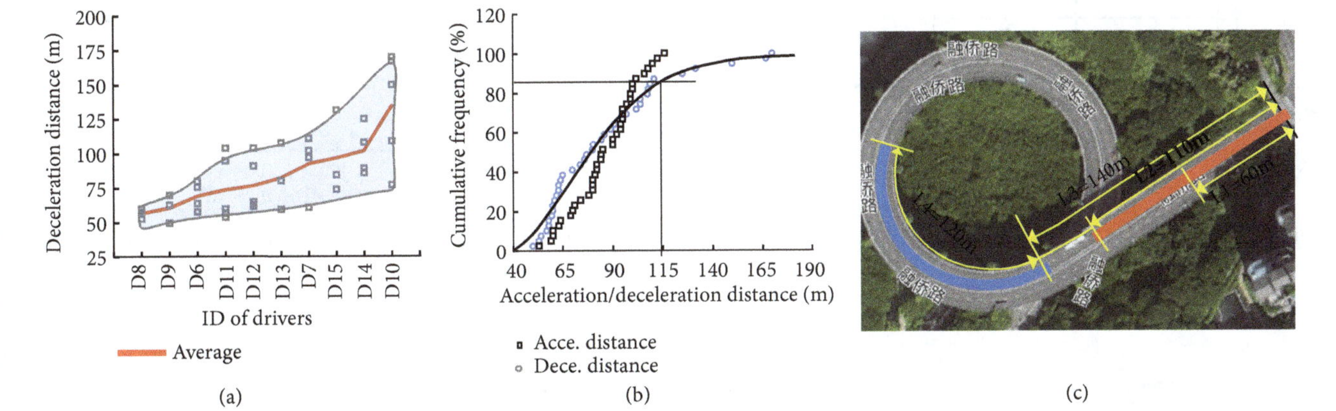

FIGURE 11: Deceleration distance on the descending ramp of the Rongqiao Helix Bridge. (a) Deceleration distances of different participants; (b) cumulative frequency curve of deceleration distance; (c) distribution area of deceleration distance.

behavior of individuals should be given equal weight to the behavior of average/overall level.

5.1. Method to Identify the Speed Patterns. Here we focus on the pattern of speed change within the range of helical ramp after drivers finish their initial acceleration and deceleration, i.e., changing patterns of speed curves of Stage II marked in Figure 5 and speed curves of Stage III marked in Figure 8. The gradient of speed, GV_1 and GV_2, was used to identify and define the speed pattern on a helical ramp, and GV_i was calculated by

$$GV_i = \frac{(V_{i+1} - V_i)}{(d_{i+1} - d_i)} \quad (3)$$

The method of getting GV_1 and GV_2 from a given speed profile was illustrated in Figure 12. Because the speed fluctuates locally along the distance, V_i was extracted from the trend line (fitting curve) of the measured data of speed. According to the shape of speed profiles in Figures 5 and 8, a broken line linked by three point can outline the trend and changing pattern of a given speed profile. Data point (GV_{1i}, GV_{2i}) would be determined and plotted in a plane coordinate after the valves of GV_1 and GV_2 were calculated.

In fact, the position of an arbitrary data point (GV_{1i}, GV_{2i}) in $GV_1 - GV_2$ coordinate is nothing more than subject to the following nine kinds, as shown in Figure 13, in which one situation corresponds to one speed pattern, i.e., the speed in the graph below the $GV_1 - GV_2$ coordinate. Therefore, we can recognize the speed pattern by using the data points (GV_{1i}, GV_{2i}).

5.2. Results of Speed Pattern Identifying. Using the method illustrated in Figure 12, data points (GV_{1i}, GV_{2i}) of all four ascending helical ramps and four descending ramps were obtained and displayed in the $GV_1 - GV_2$ coordinate, as shown in Figure 14. Comparing the quadrant distribution of (GV_{1i}, GV_{2i}) of ascending ramp against that of descending ramp, we find that there exists significant difference between speed patterns of ascending ramp and descending ramp. Taking YZ as example, for ascending ramp, twenty-seven data points are located in the first quadrant, and seventeen points are located in the fourth quadrant, while for descending ramp, of all forty-five points, twenty-one were in the fourth quadrant, and the rest were scattered in other three quadrants. The quadrant distribution of WJ' ascending ramp is similar to that of YZ' ascending ramp; i.e., patterns I, IV,

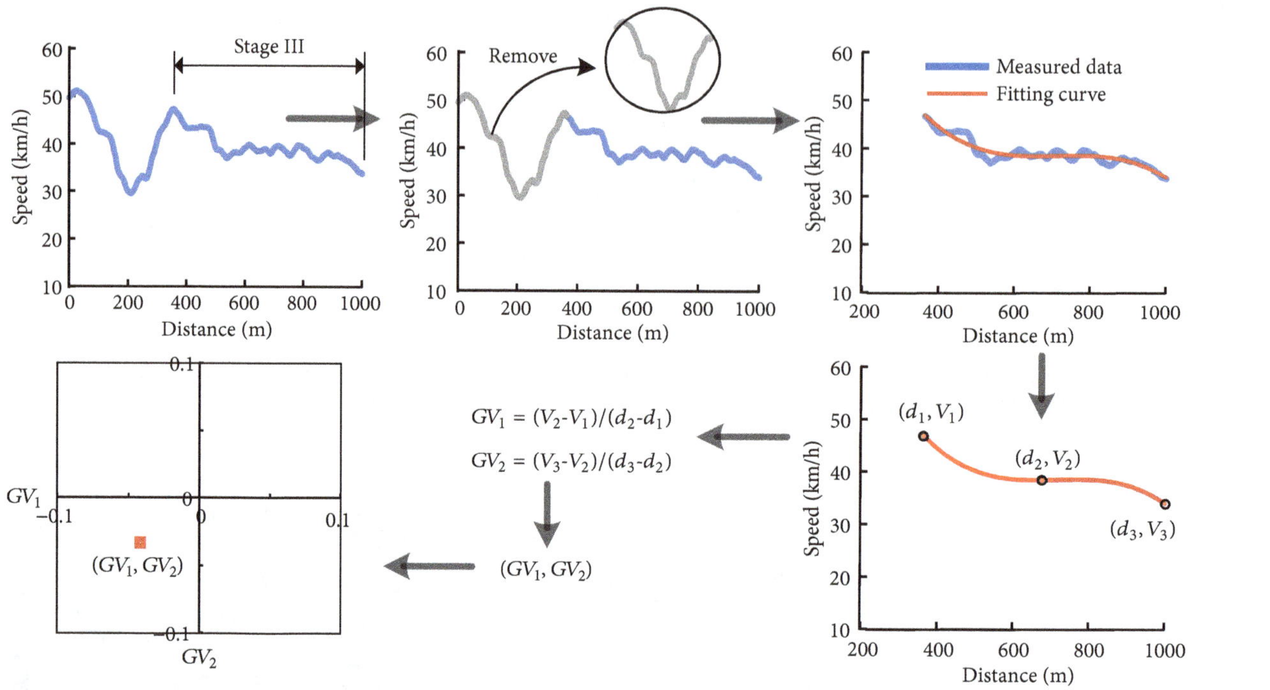

FIGURE 12: Method to get the value of GV_1 and GV_2 from a measured speed profile and show the point of (GV_1, GV_2) in a plane coordinate.

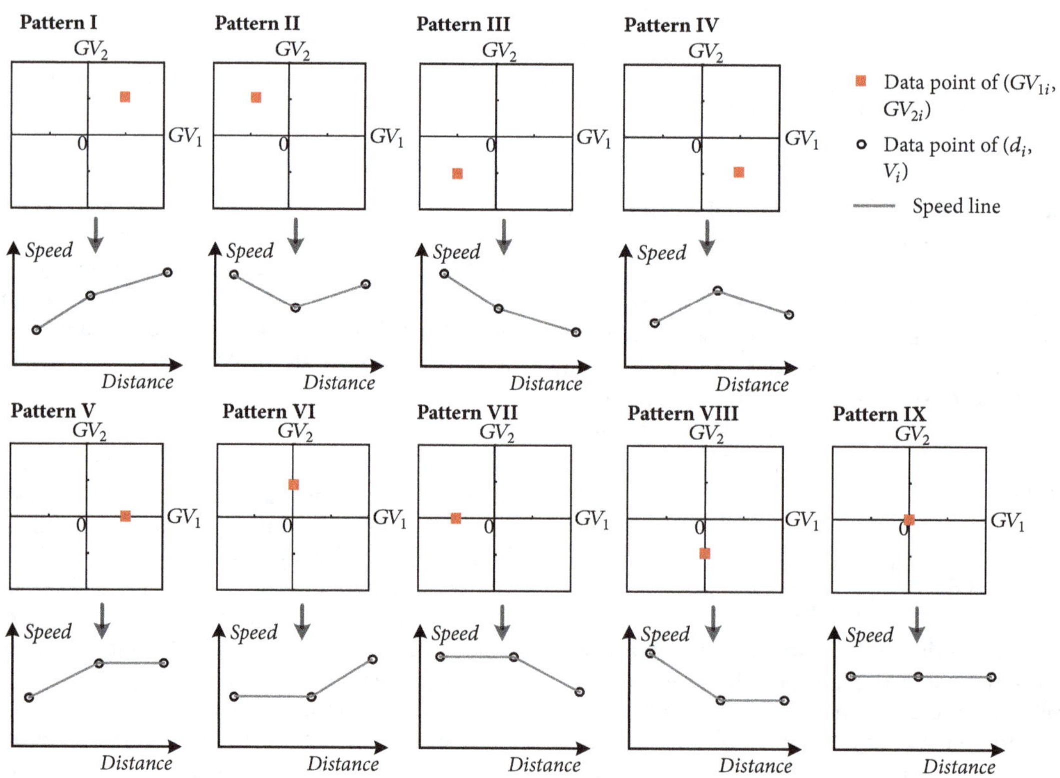

FIGURE 13: Identifying the speed pattern on helical ramp by using the position of an arbitrary data point (GV_{1i}, GV_{2i}) in $GV_1 - GV_2$ coordinate.

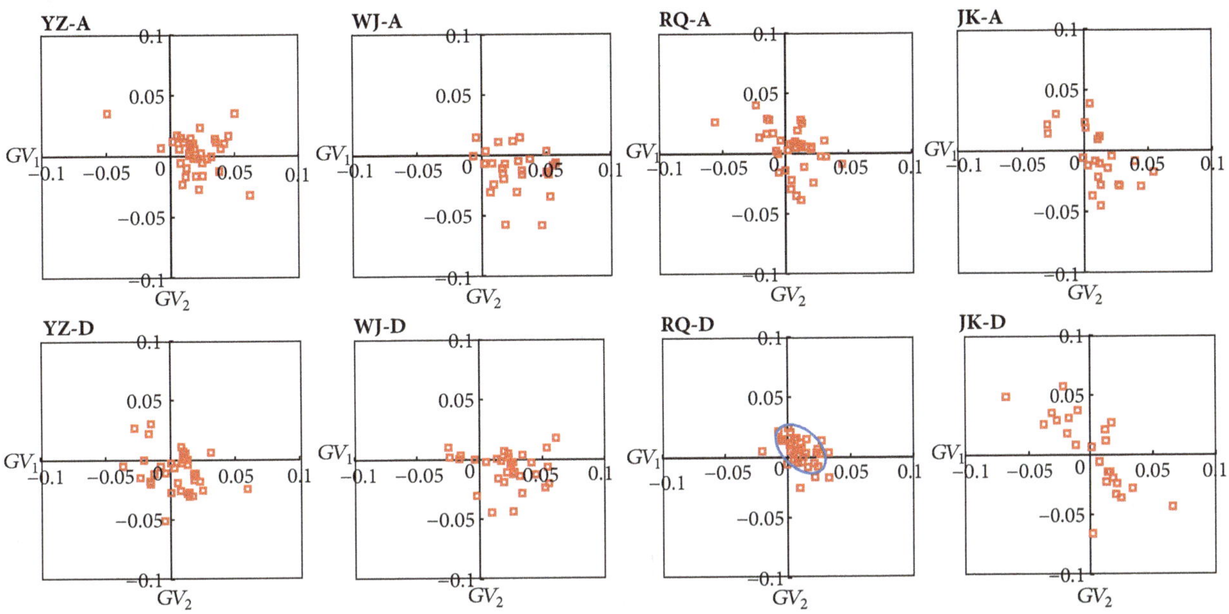

FIGURE 14: Data point (GV_{1i}, GV_{2i}) of four ascending helical ramps and four descending ramps shown in GV_1 —GV_2 coordinate, where YZ-A and YZ-D denote the ascending and descending ramp of Fuling Yangtze River Bridge, respectively; WJ-A and WJ-D denote the ascending and descending ramp of Fuling Wujiang Bridge; RQ-A and RQ-D denote the ascending and descending ramp of Rongqiao Helix Bridge; and JK-A and JK-D denote the ascending and descending lane of Jinkai ring bridge.

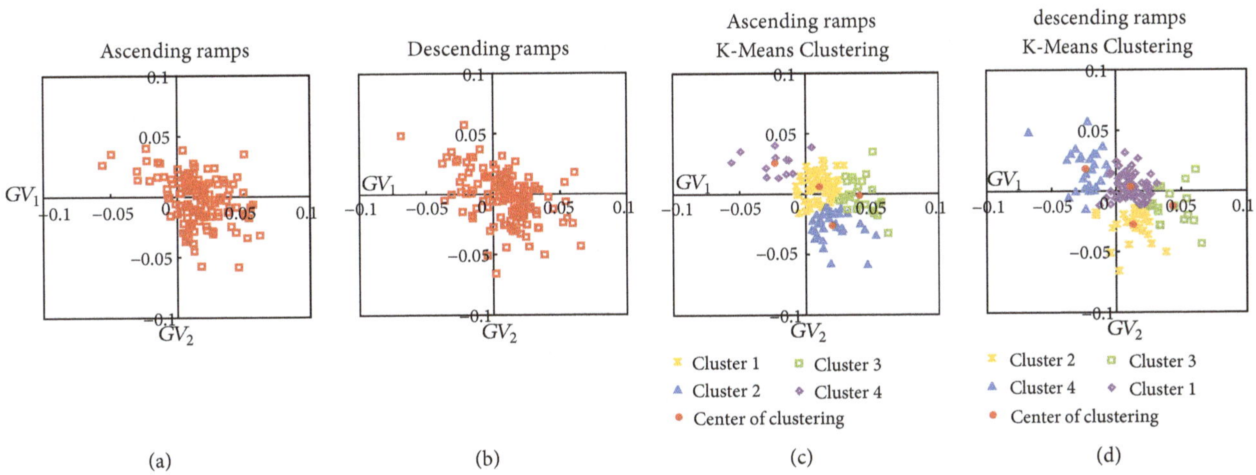

FIGURE 15: Scatter plot containing the points of (GV_{1i}, GV_{2i}) of four ascending/descending ramps. (a) Scatter plot of ascending ramps; (b) scatter plot of descending ramps; (c) four clusters for ascending ramps; (d) four clusters for descending ramps.

and V are dominant speed modes, while, for WJ' descending ramp, dominant speed patterns are IV, V, and VII.

Data points of RQ bridge and JK bridge have similar distribution characteristic; however, for RQ bridge the points' distribution is more concentrated, especially on RQ's descending lanes, which indicates that a convergent driving behavior occurs on this helical bridge.

Data points of all four ascending ramps were put together and displayed in one GV_1 —GV_2 coordinate, as shown in Figure 15(a); and Figure 15(b) for descending ramps was then obtained using the same method. In Figure 15(a) the number of data points located in the first to fourth quadrant is $\{N_1 = 53$ (39%), $N_2 = 65$ (47.8%), $N_3 = 14$ (10.3%), $N_4 = 4$ (2.9%)$\}$, while in Figure 15(b) this set of numbers is $\{N_1 = 39$ (28.3%), $N_2 = 62$ (44.9%), $N_3 = 24$ (17.4), $N_4 = 13$ (9.4%)$\}$. Therefore, pattern IV is dominant pattern for both ascending and descending ramps, followed by pattern I, and pattern III occupied the lowest share in the total. Figures 15(c) and 15(d) show four clustering results obtained by K-means arithmetic; the center of clustering distributes the quadrants except the third quadrant. Table 2 lists the coordinate values of the center of clustering for ascending/descending ramps and sample size of points divided into each cluster.

Table 2: Center of clustering for ascending/descending ramps.

	Ascending ramps				Descending ramps			
	Cluster 1	Cluster 2	Cluster 3	Cluster 4	Cluster 1	Cluster 2	Cluster 3	Cluster 4
Value of clustering center	(0.01, 0.0058)	(0.020,-0.0262)	(0.04,-0.00154)	(-0.023, 0.0255)	(0.01108, 0.00347)	(0.0131,-0.02772)	(0.0435, -0.01247)	(-0.0232, 0.0181)
sample size of each cluster	64	34	27	12	65	30	17	36

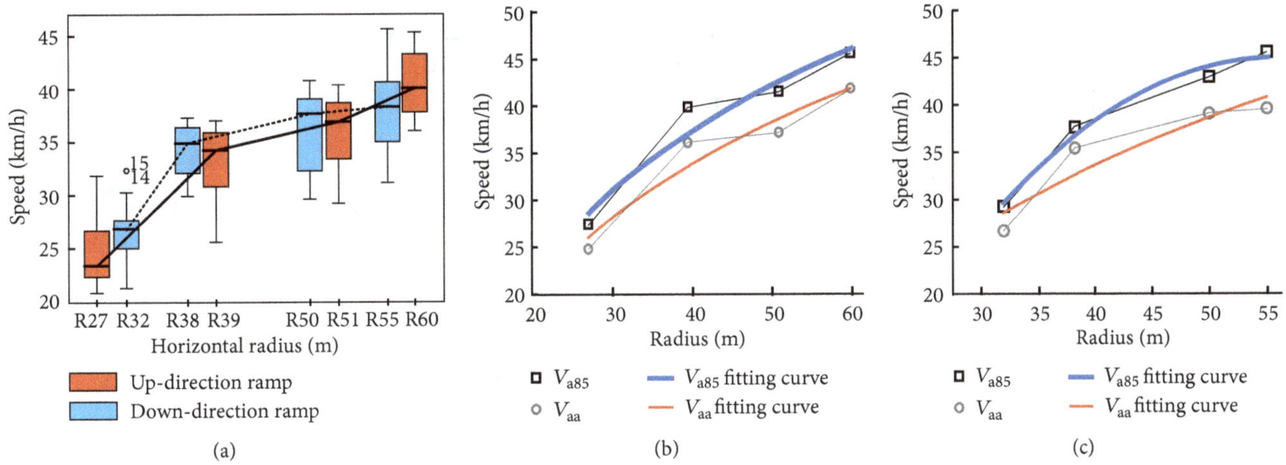

FIGURE 16: (a) Box plots of speed versus ramp radius; eigenvalue of driving speed and horizontal radius of (b) ascending ramp and (c) descending ramp.

TABLE 3: Regression model coefficients and the goodness of fit.

Object	b_1	b_2	R^2
Ascending ramp V_{a85}	21.803	-43.175	0.940
Ascending ramp V_{aa}	20.496	-41.87	0.906
Descending ramp V_{a85}	28.494	-68.198	0.962
Descending ramp V_{aa}	22.593	-49.712	0.876

6. Correlation between Speed and Ramp Radius

The box plot of the average speed on a helical ramp (the average value of a speed curve within Stage II in Figure 5 and Stage III in Figure 8), V_{ai}, was obtained using SPSS19.0, as shown in Figure 16(a). It is obvious from the figure that there is a correlation between the driving speed and horizontal radius of the helical ramp, regardless of whether it is an ascending or descending ramp; that is, the larger the radius, the higher the driving speed. The 85th percentile of V_{ai}, i.e., V_{a85}, for each helical ramp was extracted, and a broken line graph of V_{a85} changing over ramp radius was plotted, as shown in Figures 16(b) and 16(c), in addition to the quadric mean V_{aa}. From the figure, we can see a strong correlation between V_{a85} (or V_{aa}) and ramp radius. Using the least squares analysis, the regression formula for the data points in the graphs can be determined. Among various typical mathematical functions, logarithmic function has a higher goodness of fit and a simpler form, and it is expressed as follows:

$$f(x) = b_1 \cdot \ln(x) + b_2, \tag{4}$$

where x is the independent variable, namely, the horizontal radius of the centerline (m) of the bridge deck (box girder) where the traffic lane is located, and $f(x)$ is the function of the dependent variables V_{a85} or V_{aa}. The values of the regression model coefficients, b_1 and b_2, and the goodness of fit, R^2, are presented in Table 3.

7. Discussion

Based on the speed behavior of various participants measured in this study (speed profiles shown in Figures 5 and 8), we conclude that the design speed of the helical ramp of the Yangtze River Bridge is not appropriate, and the alignment elements are inconsistent with its design standard. This leads to a serious deviation between the actual driving behavior and design expectations. Therefore, the ramp radius, shoulder width, and other geometric features used in the design process should not significantly exceed the normal values recommended in the guideline; otherwise, the design speed should be increased to correspond to the design criteria of ramp geometry.

After the interchange construction is completed and opened to traffic, a proper speed limit for the ramp should be determined through an adequate field investigation and verification, rather than simply using the design speed of the ramp as the speed limit. A too low speed limit reduces the traffic capacity and the operational efficiency in ramp areas where there are no valid safety concerns. According to Figure 16, apart from R60 (ascending ramp of Rongqiao Helix Bridge), the 75th percentiles V_{ai}, V_{a75}, for other seven ramps, are less than 40 km/h. In addition, except for Jinkai Ring Bridge, the 85th percentile V_{ai}, V_{a85}, for the ramps at the other three sites, ranges from 40 to 45 km/h for the ascending path and 38 to 45 km/h for the descending path. Therefore, for one-way helical ramps and bridges with design speeds below 40 km/h, speed limit for passenger cars should be set at 40 km/h, i.e., the roundness of V_{a85} taking tens as unit. Moreover, because it is common for the drivers to decelerate in the traffic merge area (the end of the ramp), there is a high rear-end collision risk in that area when the deceleration amplitudes of the preceding and following vehicles are inconsistent. This fact also requires the attention of the designers and administrators.

For descending ramps, drivers experience an obvious deceleration when entering a helical ramp, and deceleration length and its distribution of WJ' ramp and RQ' ramp are

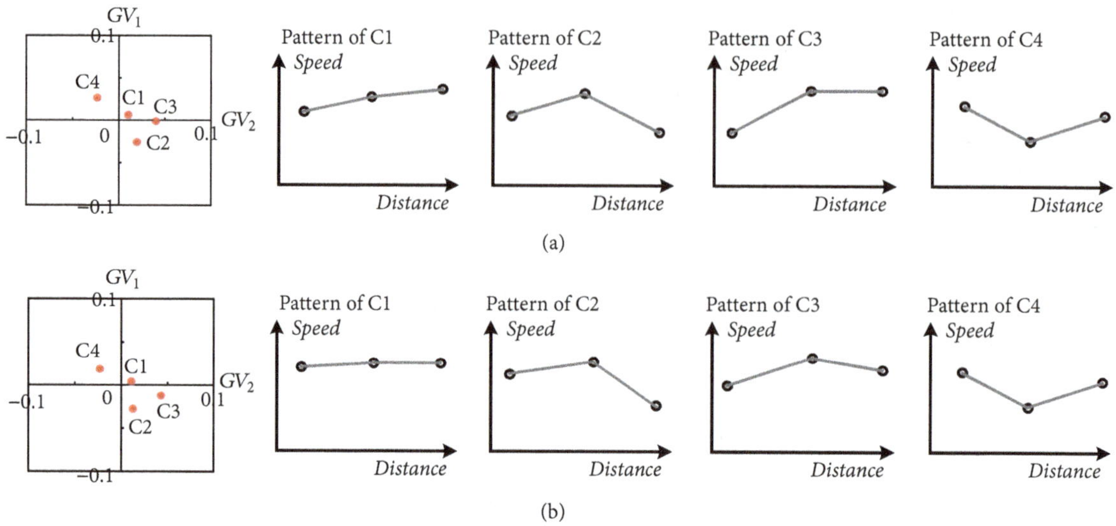

FIGURE 17: Speed patterns correspond to four clustering center. (a) Ascending ramp; (b) descending ramp.

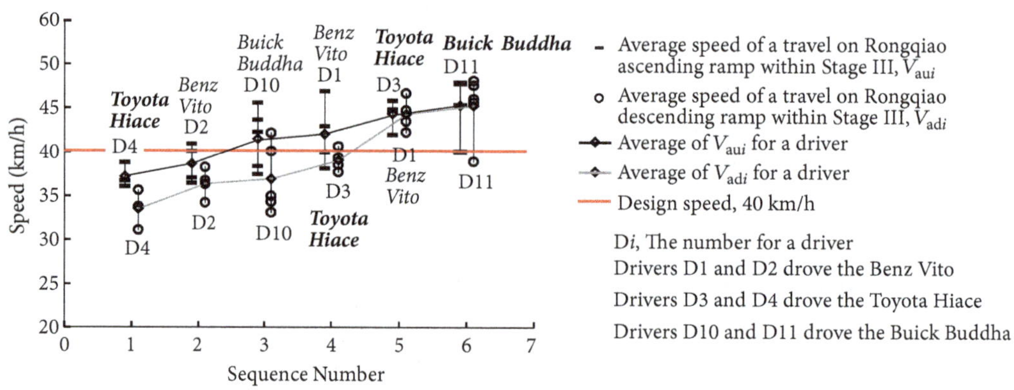

FIGURE 18: Average speed of different types of vehicles on Rongqiao Helical Bridge.

present in Section 4.2.2. The deceleration length of YZ' ramp was analyzed qualitatively, instead of getting a statistical distribution, because there are three curves between the tangent and the helix (see Figure 9); i.e., the helix is not directly connected to the tangent (mainline).

One of the main aims of this work is to get the speed pattern for helical ramps, a method of "location in four-quadrant of speed gradient (GV_{1i}, GV_{2i})" was designed to determine the speed patterns, and this method has an explicit physics meaning to explain the change in speed. In addition, four clusters by K-means algorithm are present in Figures 15(c) and 15(d); according to the location of four clustering centers in GV_1 $-GV_2$ coordinate, we can draw the four typical speed patterns, as shown in Figure 17.

The content of this study is a part of investigative work on the "identification and control of dangerous driving patterns on complex interchanges based on reverse deconstruction of multi-parameters (NSFC 51678099)," conducted by us. In the early stages of the driving experiments, we carried out driving tests on interchanges located in an urban area of Chongqing. While collecting the naturalistic data on the Rongqiao Helix Bridge, we became aware of the peculiarities of this type of ramp. Consequently, we searched for other examples of such interchanges within the jurisdiction of Chongqing and found Yangtze River Bridge and Wujiang Bridge in Fuling, a district of Chongqing. While conducting the driving tests in Fuling for the second time, we found another conspicuous object, the Jinkai Ring Bridge. Therefore, the number of participants for each experimental site was different because of the difference in the experiment period. And the sample of participants of all experimental sites is relatively small, which may reduce the statistical significance and result in a limitation of this study.

Another limitation of this paper is the two different types of vehicles adopted in the driving test, three MPVs and one minivan. Minivan (Toyota Hiace) has a higher center of gravity than the other three vehicles, which can result in a larger rolling moment that acted on the vehicle body and may influence the driving behavior. To figure out whether the effect exists, the average speeds within Stage III on Rongqiao Helical Bridge (Toyota Hiace was used on this site), V_{au} and V_{ad}, were calculated for each travel. And V_{aui} and V_{adi} for each participant were averaged again ($i = 1, 2.\ldots\ldots, N$, N is number of rounds a driver's trip on the helical bridge), and then the quadric mean speed was obtained, as shown in Figure 18.

Next, we sequenced the quadric mean speed in an ascending order and then obtained two polylines for up- and down-direction travel. In Figure 18 driver D4 drove Toyota Hiace at the lowest speed, while driver D3 of the same vehicle steered at the second-highest speed. From Figure 18 we are not able to see a correlation between vehicle type and average driving speed, and maybe the driving style and personality of drivers are the dominant factors for the difference in driving speed. And this issue will be analyzed and discussed in our future work.

8. Conclusions

In this study, we conducted field driving tests and obtained the operating parameters of vehicles traveling on helical ramps under natural driving conditions. Based on the analysis of the naturalistic data of driving speed obtained from the participants, speed characteristics and speed patterns on helical ramps were determined, as well as factors affecting speed amplitudes and fluctuations. The findings in this study can provide basic data referencing and scientific evidence to essentially improve the safety level of helical ramps in the practice of design and operation stage. The main findings of our research are as follows:

(1) When a helical ramp merged with the main route, the driver would observe the traffic on the mainline and make a judgment to adjust his/her speed. Therefore, obvious deceleration behavior exhibited by the driver was observed before the physical gore, and this was common to all participants. When drivers entered a helical ramp from a tangent, they generally decelerated, and this deceleration behavior would last for a distance equal to approximately 33–40% of the circumference of the ramp.

(2) Speed along ascending ramps shows an overall stable or a slightly rising trend, and the continuously increasing elevation in ascending ramp did not cause a remarkable speed reduction for passenger car, while speed along descending ramps, except YZ' ramp, represents an overall stable trend, and the continuously decreasing elevation along the distance did not result in an obvious increase in the traveling speed.

(3) For one-lane helical ramps, the lower the design speed, the larger the discrepancy between the measured speed and design speed. Moreover, based on the distribution of the measured speed, 20 km/h should not be recommended as the design speed for helical ramps, and the consistency between the design speed and the alignment elements value of the ramp should be increased. In addition, for ramps with design speeds no larger than 40 km/h, the speed limit should be set to 40 km/h.

(4) We designed a method of "location in four-quadrant of speed gradient (GV_{1i}, GV_{2i})" to determine the speed patterns for individuals; this method has an explicit physics meaning to explain the change in speed. Moreover, four clusters were obtained by clustering the data points of (GV_{1i}, GV_{2i}) using K-means algorithm, and four typical speed patterns were thus determined.

(5) There is a high correlation between the driving speed and the ramp radius. The 85th percentile speed and the average speed increase with the increase in the ramp radius, and predicting models were developed.

In this work, we mainly analyzed the speed characteristics on helical ramps. Other key factors, such as lateral acceleration, lane occupancy, and steering/pedal input, will be described in another article. Moreover, during the natural driving test, we collected the electrocardiographic (ECG) signals of the drivers and passengers. In our future work, we will analyze the distribution and variation characteristics of ECG signals within helical ramps to obtain the mental workload of drivers and its influence on the driving behavior.

Conflicts of Interest

The authors declare that there are no conflicts of interest regarding the publication of this article.

Acknowledgments

This research was supported by the National Natural Science Foundation of China (Grant No. 51678099) and by Applied Basic Research Program of Ministry of Transport of China (Grant No. 2015319814050).

References

[1] Y.-P. Zhao, S.-W. Yang, H.-J. Wu, Y.-F. Zhao, and B.-H. Pan, "Helix line development's minimum radius of horizontal curve in freeway tunnel," *Journal of Chang'an University*, vol. 27, no. 6, pp. 72–75, 2007.

[2] G. Yang, H. Xu, Z. Tian, and Z. Wang, "Vehicle speed and acceleration profile study for metered on-ramps in california," *Journal of Transportation Engineering*, vol. 142, no. 2, pp. 1–13, 2016.

[3] W. Zhang, "Discussion on new kind of T-shaped interchange design," *Transportation Science and Technology*, vol. no.5, pp. 145–147, 2014.

[4] Y. P. Lai, C. Yang, and G. L. Ren, "Design of spiral ramp bridge of the second FulingWujiang River Bridge," *Bridge Construction*, no. 1, pp. 47–50, 2007.

[5] S. Ivorra, D. Foti, D. Bru, and F. J. Baeza, "Dynamic behavior of a pedestrian bridge in alicante, Spain," *Journal of Performance of Constructed Facilities*, vol. 29, no. 5, pp. 1–10, 2015.

[6] W. Wang, "Foundation treatment for double-deck spiral bridge on Chongqing Rongqiao Avenue," *Building Construction*, vol. 32, no. 8, pp. 864-865, 2010.

[7] Z. Y. Zhang, X. Y. Hao, W. B. Wu, and D. Wang, "The running speed prediction model of interchange ramp," *Journal of Transportation Systems Engineering and Information Technology*, vol. 15, no. 1, pp. 93–99, 2015.

[8] C. Zhang, X. M. Yan, X. W. Li, B. H. Pan, H. J. Wang, and X. N. Ma, "Operating speed model of passenger car at single-lane exit of interchange," *China Journal of Highway and Transport*, vol. 30, no. 6, pp. 279–286, 2017.

[9] H. J. Wang and S. W. Yang, "Research of length of deceleration lane at expressway interchange," *Journal of Highway and Transportation Research and Development*, vol. 32, no. 3, pp. 124–128, 2015.

[10] Z. Guo, H. Wan, Y. Zhao, H. Wang, and Z. Li, "Driving simulation study on speed-change lanes of the multi-lane

freeway interchange," *Procedia: Social and Behavioral Sciences*, vol. 96, pp. 60–69, 2013.

[11] R. H. Wang, J. B. Hu, and X. Q. Zhang, "Analysis of the driver's behavior characteristics in low volume freeway interchange," *Mathematical Problems in Engineering*, vol. 2016, Article ID 2679516, 9 pages, 2016.

[12] A. Polus, M. Livneh, and J. Factor, "Vehicle flow characteristics on acceleration lanes," *Journal of Transportation Engineering*, vol. 111, no. 6, pp. 595–606, 1985.

[13] M. Livneh, A. Polus, and J. Factor, "Vehicle behavior on deceleration lanes," *Journal of Transportation Engineering*, vol. 114, no. 6, pp. 706–717, 1988.

[14] M. A. Ahammed, Y. Hassan, and T. A. Sayed, "Modeling driver behavior and safety on freeway merging areas," *Journal of Transportation Engineering*, vol. 134, no. 9, pp. 370–377, 2008.

[15] K. Fitzpatrick, S. T. Chrysler, and M. Brewer, "Deceleration lengths for exit terminals," *Journal of Transportation Engineering*, vol. 138, no. 6, pp. 768–775, 2012.

[16] B. Zirkel, S. Park, J. McFadden, M. Angelastro, and L. McCarthy, "Analysis of sight distance, crash rate, and operating speed relationships for low-volume single-lane roundabouts in the united states," *Journal of Transportation Engineering*, vol. 139, no. 6, pp. 565–573, 2013.

[17] P. Perco, A. Marchionna, and N. Falconetti, "Prediction of the operating speed profile approaching and departing intersections," *Journal of Transportation Engineering*, vol. 138, no. 12, pp. 1476–1483, 2012.

[18] B. Yang, "Urban interchange in chongqing-grade separation project with mountainous characteristic," *Urban Roads Bridges and Flood Control*, no. 9, pp. 46–49, 2016.

[19] G. Li and L. Sun, "Characterizing heterogeneity in drivers' merging maneuvers using two-step cluster analysis," *Journal of Advanced Transportation*, vol. 2018, Article ID 5604375, 15 pages, 2018.

[20] N. Lyu, Y. Cao, C. Wu, J. Xu, and L. Xie, "The effect of gender, occupation and experience on behavior while driving on a freeway deceleration lane based on field operational test data," *Accident Analysis & Prevention*, vol. 121, pp. 82–93, 2018.

15

On the Kinematics of Vehicles Relevant to Conflict Detection and Resolution

Luís Braga Campos[1] and Joaquim Guerreiro Marques[1,2]

[1]*CCTAE (Center for Aeronautical and Space Science and Technology), IDMEC/LAETA, Instituto Superior Técnico, Universidade de Lisboa, Av. Rovisco Pais 1, 1049-001 Lisboa, Portugal*
[2]*Escola Náutica Infante D. Henrique, Av. Engenheiro Bonneville Franco, 2770–058 Paço de Arcos, Oeiras, Portugal*

Correspondence should be addressed to Joaquim Guerreiro Marques; jmgmarques@tecnico.ulisboa.pt

Academic Editor: Jose E. Naranjo

The paper presents methods to determine the time, positions, and distance of closest approach for two vehicles following arbitrary trajectories in two or three dimensions. The distance of closest approach of two vehicles following arbitrary curved trajectories is determined by two conditions: (i) the relative velocity must be orthogonal to the relative position in order for the distance to be a nonzero extremum; (ii) the radial acceleration including centripetal terms must have a direction that increases the separation for the extremum to be a minimum. This theorem on the distance of closest approach simplifies in the case of uniform motion along rectilinear trajectories. Three examples are given: (i) the two-dimensional motion of surface vehicles changing the velocity of one of them so as to enforce a given minimum separation distance; (ii) the three-dimensional motion of two aircraft, one flying horizontally and the other climbing, changing the vertical velocity of the latter to ensure a minimum separation distance set "a priori"; (iii) the case of an aircraft flying with constant velocity in a straight line so that its closest approach to another aircraft flying in a circular holding pattern in the same plane occurs at a given time chosen "a priori".

1. Introduction

In the traffic of vehicles safety is identified with the absence of collisions or conflicts. A conflict occurs when the distance between the centroids of two vehicles is less than a safe separation distance (SSD) determined by their size. Thus (i) the absence of conflicts and (ii) the confirmation that a conflict has been resolved depend on determining the distance of closest approach (DCA) that is not less than the SSD. The conflict resolution relies (iii) on trajectory modifications that change the DCA from smaller than the SSD to larger than (or equal to) the SSD. The paper presents methods to determine the time, positions, and distance of closest approach for two vehicles following arbitrary trajectories in two or three dimensions. The two-dimensional cases include cars in road traffic, ships in sea lanes, and aircraft ground movements at an airport. The three-dimensional cases include all types of flying vehicles, like airplanes, helicopters, drones, rockets and satellites, and also submerged submarines. The differences in conflict detection and resolution (CDR) between all these types of vehicles concern the speed, size, and distances that enter as parameters in the same methods of calculation of distance and time of closest approach.

The distance and time of closest approach are essential inputs for CDR methods [1–3]. The collision risk applies to cars [4, 5], ships and submarines [6, 7], and aircraft [8, 9]. Taking as example the case of Air Traffic Management (ATM) the problem may be divided into (i) prediction of flight paths [10, 11], (ii) safety assessment [12, 13], and (iii) conflict resolution [14–16]. Collision avoidance between aircraft starts with [17] separation distances (e.g., longitudinal, lateral, and altitude) leading to high Target Level of Safety (TLS: probability of collision less than 5×10^{-9} per hour) for various aircraft encounter geometries, like level crossing [18] or climb and descent [19]. The safe separation ultimately determines airspace capacity [20]. Many of these methods assume straight trajectories or approximate curved trajectories by straight segments. While this allows a continuous trajectory, the velocity becomes discontinuous in direction at the edges and the acceleration becomes singular. The purpose of the present paper is to determine the distance and time of closest approach for arbitrary curved trajectories without approximations of any kind. This may be used for CDR methods or to assess TLS.

There are a variety of CDR methods including multiagent algorithms [21–23] that apply to vehicles moving in two [1–9, 24–27] or three [10–20] dimensions. All CDR methods (i) start with the identification of a conflict, (ii) involve trajectory changes to resolve the conflict, and (iii) end with the verification that the conflict has been resolved. The safety of traffic requires that a minimum SSD be held, for example, ensuring that the "safety volumes" around two vehicles do not penetrate. If two vehicles follow two arbitrary trajectories with position vectors $\vec{x}_1(t)$ and $\vec{x}_2(t)$, as a function of time, the modulus of the difference specifies their distance that generally varies with time:

$$D(t) = \left|\vec{x}_1(t) - \vec{x}_2(t)\right|. \tag{1}$$

A conflict is detected if at any time the distance is less than the SSD:

$$\text{Conflict: } D(t) < \delta \quad \text{for some time } t. \tag{2}$$

If a conflict is detected, then one or both trajectories must be modified to resolve the conflict. The success of the conflict resolution is checked by showing that the distance exceeds the SSD for all time:

$$\text{No Conflict: } D(t) \geq \delta \quad \text{for all time } t. \tag{3}$$

The criteria for conflict (2) or no conflict (3) are next put into a simpler form that is easier to apply.

The key concept is that of distance of closest approach D_* between two trajectories:

$$D_* \equiv D(t_*) = \left|\vec{x}_1(t_*) - \vec{x}_2(t_*)\right|, \tag{4}$$

which occurs at the time t_* of closest approach, when the vehicles are at position $\vec{x}_1(t_*)$ and $\vec{x}_2(t_*)$. There is a conflict if the distance of closest approach is less than the safe separation distance (5a) and no conflict otherwise (5b):

$$D_* < \delta \quad \text{conflict,} \tag{5a}$$

$$D_* \geq \delta \quad \text{no conflict.} \tag{5b}$$

The paper presents a method to determine (i) the distance of closest approach D_*, (ii) the time of closest approach t_*, and (iii) the positions of the two vehicles at that time $\vec{x}_1(t_*)$, $\vec{x}_2(t_*)$. The method applies to arbitrary trajectories: (a) curved or rectilinear; (b) with constant velocity, accelerated or decelerated motion. The method is deterministic and excludes external disturbances. The extension to include random disturbances can be made adding to the position vectors the deviations due to uncertainties or external effects and applying statistical methods.

Considering first arbitrary nonuniform motion a theorem is established (Section 2.1) specifying the conditions for minimum separation between two trajectories. In many traffic situations the future trajectories are not known, and the information available is only the current positions and velocities of two vehicles; if the motion is assumed to be uniform, simple formulas are obtained (Section 2.2) for the distance and time of closest approach. This in turn specifies the conditions for collision avoidance between two or any number of vehicles (Section 2.3). These conditions give a simple geometric interpretation of the two theorems (Section 2). The preceding theory is applied (Section 3) to three examples: (Section 3.1) two-dimensional collision avoidance between surface vehicles (ships, car, or airplanes on the ground at an airport) with constant velocity by choosing the modulus of the velocity of one of them; (Section 3.2) three-dimensional collision avoidance between two aircraft moving at constant velocity, one at constant altitude and the other climbing, by changing the vertical velocity of the latter; (Section 3.3) meeting a given time of closest approach between a vehicle in a holding pattern of uniform circular motion and another in uniform rectilinear motion. These three cases are sufficiently simple for analytical calculation (Section 3), illustrate the two theorems (Section 2), and substantiate the discussion (Section 4).

2. Distance of Closest Approach between Trajectories

The minimal separation for two vehicles following arbitrary trajectories is obtained by minimizing the relative distance as a function of time. This leads to two conditions to be satisfied at the time of closest approach (Section 2.1). The simplest case is that of uniform motion from given initial positions (Section 2.2). It leads to a simple criterion for collision avoidance by maintaining a minimum separation distance (Section 2.3) and is applicable to two or more vehicles. The vehicles are represented by material vehicles placed at their centroids, and their dimensions are taken into account by setting a safe separation distance to avoid collisions.

2.1. Nonuniform Motion along Curved Trajectories. Consider two vehicles moving with integrable time-dependent velocities $\vec{V}_1(t)$ and $\vec{V}_2(t)$ from initial positions $\vec{x}_{10}$ and $\vec{x}_{20}$ at time $t = 0$, so that their positions at time t are given by

$$\vec{x}_1(t) = \vec{x}_{10} + \int_0^t \vec{V}_1(t)\,dt, \tag{6a}$$

$$\vec{x}_2(t) = \vec{x}_{20} + \int_0^t \vec{V}_2(t)\,dt. \tag{6b}$$

It is necessary to find (i) the conditions for a collision, where the positions coincide with $\vec{x}_1(t_*) = \vec{x}_2(t_*)$ for some time(s) $t = t_*$; (ii) if there is no collision then the time $t = t_*$ and distance D_* of closest approach are given by (4)≡(7):

$$D_* \equiv D(t_*) = \left|\vec{x}_1(t_*) - \vec{x}_2(t_*)\right| = \left|\vec{x}_1(t) - \vec{x}_2(t)\right|_{\min}. \tag{7}$$

The relative position of the two vehicles is given at all times by:

$$\vec{x}(t) \equiv \vec{x}_1(t) - \vec{x}_2(t) = \vec{x}_{10} - \vec{x}_{20} + \int_0^t \left[\vec{V}_1(t) - \vec{V}_2(t)\right] dt. \tag{8}$$

Introducing the initial relative positions (9a) at the initial time $t = 0$ and the relative velocity (9b) at all times,

$$\vec{x}_0 \equiv \vec{x}_{10} - \vec{x}_{20}, \tag{9a}$$

$$\vec{V}(t) \equiv \vec{V}_1(t) - \vec{V}_2(t), \tag{9b}$$

the relative position at arbitrary time is given by

$$\vec{x}(t) = \vec{x}_0 + \int_0^t \vec{V}(t)\, dt = \vec{x}_0 + \vec{X}, \tag{10}$$

where the change in relative position from time $t = 0$ to time t is given by (11a):

$$\vec{X}(t) = \int_0^t \vec{V}(t)\, dt, \tag{11a}$$

$$\frac{d\vec{X}}{dt} = \vec{V}, \tag{11b}$$

and its time derivative is relative velocity (11b).

The distance between the vehicles at time t is given by

$$[D(t)]^2 = \left|\vec{x}(t)\right|^2 = \left|\vec{x}_0 + \vec{X}(t)\right|^2 = \left|x_0\right|^2 + \left|\vec{X}(t)\right|^2 + 2\vec{x}_0 \cdot \vec{X}(t). \tag{12}$$

The first derivative of (12) with regard to time,

$$2D(t)\frac{dD}{dt} = \frac{d}{dt}[D(t)]^2 = 2\left[\vec{X}(t) + x_0\right] \cdot \frac{d\vec{X}}{dt} = 2\vec{x} \cdot \vec{V}, \tag{13}$$

is zero for stationary distance between the vehicles:

$$D'(t_*) = 0 \Longrightarrow \tag{14a}$$

$$0 = \vec{x}(t_*) \cdot \vec{V}(t_*). \tag{14b}$$

Thus two cases may arise: (i) if the distance is zero at the stationary point(s) then there is collision between the vehicles and the relative velocity is arbitrary:

$$D(t_*) = 0 \Longrightarrow \tag{15a}$$

$$\vec{x}(t_*) = 0; \tag{15b}$$

(ii) if the distance is not zero then the velocity and relative position must be orthogonal:

$$D(t_*) \neq 0 \Longrightarrow \tag{16a}$$

$$\vec{V}(t_*) \perp \vec{x}(t_*). \tag{16b}$$

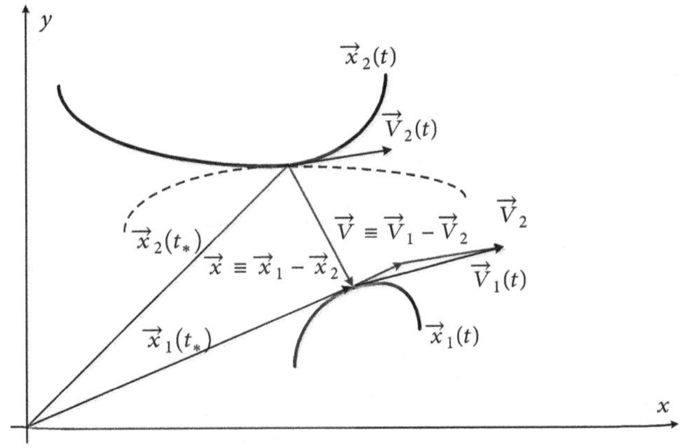

FIGURE 1: Two vehicles move along trajectories specified by the position vectors $\vec{x}_1(t)$ and $\vec{x}_2(t)$ as a function of time, so that the relative position is $\vec{x} \equiv \vec{x}_1 - \vec{x}_2$. The trajectories may be curved and described in nonuniform motion so that the velocities are functions of time $\vec{V}_1(t)$ and $\vec{V}_2(t)$ as also the relative velocity $\vec{V} \equiv \vec{V}_1 - \vec{V}_2$. The time of closest approach t_* corresponds to relative position $\vec{x}$ orthogonal $\vec{x} \perp \vec{V}$ to the relative velocity $\vec{V}$; that is, their inner product is zero $\vec{x} \cdot \vec{V} = 0$. The dotted line refers to a trajectory of the second vehicle also leading to an extremum of the relative distance but a maximum rather than a minimum.

Note that the distance of closest approach is generally not the smallest distance between the paths because the two vehicles will be at these points at different times. The condition of closest approach that the relative velocity and relative position are orthogonal (Figure 1) can be explained as follows: (i) if $\vec{V} \cdot \vec{x} < 0$ at time $t = t_*$ then the vehicles would be moving towards each other and would be closer at some later time $t > t_*$; (ii) if $\vec{V} \cdot \vec{x} > 0$ at time $t = t_*$ then the vehicles would be moving away from each other and would have been closer for some earlier time $t < t_*$. In either case (i) or (ii) the time $t = t_*$ would not be that of stationary distance. This proves by "reductio ad absurdum" that a stationary distance requires that the relative velocity be orthogonal to the relative position.

A stationary distance is a necessary but not sufficient condition for closest approach; an inflexion point in one of the trajectories would also correspond to a stationary distance but might not be a minimum; a minimum is determined by the second-order derivative of distance with regard to time:

$$2D(t)\frac{d^2D}{dt^2} + 2\left(\frac{dD}{dt}\right)^2 = \frac{d^2}{dt^2}\left\{[D(t)]^2\right\} = 2\frac{d}{dt}\left(\vec{x} \cdot \vec{V}\right) = 2\left(\vec{V} \cdot \vec{V} + \vec{x} \cdot \frac{d\vec{V}}{dt}\right), \tag{17}$$

being positive at the stationary condition. In the stationary condition (14a) ≡ (18a) the relation (17) simplifies to (18b):

$$D'(t_*) = 0: \tag{18a}$$

$$2D(t_*)D''(t_*) = 2\left\{\left|\vec{V}(t_*)\right|^2 + \vec{x}(t_*) \cdot \vec{a}(t_*)\right\} > 0, \tag{18b}$$

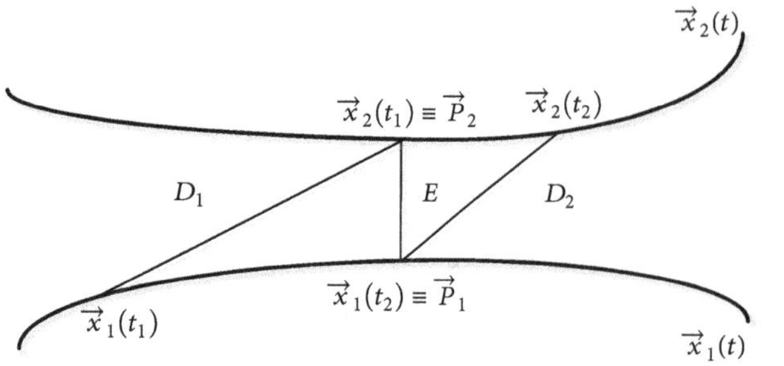

FIGURE 2: The shortest distance E between two curves or paths is along the common normal. This would be the distance of closest approach D along the two trajectories only if the two vehicles were at the closest positions (P_1, P_2) at the same time. Generally the two vehicles are at the closest positions (P_1, P_2) at different times $t \neq t_1$, so their distances (D_1, D_2) are larger than E.

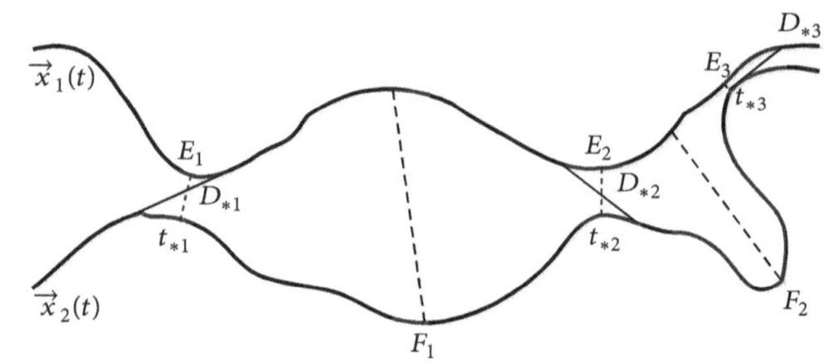

FIGURE 3: For vehicles moving along curved trajectories there may exist several times t_{*_n} of closest approach D_{*_n}, corresponding to local minima of the relative distance that usually exceeds the minimum distances E_n. The distance of closest approach is then the smallest of all local minima D_{*_n}. There may also exist local maxima F_n indicated by dotted lines.

where (19a) is the relative acceleration that must satisfy (19b) for a minimum:

$$\vec{a} \equiv \frac{d\vec{V}}{dt} \equiv \frac{d^2\vec{x}}{dt^2}: \tag{19a}$$

$$\vec{x}(t_*) \cdot \vec{a}(t_*) > -\left|\vec{V}(t_*)\right|^2. \tag{19b}$$

Thus the stationary distance will be a minimum if the acceleration is along the relative position $\vec{a} \cdot \vec{x} > 0$, because it tends to move the two vehicles away from each other. Even if the relative acceleration is from one vehicle towards the other $\vec{a} \cdot \vec{x} < 0$, the distance will still be a minimum if the centripetal acceleration is larger:

$$\vec{a} \cdot \vec{x} = -\left|\vec{a} \cdot \vec{x}\right| < 0: \tag{20a}$$

$$\frac{\left|\vec{a} \cdot \vec{x}\right|}{\left|\vec{x}\right|} < \frac{\left|\vec{V}\right|^2}{\left|\vec{x}\right|}. \tag{20b}$$

The stationary point would not be one of closest approach only if the relative acceleration projected on the relative position is opposite in sign and larger in modulus than the centripetal acceleration.

The preceding results may be summarized in a **theorem on nonuniform motion**: *two vehicles with initial positions* $\vec{x}_{10}$ *and* $\vec{x}_{20}$ *and nonuniform velocities, respectively,* $\vec{V}_1$ *and* $\vec{V}_2$, *which are integrable functions of time, will collide at an arbitrary angle if their relative distance ((8), (12)) vanishes ((15a), (15b)). If the vehicles do not collide, (i) the distance is stationary, which is a necessary condition for closest approach, if and only if ((16a), (16b)) the relative velocity (9b) is orthogonal to the relative position (8) ≡ (10); (ii) a sufficient condition for closest approach, which is that the distance is minimum, is that the relative acceleration (19a) satisfies (19b) implying that the centripetal acceleration associated with the relative curvature of the trajectories predominates over the relative acceleration projected on the relative velocity.*

The distance of closest approach D between vehicles moving two trajectories generally exceeds the smallest distance E between the two paths, because the vehicles might cross the closest points at different times; for example (Figure 2), (i) at the time t_1 when vehicle 2 is at the closest position $\vec{x}_2(t_1) = \vec{P}_2$ vehicle 1 is still behind the closest position $\vec{x}_1(t_2) \neq \vec{P}_1$:

$$D_1 \geq E \leq D_2; \tag{21}$$

(ii) by the time t_2 vehicle 1 reaches the closest position $\vec{x}_2(t_2) = \vec{P}_1$ vehicle 2 will be beyond the closest position $\vec{x}_2(t_2) \neq \vec{P}_2$. Conditions (16b) and (18b) of closest approach are local, in the sense that for small deviations to earlier or later time the relative distance increases. In the case of curved trajectories there can exist several times t_{*_n}, distances D_{*_n}, and positions $\vec{x}_1(t_{*_n}), \vec{x}_2(t_{*_n})$ of local closest approach as shown in Figure 3. In this case the distance of closest approach would be the "minimum minimorum" that is the smallest or infimum of all local distances of closest approach:

$$D_* = \inf\left(D_{*_1}, D_{*_2}, \ldots, D_{*_n}\right). \tag{22}$$

The existence of multiple local minima of the distance is not possible for straight trajectories: if the two vehicles move with constant velocity there is only one time and distance of closest approach, as shown next.

2.2. Closest Approach for Uniform Rectilinear Motion. In many traffic situations the future trajectories of vehicles are not known or can be adjusted to avoid collisions. The information available from traffic sensors is usually positions and velocities at a given time. The simplest assumption is that the velocities will be constant and the position of the centroids of the vehicles, taken as material vehicles, at time t will be

$$\vec{x}_1(t) = \vec{x}_{10} + \vec{V}_1 t, \tag{23a}$$

$$\vec{x}_2(t) = \vec{x}_{20} + \vec{V}_2 t. \tag{23b}$$

The relative position at time t is

$$\vec{x}(t) \equiv \vec{x}_1(t) - \vec{x}_2(t) = \vec{x}_0 + \vec{V} t, \tag{24}$$

where (9a) is the initial relative position and the relative velocity (9b) is constant. The relative distance at time t is given by

$$[D(t)]^2 = |\vec{x}(t)|^2 = |\vec{x}_0 + \vec{V}t|^2 = |\vec{x}_0|^2 + |\vec{V}|^2 t^2 + 2t\vec{x}_0 \cdot \vec{V}. \tag{25}$$

The second-order derivative with regard to time is always positive:

$$\frac{d^2}{dt^2}\{[D(t)]^2\} = 2|\vec{V}|^2 > 0, \tag{26}$$

and thus a stationary distance can only be a minimum distance. There is no maximum distance because a constant relative velocity leads the vehicles infinitely apart after a long time.

The first derivative of the distance

$$2D(t)\frac{dD}{dt} = \frac{d}{dt}\{[D(t)]^2\} = 2\left(\vec{x}_0 \cdot \vec{V} + |\vec{V}|^2 t\right) \tag{27}$$

vanishes at the time of closest approach:

$$D(t) \geq |D(t)|_{\min} = D(t_*): \tag{28a}$$

$$t_* = -\frac{\vec{V} \cdot \vec{x}_0}{|\vec{V}|^2} = -\frac{|\vec{x}_0|}{|\vec{V}|}\cos\alpha \tag{28b}$$

where α is the angle of the relative initial position (9a) with the constant relative velocity (9b). The position of the two vehicles at the time of closest approach is obtained by substituting (28b) in (23a), (23b):

$$\vec{x}_1(t_*) = \vec{x}_{10} - \frac{\vec{V}_1}{|\vec{V}|^2} \cdot (\vec{V} \cdot \vec{x}_0), \qquad \vec{x}_2(t_*) = \vec{x}_{20} - \frac{\vec{V}_2}{|\vec{V}|^2} \cdot (\vec{V} \cdot \vec{x}_0), \tag{29}$$

and their relative position is given by

$$\begin{aligned}\vec{x}(t_*) &= \vec{x}_1(t_*) - \vec{x}_2(t_*) \\ &= \vec{x}_{10} - \vec{x}_{20} - \frac{(\vec{V} \cdot \vec{x}_0)}{|\vec{V}|^2}(\vec{V}_1 - \vec{V}_2) \\ &= \vec{x}_0 - \frac{\vec{V}}{|\vec{V}|^2}(\vec{V} \cdot \vec{x}_0) \\ &= \frac{\vec{x}_0|\vec{V}|^2 - \vec{V}(\vec{V} \cdot \vec{x}_0)}{|\vec{V}|^2} = \frac{\vec{V} \wedge (\vec{x}_0 \wedge \vec{V})}{|\vec{V}|^2},\end{aligned} \tag{30}$$

where the double outer product of vectors was used.

The distance of closest approach (31b) is the modulus of the relative position vector (30)≡(31a):

$$\vec{x}(t_*) = \frac{\vec{V} \wedge (\vec{x}_0 \wedge \vec{V})}{|\vec{V}|^2}, \tag{31a}$$

$$D_{\min} = D(t_*) = \frac{|\vec{V} \wedge (\vec{x}_0 \wedge \vec{V})|}{|\vec{V}|^2}. \tag{31b}$$

Denoting by $\vec{N}$ the unit vector orthogonal to the plane of the initial relative position (9a) and constant relative velocity (9b) leads to the following: (i) the outer product (32a):

$$\vec{x}_0 \wedge \vec{V} = |\vec{x}_0||\vec{V}|\sin\alpha\vec{N}; \tag{32a}$$

$$|\vec{V} \wedge (\vec{x}_0 \wedge \vec{V})| = |\vec{x}_0||\vec{V}|^2 \sin\alpha; \tag{32b}$$

(ii) since the outer product involving $\vec{V}$ is orthogonal to $\vec{V}$, it follows (32b). Substituting (32b) in (31b) specifies the distance of closest approach (33):

$$D_{\min} = |\vec{x}_0|\sin\alpha > \delta, \tag{33}$$

which must exceed the safety distance to avoid a conflict between the vehicles. The distance of closest approach can also be obtained substituting the time of closest approach (28b) in the distance (25):

$$\begin{aligned}(D_{\min})^2 &= [D(t_*)]^2 \\ &= |\vec{x}_0|^2 + |\vec{V}|^2\left[\frac{\vec{x}_0 \cdot \vec{V}}{|\vec{V}|^2}\right]^2 - 2\frac{(\vec{x}_0 \cdot \vec{V})^2}{|\vec{V}|^2} \\ &= |\vec{x}_0|^2 - \frac{(\vec{x}_0 \cdot \vec{V})^2}{|\vec{V}|^2} = |\vec{x}_0|^2(1 - \cos^2\alpha) \\ &= |\vec{x}_0|^2 \sin^2\alpha,\end{aligned} \tag{34}$$

in agreement with (33).

2.3. Collision Avoidance for Several Vehicles. The preceding results with uniform motion apply to collision avoidance with two vehicles, as stated in the **theorem on uniform motion:** *two vehicles with given distinct initial positions* $(\vec{x}_{10}, \vec{x}_{20})$ *and constant velocities* $(\vec{V}_1, \vec{V}_2)$ *collide (35a) only if the initial*

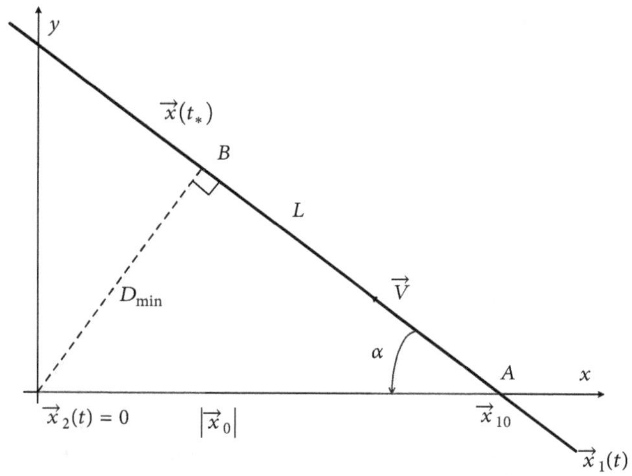

FIGURE 4: The distance of closest approach D_m is a unique minimum in the case of constant velocities when the trajectories are straight lines described in uniform motion. The distance of closest approach in this case is calculated most simply by choosing a Cartesian reference frame moving with the velocity of the second vehicle, so that it stays at the origin $\vec{x}_2(t) = 0$ for all time. The OX axis is taken through the initial position $\vec{x}_{10}$ of the first vehicle, at a distance $|\vec{x}_{10}|$. The trajectory of the first vehicle is a straight line in the direction of the relative velocity making an angle α with the x-axis. The distance of the closest approach is thus (33). The distance covered by the first vehicle is (37a), and since it travels at the relative velocity $|V|$ the time of closest approach is the ratio (37b).

relative position (9a) and the constant relative velocity (9b) are antiparallel (35b)≡(35c):

$$D_{min} = 0: \tag{35a}$$

$$\alpha = \pi, \tag{35b}$$

$$\vec{x}_0 \parallel -\vec{V}, \tag{35c}$$

$$t_* = \frac{|\vec{x}_0|}{|\vec{V}|}, \tag{35d}$$

in which case collision (28b) occurs at time (35d). If the relative velocity makes an angle α with the initial relative position, the time of closest approach is (28b) and the distance of closest approach is (33). In order that a minimum safety distance δ can be maintained at all times it is sufficient that the angle of the constant relative velocity (9b) with the relative initial position (9a) satisfies

$$\sin\alpha > \frac{\delta}{|\vec{x}_0|} = \sin\alpha_0. \tag{36}$$

In this case one vehicle "misses the other" or "passes by" at a distance larger than the minimum separation distance.

These results can be interpreted by a simple geometrical construction involving elementary Euclidean geometry (Figure 4): (i) since only relative motion is of interest the origin O is taken at the second vehicle; (ii) the initial position A of the first vehicle specifies the initial distance $|\vec{x}_0|$; (iii) through A a straight line is drawn in the direction of the constant relative velocity that makes an angle α with the relative initial position; (iv) an orthogonal line through the origin intersects at the point B of closest approach; (v) it follows that the distance of closet approach is (33); (vi) the distance travelled by the first vehicle from the initial position to the point of closest approach is (37a):

$$L = |\vec{x}_0| \cos\alpha; \tag{37a}$$

$$-t_* = \frac{L}{|\vec{V}|} = \frac{|\vec{x}_0|}{|\vec{V}|} \cos\alpha, \tag{37b}$$

(vii) since the relative velocity is constant the time of closest approach is (37b) in agreement with (28b); (viii) the minus sign arises because of the angle α is measured from the relative initial position to the constant relative velocity; (ix) at the point of closest approach the relative position (30) is orthogonal to the relative velocity:

$$\vec{x}(t_*) \cdot \vec{V} = \frac{\vec{V} \cdot \left[\vec{V} \wedge \left(\vec{x}_0 \wedge \vec{V}\right)\right]}{|\vec{V}|^2} = 0, \tag{38}$$

in agreement with the general condition ((16a), (16b)); (x) the condition (19b) for a minimum distance is met for zero relative acceleration $0 > -|\vec{V}|^2$.

The angle α in (36) would be imaginary if $\sin\alpha > 1$ but in this case $|\vec{x}_0| < \delta$ the initial separation would already be less than the minimum separation distance, and the safety criterion was violated from the start. If the safety condition is not violated from the start, the initial relative distance is larger than the safe separation distance, and a real angle α_0 exists. It suffices that the angle of the relative velocity with the relative initial position exceeds this value in modulus $|\alpha| > |\alpha_0|$ for the safe separation to be ensured at all times. The problem of safe separation between two vehicles, with a given safe separation distance δ as a chosen parameter, may be extended to several vehicles, say $n + 1$. Choosing vehicle 1 as the reference the exclusion of collision with each of the other N vehicles leads to N conditions of type (36). If these N conditions are compatible the first vehicle can avoid a collision with all the others by being steered in a constant direction. If the N conditions are incompatible a collision cannot be avoided in a straight path and a polygonal or curved path will be necessary. That may not be sufficient to avoid a collision if the traffic density is too high.

Three situations can arise concerning the time of closest approach (28b):

$$t_* > 0 \quad \text{if } \vec{V} \cdot \vec{x}_0 < 0, \tag{39a}$$

$$t_* = 0 \quad \text{if } \vec{V} \cdot \vec{x}_0 = 0, \tag{39b}$$

$$t_* < 0 \quad \text{if } \vec{V} \cdot \vec{x}_0 > 0. \tag{39c}$$

The time of closest approach is time zero (39b) if the initial relative position is orthogonal to the constant relative velocity,

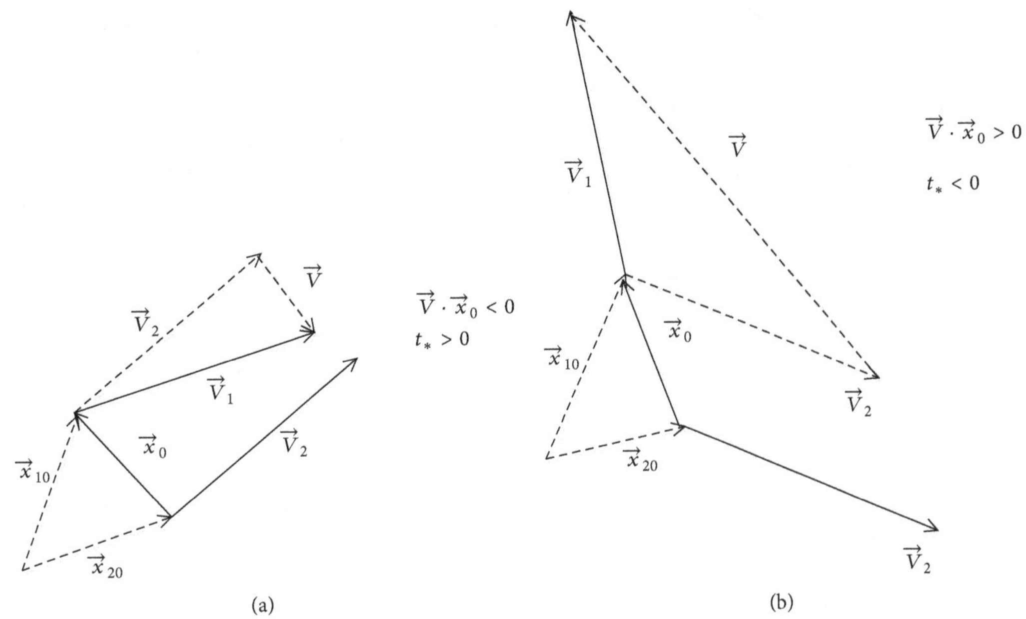

FIGURE 5: For vehicles moving with constant velocities (V_1, V_2) if the relative velocity $\vec{V} \equiv \vec{V}_1 - \vec{V}_2$ makes an obtuse angle $\vec{x}_0 \cdot \vec{V} < 0$ with the relative $\vec{x}_0 \equiv \vec{x}_{10} - \vec{x}_{20}$ initial positions then (case (a)) the trajectories converge, and the time of closest approach is in the future $t_* \geq 0$. If the relative velocity $\vec{V}$ makes an acute angle $\vec{x}_0 \cdot \vec{V} > 0$ with the relative initial position $\vec{x}_0$, the trajectories diverge, and the closest approach was in the past $t_* < 0$.

in agreement with condition (16b) of closest approach. If the vehicles move towards each other (39a) the time of closest approach is positive because (Figure 5(a)) the trajectories are converging and will be closest in the future. If the two vehicles are moving away from each other (39c) the time of closest approach is negative, because the trajectories are diverging (Figure 5(b)) and were closer in the past. Thus two situations arise: (a) for converging trajectories (40a), that is, positive time of closest approach (39a), the minimum separation distance is (31b):

$$\vec{V}\cdot\vec{x}_0 > 0: \quad D_{\min} = \left|\vec{V}\right|^{-2}\left|\vec{V}\wedge\left(\vec{x}_0\wedge\vec{V}\right)\right|, \tag{40a}$$

$$\vec{V}\cdot\vec{x}_0 \leq 0: \quad D_{\min} = \left|\vec{x}_0\right|; \tag{40b}$$

(b) for parallel or diverging trajectories (40b) the minimum separation distance is the initial separation, because the vehicles will not get closer in the future.

3. Safe Separation between Straight and Curved Trajectories

As examples of the preceding theory, the second theorem is used in the simplest case of uniform motion along straight paths (Sections 2.2 and 2.3) both in two (Section 3.1) and three (Section 3.2) dimensions motivated by collision avoidance between (i) ships or aircraft moving on ground at airports (Section 3.1)and (ii) aircraft in flight (Section 3.2). The first theorem is needed in the third (iii) case involving a curved path (Section 3.3).

3.1. Two-Dimensional Safe Separation between Ships in a Water. The first application is conflict avoidance between two ships; following nautical practice the speed is in knots (kt) in (41a) and heading (41b) in degrees:

$$\{V_1, V_2\} = \{10, 20\}\, kt, \tag{41a}$$

$$\{\varphi_1, \varphi_2\} = \{30^\circ, 150^\circ\}, \tag{41b}$$

and hence time is in hours and distance in nautical miles (nm), e.g., for the initial positions (41c,d) in a Cartesian frame:

$$\vec{x}_{10} = \{-10, 5\}\, nm, \tag{41c}$$

$$\vec{x}_{20} = \{5, -15\}\, nm. \tag{41d}$$

It is necessary to find if the minimum separation distance satisfies the safety threshold of 20 nm:

$$\left|\vec{x}_1(t) - \vec{x}_2(t)\right|_{\min} \geq \delta = 20\, nm; \tag{42}$$

if this is not the case then the modulus of the velocity of the second ship V_2 is to be modified to V_2' as little as possible so that the minimum safety distance is complied with. The problem would be similar with different values for the ground movements of two aircraft at an airport or other cases of two-dimensional motion of vehicles like cars.

The Cartesian components of the velocities of the two ships are

$$\{V_{1x}, V_{1y}\} = V_1 \{\cos\varphi_1, \sin\varphi_1\} = \{8.66, 5.00\}\, kt, \quad (43a)$$

$$\{V_{2x}, V_{2y}\} = V_2 \{\cos\varphi_2, \sin\varphi_2\} = \{-17.32, 10.00\}\, kt. \quad (43b)$$

This specifies the relative velocity (44a) and relative initial position (44b):

$$\vec{V} = \vec{V}_1 - \vec{V}_2 = \{25.98, -5.00\}\, kt, \quad (44a)$$

$$\vec{x}_0 = \vec{x}_{10} - \vec{x}_{20} = \{-15, 20\}\, nm. \quad (44b)$$

The angle between them is calculated from

$$\left|\vec{V}\right| = 26.46\, kt, \quad (45a)$$

$$\left|\vec{x}_0\right| = 25\, nm, \quad (45b)$$

$$\cos\alpha = \frac{\vec{V}\cdot\vec{x}_0}{\left|\vec{V}\right|\left|\vec{x}_0\right|} = -0.7403, \quad (45c)$$

where (45c) has two roots.

The angle is less than 180° so the root (46a) of (45c) is chosen:

$$\alpha = 137.76^\circ, \quad (46a)$$

$$D_{\min} = \left|\vec{x}_0\right| \sin\alpha = 16.80\, nm, \quad (46b)$$

$$t_* = -\frac{\left|\vec{x}_0\right|}{\left|\vec{V}\right|}\cos\alpha = 0.6995\, h, \quad (46c)$$

leading to a distance of closest approach (46b) after a time (46c). At that time the position of the two ships is

$$\vec{x}_1(t_*) = \vec{x}_{10} + \vec{V}_1 \cdot t_* = \{-3.94, 8.50\}\, nm, \quad (47a)$$

$$\vec{x}_2(t_*) = \vec{x}_{20} + \vec{V}_2 \cdot t_* = \{-7.11, -8.00\}\, nm, \quad (47b)$$

confirming the separation (46b)≡(48a)

$$\left|\vec{x}_1(t_*) - \vec{x}_2(t_*)\right| = 16.80\, nm < \delta \equiv 20\, nm, \quad (48a,b)$$

is less than the desired safe separation distance (48b) ≡(42).

In the example shown the distance of minimal approach (46b) is less than the desired safe separation distance (42)≡(48b), and thus the angle between the relative velocity and initial relative position must be decreased to α' satisfying

$$\sin\alpha' = \frac{\delta}{\left|\vec{x}_0\right|} = 0.8, \quad (49a)$$

$$\alpha' = 126.87^\circ, \quad (49b)$$

$$\cos\alpha' = -0.6. \quad (49c)$$

It is required that the desired minimum separation distance is to be obtained by changing only the modulus of velocity of the second ship from $V_2 = 20kt$ in (41a) to V_2' to be determined next; the speed of the first ship V_1 is unchanged and the headings of the two ships (φ_1, φ_2) are also maintained. The Cartesian components of the velocity of the first ship are the same (43a) and those of the second ship (43b) change to

$$\left(V_{2x}', V_{2y}'\right) = V_2' \{\cos\varphi_2, \sin\varphi_2\} = \left\{-\frac{\sqrt{3}}{2}, \frac{1}{2}\right\} V_2'. \quad (50)$$

The relative initial position (44b) remains and the relative velocity (44a) changes to

$$\vec{V}' = \vec{V}_1 - \vec{V}_2' = \left\{5\sqrt{3} + \frac{\sqrt{3}}{2}V_2', 5 - \frac{1}{2}V_2'\right\}. \quad (51)$$

The square of the modulus of the relative velocity changes to

$$\left|\vec{V}'\right|^2 = \left(5\sqrt{3} + \frac{\sqrt{3}}{2}V_2'\right)^2 + \left(5 - \frac{1}{2}V_2'\right)^2 = \left(V_2'\right)^2 + 10V_2' + 100. \quad (52)$$

The projection of the relative velocity (51) on the relative initial position (44b) changes to

$$\vec{V}' \cdot \vec{x}_0 = -15\left(5\sqrt{3} + \frac{\sqrt{3}}{2}V_2'\right) + 20\left(5 - \frac{1}{2}V_2'\right) = -29.90 - 22.99V_2'. \quad (53)$$

Its square appears in (45c) modified to

$$\left|\vec{V}'\right|^2 \left|\vec{x}_0\right|^2 \cos^2\alpha' = \left(\vec{V}' \bullet \vec{x}_0\right)^2. \quad (54)$$

Substitution of (52), (53), (45b), and (49c) in (54) leads to

$$225\left[\left(V_2'\right)^2 + 10V_2' + 100\right] = 894 + 1374.8V_2' + 528.5\left(V_2'\right)^2, \quad (55)$$

which is a quadratic equation in V_2':

$$303.5\left(V_2'\right)^2 - 875.2V_2' - 21606 = 0, \quad (56)$$

with roots:

$$V_2' = 1.442 \pm 8.560. \quad (57a,b)$$

The two roots,

$$V_{2+}' = 10.002 > V_{2-}' = -7.118, \quad (58a,b)$$

correspond to slowing down $V_{2+}' = 10kt < 20kt = V_2$ or reversing the motion $V_{2-}' = -7.1kt < 0 < 20kt = V_2$ of the second ship. They lead, respectively, to the relative velocities

$$V_+' = \left|(V_1)^2 + \left(V_{2+}'\right)^2 - 2V_1V_{2+}'\cos\alpha'\right|^{1/2} = 17.89\, kt, \quad (59a)$$

$$V_-' = \left|(V_1)^2 + \left(V_{2-}'\right)^2 - 2V_1V_{2-}'\cos\alpha'\right|^{1/2} = 8.08\, kt \quad (59b)$$

and distinct times of closet approach:

$$t_{+} = -\frac{|\vec{x}_0|}{|\vec{V'}_{+}|}\cos\alpha' = 0.8385\,h, \tag{60a}$$

$$t_{-} = \frac{|\vec{x}_0|}{|\vec{V'}_{-}|}\cos\alpha' = -1.8564\,h. \tag{60b}$$

The relative positions of the two ships at these times are

$$\begin{aligned}\vec{x}_{+} &= \vec{x}_1 - \vec{x}_{2+} = \vec{x}_{10} + \vec{V}_1 t_{+} - \vec{x}_{20} - \vec{V}'_{2+} t_{+} = \vec{x}_0 \\ &+ \left(\vec{V}_1 - \vec{V}'_{2+}\right) t_{+} = \{x_0 \\ &+ \left(V_{1x} - V'_{2+}\cos\varphi_2\right) t_{+}, y_0 \\ &+ \left(V_{1y} - V'_{2+}\sin\varphi_2\right) t_{+}\} = \{-0.48, 19.99\},\end{aligned} \tag{61a}$$

$$\begin{aligned}\vec{x}_{-} &= \vec{x}_1 - \vec{x}_{2-} = \vec{x}_{10} + \vec{V}_1 t_{-} - \vec{x}_{20} - \vec{V}'_{2-} t_{-} = \vec{x}_0 \\ &+ \left(\vec{V}_1 - \vec{V}'_{2-}\right) t_{-} = \{x_0 \\ &+ \left(V_{1x} - V'_{2-}\cos\varphi_2\right) t_{-}, y_0 \\ &+ \left(V_{1y} - V'_{2-}\sin\varphi_2\right) t_{-}\} = \{-19.63, 4.11\}.\end{aligned} \tag{61b}$$

Both solutions correspond to the desired minimum safety distance:

$$|\vec{x}_{+}| = 20.0\,km = \delta = |\vec{x}_{-}|, \tag{62}$$

obtained either by moving faster ahead (59a) or by slowing down (59b).

3.2. Three-Dimensional Safe Separation between Aircraft in Climb and at Cruise. The second example is conflict avoidance between aircraft in flight; following aeronautical practice the speed is in kilometers per hour, for the first aircraft in level flight (63a) and the second climbing (63b):

$$\vec{V}_1 = \{900, 0, 0\}\,km/h, \quad \vec{V}_2 = \{600, -200, 50\}\,km/h, \tag{63a}$$

$$\vec{x}_{10} = \{0, 0, 10\}\,km, \quad \vec{x}_{20} = \{10, 20, 5\}\,km. \tag{63b}$$

The initial positions are ((63a),(63b)) in kilometers (km) and time is in hours (h). The required safe separation distance is

$$\left|\vec{x}_1(t) - \vec{x}_2(t)\right|_{\min} \ge \delta = 15\,km; \tag{64}$$

if it is not met then only the vertical velocity of the second aircraft V_{2z} is to be changed to V'_{2z} by the minimum amount to satisfy the safe separation distance. The safety distances in (42) and (64) are examples, with standard values specified by nautical or aeronautical separation standards. For example, in aeronautics the horizontal separation may be 5 or 10 nautical miles in radar controlled airspace and 50 nautical miles in oceanic airspace and vertical separation may be 2000 or 1000 feet in altitude.

The relative initial position (65a) and the relative velocity (65b) are

$$\vec{x}_0 = \vec{x}_{10} - \vec{x}_{20} = \{-10, -20, 5\}\,km. \tag{65a}$$

$$\vec{V} = \vec{V}_1 - \vec{V}_2 = \{300, 200, -50\}\,km/h, \tag{65b}$$

and the angle between them is calculated from

$$|\vec{x}_0| = 22.9\,km, \tag{66a}$$

$$|\vec{V}| = 364\,km/h, \tag{66b}$$

$$\cos\alpha = \frac{\vec{V}\cdot\vec{x}_0}{|\vec{V}|\,|\vec{x}_0|} = -0.870. \tag{66c}$$

The angle ((67a), (67b)) leads to a distance of closest approach (67c) at the time (67d):

$$\alpha = 150.43^\circ, \tag{67a}$$

$$\sin\alpha = 0.493, \tag{67b}$$

$$D_{\min} = |\vec{x}_0|\sin\alpha = 11.3\,km, \tag{67c}$$

$$t_{*} = -\frac{|\vec{x}_0|}{|\vec{V}|}\cos\alpha = 0.0547. \tag{67d}$$

The positions of the two aircraft at this time

$$\vec{x}_1(t_*) = \vec{x}_{10} + \vec{V}_1 t_* = \{49.23, 0, 10\}\,km, \tag{68a}$$

$$\vec{x}_2(t_*) = \vec{x}_{20} + \vec{V}_2 t_* = \{42.82, 9.06, 7.73\}\,km \tag{68b}$$

confirm the minimum separation (67c) ≡ (69a):

$$\left|\vec{x}_1(t_*) - \vec{x}_2(t_*)\right| = 11.3\,km = D_{\min} < \delta = 15\,km, \tag{69a,b}$$

which is less than the required safety distance (69b) ≡ (64).

The required separation distance implies that the angle between the relative velocity and initial relative position must satisfy

$$\sin\alpha' = \frac{\delta}{|\vec{x}_0|} = 0.655, \tag{70a}$$

$$\alpha' = 139.07^\circ, \tag{70b}$$

$$\cos\alpha' = -0.7556; \tag{70c}$$

this will be met by changing only the vertical velocity of the second aircraft from $V_{2z} = 50km/h$ in (63b) to V'_{2z} determined in the sequel. The identity (54) is used again, where (i) the modulus of the relative velocity is

$$\left|\vec{V}'\right|^2 = \left|\vec{V}_1 - \vec{V}'_2\right|^2 = 300^2 + 200^2 + \left(V'_{2z}\right)^2 = 1.3 \times 10^5 + \left(V'_{2z}\right)^2; \tag{71a}$$

(ii) the inner product by the relative position is

$$\vec{V}' \cdot \vec{x}_0 = 300x_0 + 200y_0 - V'_{2z}z_0 = -7 \times 10^3 - 5V'_{2z}. \tag{71b}$$

Substituting ((71a), (71b), (66a)) and (70c) in (54) leads to a quadratic equation for the vertical velocity of the second aircraft:

$$\left(5V'_{2z} + 7 \times 10^3\right)^2 = 299.4\left[1.3 \times 10^5 + \left(V'_{2z}\right)^2\right]. \tag{72a}$$

The binomial (72a) $\equiv$ (72b)

$$274.4\left(V'_{2z}\right)^2 - 7 \times 10^4 V'_{2z} - 1.0078 \times 10^7 = 0 \tag{72b}$$

has roots

$$V'_{2z} = 127.55 \pm 230.21 = +357.76, -102.66, \tag{73a,b}$$

leading to two possible vertical velocities, one climbing and the other descending.

The corresponding relative velocities are

$$\left|\vec{V}'_+\right| = \left|\vec{V}_1 - \vec{V}'_{2+}\right| = 507.9\, km/h, \tag{74a}$$

$$\left|\vec{V}'_-\right| = \left|\vec{V}_1 - \vec{V}'_{2-}\right| = 374.9\, km/h. \tag{74b}$$

The corresponding times of closet approach are

$$t_+ = -\frac{\left|\vec{x}_0\right|}{\left|\vec{V}'_+\right|}\cos\alpha' = 0.03407h, \tag{75a}$$

$$t_- = -\frac{\left|\vec{x}_0\right|}{\left|\vec{V}'_-\right|}\cos\alpha' = 0.04615h. \tag{75b}$$

The relative positions at these times are

$$\vec{x}_+ = \vec{x}_1 - \vec{x}'_{2+} = \vec{x}_0 + \left(\vec{V}_1 - \vec{V}'_+\right)t_+ = \{0.221, -13.19, -7.19\}, \tag{76a}$$

$$\vec{x}_- = \vec{x}_1 - \vec{x}'_{2-} = \vec{x}_0 + \left(\vec{V}_1 - \vec{V}'_-\right)t_- = \{3.845, -10.77, 9.74\}. \tag{76b}$$

Both roots lead to the desired minimum separation distance:

$$\left|\vec{x}_-\right| = 15.0km = \delta = \left|\vec{x}_+\right|, \tag{77}$$

in different ways, one by descending (73b) and the other by climbing faster (73a). Next is considered a different kind of problem with one curved trajectory and time of closest approach imposed "a priori".

3.3. Minimum Distance between a Curved and a Straight Trajectory. The third example uses SI units for the velocity in meters per second (78b) of an aircraft flying in a circular path of radius (78a) in meters, using time in seconds in the centripetal acceleration (78c) that is small compared with that of gravity:

$$R = 20km = 2 \times 10^4 m, \tag{78a}$$

$$V = 216km/h = 60m/s, \tag{78b}$$

$$a = \frac{V^2}{R} = 0.18m/s^2 = 0.018g. \tag{78c}$$

The trajectory in a Cartesian reference frame with origin at the centre is

$$x_1(t) = R\cos\left(\frac{V}{R}t\right), \tag{79a}$$

$$y_1(t) = R\sin\left(\frac{V}{R}t\right) \tag{79b}$$

where $\Omega = V/R$ is the angular velocity. The second aircraft starts at an initial position (80a) and flies at the same altitude with a constant heading (80b):

$$\vec{x}_0 = \left\{-5 \times 10^4, 0\right\} m = \{-L, 0\}, \tag{80a}$$

$$\varphi = 30^\circ, \tag{80b}$$

$$t_* = 600s, \tag{80c}$$

with constant velocity u such that the time of closest approach is (80c). It is necessary to find the distance of closest approach and the corresponding positions of the two aircraft.

The trajectory of the second aircraft is

$$x_2(t) = -L + ut\cos\varphi, \tag{81a}$$

$$y_2(t) = ut\sin\varphi. \tag{81c}$$

The relative position is

$$\vec{x}(t) = \vec{x}_1(t) - \vec{x}_2(t) = \left\{R\cos\left(\frac{V}{R}t\right) + L - ut\cos\varphi, R\sin\left(\frac{V}{R}t\right) - ut\sin\varphi\right\}, \tag{82}$$

and its derivative with regard to time

$$\vec{V}(t) = \frac{d\vec{x}}{dt} = \left\{-V\sin\left(\frac{V}{R}t\right) - u\cos\varphi, V\cos\left(\frac{V}{R}t\right) - u\sin\varphi\right\} \tag{83}$$

specifies the relative velocity (83).

The condition of closest approach is ((16a), (16b)) that the relative position (82) and velocity (83) should be orthogonal:

$$\begin{aligned} 0 &= \vec{x}(t_*)\cdot\vec{V}(t_*) \\ &= u^2 t_* - Ru\cos\left(\frac{V}{R}t_* - \varphi\right) + Vut_*\sin\left(\frac{V}{R}t_* - \varphi\right) \\ &\quad - L\left[u\cos\varphi + V\sin\left(\frac{V}{R}t_*\right)\right]. \end{aligned} \tag{84}$$

The time of closest approach (80c) is satisfied if the velocity along the straight path satisfies the quadratic relation:

$$\begin{aligned} 0 &= t_* u^2 + u(Vt_*\sin\psi - R\cos\psi - L\cos\varphi) \\ &\quad - LV\sin\phi, \end{aligned} \tag{85}$$

where for the present problem ((78a), (78b); (80c), (80b))

$$\phi = \frac{V}{R}t_* = 1.8rad = 103.13^\circ, \qquad \psi = \frac{V}{R}t_* - \varphi = 73.13^\circ; \tag{86}$$

the quadratic (85)

$$0 = 6\times 10^2 u^2 - 1.4655\times 10^4 u - 2.9216\times 10^6 \tag{87}$$

has roots

$$u_\pm = 12.21 \pm 70.84 = +83.05, -58.63. \tag{88a,b}$$

The corresponding positions along the straight path at the time of closest approach are

$$\vec{x}_+ = \{-L + u_+ t_*\cos\varphi, u_+ t_*\sin\varphi\} = \left\{-6.85\times 10^3, 2.49\times 10^4\right\}, \tag{89a}$$

$$\vec{x}_- = \{-L + u_- t_*\cos\varphi, u_- t_*\sin\varphi\} = \left\{-8.05\times 10^4, -1.76\times 10^4\right\}; \tag{89b}$$

at the same time the position on the circular path is

$$\vec{x} = R\{\cos\phi, \sin\phi\} = \left\{-4.54\times 10^3, 1.95\times 10^4\right\} m. \tag{90}$$

The extrema for the relative distance,

$$\left|\vec{x} - \vec{x}_+\right| = 5.87\times 10^3 m < 8.45\times 10^4 m = \left|\vec{x} - \vec{x}_-\right|, \tag{91a,b}$$

may be local maxima or minima.

To clarify this the acceleration is needed, which is specified by the circular motion alone:

$$\vec{a} = \frac{d^2\vec{x}}{dt^2} = -\frac{V^2}{R}\{\cos\phi, \sin\phi\} = \left\{4.09\times 10^{-2}, -1.75\times 10^{-1}\right\} m/s^2. \tag{92}$$

Criterion (18b) requires also (83) relative velocities:

$$\vec{V}_+ = \{-V\sin\phi - u_+\cos\varphi, V\cos\phi - u_+\sin\varphi\} = \left\{-1.30\times 10^2, -5.52\times 10\right\} m/s, \tag{93a}$$

$$\vec{V}_- = \{-V\sin\phi - u_-\cos\varphi, V\cos\phi - u_-\sin\varphi\} = \{-7.66, 1.57\times 10\} m/s, \tag{93b}$$

through the square of the moduli:

$$\left|\vec{V}_+\right|^2 = 1.99\times 10^4 m^2/s^2 > 3.05\times 10^2 m^2/s^2 = \left|\vec{V}_-\right|^2. \tag{94a,b}$$

Acceleration (92) appears projected on the relative position ((89a), (89b); (90)):

$$\vec{x} - \vec{x}_+ = \left\{2.31\times 10^3, -5.40\times 10^3\right\} m, \tag{95a}$$

$$\vec{x} - \vec{x}_- = \left\{7.60\times 10^4, 3.71\times 10^4\right\} m, \tag{95b}$$

leading to

$$\vec{a}\cdot\left(\vec{x} - \vec{x}_+\right) = 1.04\times 10^3 m^2/s^2; \tag{96a}$$

$$\vec{a}\cdot\left(\vec{x} - \vec{x}_-\right) = -3.38\times 10^3 m^2/s^2. \tag{96b}$$

Criterion (18b),

$$\left|\vec{V}_+\right|^2 + \left(\vec{x} - \vec{x}_+\right)\cdot\vec{a} = 2.094\times 10^4 m^2/s^2 > 0, \tag{97a}$$

$$\left|\vec{V}_-\right|^2 + \left(\vec{x} - \vec{x}_-\right)\cdot\vec{a} = -3.075\times 10^3 m^2/s^2 < 0, \tag{97b}$$

shows that the first solution that has positive velocity corresponds to the closest approach and is both a local and a global minimum for the relative distance. The second solution that has negative velocity is a local maximum of the relative distance; it is not an absolute maximum because the latter corresponds to infinite distance.

4. Discussion

The problem of conflict avoidance between two vehicles involves, besides a safe separation distance chosen "a priori", 4 vectors, namely, the initial position vectors $(\vec{x}_{10}, \vec{x}_{20})$ and constant velocity vectors $(\vec{V}_1, \vec{V}_2)$ of each vehicle, leading to $4\times 2 = 8$ components in two dimensions and $4\times 3 = 12$ components in three dimensions. Since a conflict is concerned with relative motion, it might be expected that only four variables would be relevant, namely, the relative initial position (9a) and the constant relative velocity (9b). It is remarkable that the distance of closest approach (33) and hence the safety criterion involve only two combinations of these four variables: (i) initial distance (not the direction); (ii) the angle of the relative velocity with the relative initial position (not the modulus of the relative velocity). Thus a conflict can be avoided by (a) steering only one vehicle, so that the condition (36) is met, with no need to change the modulus of the velocity; (b) alternatively the heading that may be retained and conflict avoidance which is implemented by increasing the velocity to race ahead or slowing down or reversing the motion to fall behind. Other less simple combinations are to change the heading and velocity of one or both vehicles. For example, a conflict between aircraft can be resolved by changing speed (acceleration or deceleration) or heading or climb or descend rate for one or both aircraft. The present method applies to all these cases and combinations thereof.

Besides (i) the initial relative distance and (ii) angle of the relative initial position with the relative velocity, the (iii) modulus of the relative velocity also appears in the time of closest approach (28b)≡(37b). The condition of conflict avoidance between two vehicles thus requires that one of them be steered out of a dangerous angular sector for which a collision would occur. In the case of several vehicles a conflict can be avoided if the dangerous angular sectors for each vehicle do not cover all directions of straight travel. If all directions of straight travel lead to conflicts then curved trajectories must be considered. The cases of conflict between vehicles with nonuniform velocity along curved paths are addressed by the first theorem, of which the second is a particular case. The minimum distance between two curves occurs at a point with common normal (Figure 6); this would be the distance of closest approach only if the two vehicles would be at these points at the same time. In that case the relative velocity, that is tangential, would be orthogonal to the relative position, that is along the normal ((16a), (16b)); if the curves have opposite curvatures it would also be locally the point of closest approach ((19a), (19b)). However, in general the vehicles following the two paths will pass by nearest points at different times (Figure 2) and the distance of closest approach will be larger than the minimum distance between the two curves.

In all cases of uniform or nonuniform motion along straight or curved paths it remains true that an extremum of the relative distance (16a) will correspond to a relative velocity orthogonal to the relative position (16b) and that extremum will be the minimum separation distance if the relative acceleration (19a) meets the condition (19b) that may be interpreted as stating that the total acceleration, including the centripetal acceleration, must cause the vehicles to move away from each other. The latter two conditions ((16a), (16b) and (19a), (19b)) are necessary and sufficient to determine the distance of closest approach and apply the criterion of minimum safe separation distance for two vehicles or vehicles with arbitrary trajectories and velocities along them. The minimum is a local minimum for curved trajectories

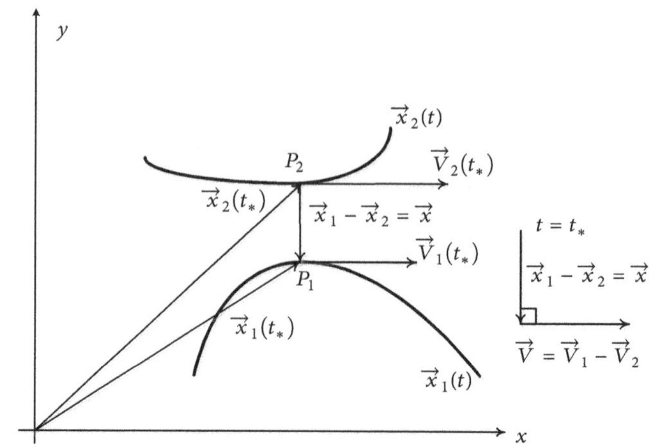

FIGURE 6: The distance between two paths $\vec{x}_1$ and $\vec{x}_2$ would be the distance of closest approach between the two trajectories $\vec{x}_1(t)$ and $\vec{x}_2(t)$ only if the two vehicles were at the closest points P_1 and P_2 at the same time t_*. In that case the relative position $\vec{x} \equiv \vec{x}_1 - \vec{x}_2$ is along the common normal and the relative velocity $\vec{V} \equiv \vec{V}_1 - \vec{V}_2$ is tangential so they are orthogonal $\vec{x} \cdot \vec{V} = 0$. This orthogonality condition applies at the time of closest approach (16b) in all cases, even if the vehicles pass at the closest points at different times.

with nonuniform velocity and an absolute minimum for straight trajectories with uniform velocity. In the latter case of straight trajectories with constant velocity (Figure 4) the maximum distance is infinity after an infinite time, and there are no local maxima. In the case of nonuniform motion along curved paths there can exist local minima D_{*_n} of the relative distance (Figure 3) as well as maxima F_*. The objective of maximizing the relative distance corresponds to (i) relative position orthogonal to the relative velocity, as for any extremum, and (ii) the local acceleration which must overcome the centripetal acceleration to increase the distance. Thus the dotted path in Figure 1 would correspond to a local maximum of the relative distance whereas the solid curve would correspond to a local minimum.

5. Conclusion

The conflict detection and resolution (CDR) required that the trajectories of two vehicles lead to a relative distance not less than the safe separation distance (SSD) at all times. This can be ensured if the distance of closest approach (DCA) is not less than the SSD. The DCA has been calculated for two arbitrary trajectories, generally curved and with accelerated motion: in this case there can exist several local minima of the relative distance between two vehicles and the DCA is the smallest of them. In the case of two vehicles with constant velocity the rectilinear trajectories lead to one time of closest approach; the maximum distance is unbounded for large time. The preceding results apply to any type of vehicle (car, ship, and aircraft) and to motion in two or three dimensions. They are illustrated by 2 examples with constant velocities: (i) conflict avoidance between two ships changing the modulus of the velocity of one of them; (ii) conflict avoidance between two aircraft changing the climb rate of one of them. A third example involves one straight and one curved trajectory and requires one velocity to be chosen to achieve a time of closest approach chosen "a priori".

List of Symbols

$\vec{a}$:	Acceleration
D:	Distance between two trajectories
D_*:	Distance of closest approach between two trajectories
R:	Radius
t_*:	Time of closest approach
$\vec{u}\cdot\vec{v}$:	Inner product or dot product of two vectors $\vec{u}$ and $\vec{v}$
$\vec{V}_1(t)$:	Velocity vector of vehicle 1
$\vec{V}_2(t)$:	Velocity vector of vehicle 2
$\vec{V}$:	Relative velocity, $\vec{V} \equiv \vec{V}_1 - \vec{V}_2$
$\vec{x}_1(t)$:	Position vector of vehicle 1
$\vec{x}_2(t)$:	Position vector of vehicle 2
$\vec{x}$:	Relative position, $\vec{x} \equiv \vec{x}_1 - \vec{x}_2$
$\vec{x}_{10} \equiv \vec{x}_1(0)$:	Initial position of vehicle 1
$\vec{x}_{20} \equiv \vec{x}_2(0)$:	Initial position of vehicle 2
$\vec{x}_0 \equiv \vec{x}(0)$:	Relative initial position, $\vec{x}_0 \equiv \vec{x}_{10} - \vec{x}_{20}$
δ:	Safe separation distance.

Units

h: Hours
nm: Nautical miles
km: Kilometers
kt: Knots.

Acronyms

ATM: Air Traffic Management
CDR: Conflict detection and resolution
DCA: Distance of closest approach
TCA: Time of closest approach
TLS: Target Level of Safety
SSD: Safe separation distance.

Conflicts of Interest

The authors declare that there are no conflicts of interest regarding the publication of this paper.

Acknowledgments

This work was supported by FCT (Foundation for Science and Technology), through IDMEC (Institute of Mechanical Engineering), under LAETA Pest-OE/EME/LA0022.

References

[1] J. K. Kuchar and L. C. Yang, "A review of conflict detection and resolution modeling methods," *IEEE Transactions on Intelligent Transportation Systems*, vol. 1, no. 4, pp. 179–189, 2000.

[2] J. Jansson, *Collision Avoidance Theory with Application to Automotive Collision Mitigation*, Linköping University, Linköping, Sweden, 2005.

[3] L. M. B. C. Campos and J. M. G. Marques, "On the combination of the gamma and generalized error distribution with application to aircraft flight path deviations," *Communications in Statistics—Theory and Methods*, vol. 33, no. 10, pp. 2307–2332, 2004.

[4] S. B. McLaughlin, J. M. Hankey, and T. A. Dingus, "A method for evaluating collision avoidance systems using naturalistic driving data," *Accident Analysis & Prevention*, vol. 40, no. 1, pp. 8–16, 2008.

[5] A. Tang and A. Yip, "Collision avoidance timing analysis of DSRC-based vehicles," *Accident Analysis & Prevention*, vol. 42, no. 1, pp. 182–195, 2010.

[6] S. McPhail, M. Furlong, and M. Pebody, "Low-altitude terrain following and collision avoidance in a flight-class autonomous underwater vehicle," *Proceedings of the Institution of Mechanical Engineers, Part M: Journal of Engineering for the Maritime Environment*, vol. 224, no. 4, pp. 279–292, 2010.

[7] Z. Lušić and M. Čorić, "Models for estimating the potential number of ship collisions," *Journal of Navigation*, vol. 68, no. 4, pp. 735–749, 2015.

[8] P. G. Reich, "Analysis of long-range air traffic systems: separation standards," *Journal of Navigation*, vol. 19, pp. 88-98, pp. 169-186, pp. 331-347, 1966.

[9] L. M. B. C. Campos, "On the probability of collision between aircraft with dissimilar position errors," *Journal of Aircraft* 38, pp. 593–599, 2001.

[10] K. Bousson, "Model predictive control approach to global air collision avoidance," *Aircraft Engineering and Aerospace Technology*, vol. 80, no. 6, pp. 605–612, 2008.

[11] J. F. Shortle, Y. Zhang, and J. Wang, "Statistical characteristics of aircraft arrival tracks," *Journal of the Transportation Research Board*, vol. no. 2177, pp. 98–104, 2010.

[12] L. F. Vismari and J. B. C. Júnior, "A safety assessment methodology applied to CNS/ATM-based air traffic control system," *Reliability Engineering & System Safety*, vol. 96, pp. 727–738, 2011.

[13] A. Narkawicz, C. Muñoz, H. Herencia-Zapana, and G. Hagen, "Formal verification of lateral and temporal safety buffers for state-based conflict detection," *Proceedings of the Institution of Mechanical Engineers, Part G: Journal of Aerospace Engineering* vol. 227, no. 9, pp. 1412–1424, 2012.

[14] J. Hu, M. Prandini, and S. Sastry, "Aircraft conflict prediction in the presence of a spatially correlated wind field," *IEEE Transactions on Intelligent Transportation Systems*, vol. 6, no. 3, pp. 326–340, 2005.

[15] A. L. Visintini, W. Glover, J. Lygeros, and J. Maciejowski, "Monte Carlo optimization for conflict resolution in air traffic control," *IEEE Transactions on Intelligent Transportation Systems*, vol. 7, no. 4, pp. 470–482, 2006.

[16] H. Emami and F. Derakhshan, "A New Method for Conflict Detection and Resolution in Air Traffic Management," in *Association for the Advancement of Artificial Intelligence. Report Ws-12-12*, pp. 37–40, 2012.

[17] L. M. B. C. Campos and J. M. G. Marques, "On Safety Metrics related to Aircraft Separation," *Journal of Navigation*, vol. 55, pp. 39–63, 2002.

[18] L. M. B. C. Campos and J. M. G. Marques, "On the probability of collision for crossing aircraft," *Aircraft Engineering and Aerospace Technology*, vol. 83, no. 5, pp. 306–314, 2011.

[19] L. M. B. C. Campos and J. M. G. Marques, "On the probability of collision between climbing and descending aircraft," *Journal of Aircraft*, vol. 44, no. 2, pp. 550–557, 2007.

[20] L. M. B. C. Campos and J. M. G. Marques, "On a dimensionless alternative to the ICAO target level of safety," *Proceedings of the Institution of Mechanical Engineers, Part G: Journal of Aerospace Engineering*, vol. 230, no. 9, pp. 1548–1557, 2016.

[21] J. K. Archibald, S. Member, J. C. Hill, N. A. Jepsen, W. C. Stirling, and R. L. Frost, "A satisficing approach to aircraft conflict resolution," *IEEE Transactions on Systems, Man, and Cybernetics, Part C: Applications and Reviews*, vol. 38, no. 4, pp. 510–521, 2008.

[22] H. Emami and F. Derakhshan, "An overview on conflict detection and resolution methods in air traffic management using multi agent systems," in *Proceedings of the 16th CSI International Symposium on Artificial Intelligence and Signal Processing, AISP 2012*, IEEE, May 2012.

[23] H. Emami and F. Derakhshan, "Multi-agent based solution for free flight conflict detection and resolution using particle swarm optimization algorithm," *UPB Scientific Bulletin, Series C: Electrical Engineering*, vol. 76, no. 3, 2014.

[24] R. Sun, F. Xie, D. Xue, Y. Zhang, and W. Y. Ochieng, "A Novel Rear-End Collision Detection Algorithm Based on GNSS Fusion and ANFIS," *Journal of Advanced Transportation*, vol. 2017, Article ID 9620831, 10 pages, 2017.

[25] V. Lücken, N. Voss, J. Schreier et al., "Density-based statistical clustering: enabling sidefire ultrasonic traffic sensing in smart cities," *Journal of Advanced Transportation*, vol. 2018, Article ID 9317291, 15 pages, 2018.

[26] H. Bai, J. Shen, L. Wei, and Z. Feng, "Accelerated lane-changing trajectory planning of automated vehicles with vehicle-to-vehicle collaboration," *Journal of Advanced Transportation*, vol. 2017, Article ID 8132769, 11 pages, 2017.

[27] M. M. Morando, Q. Tian, L. T. Truong, and H. L. Vu, "Studying the safety impact of autonomous vehicles using simulation-based surrogate safety measures," *Journal of Advanced Transportation*, vol. 2018, Article ID 6135183, 11 pages, 2018.

16

Shared Autonomous Vehicles Effect on Vehicle-Km Traveled and Average Trip Duration

Ana T. Moreno, Andrzej Michalski, Carlos Llorca, and Rolf Moeckel

Department of Civil, Geo and Environmental Engineering, Technical University of Munich, Munich 80333, Germany

Correspondence should be addressed to Ana T. Moreno; ana.moreno@tum.de

Academic Editor: Alain Lambert

Intermediate modes of transport, such as shared vehicles or ride sharing, are starting to increase their market share at the expense of traditional modes of car, public transport, and taxi. In the advent of autonomous vehicles, single occupancy shared vehicles are expected to substitute at least in part private conventional vehicle trips. The objective of this paper is to estimate the impact of shared autonomous vehicles on average trip duration and vehicle-km traveled in a large metropolitan area. A stated preference online survey was designed to gather data on the willingness to use shared autonomous vehicles. Then, commute trips and home-based other trips were generated microscopically for a synthetic population in the greater Munich metropolitan area. Individuals who traveled by auto were selected to switch from a conventional vehicle to a shared autonomous vehicle subject to their willingness to use them. The effect of shared autonomous vehicles on urban mobility was assessed through traffic simulations in MATSim with a varying autonomous taxi fleet size. The results indicated that the total traveled distance increased by up to 8% after autonomous fleets were introduced. Current travel demand can still be satisfied with an acceptable waiting time when 10 conventional vehicles are replaced with 4 shared autonomous vehicles.

1. Introduction

Autonomous vehicles (AVs), or driverless cars, have the potential to make a revolution on travel behavior and urban development. While AVs might reduce traffic congestion by improving network capacity and traffic efficiency, they could also increase traffic volumes due to induced travel demand because of improved convenience of travel or because of empty trips of shared autonomous vehicles to pick up the next passenger or to find a parking place [1, 2]. Traditional modes of car, public transport, and taxi are slowly beginning to lose market share to intermediate modes such as shared taxis (e.g., Cabify, Lyft, or Uber) or shared vehicles (e.g., Car2Go, DriveNow, or ZipCar) [3]. When shared AVs (SAVs) are used individually (no ride sharing), they can be assimilated as autonomous taxis. In this sense, there has been a growing interest in how SAVs can operate in urban areas and how they would affect urban mobility.

City-wide simulations of complete replacement of private conventional vehicles (PCVs) with SAVs in Berlin, Lisbon, and Austin indicated that one SAV could replace the demand served by ten PCVs [4–6]. In a region-wide study for the greater Munich metropolitan area, the substitution rates were at least equal to 3-for-10 to maintain reasonable waiting times. In that study, AT use was randomly selected at predefined penetration rates of 20% and 40% [7]. Because of empty trips, vehicle-kilometers traveled (VKT) increased between 4% and 51%, depending on the penetration rate of ATs and the case study.

Previous studies on the impact of SAVs on urban mobility had to simplify various assumptions, such as penetration rates, willingness to use SAVs, or limiting the study area. Milakis et al. [8] expected that the percentage of AVs in vehicle fleets will be less than 11% in 2030 and between 7% and 61% in 2050. Therefore, conventional and autonomous vehicles are probably going to coexist in the next few decades. To forecast the demand side, penetration rates should consider the willingness to use the technology as shared vehicle (single occupancy or ride sharing). Additionally, limiting the study area to single cities can be a significant shortcoming in polycentric regions, particularly for commute trips and long-distance trips.

The objective of this paper is to estimate the impact of SAVs on average trip duration and VKT in the greater Munich metropolitan area, taking into account the willingness to use SAVs.

The organization of the paper is as follows. Section 2 reviews research related to regional impacts of AVs and subjective perceptions by users, after which Section 3 presents the methodology. Results and analyses are discussed in Section 4. Finally, conclusions are presented in Section 5.

2. Literature Review

Researchers have developed frameworks for managing SAVs fleets and their daily operations [9–11] or determined the driving forces for their implementation in large cities [12]. Only rarely it has been attempted to model travel behavior with SAVs at a regional scale. Notable exceptions are the transport analyses in Berlin [4], Lisbon [6, 13], downtown Austin [5], and the metropolitan area of Munich [7] and the optimization of the New York City taxicab fleet using SAV real-time ride sharing [14]. On the demand side, some studies had modeled acceptance and use of SAVs [15–18], even though many others had used stated preference surveys to evaluate the willingness to pay extra for privately owned AVs, as summarized by Becker and Axhausen [19].

2.1. Effects of Shared Autonomous Vehicles on Urban Mobility. Bischoff and Maciejewski [4] tested a complete fleet of shared autonomous vehicles in Berlin using simulations in MATSim with the extension RoboTaxi. MATSim is a multiagent transport simulation software that allows performing large-scale agent-based transport simulations. The model was run for a typical weekday in Berlin 2008 assuming the modal split between auto and transit. The scenarios considered a 100% penetration rate of shared autonomous vehicles. The service assumption was no advance bookings and individual rides (no ride sharing). The initial fleet was distributed in the morning proportionally to population density. The results indicate that waiting times were distributed differently in the city center and the outskirts. While waiting times were quite stable during the day in the city center, some regions had average waiting times of almost 20 minutes. Those regions also had an overall low share of trips and longer pickup distances (from 1 km to 5 km) and produced an empty ride share of up to 45%. The results indicate that 1 SAV can serve the same demand as 11 PCVs (substitution rate of 1-for-11).

Fagnant and Kockelman [5] simulated one typical day in downtown Austin in 2010 using MATSim. All personal trips were dispatched using PCVs. Departure times were distributed using Seattle's household travel diaries. Average travel speeds across the network by hour were obtained from MATSim. An independent module was used to model SAVs, including their trip assignment, movement, relocation, and initial fleet generation. They considered single occupancy and ride sharing, with an increase of travel time by 20% and 40%. Ride sharing reduced the number of extra VKTs from 8.7% to 1.5% due to less and shorter empty trips. The percentage of shared miles increased from 4.8% to 11.2% after allowing 40% longer travel times. Average waiting time was lower than 2 minutes with a substitution rate of 1-for-10.

The International Transport Forum at the OECD [6] simulated the synthetic trips of one average day in Lisbon. All trips with distances below 1 km were assumed to be traveled by foot or bicycle (18% of trips), while trips with origins and destinations close to metro stations were directly assigned to metro (22% of trips). Auto trips were partially substituted by single occupancy SAVs and ride sharing SAVs. They tested nine scenarios varying the mode of AV technology (single occupancy or ride sharing), penetration rate (50% or 100% of AV technology), presence of metro (yes or no), and the base scenario with 100% of PCVs. Fleet size was assessed by calculating the required number of AVs to deliver the same level of mobility as with private conventional cars in terms of travel origins, destinations, and length of trip. Substitution rates 1-for-10 and 2-for-10 were obtained for the scenario of 100% SAVs and 50% SAVs, respectively. Travel volumes increased between 6% and 51%, depending on the AV technology and the penetration rate. For 50% of penetration rate, VKTs increased by 31% with ride sharing and 51% with single occupancy. The impact on v/c ratio was little, from an average of 42% on the base scenario to 45% on the worst case. Finally, average waiting times were between 3 and 4 minutes.

In a follow-up paper, Martinez and Viegas [13] examined the changes in mobility and CO2 emissions in Lisbon in a scenario where taxi-buses and shared taxis modes replace private conventional cars, buses, and taxis. Metro and walking are kept as modes. The trip origins and destinations are the same as in the previous study [6]. For the current scenario mode choice, they derived a nested-logit model based on age and income, land use and mode availability, and transport operation attributes. For the future scenario, the mode choice process had sequential rules due to the absence of stated preferences for shared mobility solutions. Walking, metro, shared-taxi, and taxi-bus selection was tested step by step. Users selected taxi-bus only if they had not selected the previous options. Taxi-buses should be requested 30 minutes in advance, while shared taxis had real-time booking system and performed door-to-door services. If taxi-buses could not supply the demand in a reasonable waiting time, users were upgraded to shared-taxi. The results from one average day indicate that carbon emissions and VKTs would be reduced by 40% and 30%, respectively. The substitution rate from current private conventional vehicles to AVs fleet was 3-for-100 to produce the same mobility, showing an impressive reduction.

Llorca et al. [7] simulated the use of SAVs to substitute partially the travel demand for commute trips in the greater Munich metropolitan area. Commute trips were generated based on the synthetic population. A two-step mode choice model was applied. First, individuals selected among current transport modes: auto, public transport, bicycle, and walking. Secondly, individuals that selected auto mode opted to change their private conventional vehicle by a SAV. This selection was random and the threshold was equal to the predefined penetration rate. Eleven scenarios were considered varying penetration rate (0, 20, and 40%) and substitution rate

(1-for-10 to 5-for-10). The results from the simulations in MATSim showed that the total VKTs increased up to 5% and 10%, for penetration rates of 40 and 20%, respectively. Moreover, the presence of SAVs increased not only SAV average trip duration due to waiting times, but also PCV average trip duration due to SAV empty trips. They suggested substitution rates of 3-for-10 to provide waiting time shorter than 8 minutes.

2.2. Autonomous Vehicle Acceptance and Use. Zmud and Sener [15] and Bansal et al. [16] conducted two state preference online surveys in the Austin metropolitan area. The objective was to gather how likely people are to use AVs, which factors influence their willingness to use them, and how that might change travel behavior in the future. They gathered 556 and 347 responses online, respectively. Zmud and Sener [15] concluded that personality and psychology of the individual, specially self-perception of risk and usage of new technologies, are currently much more important than the demographic characteristics. On the other hand, Bansal et al. [16] found statistical significance among demographic characteristics. Specifically, their results indicate that full-time male workers living in urban areas, without children, are more likely to use shared AVs more frequently. Holding a driver license decreased the likelihood of using shared AVs.

On follow-up stage, Zmud and Sener conducted face-to-face interviews with 44 persons to assess the perceptions of self-driving and their potential usage of AVs. They indicate that "the intent to use is an importance concept because the technology is not yet on the market. It is not until the product becomes tangible and drivers have an opportunity to experience it "for real" that they can form judgments and provide reliable and valid responses to questions pertaining to actual use". Most of the 44 respondents would rather use autonomous vehicles as private cars (59%), compared to 41% that would use them as shared vehicle. There were no age or income difference among individuals interested in shared vehicles. Additionally, 66% of the respondents indicated that their annual mileage would not change with AVs.

Krueger et al. [17] identified the characteristics of users who are likely to adopt shared AVs using a stated choice online survey. The respondents were asked to indicate whether they would switch to a SAV on a trip they recently undertook. A total of 435 residents in major metropolitan areas of Australia completed the survey. Individuals that use all four modes relatively frequently, as well as individuals that travel by car, public transport, and walk frequently are more likely to use shared AVs. Moreover, the willingness to use SAVs was higher for shopping and medical trips. Demographic characteristics were not statistically significant.

Finally, Haboucha et al. [18] conducted a stated preference survey in Israel and North America. The online survey gathered 721 responses. They concluded that only 75% of individuals would be willing to use SAVs even if the service was completely free. Among their sociodemographic characteristics, young males with higher education will be more likely to use SAVs.

3. Materials and Methods

3.1. Study Area. The study area was the greater Munich metropolitan area. It included the cities of Munich, Augsburg, Ingolstadt, Landshut, and Rosenheim and their respective suburbs to cover the large commuting shed of the Munich region (Figure 1). The study area covers 150 by 150 km. The delineation of the study area was carried out using the share of commuters' flow into central cities and among cities. This metropolitan area includes a population of 4.5 million living in 2.1 million households and 14 million person-trips every day.

After delineating the study area, the municipalities were divided into traffic analysis zones (TAZs) using a gradual raster-based zone system [20]. The total number of TAZs was 4,950, and the finest resolution was 200 by 200 m. The zone system nested TAZs within the municipal regions respecting municipal boundaries, which allowed allocating the population according to the census division (Figure 2).

3.2. Stated Preference Online Survey. Stated preferences devoted to AV technology were gathered through an online survey. A 19-question survey was created in Google Forms and placed on the research group website. It was designed in four main parts:

(1) Description of AVs and General Knowledge about the Technology. Individuals were asked about how informed they were about the development of AVs and how they subjectively perceived the spread of information about AVs.

(2) Travel Behavior Characteristics on an Average Day. Individuals reported their number of trips (from one origin to one destination), the used modes of transport, the most important mode of transport, how much time they spent traveling daily, and how safe they perceive conventional vehicles and autonomous vehicles.

(3) Willingness to Use AVs. Individuals selected which modes they would consider using AVs (multiple choice), for which situations they could imagine using AVs (multiple choice), which increase on travel time they would be willing to accept with AVs, if they would consider purchasing a private AV (scale), and how much extra they would be willing to pay to own one.

(4) Sociodemographic Profile. Individuals provided their age range, gender, household size, household income, car ownership, and county of residence.

The average time to respond to the survey was below 10 minutes. The survey was offered in both German and English. The survey targeted respondents in the Munich metropolitan area. The link to the survey was posted at the research group website, website of the city of Fürstenfeldbruck and the monthly newsletter of Fürstenfeldbruck, Facebook group pages for Munich-region activities, and MotorForum website in Munich.

The survey was available online for six weeks along March and April 2017 and gathered 86 responses. In order to

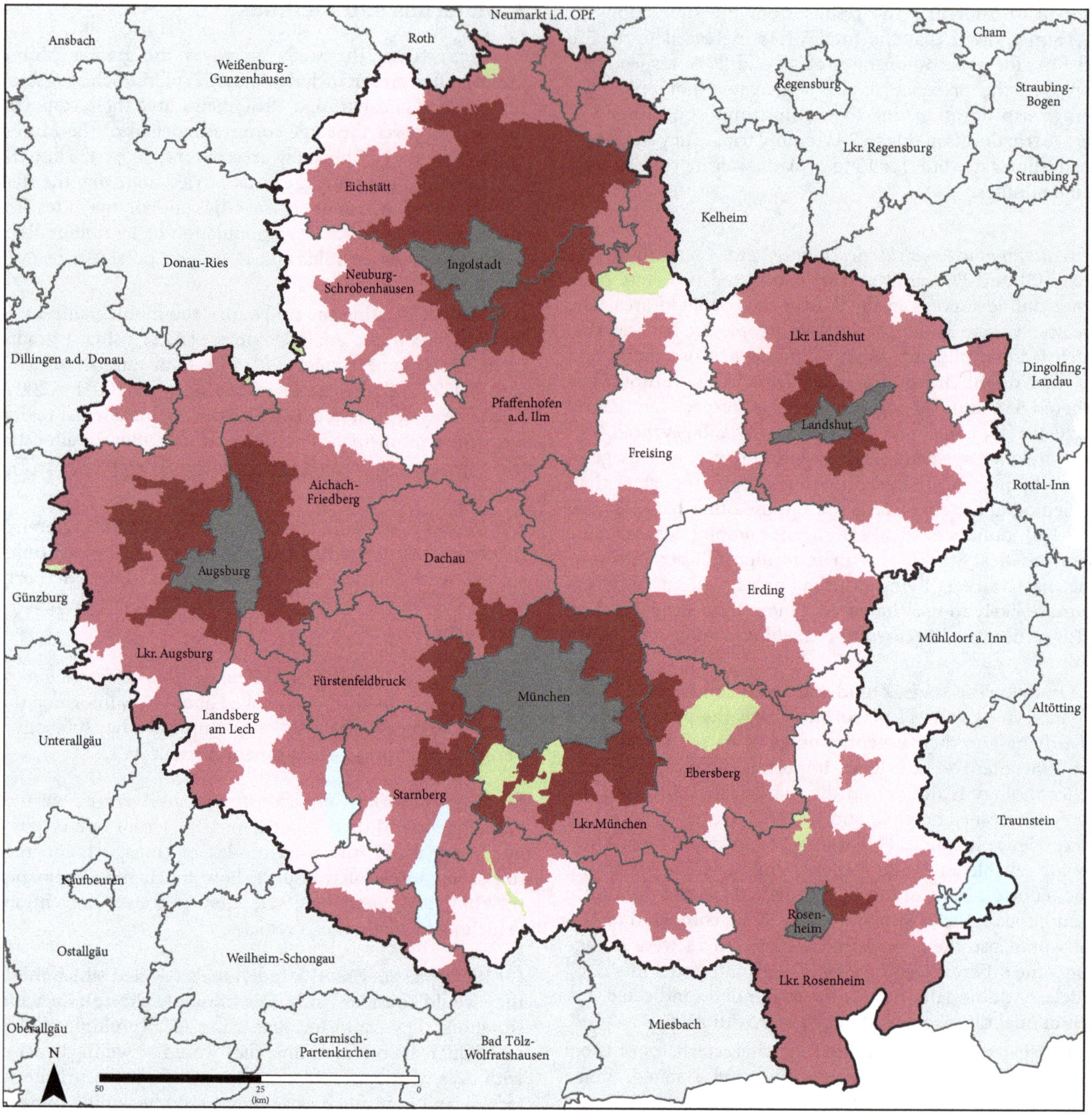

Percentage of commuters in the core region out of total commuters per municipality

- Percentage < 25% or outside of the study area
- <25,0
- 25,0–49,9%
- 50,0–78,4%

FIGURE 1: Delineation of the Greater Munich metropolitan area.

increase the sample, 20 on-street surveys were conducted in Fürstenfeldbruck with the same questions as on the online survey. The total sample was 106. English- and German-speaking respondents were evenly distributed, and around 70% of respondents were men. 98% of English-speaking respondents were younger than 35 years old and only 19% owned a car. Conversely, 62% of German-speaking respondents owned a car and covered a wider spectrum of age (18–65 years old).

3.3. Travel Demand. Travel demand was generated based on the synthetic population of the land use model SILO [21,

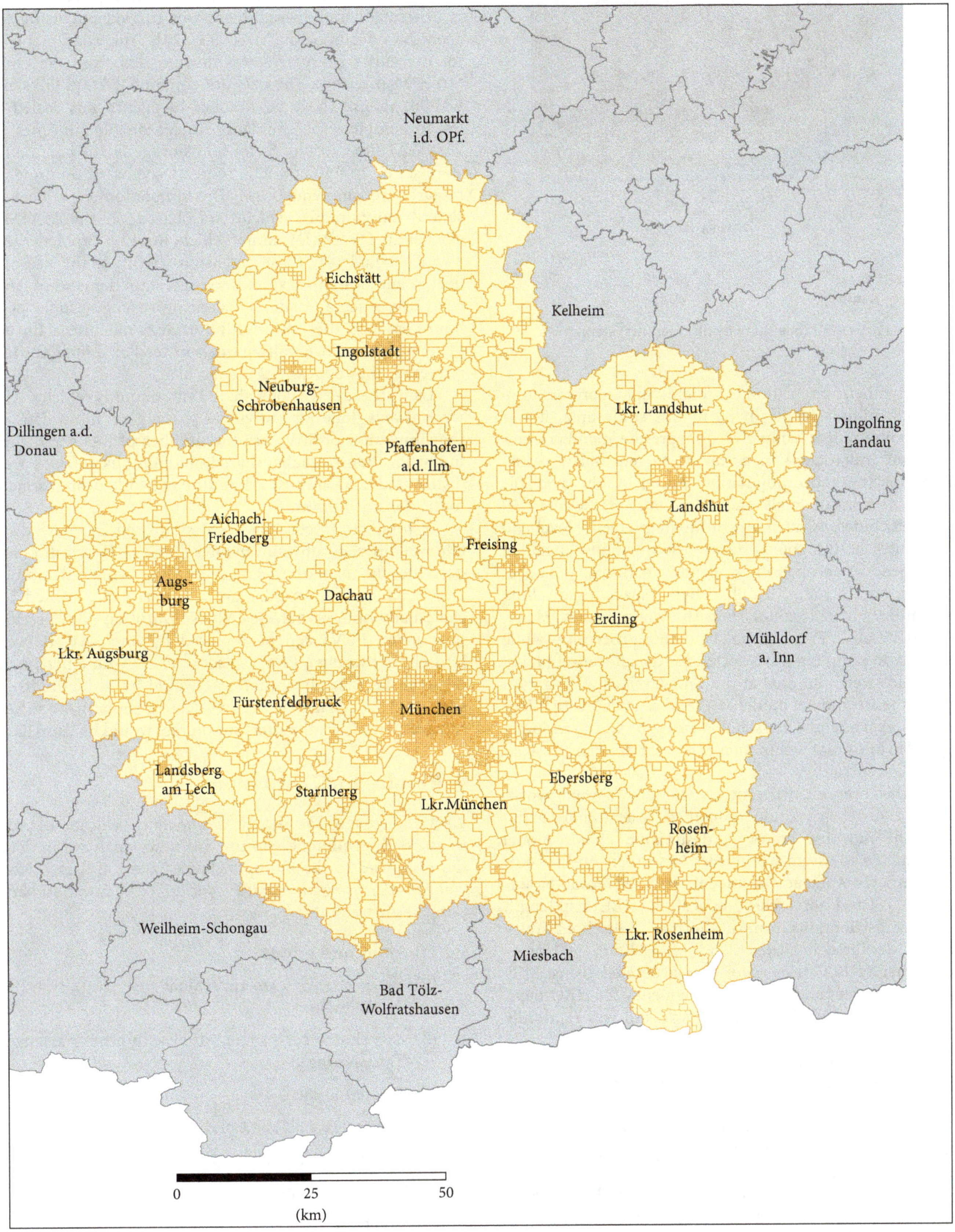

FIGURE 2: Greater Munich metropolitan area: counties and traffic analysis zones.

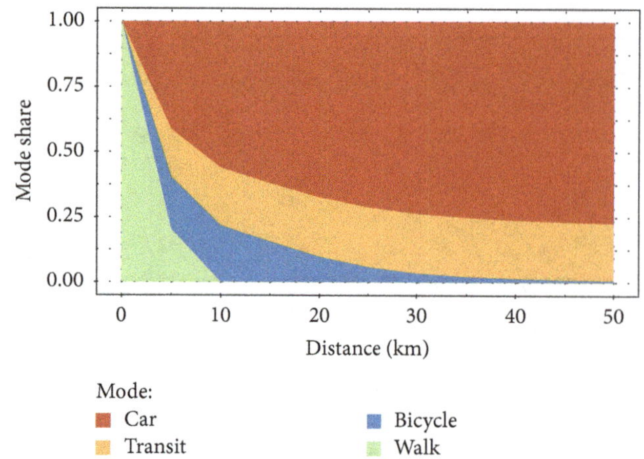

FIGURE 3: Mode share by distance: first step.

22]. The synthetic population is a microscopic representation of the 4.5 million people living in this study area. It was created combining a sample of disaggregate members of a population (microdata) in a way as to match key distributions for the entire population. Based on the available data in the control totals and microdata, we selected 59 control attributes: 46 attributes at the municipality level and 13 attributes at the county level. The attributes included age by gender, occupation by gender, nationality, household size, tenure status, dwelling living space, or dwelling type (http://www.silo.zone). For the optimization phase, we used Iterative Proportional Updating procedure, while Monte Carlo sampling was selected for the allocation phase. Households were allocated at the TAZ level, as well as jobs. Because the assignment model MATSim can work with micro locations as trip ends, households and jobs were randomly assigned to one point within their corresponding TAZ instead of the TAZ centroid.

For this exercise, we sent all workers to their workplace and all tertiary students to their educational facility during the morning hours. A destination choice model was calibrated to assign workplaces to workers and schools to students. To validate the model, we used home-based-work (HBW) and home-based-education (HBE) trip length frequency distributions from the German Household Travel Survey—Mobilitaet in Deutschland (MiD 2008) [23]. Home-based-other (HBO) trips were also generated using MiD observed trip rates and trip length frequency for HBO trips. Time of day depended on the job type and was randomly assigned for HBO trips.

Mode was selected in two steps. At the first step, the available modes were four: auto; public transport (metro, tram, and bus); bicycle; and walk. Mode was selected according to distance, with short trips being dominated by nonmotorized modes and the longest trips being dominated by auto (Figure 3).

At the second step, auto users selected whether to travel with private PCV or one SAV. The probability of one individual to select SAV was predicted using the logit model from the stated preference survey. The logit model will be defined in Section 4.1.

3.4. Traffic Simulations in MATSim. Once travel demand is generated, traffic assignment was carried out using the agent-based simulator MATSim [24]. The road network for the entire study area was downloaded from the Open Street Map service. The network contains 138,000 links and 68,000 nodes. The total number of agents was scaled to 5% to complete the simulation in reasonable runtimes [7, 25]. MATSim capacity factor and storage factor parameters were set up to 0.10 and 0.18, respectively, based on previous research recommendations. The simulation of a fleet of shared autonomous vehicles was based on the MATSim extensions DVRP (Dynamic Vehicle Routing Problem) [25] and RoboTaxi [4]. These extensions optimize the use of a fleet of autonomous taxis to serve a certain demand of trips, based on a demand-supply balancing strategy. Under peak hour demands, this optimization strategy maximizes the use of vehicles to reduce waiting times instead of serving the taxi request as they appear.

Using these extensions, shared autonomous vehicles' trips were always done for individual passengers. Input data for the simulation of SAVs are the number of SAVs and their initial location. Initially, shared autonomous vehicles were distributed randomly over the entire network, proportionally to the density of links.

A total of 11 scenarios were simulated to analyze the effect of different substitution rates (Table 1). Penetration rate describes the share of trips made by SAVs, while substitution rate assumes that vehicles are not personally owned but shared autonomous vehicles. The substitution rate describes how many PCVs can be replaced by one SAV. If every household owning a PCV replaces the car with a personally owned SAV, the substitution rate would be 10-for-10. If—as suggested by scenarios in Berlin [4] and Lisbon [6]—one SAV replaces ten PCVs, the substitution rate would be 1-for-10.

4. Results and Discussion

4.1. Willingness to Use Shared Autonomous Vehicles. In order to assess the willingness to use SAVs, we analyzed in detail the responses to the multiple choice question "*How would you consider using automated vehicles?*". The question had 8 possible answers:

(1) As my own private car.

(2) As a shared car—individual use (e.g., Car2Go, DriveNow).

(3) As a shared car—shared used with other passengers (ride sharing).

(4) As a rental car.

(5) As a taxi (e.g., Uber, Cabify).

(6) As a delivery service.

(7) I won't use them.

(8) Other use.

Individuals that selected either AVs as taxi and/or shared vehicle-individual use were considered willing to use SAVs.

TABLE 1: Simulated scenarios (agents, trips, and vehicles already scaled to 5%).

Scenario	Substitution rate[1]	SAVs	Penetration rate[2]	PCV trips	AV trips	Number of agents
0 - base	No SAVs	0	0	62,177	0	53,000
1	1-for-10	1,523	24.5%	46,941	15,236	53,000
2	2-for-10	1,692				
3	3-for-10	1,904				
4	4-for-10	2,176				
5	5-for-10	2,539				
6	6-for-10	3,048				
7	7-for-10	3,809				
8	8-for-10	5,078				
9	9-for-10	7,618				
10	10-for-10	15,236				

[1]Number of AVs that replace ten PCVs. [2]Share of PCV trips that is replaced by AV trips.

TABLE 2: Logit model for willingness to use shared autonomous vehicles.

Coefficients	Estimate	Error	t-value	Pr(>|t|)	Significance
Intercept	−3.4893	0.82455	−4.2317	0.000023	* * *
Male	1.2622	0.52426	2.4076	0.016058	*
Female	(Reference)				
3-4 trips per day	1.5181	0.46973	3.2319	0.001230	* *
0–2 or 4+ trips per day	(Reference)				
<35 years & owns car	2.2993	0.76888	2.9904	0.002786	* *
<35 years & does not own car	1.5892	0.66211	2.4003	0.016384	*
>35 years & does not own car	1.8386	1.01506	1.8113	0.070097	.
>35 years & owns car	(Reference)				

Note. Signif. codes: 0*** 0.001** 0.01* 0.05˙ 0.1; log-likelihood: −57.709; McFadden R^2: 0.19779; likelihood ratio test: chisq = 28.458 (p value = 2.9618e − 05).

41.5% of the respondents were willing to use AVs as SAVs, while 58.5% were not willing to use them as SAVs.

A logit model was applied to estimate the willingness to use autonomous vehicles as SAVs. We used the mlogit package in *R* [26]. Age and language had a correlation higher than 0.5 and therefore were not included simultaneously on the statistical analysis. No other pair of variables showed correlations higher than 0.5. The tested independent variables were age, gender, household size, number of trips per day, time spent traveling per day, car ownership, main transport mode, and the correlations among variables. Stepwise selection was followed to estimate the model with only significant variables, at the 90% level of confidence. Table 2 shows the summary of the final model, with McFadden R^2 equal to 19.8%.

The statistically significant variables are gender, number of trips per day, and the correlation between age and car ownership. Males are more likely to travel with SAVs than females. The least prone to travel with SAVs are individuals older than 35 years that own a conventional car. Other age groups and car ownership status show higher probability of using SAVs, being the most likely to use SAVs individuals younger than 35 years that own a conventional car. In this sense, young car owners will be more willing to experiment with new driving options. Finally, individuals that travel between 3 and 4 trips per day are more likely to use SAVs than less or more mobile individuals. It is expected that individuals that engage in more trips per day relay on a single mode of transport more available or reliable, while less mobile individuals can be hesitant to change their mode of transport. The results agree with previous research [16, 18].

Table 3 shows the resulting modal split on the base scenario without SAVs and the remaining scenarios with SAVs, by trip purpose. The modal split on the base scenario was calibrated using MiD 2008 data. Half of working and educational trips (HBW, HBE) are made by auto, while 18.6% are made in public transport. For HBO trips, most of them rely on nonmotorized modes (71.5%), given the shorter trip lengths. In all trip purposes, the share of SAVs on the auto mode is 26.3%.

4.2. Effect of SAVs on Average Trip Duration and VMTs. Average trip duration was compared across different scenarios, individually for SAVs and PCVs, and for all the vehicles (Figure 4). The trip duration was the sum of the driving time and waiting time (with the latter being 0 for PCV trips). Average trip duration of SAV trips was strongly related to the substitution rate, with very long trips (caused by longer wait times) when it was 3-for-10 or lower. With higher substitution rates than 3-for-10, average trip duration decreased at a constant rate. For substitution rates higher than 8-for-10, the average trip duration of SAVs was very close to PCVs. When

Table 3: Modal split by trip purpose and scenario.

Mode	HBW + HBE		HBO	
	Without SAVs	With SAVs	Without SAVs	With SAVs
Private conventional vehicle (%)	51.1	37.7	20.7	16.1
Shared autonomous vehicles (%)	-	13.4	-	4.6
Public transport (%)	18.6	18.6	7.8	7.8
Bicycle (%)	13.2	13.2	6.5	6.5
Walk (%)	17.1	17.1	65.0	65.0
Total (%)	100	100	100	100

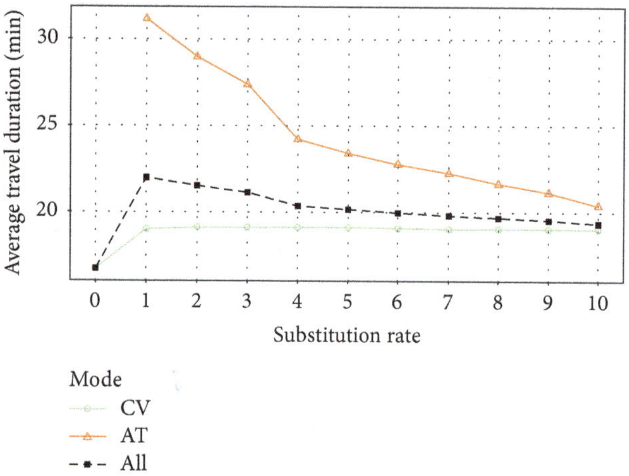

Figure 4: Average travel time by mode depending on substitution rate.

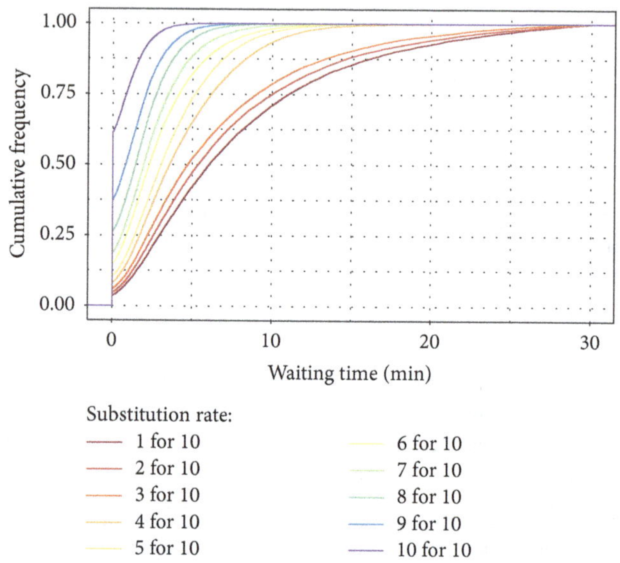

Figure 5: Waiting time cumulative distribution function by substitution rate.

the substitution rate was 3-for-10, the average waiting time for SAV travelers was around 8 min, and it was less than 5 minutes for substitution rates of 4-for-10 or higher.

The presence of SAVs triggered an increase of the average trip duration of PCVs. The increase (of up to a 14% with respect to the base scenario, from 16.7 to 19.0 minutes) was caused by additional traffic related to empty trips. The increase of trip duration was slightly higher when the substitution rate of AV trips was between 3-for-10 and 6-for-10: for higher substitution rates, the amount of empty trips is smaller while, for smaller substitution rates, the amount of vehicles in the network is smaller.

As a result of increasing both average travel times of SAVs and PCVs, the average trip duration of all modes also increased. As can be observed in Figure 2, the presence of SAVs showed in all scenarios an increase of the average trip duration, caused by increased congestion and added waiting times for SAVs. When the substitution rate was low (1-for-10), the average trip duration of all trips increased 31%, from 16.7 minutes to 22.0 minutes. For substitution rate of 3-for-10, suggested by Llorca et al. [7] for the Munich region, it still had an increase on average trip duration of 21%, from 16.7 to 21.1 minutes.

It is unknown territory how we will perceive "on the average" the waiting time for a SAV. While in some situations people could spend waiting time doing other things (such as preparing for the trip at home), the same person could face another situation on which waiting time is bothersome (such as finishing a doctor's appointment and wanting to be picked up). In this sense, average waiting time would not be the most appropriate performance measure. Waiting time cumulative distribution function could be used instead (Figure 5). It indicates the percentage of users served within a particular waiting time, similar to headways in public transit lines. As observed, substitution rates lower than 3-for-10 provided longer waiting times which ranged between 0 and 30 minutes. For substitution rates of 4-for-10 or higher, the range of waiting times is between 0 and 10 minutes with a confidence of 95%. It implicates that the maximum waiting time is lower than 10 minutes, providing a more reliable service.

Figure 6 shows that all the scenarios with SAV involved an increase of the VKT with respect to the base scenario. The results showed an increase of the total driven kilometers in the entire area in all the scenarios with AVs because of empty trips. The maximum increase was found for the lowest substitution rates: 8% increase for 1-for-10 and 7% increase for 3-for-10. This increase drops for substitution rates below 4-for-10 and is less than 4%.

The results agree with Fagnant and Kockelman [5] findings, where extra VKT was around 8.7% for single occupancy SAVs, and are smaller than the estimates for Lisbon of extra 51% VKT [6]. Nevertheless, they are greater than the

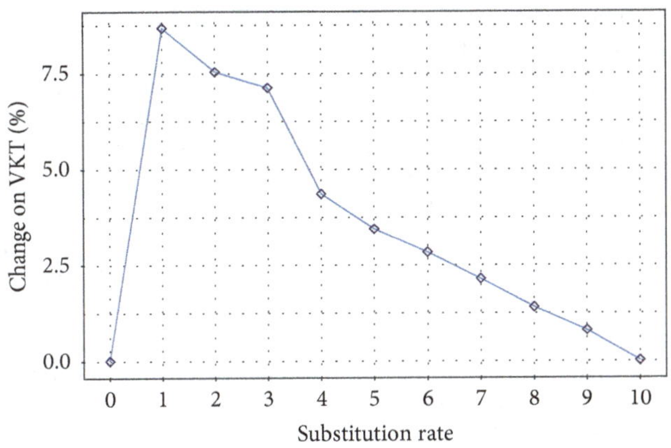

FIGURE 6: Change on VKTs with respect to base scenario depending on substitution rate.

estimates for the Munich region with 20% of penetration rate (less than 5% extra VKT). It could be because previous SAVs users were randomly distributed along the network but now are more clustered due to their travel behavior. Therefore, the same amount of trips by SAV could require longer empty trips. In this sense, ride sharing would be more efficient to reduce extra VKTs in our study area.

5. Conclusions

The results and conclusions are plausible within the simulation range and for the considered assumptions. In this research, we assumed that the willingness to use AVs as SAVs is a proxy of the actual use. It is readily admitted that the current lack of experience to travel in an AV limits the reliability and validity of survey respondent on the actual use of AVs. The limited sample of the survey is likely to have introduced a bias to the results. However, those surveys help making modeling assumptions less arbitrary when AVs need to be analyzed in transport modeling. To prepare cities to the upcoming arrival of AVs, transport models need to be built today to analyze the impact of this technology. Given the lack of observed data, surveys in conjunction with travel behavior theory is the most promising approach to reasonable model in the unknown future.

The described implementation makes the simplifying assumption that travel demand remained unchanged after AVs were introduced. Given the fundamental change AVs provide to the transport system, this is rather unlikely. Mobility-as-a-Service (MaaS) provider claims that transportation in shared AVs will be less expensive than current MaaS rates. A transport option that is cheaper and more convenient is likely to increase in demand. This induced demand generated by AVs has not been quantified with evidence yet. Transportation planners need to be aware that induced demand will make it rather unlikely that AVs will solve congestion problems. Transportation modelers need to find ways to explicitly model suppressed and induced travel demand [27].

It is planned to further improve the simulation of AVs by integrating the described model with a microscopic travel demand model and a land use model. The microscopic travel demand model MITO simulates travel behavior individually for every household [28, 29]. It is planned to model not only the activity, for which a certain trip is conducted, but also the activity done while traveling. This is relevant for transit travelers already today, as someone might do some work while commuting and thereby shorten the hours spent in the office. With AVs, conducting activities while traveling is expected to occur much more often. The integration with the land use model SILO [21] will allow reflecting the impact of AVs on location housing choice. Some households might choose to move further away from their workplace, as commuting will become more convenient with AVs. Others might decide to move closer towards downtown, as lack of parking space could be eliminated with shared AVs or private AVs that park outside of town. Modeling the impact of AVs on household location choice and firm location choice will be a key improvement to reasonably assess travel demand in the future. In the long-run, it is intended to account for improved speed flow relationships of automated highways, modal shifts between existing and new transportation modes, changes in value of time when traveling in AVs, and induced travel demand.

Conflicts of Interest

The authors declare that there are no conflicts of interest regarding the publication of this paper.

Acknowledgments

The research was completed with the support of the Technische Universität München-Institute for Advanced Study, funded by the German Excellence Initiative and the European Union Seventh Framework Programme under Grant Agreement no. 291763. This work was supported by the German Research Foundation (DFG) and the Technical University of Munich (TUM) in the framework of the Open Access Publishing Program. The authors thank Matthias Fiedler, from the Department of Transport Planning and Mobility Management in the city of Fürstenfeldbruck, for his support on the distribution of the survey on the city website and the monthly magazine.

References

[1] D. Milakis, M. B. Snelder, B. Van Arem, G. Van Wee, Homem., and D. A. Correia, "Exploring Plausible Futures of Automated Vehicles in the Netherlands: Results from a Scenario Analysis," in *Proceedings of the Automated Vehicles Symposium*, San Francisco, 2015.

[2] A. T. Moreno, "Autonomous Vehicles: Implications on an Integrated Land-Use and Transport Modelling Suite," in *Proceedings of the 11th AESOP Young Academics Conference*, Munich, 2017.

[3] M. P. Enoch, "How a Rapid Modal Convergence into a Universal Automated Taxi Service Could Be the Future for Local Passenger Transport," *Technology Analysis Strategic Management*, vol. 27, no. 8, pp. 910–924, 2015.

[4] J. Bischoff and M. Maciejewski, "Autonomous Taxicabs in Berlin - A Spatiotemporal Analysis of Service Performance," *Transport Research Procedia*, vol. 19, pp. 176–186, 2016.

[5] D. J. Fagnant and K. M. Kockelman, "Dynamic ride-sharing and fleet sizing for a system of shared autonomous vehicles in Austin, Texas," *Transportation*, pp. 1–16, 2016.

[6] OECD, "Urban Mobility System Upgrade: How Shared Self-Driving Cars Could Change City Traffic," in *Corporate Partnership Board Report*, 2015.

[7] C. Llorca, A. T. Moreno, and R. Moeckel, "Effects of Shared Autonomous Vehicles on the Level of Service in the Greater Munich Metropolitan Area," in *Proceedings of the mobil.TUM 2017*, Munich, 2017.

[8] D. Milakis, M. Snelder, B. Van Arem, B. Van Wee, and G. H. De Almeida Correia, "Development and transport implications of automated vehicles in the Netherlands: Scenarios for 2030 and 2050," *European Journal of Transport and Infrastructure Research*, vol. 17, no. 1, pp. 63–85, 2017.

[9] M. Kümmel, F. Busch, and D. Z. W. Wang, "Framework for automated taxi operation: The family model," *Transportation Research Procedia*, vol. 22, pp. 529–540, 2016.

[10] H. Billhardt, A. M. Fernandez, S. Lujak et al., "Multi-Agent Systems and Agreement Technologies," in *Proceedings of 13th European Conference, EUMAS, 2015, and Third International Conference*, Athens , Greece, 2015.

[11] M. Ramezani and M. Nourinejad, "Dynamic modeling and control of taxi services in large-scale urban networks: A macroscopic approach," *Transportation Research Procedia*, vol. 23, pp. 41–60, 2017.

[12] W. E. Walker and V. A. Marchau, "Dynamic adaptive policy-making for the sustainable city: The case of automated taxis," *International Journal of Transportation Science and Technology*, vol. 6, no. 1, pp. 1–12, 2017.

[13] L. M. Martinez and J. M. Viegas, "Assessing the impacts of deploying a shared self-driving urban mobility system: An agent-based model applied to the city of Lisbon, Portugal," *International Journal of Transportation Science and Technology*, vol. 6, no. 1, pp. 13–27, 2017.

[14] J. Alonso-Mora, S. Samaranayake, A. Wallar, E. Frazzoli, and D. Rus, "On-demand high-capacity ride-sharing via dynamic trip-vehicle assignment," *Proceedings of the National Acadamy of Sciences of the United States of America*, vol. 114, no. 3, pp. 462–467, 2017.

[15] J. P. Zmud and I. N. Sener, "Towards an understanding of the travel behavior impact of autonomous vehicles," *Transportation Research Procedia*, vol. 25, pp. 2500–2519, 2017.

[16] P. Bansal, K. M. Kockelman, and A. Singh, "Assessing public opinions of and interest in new vehicle technologies: An Austin perspective," *Transportation Research Part C: Emerging Technologies*, vol. 67, pp. 1–14, 2016.

[17] R. Krueger, T. H. Rashidi, and J. M. Rose, "Preferences for shared autonomous vehicles," *Transportation Research Part C: Emerging Technologies*, vol. 69, pp. 343–355, 2016.

[18] C. J. Haboucha, R. Ishaq, and Y. Shiftan, "User preferences regarding autonomous vehicles," *Transportation Research Part C: Emerging Technologies*, vol. 78, no. Part C, pp. 37–49, 2017.

[19] F. Becker and K. Axhausen, "Literature review on behavioral experiments for autonomous vehicles," in *Proceedings of the Annual Meeting of the Transportation Research Board*, Washington, D.C., USA, 2017.

[20] J. Molloy and R. Moeckel, "Automated Design of Gradual Zone Systems. An iterative algorithm to design optimally sized spatial zones suitable for spatial modeling, while respecting municipal boundaries," *Open Geospatial Data, Software and Standards*, vol. 2, no. 19, pp. 1–10, 2017.

[21] R. Moeckel, "Constraints in household relocation: Modeling land-use/transport interactions that respect time and monetary budgets," *Journal of Transport and Land Use*, vol. 10, no. 1, pp. 211–228, 2013.

[22] A. T. Moreno and R. Moeckel, "Population synthesis handling three geographical resolutions," *ISPRS International Journal of Geo-Information*, vol. 7, no. 5, p. 174, 2018.

[23] R. Follmer, B. Lenz, P. Loecker, and M. Sigismund, in *Proceedings of the Annual Meeting of the Transportation Research Board*, Washington, D.C., USA, 2012.

[24] A. Horni, K. Nagel, and K. W. Axhausen, "The Multi-Agent Transport Simulation," in *MATSim*, Ubiquity Press, London, UK, 2016.

[25] M. Maciejewski, J. M. Salanova, J. Bischoff, and M. Estrada, "Large-scale microscopic simulation of taxi services. Berlin and Barcelona case studies," *Journal of Ambient Intelligence and Humanized Computing*, vol. 7, no. 3, pp. 385–393, 2016.

[26] Y. Croissant, "Estimation of Multinomial Logit Models in R: The Mlogit Packages: an Introductory Example," in *Data Management*, p. 73, 2003.

[27] R. Kitamura, S. Fujii, and E. I. Pas, "Time-use data, analysis and modeling: Toward the next generation of transportation planning methodologies," *Transport Policy*, vol. 4, no. 4, pp. 225–235, 1997.

[28] A. Moreno and R. Moeckel, "A Framework to Respect Travel Time Budgets as Constraint for Microscopic Destination Choice Modeling," in *Proceedings of the the Annual Meeting of the Transportation Research Board*, Washington, D.C., USA, 2017.

[29] A. Moreno, A. Michalski, C. Llorca, and R. Moeckel, "Autonomous Taxis Effect on Vehicle-Km Traveled and Average Trip Duration in the Greater Munich Metropolitan Area," in *Proceedings of the the Annual Meeting of the Transportation Research Board*, Washington, D.C., USA, 2018.

17

Changes in Trust after Driving Level 2 Automated Cars

Francesco Walker,[1] Anika Boelhouwer,[1] Tom Alkim,[2] Willem B. Verwey,[1] and Marieke H. Martens[1]

[1] *University of Twente, Netherlands*
[2] *Rijkswaterstaat, Ministry of Infrastructure and Water Management, Netherlands*

Correspondence should be addressed to Francesco Walker; f.walker@utwente.nl

Academic Editor: Jay Katupitiya

Overtrust and undertrust are major issues with partially automated vehicles. Ideally, trust should be *calibrated* ensuring that drivers' subjective feelings of safety match the objective reliability of the vehicle. In the present study, we examined if drivers' trust toward Level 2 cars changed after on-road experience. Drivers' self-reported trust was assessed three times: before having experience with these vehicles, immediately after driving two types of vehicles, and two weeks after the driving experience. Analysis of the results showed major changes in trust scores after the on-road driving experience. Before experiencing the vehicles, participants tended to overestimate the vehicle capabilities. Afterwards they had a better understanding of vehicles' limitations, resulting in better calibrated trust.

1. Introduction

Advanced driver assistance systems (ADAS) are technologies that support the human driver in the driving task. This is done by providing information and warnings, and automating demanding and repetitive tasks [1–3]. Several modern vehicles are equipped with a combination of ADAS. Examples include Volvo's Adaptive Cruise Control with Steer Assistance, BMW's Traffic Jam Assistant, Toyota's Automated Driving Highway Assistant, and Tesla's Autopilot [4]. The Society of Automotive Engineers has formally classified these vehicles as "Level 2", or vehicles equipped with "partial automation" [5]. With SAE Level 2, or partial automation, the driver is still responsible for deciding when and whether to engage the automated system. Furthermore, with these systems, the driver should always monitor the driving environment and be ready to respond to system failures [5].

The benefits of partial automation will not be realized until Level 2 vehicles are widely accepted and used on the road [6]. If drivers do not trust the automated system, it is likely that they will overrule the vehicle's decisions or not activate the partially automated system at all. This will reduce the potential positive safety impact of the technology [7]. In their extension of the Technology Acceptance Model (TAM), Choi and Ji [8] showed that perceived usefulness and intention to use the technology were strongly mediated by trust [8]. A similar conclusion was reached by Ghazizadeh and colleagues [9]. Through their Automation Acceptance Model, the authors suggested that trust is, in fact, one of the main factors affecting users' acceptance of automation technology.

In line with these considerations, reports show that although driver awareness of automation technology is increasing, usage is still limited [10]: A survey conducted in 2017 by the news agency AAA showed that while drivers of the United States look for autonomous technologies in their next car, only 25% would feel safe actually driving an autonomous vehicle. Despite these results, it is also possible for drivers to overestimate the capabilities of the vehicle, relying excessively on SAE Level 2 functions even under unsafe conditions or outside of the Operational Design Domain (ODD). With partially self-driving (i.e., Level 2) cars, as with other automated systems, the goal of human factors research should NOT be to maximize drivers' trust, but to *align drivers' subjective perceptions of safety with the actual reliability of the vehicle*, thus avoiding over- and underestimation of the car's capabilities (e.g., [11, 12]). This is commonly known as "trust calibration" (e.g., [11, 13]). Notably, trust calibration cannot be achieved if drivers are unable to discriminate situations in which the vehicle will behave reliably from situations in which it will not [14]. In other words, to align drivers' subjective feelings of safety with the actual reliability of the vehicle, drivers need to know how the system behaves in a variety of situations. The ability to discriminate between different situations is commonly known as "resolution" [14].

In general, paraphrasing Ghazizadeh and colleagues, trust affects use, and use affects trust [9]. Importantly, the relationship between trust and use is not always positive: Drivers might not trust the technology but still use it, and vice versa. Furthermore, the relationship between trust and use is mediated by drivers' initial trust levels and the objective reliability of the technology. Thus, calibrating drivers' trust in Level 2 vehicles requires a better understanding of their initial trust toward the vehicles, and how this is influenced by vehicle experience. Most of what we know about trust in automation comes from studies of automated aircraft or production systems [6]. These studies indicate that user trust in automation is strongly affected by system performance, unexpected failures of the automated system [15], and predictability of system behaviour [6, 11].

In the automotive domain, several studies have analysed drivers' trust in relation to Adaptive Cruise Control (ACC). ACC is an Advanced Driving Assistance System that automatically regulates speed and headway distance. Rajaonah and colleagues [16] used a simulated vehicle equipped with ACC to analyse driver behaviour when a truck suddenly cut into the driver's lane. Trust was measured via self-reports. The study identified two groups of drivers: One group that started braking before the truck pulled into their lane—correct anticipatory behaviour—and a second group that started braking only when the truck was already in their lane. Given that drivers should not rely on ACC during hazardous situations, the latter behaviour suggests overtrust in the system [16]. The group that correctly disengaged the ACC before the cut-in event reported high trust in their interaction with the ACC system, but not in the system itself. This result suggests that this group had a clear understanding of the ACC's limitations, which may allow appropriate trust calibration [16]. A survey conducted by Dickie and Boyle [17] showed that drivers who are unaware of the way ACC works were unsure about how to use it, tending to overestimate the capabilities of the system. The authors argue that the inappropriate expectations of "unaware" and "unsure" drivers lead to poor trust calibration (see also [13]).

A driving simulator study by Gold and colleagues [6] suggests that driver experience with SAE Level 3 vehicles can influence drivers' trust in the system. In such vehicles the system, and not the driver, is responsible for monitoring the driving environment [5]. Thus, in Level 3 vehicles drivers can perform all sorts of activities while the car takes care of the driving task. In Gold et al.'s [6] study participants had been briefed that the vehicle could cope with most driving scenarios. Thus, monitoring the system was unnecessary. Nevertheless, participants were also told that the vehicle could not cope with every driving scenario and that the system would emit a warning signal when it could not cope. Participants had seven seconds to take back control of the vehicle. Drivers experienced three take-over scenarios during a twenty-minute simulated drive. Interestingly, drivers' clearer understanding of the vehicles' limitations (i.e., take-over scenarios) due to on-hand experience leads to a decreased perception of safety advantages, but also to an increase in trust. In line with these results, a study investigating drivers' experience with Forward Collision Warnings showed that a learning period helped drivers to achieve a better understanding of system behaviour, ultimately improving their interactions with the system and increasing their trust [18].

Overall, the literature suggests that trust strongly affects how drivers interact with SAE Level 2 technology and that a clear understanding of vehicles' limitations is important for trust calibration [6, 13, 16–19]. As pointed out by Muir [20], trust can be built by giving to the user the chance to experience the system in a variety of situations. Notably, by doing so, on-road experience with real Level 2 vehicles can increase drivers' resolution [14]. Improving this ability could play a key role in calibrating drivers' trust.

Previous work on drivers' trust in automation has been mostly carried out in driving simulators and analysed the effect of specific ADAS (e.g., ACC, Forward Collision Warning System) or, as in the case of Gold et al. [6], systems that were not yet available on the market. They thus provide no information about drivers' trust after experiencing the systems on real roads, during a variety of conditions, and when driving different vehicles equipped with similar automated functions during the same driving session.

In our study, we explored how on-road experience with Level 2 vehicles influences drivers' trust in automation. In line with Gold et al. [6] and Koustanaï et al.'s [18] results, we expected that real-life experience would give drivers a better understanding of system behaviour, thereby improving trust calibration. In the study, therefore, participants were asked to fill in a trust questionnaire before, immediately after, and two weeks after driving two different Level 2 vehicles on motorway, urban and rural roads. The impact of the driving experience on participants' trust was assessed through the comparison of the pre- and post-trust scores. The "2-week" measurement was included to measure if changes in trust would fade after two weeks. Features such as age, gender, and views of new technology were also taken into account. To our knowledge, this is the first time trust scores have been collected before and after on-road experience with real Level 2 vehicles.

2. Method

2.1. Participants. Participants in the study were all employees of Rijkswaterstaat, the Dutch Ministry of Infrastructure and Water Management. None of the participants had driven or been driven (as a passenger) in a Level 2 vehicle before joining the study. Drivers were asked to fill in the same questionnaire three times (see "Procedure" section). Only participants who completed all three questionnaires (i.e., measurement 0, measurement 1, and measurement 2) were included in the analysis. This resulted in a sample of 106

Figure 1: Picture following description of scenario 1.

Figure 2: Picture following description of scenario 2.

participants of which 77.4% (N = 82) were male. This group had a mean age of 43.77 years (SD = 10.14). The youngest participant was 24 years old and the oldest 62.

2.2. Questionnaire. The questionnaire was composed of 34 items. Of these, 4 concerned participants' age, gender, travelling profile, and attitudes toward new technology. A subsection of the questionnaire, composed of 12 items, investigated drivers' trust toward Level 2 cars in twelve different scenarios. This was the focus of the current study.

In this section of the questionnaire, participants were first asked to indicate which driving scenarios they thought partially automated cars could or could not handle. Then, after experiencing the vehicles, participants were asked to imagine driving a Level 2 car in these different scenarios. More specifically, drivers were asked to indicate to what extent they *trusted* the system (i.e., the Level 2 car) to cope with each scenario. This was done through the statement "I trust that the system can handle the situation, thus, I do not need to take over control". Drivers were asked to indicate, through a 5-point Likert scale (i.e., 1 indicating "I fully disagree" and 5 indicating "I fully agree"), how much they agreed with the statement. Each of the 12 items was accompanied by an image illustrating the scenario (e.g., Figures 1 and 2). Scenarios were presented in Dutch and translated into English for this manuscript. The 12 scenarios were as follows.

(1) You are driving on a highway with an average amount of traffic. The system is set to automatically keep distance from the vehicle in front, and to keep within the lane (see Figure 1).

(2) You drive on the highway and the vehicle in front brakes hard (see Figure 2).

(3) A car overtakes you in the right lane.

(4) You drive on the highway with a lot of rain, so the visibility is poor.

(5) You drive in a work area with yellow road marking in addition to the normal white road marking.

(6) A car merges in from the right lane and has a much lower speed than yours.

(7) A deer suddenly crosses the highway – a situation in which normally you would brake.

(8) The lane where you drive ends, you need to move to the lane to your right.

(9) You are approaching a curve on the highway.

(10) You are driving on the highway behind a motorcycle instead of behind a car.

(11) You are driving on the highway and the speed limit decreases. Your speed should adapt to the new speed limit.

(12) You are driving at night on the highway.

2.3. Vehicles. The following Level 2 vehicles, all 2016 models, were selected for the study: Two Teslas (a Model S and a Model X), two Mercedes (an E-200 and an E-350-E), and two Volvos (a XC90 and a V90). All the vehicles were equipped with similar SAE Level 2 functions. More detailed information on the vehicle features described in this section can be retrieved from Tesla, Mercedes, and Volvo official websites.

2.3.1. Tesla Model S and Model X. Tesla's combination of Advanced Driving Assistance Systems (ADAS) falls under the name of *Autopilot*. Tesla's Autopilot features Automatic Emergency Braking (AEB), Lane Keeping capabilities (i.e., "Autosteer"), and ACC. Autopilot also allows the vehicle to automatically change lanes when prompted by the driver's tap on the turn signal lever. The 2016 version of Tesla's Autopilot could already detect pedestrians, a feature later significantly improved through the v8.0 software update.

2.3.2. Mercedes E-200 and E-350-E. Mercedes's combination of ADAS is known as *Drive Pilot*. Drive Pilot features Active Break Assist, ACC (i.e., "Distronic Cruise Control"), and Lane Keeping functions (i.e., "Active Lane Keeping Assist"). Other ADAS include Speed Limit Pilot. This is a subfunction of Drive Pilot that detects speed limits and adjusts the vehicle's speed accordingly. Like the Tesla, the Mercedes E class comes with Active Lane Change Assist, allowing the driver to initiate an automatic lane change by lightly touching the turn control. Drive Pilot does not provide pedestrian detection. Pedestrian and animal recognition was implemented in the 2017 models through the *Intelligent Drive* software update. However, the study reported here only used 2016 models.

2.3.3. Volvo XC90 and V90. *Pilot Assist* is the name chosen by Volvo to describe the combination of ADAS in their XC90 and V90 2016 models. Pilot Assist features ACC, AEB, and Lane Keeping functions. Volvo vehicle user manuals clearly specify that Pilot Assist is not a collision warning system: Even if Pilot Assist can clearly detect leading vehicles, it cannot detect pedestrians, cyclists, animals, and motorcycles. *Pilot Assist II*, implemented in the Volvo XC90 2017 model (thus, not available for this study), features pedestrian, cyclist, and Large Animal Detection functions. Unlike Tesla's Autopilot and Mercedes's Drive Pilot, Volvo's Pilot Assist does not support lane changing.

2.4. Procedure. Five separate driving sessions were organised. Each participant joined only one of the five sessions. Participants were asked to fill in the same questionnaire (see "Questionnaire" section) three times: before the session ("measurement 0"); immediately after the session ("measurement 1"); and two weeks after the session ("measurement 2"). In each session, participants drove and were also passengers in two of the Level 2 vehicles described earlier (see "Vehicles" section). Each vehicle carried a driver, two passengers in the back seats (also participants in the study), and a PRO DRIVE expert or car dealer expert in the front passenger seat. The expert explained the functionalities of the vehicle and made sure that overreliance in automation would not lead to dangerous situations. The expert also prompted participants to use the vehicle's Level 2 functions. In each driving session, participants drove for ~ 20 minutes on a predefined route including a motorway, an urban road, and a rural road.

2.5. Analysis. A Bonferroni correction was applied to reduce the likelihood of Type I Errors. The corrected p value was calculated by dividing the alpha-value ($\alpha_{original}$ = .05) by the number of scenarios (12): ($\alpha_{altered}$ = .05/12) =.004 for each scenario. A Friedman test was used to assess differences between measurement 0 (m0), measurement 1 (m1), and measurement 2 (m2). If differences between the three measurements were found, a Wilcoxon Signed Ranks test was used to assess differences between m0 and m1, m0 and m2, and m1 and m2. If the Friedman test showed differences between the three measurements, a Mann–Whitney U test was performed to analyse the effect of gender. A Kruskal-Wallis test was used to analyse innovation and age effects.

2.6. Exclusions. Participants who did not fill in the questionnaire at all three time points (i.e., m0, m1, and m2) were excluded from the analysis. 60 participants were excluded on these grounds. Missing values for any of the 12 items were replaced by the mean of the other participants' responses. Missing values were only present at m1. In no case were there more than four missing values for a single scenario.

3. Results

3.1. Participants. None of the participants selected for this study had ever experienced a Level 2 car, as a driver or a passenger. Of the final sample of 106 participants, 46.7% of the drivers reported that they used their car more often than public transport, 37.1% reported that they used public transport more than the car, and 16.2% reported equal use. None of the respondents reported exclusive use of public transport or the car. Questioned about their attitudes to new technology, 1.90% viewed themselves as innovators, 20.8% as early adopters, 48.10% as early majority, 28.30% as late majority, and 0.9% as laggards. For the purposes of the subsequent analysis, these categories were condensed into three innovation profiles: "innovators/early adopters", "early majority", and "late majority/laggards". A Mann–Whitney test showed a significant main effect of gender on innovation profile, with men considering themselves more innovative than women (U = 736.50; p = .043). Ages were categorized into three groups: <35, 35-50, 50+. Questionnaire responses showed no significant effect of age, gender, or innovation profile (data not shown).

3.2. Responses to Questionnaire. For seven of the twelve scenarios, Wilcoxon Signed Ranks tests showed significant differences (i.e., p <= .004) between m0 and m1 (see Figure 3 and Table 1). In scenarios 1 and 2 (see Table 1), trust *increased* after vehicle experience. All the Level 2 vehicles included in this study could cope with scenarios 1 and 2. In scenario 1 (ACC and Lane Keeping), the high scores reported at m0 (M = 4.21) showed relatively high trust for the ACC and Lane Keeping functions of the systems. The increase in trust reported at m1 (M = 4.62) suggests that the vehicles' ACC and Lane Keeping functions exceeded participants' expectations. Drivers' responses to scenario 2 (brake preceding vehicle) suggest that experiencing this scenario also had a positive impact on participants' trust ratings.

In scenarios 4, 7, 8, 9, and 12, Wilcoxon Signed Ranks tests showed significant *decreases* in drivers' trust after vehicle experience (see Figure 3 and Table 1). In scenario 4 (rain), drivers' trust decreased after they had driven the vehicles, suggesting an initial overestimation of the vehicles' capabilities. Bad weather still represents a challenge for Level 2 technology, for instance, because it makes it more difficult to detect road markings. Even though participants did not directly experience this scenario, the drop in trust from m0 (M = 2.96) to m1 (M = 2.57) can probably be explained by a clearer understanding of vehicles' limitations. Scenario 7 (crossing deer) implies that the vehicle should automatically brake in response to a crossing deer. Animal detection currently represents a big challenge for automakers, and none of the 2016 vehicles selected for the study were equipped with this feature. Even though none of the drivers experienced the scenario on the road, the reported decrease in trust from m0 (M = 2.69) to m1 (M = 2.20) suggests a better understanding of the systems' limitations after the driving experience. None of the Level 2 vehicles were capable of handling scenario 8 (merging from left lane). As with scenario 7, drivers' decrease in trust from the pre- to the postmeasurements suggests an initial overestimation of the vehicles' capabilities, but also a clearer understanding of the systems' limitations after on-road experience with the vehicles. In scenario 9 (curve on motorway), even when Lane Keeping functions were activated, the selected Level 2 vehicles would occasionally stray out of lane when facing tight curves. Most drivers experienced this limitation, and their trust decreased from m0 (M = 3.63) to m1 (M = 2.81). Drivers did not have the chance to experience scenario 12 (driving at night). Here, drivers' trust decreased from the pre- (M = 3.61) to the post- (M = 3.28) measurement. In scenario 10 (driving behind a motorbike), no significant difference was found between m0 and m1. However, the Wilcoxon Signed Ranks test showed a significant difference between m0 (M = 3.5) and m2 (M = 3.15).

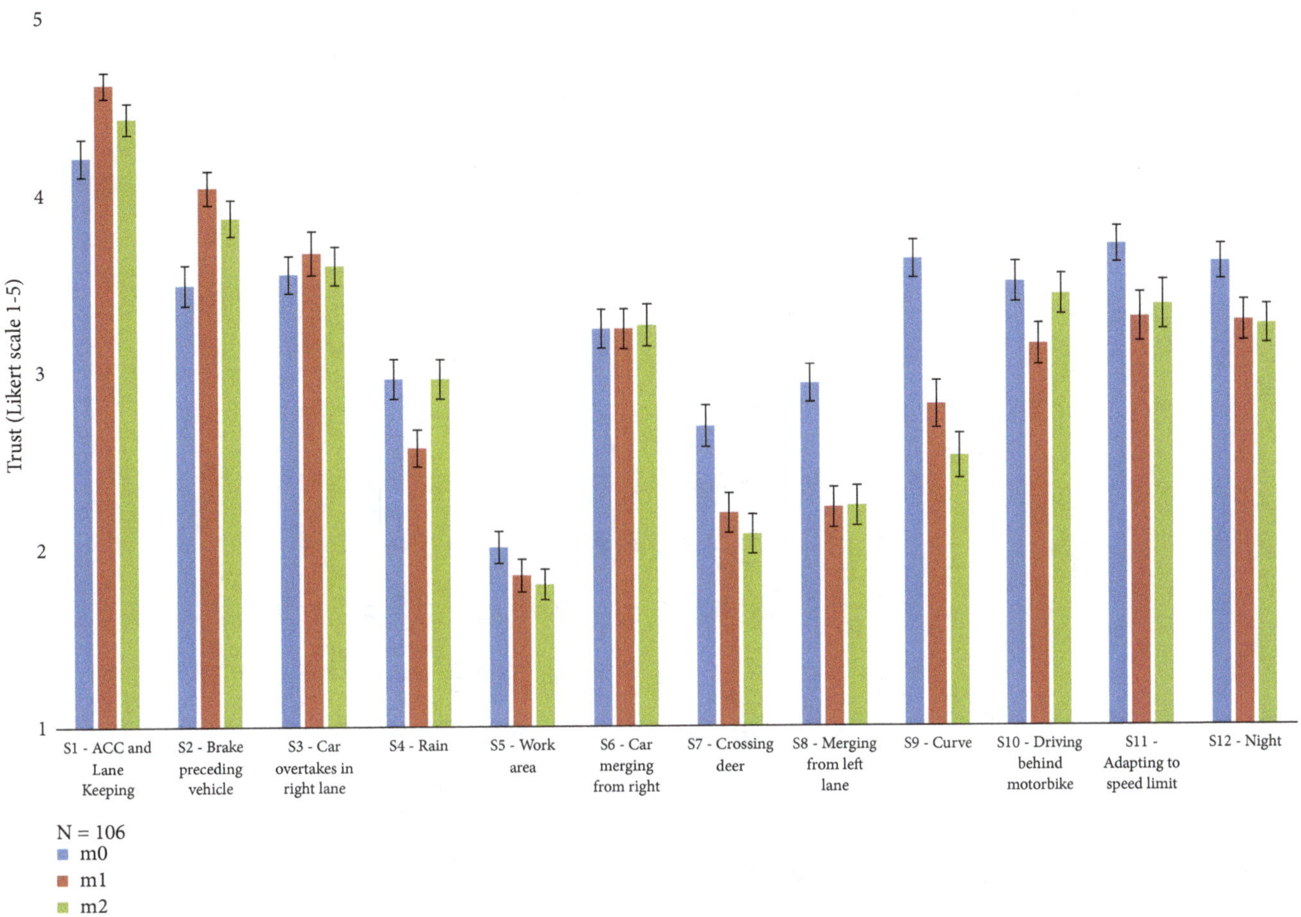

FIGURE 3: **Questionnaire results. Scenarios (S) one to twelve are displayed from left to right, respectively.** For scenarios 1 and 2, drivers' trust significantly increased after experiencing the vehicles. Conversely, for scenarios 4, 7, 8, 9, and 12, drivers' trust significantly decreased after experiencing the vehicles. No significant trust changes between pre- and postmeasurements were found for scenarios 3, 5, 6, and 11. For scenario 10, significant differences were found between the pre- and the "2-week" measurement (i.e., m2), but not between m0 and m1. $\alpha = .004$. Error bars were calculated based on the standard error of the mean.

A Friedman test showed no significant differences across measurements for scenarios 3, 5, 6, and 11 (see Figure 3 and Table 1). The relatively high trust scores observed for scenario 3 (car overtakes in right lane) indicate that drivers trusted the vehicles to cope with this scenario and that this did not change after their experience with the vehicles. A car overtaking in the right lane represents a dangerous situation. Nevertheless, Autopilot, Drive Pilot, and Pilot Assist would be able to detect the oncoming vehicle. Even if drivers did not directly experience scenario 5 (work area), the selected Level 2 vehicles do not perform well with mixed white and yellow road markings.

In scenario 6 (car merging from right), the absence of an effect across measurements suggests that participants are unsure of how the systems can handle the scenario. In scenario 11 (adapting to speed limit), the absence of any significant difference between the three measurements is probably due to the fact that most participants did not have the chance to experience the scenario. In reality, of the vehicles used in the study, only the Mercedes E class was capable of adapting to speed limits.

4. Discussion

While engineering developments are moving fast, there is still little understanding of how drivers interact with automated driving technology [7]. The study of drivers' trust in automation represents a fundamental challenge for current and future human factors research. Importantly, the goal should not be to maximize users' trust toward the technology, but to align drivers' subjective feelings of safety with the actual reliability of the vehicle. In this respect, a decrease in trust is not a negative outcome *per se*: If, for example, vehicles are incapable of handling a specific driving scenario (e.g., deer crossing the road), decreases in trust suggest an improvement in trust calibration. More generally, if drivers overestimate the capabilities of the system, a better understanding of vehicles' limitations after vehicle experience can lead to a decrease in trust. Conversely, if drivers underestimate the capabilities of the system, a better understanding of the vehicles' functionalities can lead to increased trust [11].

Bearing these considerations in mind, determining the "correct" level of trust for particular driving scenarios is a

Table 1: **Questionnaire results.** A Friedman test was performed for an analysis of the variances of m0 (measurement 0), m1 (measurement 1), and m2 (measurement 2). Wilcoxon Signed Ranks test was used for pairwise comparisons (i.e., m0 versus m1, m0 versus m2, and m1 versus m2); *significant ($p <= .004$).

								N = 106
Scenario	**Missing values**	**Friedman test**	**M0 (M, SD)**	**M1 (M, SD)**	**M2 (M, SD)**	**M0 vs m1**	**M0 vs m2**	**M1 vs m2**
(1) You are driving on a highway with an average amount of traffic. The system is set to automatically keep distance from the vehicle in front, and to keep within the lane.	4	$\chi^2 = 23.096$, $p < .001^*$	M = 4.21, SD = 1.09	M = 4.62, SD = 0.76	M = 4.43, SD = 0.92	Z = -3.927, $p < .001^*$	Z = -2.511, p = .012	Z = -2.360, p = .018
(2) You drive on the highway and the vehicle in front brakes hard.	3	$X^2 = 26.315$, $p < .001^*$	M = 3.49, SD = 1.18	M = 4.04, SD = 0.99	M = 3.87, SD = 1.06	Z = -4.014, $p < .001^*$	Z = -3.390, $p = .001^*$	Z = -2.024, p = .043
(3) A car overtakes you in the right lane.	1	$X^2 = 2.537$, p = .281	M = 3.55, SD = 1.07	M = 3.67, SD = 1.27	M = 3.60, SD = 1.12	Z = -1.065, p = .287	Z = -.444, p = .657	Z = -.676, p = .499
(4) You drive on the highway with a lot of rain, so the visibility is poor.	0	$X^2 = 6.483$, p = .039	M = 2.96, SD = 1.18	M = 2.57, SD = 1.07	M = 2.71, SD = 1.15	Z = -3.390, $p = .001^*$	Z = -2.212, p = .027	Z = -1.484, p = .138
(5) You drive in a work area with yellow road marking in addition to the normal white road marking.	2	$X^2 = 8.130$, p = .017	M = 2.01, SD = 0.91	M = 1.85, SD = 0.95	M = 1.80, SD = 0.89	Z = -1.883, p = .06	Z = -2.211, p = .027	Z = -0.668, p = .504
(6) A car merges in from the right lane and has a much lower speed than yours.	1	$X^2 = 0.087$, p = .957	M = 3.24, SD = 1.14	M = 3.24, SD = 1.15	M = 3.26, SD = 1.23	Z = -.092, p = .926	Z = -.327, p = .744	Z = -.235, p = .814
(7) A deer suddenly crosses the highway – a situation in which normally you would brake.	1	$X^2 = 19.196$, $p < .001^*$	M = 2.69, SD = 1.21	M = 2.20, SD = 1.16	M = 2.08, SD = 1.14	Z = -3.437, $p = .001^*$	Z = -4.026, $p < .001^*$	Z = -1.233, p = .218
(8) The lane where you drive ends, you need to insert the lane to your right.	0	$X^2 = 31.848$, $p < .001^*$	M = 2.93, SD = 1.10	M = 2.23, SD = 1.18	M = 2.24, SD = 1.17	Z = -4.658, $p < .001^*$	Z = -4.641, $p < .001^*$	Z = -0.329, p = .742
(9) You are approaching a curve on the highway.	4	$X^2 = 48.771$, $p < .001^*$	M = 3.63, SD = 1.10	M = 2.81, SD = 1.37	M = 2.52, SD = 1.30	Z = -5.239, $p < .001^*$	Z = -6.326, $p < .001^*$	Z = -2.556, p = .011
(10) You are driving on the highway behind a motorbike instead of behind a car.	1	$X^2 = 13.076$, $p = .001^*$	M = 3.5, SD = 1.17	M = 3.43, SD = 1.20	M = 3.15, SD = 1.17	Z = -0.578, p = .564	Z = -3.003, $p = .003^*$	Z = -2.970, $p = .003^*$
(11) You are driving on the highway and the speed limit decreases. Your speed should adapt to the new speed limit.	3	$X^2 = 2.299$, p = .317	M = 3.71, SD = 1.06	M = 3.30, SD = 1.41	M = 3.37, SD = 1.42	Z = -2.402, p = .016	Z = -2.253, p = .024	Z = -.405, p = .686
(12) You are driving at night on the highway.	0	$X^2 = 13.292$, $p = .001^*$	M = 3.61, SD = 1.02	M = 3.28, SD = 1.18	M = 3.26, SD = 1.13	Z = -2.989, $p = .003^*$	Z = -3.214, $p = .001^*$	Z = -0.180, p = .857

challenging task, particularly in realistic driving conditions where road conditions change continuously and unexpected issues can compromise the safety of the driver at any moment. Nonetheless, observed changes in drivers' trust after vehicle experience were in the direction we would expect if drivers had improved their understanding of vehicles' limitations and capabilities.

More specifically, the results of our study suggest an improvement in trust calibration in 7/12 scenarios. In two scenarios (i.e., car overtakes in right lane; work area) drivers already showed correct trust calibration and this was maintained after the driving experience. In four scenarios, including the two in which trust was already well calibrated, the changes observed were not significant.

Notably, improved trust calibration involved both increases and decreases in trust. This is a key finding, showing that on-road driving experience positively influenced drivers' trust by increasing participants' ability to discriminate between different driving situations. Importantly, this was also true for scenarios (e.g., crossing deer) that were not directly experienced by any of the participants. Thus, vehicle experience leads to more reliable inferences concerning both experienced and unexperienced scenarios.

More in detail, after vehicle experience, drivers showed *increased trust* in vehicles' ACC and Lane Keeping capabilities, which are normally considered to be reliable. In the other five scenarios where we observed a change in trust, we observed a *reduction in trust*, again reflecting a better understanding of the vehicles' capabilities. Specifically, drivers seem to have understood that weather conditions can have a major impact on the reliability of the system, that they cannot rely on the emergency braking system when a deer suddenly crosses the highway, that current Level 2 vehicles do not have the capability to merge autonomously, that partially automated vehicles with Lane Keeping functionality can stray out of lane when facing tight curves, and that drivers should not blindly rely on automation in low visibility conditions.

When asked if the system could work optimally when driving behind a motorcycle (instead of behind a car), drivers' trust did not change immediately after experiencing the vehicles. However, in the "2-week" measurement, drivers' reported trust was significantly lower. In a recent study by Lenkeit [21], motorcycles were inadequately detected in more than 40% of cases (against a ~4% failure rate for detection of cars). In sum, vehicle detection systems are still far from perfect. Even if this scenario was not directly experienced by our participants, drivers' trust was better aligned to the actual reliability of the systems after their on-road driving experience. The fact that participants' decrease in trust occurred only two weeks after experiencing the vehicles could indicate that they needed more time to process this specific scenario. Nevertheless, since trust did not change immediately after vehicle experience, we cannot decisively conclude that the driving session was the main factor leading to improved trust calibration.

Participants' initial trust scores (i.e., before their driving experience) suggest a general overestimation of vehicles' capabilities. Drivers had no experience with the selected systems and their expectations of system behaviour were probably influenced by the media. "Self-driving cars" are currently a hot topic, leading to increased awareness of the presence and promises of the technology [10, 22]. News agencies, dealers, and manufacturers often suggest that fully autonomous vehicles will soon be cruising our roads (e.g., Tesla's website opens with the statement "Full Self-Driving Hardware on All Cars"). In reality, many analysts believe that full automation is still decades away (e.g., [23–25]). Thus, the initial overestimation of vehicles' capabilities seen in our participants might be the result of erroneous beliefs concerning the status of available driving technology.

In the scenarios where we found differences between the pre- and postmeasurement, no differences were found between the post- and the "2-week" measurement (see Table 1). This suggests that the impact of the driving experience on drivers' trust was relatively stable. No differences between the pre- and the "2-week" measurement were found when drivers were questioned about the ACC and Lane Keeping functions of the Level 2 vehicle (i.e., scenario 1), or when they were asked if the system could function optimally in low visibility conditions due to heavy rain (i.e., scenario 4). In these two scenarios, the driving experience had a direct positive impact on drivers' trust, but the effect did not seem to last. In the other scenarios, where differences between the pre- and postmeasurements were found, we also found differences between the pre- and the "2-week" measurement.

Our study presents several limitations that should be acknowledged. Participants drove two different vehicles before completing the post questionnaires (i.e., immediately after the driving session and two weeks after the driving session). These vehicles were all categorized as Level 2 and presented comparable Level 2 functions. Nevertheless, we cannot assess from the questionnaire data if one vehicle influenced trust scores more than the other. The event started with a presentation, in which the concept of partial automation was introduced to the participants for the first time. Furthermore, during every drive, a vehicle expert explained the Level 2 functions available in the vehicle and prompted the drivers to use them. We cannot assess if trust scores were mainly influenced by the participants' on-hand driving experience, by the initial presentation, by the experts' feedback, or by a combination of these factors. Finally, data was collected during five separate driving sessions. Varying weather conditions between sessions may have affected drivers' trust scores.

To our knowledge, this study represents the first attempt to measure changes in trust after drivers' experience with real Level 2 cars. Overall, our results suggest that hands-on experience with Level 2 vehicles helps trust calibration. Drivers interested in purchasing a Level 2 vehicle should be given the chance to drive the car along with someone (e.g., car dealer) who is fully aware of its functionalities. Such a procedure is likely to improve driver's understanding of the vehicles' capabilities and contribute to trust calibration even in situations they did not encounter during their on-road experience.

5. Conclusions

In this study we analysed, for the first time, participants' trust before (i.e., premeasurement), immediately after (i.e., postmeasurement), and two weeks after (i.e., "2-week" measurement) on-road driving experience with SAE Level 2 semiautomated vehicles. The results showed that the real-life driving experience led to improvements in trust calibration. Drivers' ability to more precisely differentiate, after experience, between situations where the automation performed well from situations where it did not, played a key role in calibrating participants' trust. These improvements were experienced by all drivers, independent of age, gender, and innovation profile. We conclude that carefully organised instruction and on-road experience can make crucial contributions to safe driving, while simultaneously improving sales, encouraging uptake, and helping to realize the potential benefits of self-driving technology for drivers and society as a whole.

Conflicts of Interest

The authors declare that there are no conflicts of interest regarding the publication of this article.

Acknowledgments

This research was completely funded by Rijkswaterstaat, The Ministry of Infrastructure and Water Management, the Dutch road authority. The authors thank Rijkswaterstaat for providing the opportunity to conduct this research.

References

[1] A. Tigadi, R. Gujanatti, and A. Gonchi, "Advanced driver assistance systems," *International Journal of Engineering Research and General Science*, vol. 4, no. 3, pp. 151–158, 2016.

[2] K. A. Brookhuis, D. De Waard, and W. H. Janssen, "Behavioural impacts of advanced driver assistance systems - an overview," *European Journal of Transport and Infrastructure Research*, vol. 1, no. 3, pp. 245–253, 2001.

[3] O. Gietelink, J. Ploeg, B. de Schutter, and M. Verhaegen, "Development of advanced driver assistance systems with vehicle hardware-in-the-loop simulations," *Vehicle System Dynamics*, vol. 44, no. 7, pp. 569–590, 2006.

[4] J. C. F. de Winter, R. Happee, M. H. Martens, and N. A. Stanton, "Effects of adaptive cruise control and highly automated driving on workload and situation awareness: a review of the empirical evidence," *Transportation Research Part F: Traffic Psychology and Behaviour*, vol. 27, pp. 196–217, 2014.

[5] S. A. E. Taxonomy, "Taxonomy and definitions for terms related to on-road motor vehicle automated driving systems," Tech. Rep., SAE International, 2014.

[6] C. Gold, M. Körber, C. Hohenberger, D. Lechner, and K. Bengler, "Trust in automation–before and after the experience of take-over scenarios in a highly automated vehicle," *Procedia Manufacturing*, vol. 3, pp. 3025–3032, 2015.

[7] M. H. Martens and A. P. van den Beukel, "The road to automated driving: dual mode and human factors considerations," in *Proceedings of the 16th International IEEE Conference on Intelligent Transportation Systems: Intelligent Transportation Systems for All Modes (ITSC '13)*, pp. 2262–2267, IEEE, October 2013.

[8] J. K. Choi and Y. G. Ji, "Investigating the Importance of Trust on Adopting an Autonomous Vehicle," *International Journal of Human-Computer Interaction*, vol. 31, no. 10, pp. 692–702, 2015.

[9] M. Ghazizadeh, J. D. Lee, and L. N. Boyle, "Extending the Technology Acceptance Model to assess automation," *Cognition, Technology & Work*, vol. 14, no. 1, pp. 39–49, 2012.

[10] N. Trübswetter and K. Bengler, "Why should I use ADAS? Advanced driver assistance systems and the elderly: knowledge, experience and usage barriers," in *Proceedings of the 7th International Driving Symposium on Human Factors in Driver Assessment, Training and Vehicle Design*, pp. 17–20, Bolton Landing, NY, USA, June 2013.

[11] B. M. Muir, "Trust between humans and machines, and the design of decision aids," *International Journal of Man-Machine Studies*, vol. 27, no. 5-6, pp. 527–539, 1987.

[12] E. J. de Visser, M. Cohen, A. Freedy, and R. Parasuraman, "A design methodology for trust cue calibration in cognitive agents," in *International Conference on Virtual, Augmented and Mixed Reality*, pp. 251–262, Springer, Cham, Switzerland, 2014.

[13] J. D. Lee and K. A. See, "Trust in automation: designing for appropriate reliance," *Human Factors*, vol. 46, no. 1, pp. 50–80, 2004.

[14] M. S. Cohen, R. Parasuraman, D. Serfaty, and R. Andes, *Trust in Decision Aids: A Model and a Training Strategy*, Cognitive Technologies Inc, Arlington, Va, USA, 1997.

[15] J. Lee and N. Moray, "Trust, control strategies and allocation of function in human-machine systems," *Ergonomics*, vol. 35, no. 10, pp. 1243–1270, 1992.

[16] B. Rajaonah, F. Anceaux, and F. Vienne, "Trust and the use of adaptive cruise control: A study of a cut-in situation," *Cognition, Technology & Work*, vol. 8, no. 2, pp. 146–155, 2006.

[17] D. A. Dickie and L. N. Boyle, "Drivers' understanding of adaptive cruise control limitations," *Proceedings of the Human Factors and Ergonomics Society Annual Meeting*, vol. 53, no. 23, pp. 1806–1810, 2009.

[18] A. Koustanaï, V. Cavallo, P. Delhomme, and A. Mas, "Simulator training with a forward collision warning system: effects on driver-system interactions and driver trust," *Human Factors*, vol. 54, no. 5, pp. 709–721, 2012.

[19] M. Körber, E. Baseler, and K. Bengler, "Introduction matters: Manipulating trust in automation and reliance in automated driving," *Applied Ergonomics*, vol. 66, pp. 18–31, 2018.

[20] B. M. Muir, "Trust in automation: Part I. Theoretical issues in the study of trust and human intervention in automated systems," *Ergonomics*, vol. 37, no. 11, pp. 1905–1922, 1994.

[21] J. F. Lenkeit, "Preliminary study of the response of forward collision warning systems to motorcycles," in *Proceedings of the 11th International Motorcycles Conference*, Cologne, Germany, October 2016.

[22] V. Hahanov, W. Gharibi, K. L. Man et al., "Cyber-physical technologies: hype cycle," in *Cyber Physical Computing for IoT-driven Services*, pp. 259–272, Springer, Cham, Switzerland, 2018.

[23] P. Bansal and K. M. Kockelman, "Forecasting Americans' long-term adoption of connected and autonomous vehicle technologies," *Transportation Research Part A: Policy and Practice*, vol. 95, pp. 49–63, 2017.

[24] A. Lumiaho and F. Malin, "Road transport automation: road map and action plan 2016-2020," Research Reports, Finnish Transport Agency, 2016.

[25] T. Litman, *Autonomous vehicle implementation predictions*, Victoria Transport Policy Institute, 2018.

Permissions

All chapters in this book were first published in JAT, by Hindawi Publishing Corporation; hereby published with permission under the Creative Commons Attribution License or equivalent. Every chapter published in this book has been scrutinized by our experts. Their significance has been extensively debated. The topics covered herein carry significant findings which will fuel the growth of the discipline. They may even be implemented as practical applications or may be referred to as a beginning point for another development.

The contributors of this book come from diverse backgrounds, making this book a truly international effort. This book will bring forth new frontiers with its revolutionizing research information and detailed analysis of the nascent developments around the world.

We would like to thank all the contributing authors for lending their expertise to make the book truly unique. They have played a crucial role in the development of this book. Without their invaluable contributions this book wouldn't have been possible. They have made vital efforts to compile up to date information on the varied aspects of this subject to make this book a valuable addition to the collection of many professionals and students.

This book was conceptualized with the vision of imparting up-to-date information and advanced data in this field. To ensure the same, a matchless editorial board was set up. Every individual on the board went through rigorous rounds of assessment to prove their worth. After which they invested a large part of their time researching and compiling the most relevant data for our readers.

The editorial board has been involved in producing this book since its inception. They have spent rigorous hours researching and exploring the diverse topics which have resulted in the successful publishing of this book. They have passed on their knowledge of decades through this book. To expedite this challenging task, the publisher supported the team at every step. A small team of assistant editors was also appointed to further simplify the editing procedure and attain best results for the readers.

Apart from the editorial board, the designing team has also invested a significant amount of their time in understanding the subject and creating the most relevant covers. They scrutinized every image to scout for the most suitable representation of the subject and create an appropriate cover for the book.

The publishing team has been an ardent support to the editorial, designing and production team. Their endless efforts to recruit the best for this project, has resulted in the accomplishment of this book. They are a veteran in the field of academics and their pool of knowledge is as vast as their experience in printing. Their expertise and guidance has proved useful at every step. Their uncompromising quality standards have made this book an exceptional effort. Their encouragement from time to time has been an inspiration for everyone.

The publisher and the editorial board hope that this book will prove to be a valuable piece of knowledge for researchers, students, practitioners and scholars across the globe.

List of Contributors

Yang Li, Dihua Sun, Min Zhao, Dong Chen and Senlin Cheng
Key Laboratory of Dependable Service Computing in Cyber Physical Society of Ministry of Education, Chongqing University, Chongqing 400044, China
School of Automation, Chongqing University, Chongqing 400044, China

Fei Xie
China Automotive Engineering Research Institute Co., Ltd., Chongqing 401122, China

Chun-Cheng Chen
Graduate School of Technological and Vocational Education, National Yunlin University of Science and Technology, Yunlin 64002, Taiwan

Shun-Chang Chang
Department of Mechanical and Automation Engineering, Da-Yeh University, Changhua 51591, Taiwan

Hongxing Zhao, Ruichun He and Changxi Ma
School of Traffic and Transportation, Lanzhou Jiaotong University, Lanzhou 730070, China

Li Li, Ping Wang and Guiping Wang
School of Electronics & Control Engineering, Chang'an University, Middle Section of Nan Er Huan, Xi'an, Shaanxi 710064, China

Dong Zhang
School of Transportation & Logistics, Dalian University of Technology, No. 2 Linggong Rd., Gan Jingzi District, Dalian, Liaoning 116024, China

Zhigang Xu
School of Information Engineering, Chang'an University, Middle Section of Nan Er Huan, Xi'an, Shaanxi 710064, China

Jianyou Zhao and Jing Liu
School of Automobile, Chang'an University, Xi' an 710064, China

Jing Liu and Ping He
School of Mechanical and Electrical Engineering, Anhui Jianzhu University, Hefei 230601, China

Liping Yang
School of Automobile and Traffic Engineering, Hefei University of Technology, Hefei 230009, China

Mudasser Seraj, Jiangchen Li and Zhijun Qiu
Department of Civil and Environmental Engineering, University of Alberta, Edmonton, Alberta, Canada

Zhijun Qiu
Intelligent Transportation System Research Center, Wuhan University of Technology, Wuhan, Hubei, China

Mark Mario Morando, Long T. Truong and Hai L. Vu
Monash Institute of Transport Studies, Department of Civil Engineering, Monash University, Melbourne, VIC, Australia

Qingyun Tian
School of Civil and Environmental Engineering, Nanyang Technological University, Singapore

Sehyun Tak
Department of The Fourth Industrial Revolution and Transport,The Korea Transportation Institute (KOTI), Republic of Korea

Sunghoon Kim, Donghoun Lee and Hwasoo Yeo
Department of Civil and Environmental Engineering, Korea Advanced Institute of Science and Technology, Republic of Korea

Chao Li, Mohammad Karimi and Ciprian Alecsandru
Department of Building, Civil and Environmental Engineering, Concordia University, Montréal, QC, Canada H3G 2W1

Xiaobo Liu, Lijuan Lai, You Kong and Scott Le Vine
School of Transportation and Logistics, Southwest Jiaotong University, Chengdu, China

Scott Le Vine
Department of Geography, State University of New York (SUNY) at New Paltz, New Paltz, NY, USA

Scott Le Vine
Department of Civil and Environmental Engineering, Imperial College London, South Kensington SW7 2AZ, UK

Johannes Masino, Jakob Thumm, Michael Frey and Frank Gauterin
Institute of Vehicle System Technology, Karlsruhe Institute of Technology, Germany

Guillaume Levasseur
Artificial Intelligence Research Laboratory, Université Libre de Bruxelles, Belgium

Ralf Mikut and Markus Reischl
Institute for Automation and Applied Informatics, Karlsruhe Institute of Technology, Germany

Sina Nordhoff, Bart van Arem and Riender Happee
Department of Transport & Planning, Delft University of Technology, Delft, Netherlands

Sina Nordhoff
Innovation Centre for Mobility and Societal Change, Berlin, Germany

Joost de Winter
Department of Biomechanical Engineering, Delft University of Technology, Delft, Netherlands

Miltos Kyriakidis
ETH Zurich, Future Resilient Systems, Singapore ETH Centre, Singapore

Riender Happee
Department of Cognitive Robotics, Delft University of Technology, Delft, Netherlands

Zhan Wang and Alain Lambert
Laboratoire de Recherche en Informatique (LRI), CNRS, Univ Paris-Sud, Université Paris-Saclay, 91403 Orsay, France

Jin Xu
Professor, Chongqing Key Laboratory of "Human - Vehicle -Road" Cooperation & Safety for Mountain Complex Environment, Chongqing Jiaotong University, No. 66, Xuefu Ave., Chongqing 400074, China

Jinghou Fu and Xiaoming Liu
Postgraduate Student, College of Traffic and Transportation, Chongqing Jiaotong University, No. 66, Xuefu Ave., Chongqing 400074, China

Yiming Shao
Professor, College of Traffic and Transportation, Chongqing Jiaotong University, No. 66, Xuefu Ave., Chongqing 400074, China

Luís Braga Campos
CCTAE (Center for Aeronautical and Space Science and Technology), IDMEC/LAETA, Instituto Superior Técnico, Universidade de Lisboa, Av. Rovisco Pais 1, 1049-001 Lisboa, Portugal

Joaquim Guerreiro Marques
Escola Náutica Infante D. Henrique, Av. Engenheiro Bonneville Franco, 2770-058 Paço de Arcos, Oeiras, Portugal

Ana T. Moreno, Andrzej Michalski, Carlos Llorca and Rolf Moeckel
Department of Civil, Geo and Environmental Engineering, Technical University of Munich, Munich 80333, Germany

Francesco Walker, Anika Boelhouwer, Willem B. Verwey and Marieke H. Martens
University of Twente, Netherlands

Tom Alkim
Rijkswaterstaat, Ministry of Infrastructure and Water Management, Netherlands

Index

S

T

V

W

www.ingramcontent.com/pod-product-compliance
Lightning Source LLC
LaVergne TN
LVHW080918210926
848485LV00004B/676